# The Healthy Indoor Environment

Despite policy directives, standards and guidelines, indoor environmental quality is still poor in many cases. *The Healthy Indoor Environment* aims to help architects, building engineers and anyone concerned with the wellbeing of building occupants to better understand the effects of spending time in buildings on health and comfort. In three clear parts dedicated to mechanisms, assessment and analysis, this book looks at different indoor stressors and their effects on wellbeing in a variety of scenarios with a range of tools and methods.

The book supports a more holistic way of evaluating indoor environments and argues that a clear understanding of how the human body and mind receive, perceive and respond to indoor conditions is needed. At the national, European and worldwide level, it is acknowledged that a healthy and comfortable indoor environment is important both for the quality of life, now and in the future, and for the creation of truly sustainable buildings. Moreover, current methods of risk assessment are no longer adequate: a different view on indoor environment is required.

Highly illustrated and full of practical examples, the book makes recommendations for future procedures for investigating indoor environmental quality based on an interdisciplinary understanding of the mechanisms of responses to stressors. It forms the basis for the development of an integrated approach towards the assessment of indoor environmental quality.

**Philomena M. Bluyssen** started as a full Professor of Indoor Environment at the Delft University of Technology in 2012 after more than two decades working for TNO in Delft, The Netherlands. She has written more than 170 publications and won the Choice Award for Outstanding Academic Titles of 2010 for *The Indoor Environment Handbook* (also published by Earthscan from Routledge).

'The author of this book does not start from the more conventional points of departure, like the possibilities provided by new theories or technologies, nor does the book start from the most problematic aspects in assessing Indoor Environmental Quality. It starts where it should start: with people, us building occupants. The occupants are put central in this multi-disciplinary quest on how to analyse and assess IEQ in order to ameliorate conditions in buildings. From this perspective it gives an overview of all possible approaches. Therefore it is highly recommended to everyone interested to know how to assess occupants' well-being in buildings.'

*Professor Mieke Oostra, Hanzehogeschool*
*Groningen University of Applied Sciences*

'"Everybody" knows that in some buildings we feel good and in others we do not. We know a lot about the determinants of indoor environment, and yet we are unable to predict which buildings we will thrive in. Professor Bluyssen's book offers a not-so-common occupant-centric point of view. In nine well-documented chapters, the reader is lead through the background of the various disciplines needed to understand indoor environment. All this wealth of information is accompanied by contextual glimpses and a personal touch, conveying the passion with which the book was conceived and executed. Enjoyable reading, all the time reminding us that "our" scientific discipline lives in a wider context, and that after all, buildings are built for people, not the other way around.'

*Dr Alena Bortonava, Center for Ecology and Economics*
*at the Norwegian Institute for Air Research*

# The Healthy Indoor Environment

## How to assess occupants' wellbeing in buildings

*Philomena M. Bluyssen*

Routledge
Taylor & Francis Group

LONDON AND NEW YORK

from Routledge

First published by Earthscan in the UK and USA in 2014

For a full list of publications please contact:
Earthscan
2 Park Square, Milton Park, Abingdon, Oxon OX14 4RN
605 Third Avenue, New York, NY 10017

First issued in paperback 2021

*Earthscan is an imprint of the Taylor & Francis Group, an informa business*

*British Library Cataloguing in Publication Data*
A catalogue record for this book is available from the British Library

*Library of Congress Cataloging-in-Publication Data*
Bluyssen, Philomena M.
    The healthy indoor environment: how to assess occupants' wellbeing
in buildings/Philomena M. Bluyssen.
    pages cm
Includes bibliographical references and index.
    1. Buildings—Environmental engineering. 2. Architecture—Human factors.
3. Architecture—Psychological aspects. 4. Indoor air quality. I. Title.
TH6021.B58 2014
720.28′6—dc23
2013011293

ISBN-13: 978-1-03-209908-8 (pbk)
ISBN-13: 978-0-415-82275-6 (hbk)

Typeset in Sabon and Helvetica Neue by
Florence Production Ltd, Stoodleigh, Devon, UK

### Middle of night

Sometimes I wake up in the in the middle of the night
Turning and turning
Thinking and thinking
From this to that, from that to this
I tend to solve the most complex issues half asleep
Unfortunately the next day
I don't remember a thing
Only the notion that I didn't sleep well
And that I did solve something
But what?

# Contents

# Illustrations

## Figures

# Tables

# Boxes

# Preface

How to achieve a healthy indoor environment has been an issue among architects, engineers and scientists for centuries. However, it was not until the early decades of the twentieth century that the first relationships between parameters describing heat, lighting and sound in buildings and human needs were established. In fact, in the last hundred years much effort has been put into the management of the indoor environment with the goal of creating healthy and comfortable conditions for the people living, working and recreating in them for more than 90 per cent of their time. Nevertheless, enough health problems and comfort complaints still occur to trigger more research and development.

Relationships between indoor building conditions and the wellbeing of occupants are complex. Many indoor stressors can exert their effects additively or through complex interactions. It has been shown that exposure to these stressors can cause both short-term and long-term effects. Relevant relations between measurements of chemical and physical indoor environmental parameters and effects have been difficult to establish.

Over the years, control of indoor environmental factors has focused merely on the prevention or cure of their observed physical effects in a largely isolated way – i.e. trying to find separate solutions for thermal quality, lighting quality, sound quality and air quality with models that in general consider only physical conditions and address only one parameter at a time. Many control strategies for these parameters have been implemented in order to minimize or prevent possible diseases and disorders. Only in the last decades of the twentieth century was an attempt made, through epidemiological studies, to approach the indoor environment in a holistic way. The scientific approach towards evaluating and creating a healthy and comfortable indoor environment developed from a component-related to a bottom-up holistic approach (that tried simply to add the different components). Performance concepts and indicators emerged, including not only environmental parameters, but also possible associated variables such as the characteristics of buildings. New methods of investigating indoor environmental quality from different perspectives were introduced. Nevertheless, control strategies were still focused on a component basis. Even though these control strategies are currently being applied, complaints and symptoms related to the indoor environment still occur.

Our current standards are focused mainly on single-dose responses. With the exception of health-threatening stimuli, the complexity and number of indoor environmental parameters, as well as lack of knowledge, make a performance assessment using only threshold levels for single parameters difficult and even meaningless. Most standards are based on averaged data and do not take into account the fact that buildings, individuals and their activities may differ widely and change continuously; not every person receives, perceives and responds in the same way. This is due to physical, physiological and psychological differences, but also to differences in personal experience, context and situation. Considering both the numerous indoor stimuli and the lack of a solid scientific basis, it appears implausible to make the final and complex integrating step.

To increase the chance of successful assessment of cause–effect relationships in future indoor environmental quality (IEQ) investigations, there seems to be a need to improve the procedures applied to gather the relevant information. First, it is essential to understand the mechanisms behind how and why people respond to external stressors. The next step is then to determine which parameters or indicators can be used to explain these responses and how to assess them. Only when the picture is more clear can procedures be improved in such a way that the chances to successfully assess the effects caused by different stressors (or combinations of stressors) increase. The following questions seem thus important to discuss in order to get a better picture:

- How and why do people respond to external stressors?
- How can the health and comfort (wellbeing) of people in the indoor environment be evaluated?
- What do we need to assess and/or predict the effects or responses?

This book presents an attempt to answer these questions and to recommend an adapted procedure based on information from previous studies and available information from different disciplines.

This book comprises three parts:

- Part I, 'Mechanisms', in which the mechanisms of response to external stressors are explored, including aspects such as acute/long-term effects, psychological/physical stressors, personal factors and history. As a first step a model of the human body is outlined, which acknowledges that we are dealing with an extensive number of variables (person variables, situation variables, activity variables, history (previous exposures and experiences), time period, future expectations, etc.), which together determine how a person is feeling in a certain situation, during a certain activity at a certain point in time.
- Part II, 'Assessment', in which different assessment techniques of indicators and parameters (from responses) to external stressors are inventoried and discussed, including aspects such as study design, statistical relevance, different assessment techniques (e.g. questionnaires, observation, medical examination), and building life-cycle and intervention. Based on the human model established in Part I, the second step is to determine which techniques and procedures are available to extract 'relevant' information from the following sources:
  - People and the human systems: different types of information on how our body copes with stress and responds to stress.
  - The indoor environment in the built environment: characteristics and processes that can cause and influence physical stressors and factors.
  - The psychosocial environment: characteristics and processes that can cause and influence the psychosocial stressors and factors.
- Part III, 'Analysis', in which current available performance indicators for healthy and comfortable indoor environments, as well as the available information on cause–effect relations for different scenarios (schools, homes and offices), is presented. In the third step it is determined what is needed to assess and/or predict the effects or responses of people in their indoor environment, applying different levels of assessment (at the level of the occupants, of the dose or environmental parameter, and of the building and its components).

Each part begins with an introductory chapter (Chapters 1, 4 and 7) that provides a summary of the information presented in the following chapters.

The focus is on non-industrial buildings (e.g. offices, schools and homes), and on non-infectious diseases and disorders potentially caused by chemical, physical and biological exposures in indoor environments.

# Acknowledgments

After the successful, award-winning *Indoor Environment Handbook*, I got inspired to evolve my ideas on the assessment of IEQ in more depth. I wrote a few first pages on the contents and, even before those were reviewed, I seem to have convinced my publisher and received the unofficial 'go' to start writing in August 2011, during the summer holidays.

'Writing in your free time?' I hear you ask. 'You must be either bored, or really fond of writing!' I guess it is a little bit of both. I have a hard time doing nothing, and, yes, the writing of the *Indoor Environment Handbook* was so much fun, that I knew I would write another one sometime. To write this book, however, I didn't have so much obligatory sitting at home time: I didn't have to go through several revalidations of my knee, as was the case with my first book. Moreover, in the last year, next to my full time research job, I successfully applied for a full professorship at the Delft University of Technology.

Considering the above, it is without doubt that my family suffered the most from my time consuming 'hobby'. Without their support and endurance, it wouldn't have been possible to spend all the 'free' hours I could find on this book. Especially my husband Darell Meertins should receive an award for his patience. A special thanks also to my two sons, Anthony and Sebastian, who not only gave me the opportunity, but also gave me a couple of their drawings to use! They clearly inherited some of the artistic qualities of my mother, who also contributed with one of her finest paintings.

Then I need to thank a number of friends and colleagues, who helped me with the writing itself. First of all, Juliette Clipper, my golf mate, who has undertaken a second study 'Medicine'. She reviewed the first part on mechanisms of the human body and encouraged me to continue with the rest of the book. Second, Alena Bartonova, my dear Czech friend from Norway who gave me very good advice on Part II, and with whom I coordinated work package data management for the European project SINPHONIE. And last, but not least, Mieke Oostra, ex-colleague and close friend, who not only reviewed Part III, but with whom I had intense discussions on the ideas presented.

Finally, I would like to thank several persons from the consortia of the European projects SINPHONIE and OFFICAIR, for helping and encouraging me to shape the ideas on IEQ presented in the underlying publication: Corinne Mandin (CSTB in France), Paolo Carrer (University of Milano, Italy), Eduardo de Oliveira Fernandes (University of Porto, Portugal), John Bartzis (University of Western Macedonia, Greece), Eva Csobod (Regional Environmental Center, Hungary), Yvonne de Kluizenaar and Michel Böhms (TNO, The Netherlands).

Enjoy reading!

Philomena M. Bluyssen
February 2013

# Symbols, acronyms and abbreviations

## Symbols

| | |
|---|---|
| 8-OHdG | 8-hydroxydeoxyguanosine |
| $a_{H^+}$ | the activity of hydrogen ions in units of mol/L (molar concentration) |
| A | adenine |
| AB | attributable burden: number of persons with a certain disease as a result of exposure to a risk factor, not corrected for co-morbidity |
| Ar | argon |
| AR | attributive risk: risk of getting a specific disease as a result of exposure to a certain factor |
| C | cytosine |
| C | concentration of risk factor |
| Ca | calcium |
| Cl | chlorine |
| CO | carbon monoxide |
| $CO_2$ | carbon dioxide |
| Cr | chromium |
| Cu | copper |
| D | duration of disease: for morbidity duration is one year in case of prevalence numbers, for mortality the duration of time lost due to premature mortality |
| e | error |
| F | between conditions variance/error variance |
| F | fraction of population exposed to the risk factor |
| Fe | iron |
| G | guanine |
| $H^+$ | hydrogen ion |
| Hb | hemoglobin |
| $HbO_2^-$ | oxyhemoglobine |
| $H_2CO_3$ | carbonic acid |
| $HCO_3^-$ | bicarbonate |
| $H_2O$ | water |
| $H_2O_2$ | hydrogen peroxide |
| He | helium |
| Hg | mercury |
| HHb | deoxyhemoglobine |
| K | potassium |
| Kr | krypton |
| L | allowable error or estimated confidence interval |
| mmHg | millimeters of mercury; unit for blood pressure |
| n | size of sample |
| Na | sodium |
| NaCl | sodium chloride |

| | |
|---|---|
| Ne | neon |
| NO | nitric oxide |
| $NO_2$ | nitrogen dioxide |
| $NO_x$ | nitrogen oxides |
| $O_2$ | oxygen |
| $O_2{}^{\bullet-}$ | superoxide |
| $O_3$ | ozone |
| OH$\bullet$ | hydroxyl radical |
| P | phosphate |
| P | base prevalence for morbidity; number of deaths for mortality |
| p | percentage choosing yes |
| p | significance level |
| pH | acidity or basicity of an aqueous solution |
| $PN_2$ | partial pressure of nitrogen |
| $PO_2$ | partial pressure of oxygen |
| q | 1– p; percentage choosing no |
| $\rho$ | Spearman correlation coefficient |
| r | Pearson correlation coefficient |
| R$\bullet$ | alkyl radical |
| RH | relative humidity (per cent) |
| Rn | radon |
| $RO_2$ | peroxyl radical |
| ROOH | hydroperoxides |
| RR | relative risk |
| RR' | adjusted relative risk |
| $\sigma$ | standard deviation; the square root of the variance |
| S | standard deviation |
| S | severity: reduction in capacity due to morbidity using severity weights (between 0 (perfect health) and 1 (death), determined by experts) |
| T | thymine |
| T | temperature (°C) |
| $t_{.05}$ | t-distribution of 95 per cent confidence interval |
| T3 | tri-iodothyronine |
| T4 | thyroxine |
| V | vanadium |
| X | estimated mean or average |
| $X^2$ | chi-square |
| Xe | xenon |

# Acronyms and abbreviations

| | |
|---|---|
| ACTH | adrenocortico tropin hormone |
| ADH | antiduretic hormone or vasopressin |
| ADP | adenosine diphosphate |
| ANOVA | analysis of variance |
| ANS | autonomic nervous system |
| ASD | autism spectrum disorder |
| ASHRAE | American Society of Heating, Refrigerating and Air-Conditioning Engineers |
| ATD | automated thermal desorption |

| | |
|---|---|
| ATP | adenosine triphosphate |
| BBB | blood–brain barrier |
| BBzP | n-butyl benzyl phthalate |
| BCAA | branched chain amino acid |
| BMI | body mass index |
| BMR | basal metabolic rate |
| BPS | boredom proneness scale |
| BREEAM | BRE Environmental Assessment Method |
| cAMP | cyclic adenosine monophosphate |
| CART | classification and regression tree |
| CASBEE | Comprehensive Assessment System for Built Environment Efficiency |
| CAVI | cardio-ankle vascular index |
| CBE | Center for the Built Environment |
| CgA | chromogranin A |
| CHD | coronary heart disease |
| CIBSE | Chartered Institution of Building Services Engineers |
| CNS | central nervous system |
| Co-A | co-enzyme A |
| COPD | chronic obstructive pulmonary disease |
| COPE | Cost-effective Open Plan |
| CPD | Construction Products Directive |
| CRH | corticotropin-releasing hormone |
| CRP | C-reactive protein |
| CS | conditioned stimulus |
| CT | computerized tomography |
| CTD | cumulative trauma disorder |
| CVD | cardiovascular disease |
| DALY | disability adjusted life year |
| DASS | depression anxiety stress scales |
| DBH | dampness in buildings and health |
| DBMS | Database Management System |
| DDT | dichlorodihenyltrichloroethane |
| DEHP | di(2-ethylhexyl)phthalate |
| DEP | diethyl phthalate |
| DES | differential emotions scale |
| DHEA | dehydroepiandrosterone |
| DIBP | diisobutyl phthalate |
| DNA | deoxyribonucleic acid |
| DnBP | di-n-butyl phthalate |
| DPB | dibutyl phthalate |
| DRM | day reconstruction method |
| DTIE | Division of Technology, Industry and Economics |
| DTT | dichlorodiheny-richloroethane |
| DV | dependent variable |
| EBC | exhaled breath condensate |
| EC | European Commission |
| ECG | electrocardiogram |
| EDC | endocrine-disrupting chemical |
| EDR | electrodermal response |

| | |
|---|---|
| EEG | electro-encephalogram |
| ELISA | enzyme-linked immunosorbent assay |
| EMG | electromyography |
| EPIQR | Energy Performance Indoor air Quality Retrofit |
| ERI | effort–reward imbalance |
| ESM | experience sampling method |
| ESRM | experience sampling reconstruction method |
| $ETCO_2$ | end-tidal partial $CO_2$ |
| ETS | environmental tobacco smoke |
| FA | fatty acids |
| FEF | forced expiratory flow |
| FeNO | fractional concentration of exhaled nitric oxide |
| FEV1 | forced expiratory volume in one second |
| FMD | flow-mediated dilation |
| fMRI | functional magnetic resonance imaging |
| FSH | follicle stimulating hormone |
| FVC | forced vital capacity |
| GABA | gamma-aminobutyric acid |
| GC-MS | gas chromatography-mass spectrometry |
| GDP | G protein diphosphate |
| GH | growth hormone |
| GHG | greenhouse gas |
| GHQ | general health questionnaire |
| GIH | growth inhibiting hormone (somatostatin) |
| GJIC | gap junctional intracellular communication |
| GnRH | gonadotropin-releasing hormone |
| GPP | green public procurement |
| GRH | growth hormone-releasing hormone |
| GSH | reduced glutathione |
| GSH-PX | glutathione peroxidase |
| GSR | galvanic skin response |
| GTP | G protein triphosphate |
| HDL | high-density lipoprotein |
| HESE | Health Effect of School Environment |
| HO-1 | heme oxygenase-1 |
| HOPE | Health Optimisation Protocol for Energy-efficient Buildings |
| HPA | hypothalamus–pituitary–adrenal |
| HPG | hypothalamic–pituitary–gonadal |
| HPLC | high-performance liquid chromatography |
| HPT | hypothalamic–pituitary–thyroid |
| HRV | heart rate variation |
| HVAC | heating, ventilating and air conditioning |
| IAQ | indoor air quality |
| ICT | Information and Communication Technology |
| IECH | indoor environment and children's health |
| IEI | idiopathic environmental intolerance |
| IEQ | indoor environment quality |
| IFN | interferon |
| Ig | immune-gobulins |

| | |
|---|---|
| IL | interleukin |
| IR | infrared |
| IRT | infrared thermography |
| IV | independent variable |
| LARES | Large Analysis and Review of European Housing and Health Status |
| $LC_{50}$ | lethal concentration, 50 per cent |
| LCA | life-cycle assessment |
| $LCt_{50}$ | lethal concentration and time |
| LCD | life-cycle design |
| $LD_{50}$ | the median lethal dose, abbreviation for 'lethal dose, 50 per cent' |
| LDH | lactaat dehydrogenase |
| LDL | low density lipoprotein |
| LEED | Leadership in Energy and Environmental Design |
| LH | luteinizing hormone |
| LOAEL | lowest-observed-adverse-effect level |
| MBP | mono n-butyl phthalate |
| MCP | moncyte chemoattractant protein |
| MCS | multiple chemical sensitivity |
| MEF | maximal expiratory flow |
| MEG | magnetoencephalography |
| MEHP | mono-2-ethyl phthalate |
| MIP | macrophage inflammatory protein |
| miRNA | microRNA |
| MONAVA | multivariate analysis of variance |
| MR | metabolic rate |
| MRI | magnetic resonance imaging |
| mRNA | messenger RNA |
| MS | mass spectrometry |
| mtDNA | mitochondrial DNA |
| NA | negative affect |
| NAL | nasal lavage |
| NEO-PI | Neuroticism, Extraversion and Openess Personality Inventory |
| NIF | non-imaging forming |
| NK cell | natural killer cell |
| NMR | nuclear magnetic resonance |
| NOAEL | no-observed-adverse-effect level |
| NOTEL | no observed transcriptional effect level |
| OIE | open information environment |
| OSDI | Ocular Surface Disease Index |
| OWL | Web Ontology Language |
| PA | positive affect |
| PAD | pleasure–arousal–dominance |
| PAH | polyaromatic hydrocarbons |
| PANAS | Positive and Negative Activation Scale |
| PAS | primary attentional system |
| PAT | peripheral arterial tonometry |
| PBDE | polybrominated diphenyl ether |
| PC | personal computer |
| PCA | principal component analysis |

| | |
|---|---|
| PCB | polychlorinated biphenyl |
| PCR | polymerase chain reaction |
| PDA | personal digital assistant |
| PEF | peak expiratory flow |
| PEP | pre-prejection period |
| PET | positron emission tomography |
| PFC | perflurorinated chemical |
| PIH | prolactin-inhibiting hormone |
| PM | particulate matter |
| PMS | premenstrual syndrome |
| PNS | peripheral nervous system |
| POE | post-occupancy evaluation |
| POMS | profile of mood states |
| PPD | percentage of people disatisfied |
| PRH | prolactin-releasing hormone |
| PRS | perceptual representation system |
| PTF | pre-corneal tear film |
| PTH | parathyroid hormone |
| PUFA | polyunsaturated fatty acid |
| PVC | polyvinylchloride |
| PWA | pulse wave amplitude |
| QALY | quality adjusted life-year |
| QPCR | quantitative polymerase chain reaction |
| R | ribosomes |
| R | reproducibility |
| rCBF | regional cerebral blood flow |
| RCDC | reducing agent compatible detergent compatible |
| RCT | randomized controlled trials |
| REM | rapid eye movement |
| RH | relative humidity |
| RHP-axis | retino-hypothaliamic-pineal axis |
| RNA | ribonucleic acid |
| ROS | reactive oxygen species |
| RSI | repetitive strain injury |
| S | sympathetic |
| SAA | salivary $\alpha$-amylase activity |
| SAD | seasonal affective disorder |
| SAM | self-assessment manikin |
| SAS | secondary attentional system |
| SBA | Sustainable Building Alliance |
| SBCI | Sustainable Buildings and Climate Initiative |
| SBS | Sick Building Syndrome |
| SBUT | self-reported break up time |
| SCN | supra-chiasmatic nuclei |
| SDR | social desirable response |
| SEM | structural equation model |
| SES | social economical status |
| SMPS | scanning mobility particle sizer |
| SNS | sympathetic nervous systems |

| | |
|---|---|
| SOA | secondary organic aerosols |
| SPECT or SPET | single-photon emission (computerized) tomography |
| SPN | stochastic petri network |
| $SPO_2$ | blood oxygen saturation |
| SQUID | superconducting quantum interference device |
| STAI | state–trait anxiety inventory |
| STICSA | state–trait inventory for cognitive and somatic anxiety |
| SVOC | semi-volatile organic compounds |
| SWB | subjective wellbeing |
| SWOP | semantic web-based open engineering platform |
| T-cell | thymus cell |
| TG | triglycerides or triagcylglycerols |
| TGF | transforming growth factor |
| TLR | toll-like receptor |
| TMS | transcranial magnetic stimulation |
| TNF-$\alpha$ | tumor necrose factor |
| TRH | thyroid-releasing hormone |
| tRNA | transfer RNA |
| TSH | thyroid-stimulating hormone |
| UFP | ultra fine particles |
| US | unconditioned stimulus |
| USGBC | US Green Building Council |
| UV | ultraviolet |
| VDU | video display unit |
| VOCs | volatile organic compounds |
| WAIS-III | Wechsler Adult Intelligence Scale |
| WASO | time awake after sleep onset |
| WHO | World Health Organization |
| WMS-II | Wechsler Memory Scale |
| YLD | years lived with disability |
| YLL | years of life lost |

# Part I

# Mechanisms

# 1

# Human model

*In this chapter a synthesis of knowledge on the fundamentals of how the human body and mind respond to external stressors is presented. Depending on how the body and mind cope with acute or chronic stress, stressors (both psychosocial and physical) may cause an imbalance or disturbance of the human systems, which immediately or over time may cause physiological, physical and psychological changes. The human bodily processes in which these changes have an effect and can lead to diseases and disorders are described in Chapter 2. Chapter 2 also shows how the three major human systems regulate and control these processes and how they cope with these external stressors. The possible stress response mechanisms, including stressors, responses and factors of influence, are dealt with in Chapter 3.*

## 1.1 Introduction

Human exposure to environmental factors (such as indoor air compounds) occurs mainly through the senses. Receptors in our nervous system receive sensory information as sensations via the eyes, ears, nose and skin, enhanced by bodily processes such as inhalation, ingestion and skin contacts. Most of us are familiar with several reactions of the human body to certain stimuli, such as sweating when warm, closing/narrowing the eyes against a sharp light, covering the ears against a loud noise, temporarily stopping breathing while in the presence of a bad smell, allergic reactions to pollen or even certain inflammation and infection defence mechanisms of the immune system upon injury of the epithelium (the 'skin' of an organ).

In addition to the stimuli that can be processed by our sensory system, the environment affects us in other ways, which are not always recognizable to us and which we are not (immediately) conscious of. The latter stimuli can cause changes in our physiological and psychological state. These changes can be harmful to our physical state of wellbeing in the long-term even though the exposed levels are well below current set threshold levels. External stress factors seem to be able to result in both mental and physical effects (Vroon, 1990; Kapit *et al.*, 2000; Bonnefoy *et al.*, 2004; Fisk *et al.*, 2007; Lewtas, 2007; Houtman *et al.*, 2008; Babisch, 2008).

Already Hippocrates (born *c.*460 BC) believed that when the bodily fluids are out of balance this can eventually lead to diseases (Nicolle and Woodriff Beirne, 2010). He introduced the so-called *humoral* theory, based around four humours (from Greek: 'chymos', meaning 'juice' or metaphorically 'flavour'): blood, black bile, yellow bile and phlegm related to respectively the liver, spleen, gall

---

### Box 1.1 Pasteurization and the Germ theory

Pasteurization is named after the French chemist and microbiologist Louis Pasteur (1822–1895). Together with Robert Koch (1843–1910), he is seen as one of the 'fathers of bacteriology and the Germ theory'. Unlike sterilization, pasteurization is not intended to kill all microorganisms. It is aimed at reducing the number of viable pathogens, so that they are unlikely to cause disease. In the Germ theory of disease, also called the pathogenic theory of medicine, microorganisms are the cause of many diseases. Pasteur demonstrated that growth of bacteria in nutrient broths is not caused by spontaneous generation but rather by biogenesis (*Omne vivum ex vivo* – all life is from life).

(Madigan *et al.*, 2005)

---

bladder and the brain/lungs. According to him, in order to be healthy, those four humours should be in balance. Additionally, Hippocrates was also in favour of including family history and the environment in his examination routine. Illness was thus the product of environmental factors, diet and living habits. Unfortunately, it was not until the nineteenth century, when the *Germ theory* was proposed in 1865 (see Box 1.1), that it became accepted that environmental factors such as unsanitary conditions, or contaminated water or air could cause diseases.

Around the same time Claude Bernard introduced his ideas on the importance of the state of the body (the *milieu interieur*) in disease, rather than any invading organism (Bernard, 1865). This formed the basis for the principles of *homeostasis* (in Greek, *homo* means 'same', and 'stasis' means 'stable', thus meaning 'remaining stable by staying the same') introduced by Walter Cannon (1963 [1932]). Cannon defined homeostasis 'as a condition which may vary but which is relatively constant', or 'The coordinated physiological processes which maintain most of the steady states in the organisms' (in Cannon, 1963: p. 24). He claimed that constancy is the goal of all internal mechanisms.

Much later, in 1988, Sterling and Eyer proposed the theory of *allostasis*: remaining stable by changing (in Greek, *allo* means 'different' or 'varying'). According to them, but also others, homeostasis alone cannot fully explain physiological regulation (Schulkin, 2004). Homeostasis cannot explain chronic conditions such as hypertension, obesity and diabetes. According to them the main goal of internal mechanisms cannot be constancy (regulation to one set point), but survival to reproduce under natural selection (Sterling, 2004). Many scientists have pointed out weaknesses in the homeostatic perspective. Many things within the 'internal milieu' are constantly changing and adapting. Some of the changes are programmed, such as circadian or seasonal rhythms, or the physiological changes associated with pregnancy. Others are acute responses to challenges. But this is not an inherent contradiction of homeostasis. The term allostasis was proposed for physiological processes that are not homeostatic and that oppose, at least temporarily, stability (Power, 2004).

Through allostasis – the ability to achieve stability through change – the bodily and regulatory systems protect the body by responding to internal and external stress (McEwen, 1998). This accommodation to stress, McEwen named 'Allostatic load', the wear and tear that results from chronic over- or under-activity of those systems. He defined four basic conditions that can lead to allostatic load (McEwen, 2004: p. 71):

- Too much stress in the form of repeated novel events.
- Failure to adapt to the same stressor.
- Failure to shut off response after the stressful situation is over.
- Inadequate response that allows other systems to become overactive (compensatory activities).

Parallel to the homeostasis–allostasis theories, additional views have been introduced. Antoine Béchamps (1816–1908) was convinced that harmful bacteria (he named them microzymas – 'micro' meaning 'small' and 'zymas' referring to a special class of immortal 'enzymes') arise because the body is out of balance; they do not cause the imbalance, as the Germ theory suggests (Béchamps, 1912). Later it was proposed that the microzymas Béchamps was referring to might have been 'RNA' (ribonucleic acid) or 'microsomes' (Nicolle and Woodriff Beirne, 2010). The elucidation of the double helix structure of DNA (deoxyribonucleic acid) in 1953 (Watson and Crick) has resulted in enormous steps forward in explaining causes and development of diseases.

To be more successful in determining the health and comfort effects of certain indoor environmental aspects it seems essential to understand the mechanisms behind how and why people respond to external stressors. We need to know how the human body and mind receive, perceive and respond to certain environmental conditions. Not every person receives, perceives and responds in the same way, due to physical, physiological and psychological differences but also due to differences in history, context and situation.

In the beginning of the twenty-first century, we find that response mechanisms of the bodily systems that keep our body and mind healthy are being studied from different angles (e.g. psychology, physiology, genetics and neurology), all adding to the further explanation of how the human body and its systems perceive, control and respond to environmental stressors in order to keep its balance.

---

**Box 1.2  A human body and mind as a complex system**

Complexity theory examines systems that are capable of becoming chaotic and are open to receiving input from outside themselves. . . . Thinking in system terms requires that we focus on the relationships among the elements that interact to compose the system. An example of a complex system is a cloud – a collection of water molecules capable of random distribution (it can be chaotic), and which receives light and energy such as wind and heat from outside itself (it is open). Human lives also meet these criteria – we are open systems capable of chaotic behaviour.

(Siegel, 2011: p. 68)

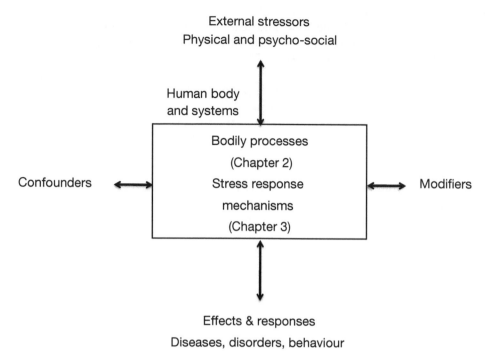

**Figure 1.1** *Human model: external stressors initiate several stress response mechanisms in which several bodily systems can be involved, under the influence of confounders and modifiers*
Source: Bluyssen.

In part I of the underlying book an attempt is made to answer therefore the following question:

How and why do people respond to external stressors?

Information from previous studies and available information from different disciplines is used to explain better the human model illustrated in Figure 1.1. Depending on how the body and mind cope with acute or chronic stress, stressors (both psychosocial and physical) may initiate (several) stress mechanisms in or with the bodily human systems, which immediately or over time may cause physiological, physical and psychological changes (effects and responses). Personal factors, such as state and traits, but also previous exposures and circumstances and other factors (confounders and modifiers), may influence the perception of, coping with and responses to those stressors.

## 1.2  Human systems and effects

### 1.2.1  Indoor-environment-related health effects

Recent studies have indicated that indoor building conditions may be associated with mental health effects (Houtman *et al.*, 2008), illnesses that take longer to manifest (e.g. cardiovascular disease and lung cancer (Lewtas, 2007; Babisch,

2008)), a variety of asthma-related health outcomes (Fisk *et al.*, 2007) and obesity (Bonnefoy *et al.*, 2004; WHO, 2011a) (see Boxes 1.3, 1.4 and Table 1.1). Additionally, in office buildings, a whole range of effects have been associated with indoor environmental stressors such as Sick Building Syndrome (SBS), building-related illnesses (e.g. humidifier fever and Legionnaires' disease) and productivity loss (Bluyssen, 2009a). People in the Western world in general spend 80–90 per cent of their time indoors (e.g. at home, at school, at the office, etc.). And the increased asthma prevalence in most countries in the past decades – it has become the first chronic disease of childhood (Eder *et al.*, 2006) – seems to put a finger to the indoor environment of schools and homes.

---

## Box 1.3  Major diseases associated with indoor building conditions

*Cancer*:   Cancer is a leading cause of death worldwide, accounting for 7.6 million deaths (around 13% of all deaths – 57 million) in 2008. Lung, stomach, liver, colon and breast cancer cause the most cancer deaths each year. Tobacco use is the most important risk factor for cancer, causing 22 per cent of global cancer deaths and 71 per cent of global lung cancer deaths (WHO, 2012a). Radon is considered to be the second cause of lung cancer. Long-term exposure to low concentrations of particulate matter (e.g. combustion particles, cleaning products) is associated with mortality (Carrer *et al.*, 2008). Annual cancer death is projected to increase by 4 million (WHO, 2011a).

*Cardiovascular diseases (CVDs)*:   17.3 million people died from CVDs in 2008, representing 30 per cent of all global deaths. Of these deaths, an estimated 7.3 million were due to coronary heart disease and 6.2 million were due to stroke. By 2030, almost 23.6 million people will die from CVDs, mainly from heart disease and stroke. Behavioural risk factors are responsible for about 80 per cent of coronary heart disease and cerebrovascular disease. Stress is also an important determinant (WHO, 2011b). Exposure to particulate matter indoors is associated with cardiovascular diseases, lung cancer, asthma, etc. (Lewtas, 2007). Annual cardiovascular disease is projected to increase by 6 million (WHO, 2011a).

*Chronic respiratory diseases*:   4.2 million people died from respiratory diseases in 2008 including asthma and COPD (chronic obstructive pulmonary disease) (WHO, 2010). 235 million people currently suffer from asthma. It is the most common chronic disease among children (WHO, 2011c). Building dampness and moulds are associated with 30–50 per cent increase in a variety of asthma-related health outcomes (Fisk *et al.*, 2007). More than 3 million died of COPD in 2005 (equal to 5 per cent of all deaths globally that year), and it is predicted that in the next 10 years this death rate will increase by 30 per cent if nothing is undertaken (WHO, 2011d). Active smoking is the most important risk factor, but other noxious particles and gases seem to be possible risk factors as well (e.g. particles, moulds) (Carrer *et al.*, 2008).

*Obesity*:   In 2008, 1.5 billion adults, 20 and older, were overweight. Of these over 200 million men and nearly 300 million women were obese. Nearly 43 million children under the age of five were overweight in 2010 (WHO, 2011e). Obesity (leading to increases in diabetes and CVDs) has been related to indoor conditions (Bonnefoy *et al.*, 2004).

*Diabetes*:   346 million people worldwide have diabetes. In 2004, an estimated 3.4 million people died from consequences of high blood sugar. WHO projects diabetes deaths will double between 2005 and 2030. Healthy diet, regular physical activity, maintaining a normal body weight and avoiding tobacco use can prevent or delay the onset of type 2 diabetes (WHO, 2011f).

*Depression*:   Depression is common, affecting about 121 million people worldwide. Depression is among the leading causes of disability worldwide and the fourth leading contributor to the global burden of disease (WHO, 2011g). A shift from physical complaints towards mental illness (depression) related to indoor environment has been observed (Houtman *et al.*, 2008).

---

### Box 1.4  How many people become sick each year and are sick at any given time?

Diarrhoeal disease (in 2004: 4,620.4 million) and lower respiratory infections (in 2004: 429.2 million) (both episodes of illness) are respectively the first and second most common cause of illness globally (not including upper respiratory tract infections such as common cold, and allergic rhinitis (hay fever)).

Worldwide, at any given moment, more individuals have iron-deficiency anaemia than any other health problem (in 2004: 1,159.3 million). Other very common conditions, with varying levels of severity, include asthma (234.9 million), arthritis (175.1 million), vision (incl. blindness: 315.1 million) and hearing (636.6 millions) problems, migraine (324.1 million), major depressive episodes (151.2 million) and intestinal worms (150.9 million).

Hearing loss, vision problems and mental disorders were the most common cause of disability in 2004!

(WHO, 2008)

---

**Table 1.1** *The leading causes of burden of disease in 2004 and projected in 2030*

|          | 2004                             | 2030 (projected)              |
|----------|----------------------------------|-------------------------------|
| Rank 1   | Lower respiratory infections     | Unipolar depressive disorders |
| Rank 2   | Diarrhoeal diseases              | Ischaemic heart disease       |
| Rank 3   | Unipolar depressive disorders    | Road traffic accidents        |
| Rank 4   | Ischaemic heart disease          | Cerebrovascular disease       |
| Rank 5   | HIV/AIDS                         | COPD                          |
| Rank 6   | Cerebrovascular disease          | Lower respiratory infections  |
| Rank 7   | Prematurity and low birth weight | Hearing loss, adult onset     |
| Rank 8   | Birth asphyxia and birth trauma  | Refractive errors             |
| Rank 9   | Road traffic accidents           | HIV/AIDS                      |
| Rank 10  | Neonatal infections and other    | Diabetes mellitus             |

Source: adapted from WHO, 2008, p. 51, figure 27.

## 1.2.2  Human systems

A number of human systems are available to the human body to make certain our body can maintain its balance and respond adequately to external and internal stressors and needs. Those systems can be divided into systems that take care of the basic bodily processes and those that regulate or control the basic bodily and other processes required to keep the body healthy and comfortable.

Basic bodily processes comprise of uptake and digestion of food, exchange of metabolic products via the blood circulation, exchange of gasses via the respiratory system, excretion of waste products, movement via the skeletal and muscular system, and reproduction. Exchange of metabolic products between the cells of the human body and the blood circulation are essential to *survive* and *grow*. For uptake and digestion of food, the digestive system and the liver are mainly responsible. *The digestive system* breaks down food into proteins, vitamins, minerals, carbohydrates and fats (required for energy, growth and repair), after which it is transported via the circulatory system. *The circulatory system*, comprising the heart, arteries and veins, is the body's transport system. *The respiratory system* (which includes the lungs) removes carbon dioxide and brings oxygen into the body required for *metabolism*. And for movement of the body the *skeletal and muscular system* is used. The secretory and digestive system (e.g. liver, kidneys and skin) takes care of the excretion of waste and foreign matter, the water and salt balance, and the exchange of heat with the external environment.

Three main control systems regulate the processes required to keep our body and mind healthy: the nervous system, the immune system and the endocrine system. External stress factors can influence all three control systems of the human body and can result in both mental and physical effects (Vroon, 1990; Kapit *et al.*, 2000). Our emotions and evaluations are controlled by our limbic system (part of the central nervous system) and other parts of the brain, and the autonomic nervous system keeps the parasympathetic and sympathetic activity in balance. The defence of our human body against (potential) disease (e.g. irritation, allergy, infection, toxicity) caused by stimuli from the environment are controlled (or, better, fought against) by the *immune system* that produces cytokines, which are transported by the lymphoid system. The *endocrine system* receives and sends information via blood vessels to endocrine glands that produce specific hormones, and provides boundary conditions for 'control' of environmental stimuli by our immune as well as our limbic system. These systems are thus pretty much intertwined and interactions occur. In a neuro-hormonal system two sources of stimuli are used in the control: other brain areas mediating exogenous (environmental) stimuli and stresses as well as endogenous rhythms, and feedback signals from the target hormones in the plasma. This results in a dynamic control, adjusting its operation to the needs of the body. Interactions between the endocrine system and the immune system are also not uncommon. Hormones can modulate the sensitivity of the immune system.

## 1.3 Stressors and response mechanisms

### 1.3.1 Stressors

Stress may be defined as '*a physical, chemical or emotional factor that causes bodily or mental tension and may be a factor in disease causation*' (Merriam-Webster Dictionary, 2011). A stressor or stress can trigger a mechanism or several mechanisms, and can cause an effect (or multiple effects) immediately (within seconds), or in the medium term (within minutes to hours) or long term (days to years). When exposed to this stressor, stress or stress situation continuously (chronically) or repeatedly, this may lead to an imbalance of the bodily systems, characterized by changes in production of hormones and cytokines and other physiological processes to restore the balance.

Stressors can be divided into stressors that are caused by the physical environment and those that are caused by the psychosocial environment: physical and psychosocial stressors (see Table 1.2). Psychosocial stressors can be categorized into discrete stressful events or stressors and chronic stressors.

**Table 1.2** *Physical and psychosocial stressors*

| Category of stressors | Stressor |
|---|---|
| **Psychosocial stressors** | |
| *Discrete stressful events or stressors* | |
|    – Major life events | *Positive events*: e.g. marriage, birth of a child. |
| | *Negative events*: e.g. job loss, death of family member. |
|    – Minor daily events | e.g. oral speech, catching the bus on time. |
| *Chronic stressors* | |
|    – Community stressors | *Locally*: e.g. crowding (neighbourhood), social disorganisation, neighbourhood quality, safety (crime and violence) and poverty. |
| | *Nationally*: e.g. fear and economic deprivation, social support, racial segregation and income inequality. |
| | *Worldwide*: e.g. fear for climate and energy related problems, epidemic diseases and economic depression. |
|    – Individual stressors | *At work*: e.g. working hours, high demands and low control, flexible versus same working place and working pattern. |
| | *At home*: e.g. financial stress, access to health care, family turmoil, marital problems, cold-harsh parenting and separation from family. |
| | *During commuting conditions*: e.g. travel time, queuing and transportation means. |
| **Physical stressors** | |
| *Noise or acoustical quality* | Influenced by outside and indoor noise as well as vibrations. |
| *Indoor air quality* | A complex phenomenon comprising odour, indoor air pollution, fresh air supply, etc. |
| *Visual or lighting quality* | Determined by view, illuminance, luminance ratios, reflection and other parameters. |
| *Thermal comfort or indoor climate* | Comprising parameters as moisture, air velocity and temperature. |

Source: Cohen *et al.*, 1995; Gee and Payne-Sturges, 2004; deFur *et al.*, 2007; Pieper, 2008; Bluyssen, 2009a.

*Discrete stressful events or stressors* are stressors with a beginning and ending, comprising major and minor life events. Major life events can be positive (e.g. a marriage or a birth) or negative (e.g. job loss or death) (Cohen *et al.*, 1995). Minor daily events causing stress can be giving an oral speech or catching the train on time (Pieper, 2008). *Chronic psychosocial stressors*, characterized by their continuous presence or high frequency, comprise stressors in the social environment (community), such as crowding, racial segregation or poverty, and individual stressors such as working hours, financial stress or travel time (Gee and Payne-Sturges, 2004).

The physical stressors caused by the indoor environment considered in the underlying publication are *noise or acoustical quality*, *indoor air quality*, *visual or lighting quality*, and *thermal quality* (see Bluyssen, 2009a, for details). In addition, *ergonomics*, such as the dimensions and sizes of the space, tools, furniture, and so on, play an important role in the total body perception. Although significant, this will not be discussed in this book. It is mentioned only briefly in Section 2.3 (in relation to movement, skeleton and muscles).

## 1.3.2 Stress response mechanisms

Several stress mechanisms have been or are being investigated. Roughly those mechanisms can be divided into two categories: mechanisms originating with the endocrine system (anti-stress mechanism, disturbance of sleep–awake rhythm and endocrine disruption) and mechanisms originating with the immune system (oxidative stress, inflammation and cell death and changes), afterwards affecting other bodily systems (see Boxes 1.5 and 1.6). While the psychosocial stressors are mainly involved in the HPA-axis mechanism and to some extent in the disruption of sleep–awake rhythm, the physical stressors have in general more relations.

---

### Box 1.5  Stress response mechanisms in general originating with the endocrine system

*Anti-stress mechanism*:   With prolonged stress, caused by external stressors (physical or psychosocial), the production of corticotrophin may increase causing an increased production of anti-stress glycocorticoids (e.g. cortisol) via the HPA-axis (hypothalamus–pituitary–adrenal-axis) (McEwen, 1998). These hormones can lead to physiological responses such as increases of heart rate, ventilation (breathing), myocardial contraction force, arterial vasodilation to working muscles, vasoconstriction to nonworking muscles, or dilating pupils and bronchi. The effects of chronic stress have been related to a chronic *imbalance in hormones* released during stress, causing or contributing to in the long-term degradation of the immune system, hypertension, vascular disorders, changes in carbohydrate and fat metabolism *(obesity)*, anxiety, depression, heart disease, fatigue, allergies and asthma.

*Disturbance of sleep–awake rhythm*:   Improper lighting, noise pollution during the night, and even thermal discomfort during night, can lead to sleep disturbances (Muzet *et al.*, 1983; Brainard *et al.*, 2001; Boyce, 2003; Muzet, 2007). Under the influence of light, the hypothalamus signals to the pineal body to produce melatonin via the RHP-axis (retino–hypothalamic–pineal–axis), a hormone that makes us want to sleep. If exposed to light during the night, the production of the antioxidant melatonin is immediately stopped, alertness and core body temperature is increased and sleep is distorted. Consequences of distorted sleep can include impaired alertness, memory, performance, and disturbed endocrine functions, and upset gastrointestinal function (Klerman, 2005), but also (winter) depression and even cancer, obesity and diabetes have been related to circadian disruption (Duffy and Wright, 2005; Stevens and Rea, 2001; Stevens *et al.*, 2007; Straif *et al.*, 2007).

*Endocrine disruption*:   In the indoor environment, inhalation of air (Adibi *et al.*, 2008), ingestion of house dust (Bornehag *et al.*, 2005a) and uptake via skin (Weschler and Nazaroff, 2012) have been considered important pathways for endocrine-disrupting chemicals (EDCs) such as phthalates and flame retardants. EDCs may alter endocrine function by affecting the availability of a hormone at a target issue or the cellular response to a hormone (e.g. receptor binding) (Diamanti-Kandarakis *et al.*, 2009; Marty *et al.*, 2011). Along with endocrine effects, EDCs can have neurobiological and neurotoxic effects. Besides those neuroendocrine systems, the cardiovascular and the respiratory system are also a target of environmental chemicals that interfere with intracellular signalling of hormonal and inflammatory pathways (Newbold *et al.*, 2007). Exposure to EDCs may play a role in several diseases such as reproduction disorders including malformation of genitals, asthma and allergy, neurodevelopment disorders (e.g. autism), overweight/obesity, and diabetes among others (Walker and Gore, 2007; Bornehag, 2009).

## Box 1.6  Stress response mechanisms in general originating with the immune system

*Oxidative stress*:   Oxidative stress can be caused by air pollution (lungs, eyes and brain, all organs) (Ayres *et al.*, 2008), light (skin and eyes) (Boyce, 2003; Halliwell and Gutteridge, 2007) and noise (ears) (Seidman and Standring, 2010). Oxidative stress occurs when free radicals (which steal electrons) overwhelm antioxidant defences (Kelly, 2003). Removal of electrons from cells through oxidation can create highly reactive oxygen species (ROS). Oxidative stress can damage cellular proteins, membranes and genes and can lead to systemic inflammation. Damage to circulating triglycerides has been implicated in cardiovascular disease. Oxidation of LDL (low density lipoprotein) cholesterol has been shown to incite inflammatory mechanisms that are detrimental to vascular cells. ROS can also damage cellular DNA, initiating changes implicated in cancer (Halliwell and Gutteridge, 2007). At a higher level of oxidative stress, cytotoxic effects may induce cell death (Li *et al.*, 2008).

*Inflammation*:   Inflammation is the body's response to the cause of the infection (an allergic reaction). Two inflammatory conditions are asthma (usually caused by an allergy to an ingested substance) and pneumonia, an inflammatory reaction to the invading microorganisms in the lungs (Nadler, 2007). The key exposure indoor environmental agents for asthma include: microbial agents (e.g. endotoxin of gram negative bacteria, fungal spores and fragments, bacterial cells, spores and fragments, microbial metabolites, and allergens such as house dust mites, pet and fungal allergens); chemical agents (e.g. formaldehyde, aromatic and aliphatic chemical compounds, phthalates, and indoor chemistry products); and particles (e.g. ETS (environmental tobacco smoke), indoor ultrafine particulate matter, wood or oil smoke, soot and exhaust) (Carrer *et al.*, 2008).

*Cell death and cell changes*:   Both cell death and changes to cells are processes that are a normal part of cell processes. External stressors can cause unwanted or unintended cell death and cell changes. Necrosis (passive form of cell death associated with cell swelling) can occur with exposure to an intensive physical or chemical insult, for example, after traumatic noise exposure or inhalation of toluene. Apoptosis (active form of cell death) can be initiated at the wrong time causing crucial cells to die, for example, during a stroke, heart failure or autoimmune disease. Cell changes (mutation or altered gene expression) can occur through damage (e.g. by phthalates), inhibition or mediation of certain functions (e.g. lead competing with calcium), epigenetic changes, cumulative effects from certain exposures (e.g. dioxins), or exposure to radioactive substances and certain volatile organic compounds (VOCs; e.g. formaldehyde and benzene) (Miller and Ho, 2008; Ash, 2010; Müller and Yeoh, 2010; Puri and Lynam, 2010).

*Noise* is typically defined as an unwanted sound or combinations of sounds that may adversely affect people (Seidman and Standring, 2010). The mechanisms of physiological damage from noise are not completely understood, but several mechanisms have been demonstrated:

- Traffic noise exposure has been associated with changes in stress hormone levels and with cardiovascular changes (e.g. Babisch *et al.*, 2001; Babisch, 2006), and has been related to the parasympathetic and sympathetic balance (Graham *et al.*, 2009).
- Road traffic noise has been identified as a major cause of sleep disturbance (Berglund *et al.*, 1999a; Muzet, 2007).
- Oxidative stress in chronic noise exposure leading to noise-induced hearing loss (Henderson *et al.*, 2006; Darrat *et al.*, 2007; Seidman and Standring, 2010).

- Acoustic trauma causing mechanical disruption of the cochlea, which may result in permanent hearing loss (Henderson *et al.*, 2006).

*Indoor air pollutants* can:

- Be odorous and lead to annoyance or pleasure (Herz, 2002; Köster, 2002).
- Stimulate the trigeminal nerve endings in nose and eyes, causing irritation (Bluyssen, 2009a).
- Disrupt endocrine function (Bornehag *et al.*, 2005a; Newbold *et al.*, 2007; Walker and Gore, 2007; Adibi *et al.*, 2008; Marty *et al.*, 2011).
- Cause oxidative stress (Pope and Dockery, 2006; Ayres *et al.*, 2008), inflammatory and allergic responses (Nadler, 2007; Carrer *et al.*, 2008).
- Induce cell alterations and even cell death (Goyer, 1995; Henderson *et al.*, 2006; Beil, 2008; Miller and Ho, 2008; Müller and Yeoh, 2010; Ash, 2010).

*Radiation (light)* wrongly used or exposed can be looked upon from several angles:

- Visual discomfort, which can lead to eyestrain (Boyce, 2003).
- Improper lighting that can cause disturbance of the circadian rhythm (Cajochen *et al.*, 2000; Brainard *et al.*, 2001; Duffy and Czeisler, 2009).
- Damage of both eye and skin through both photochemical and thermal mechanisms (Boyce, 2003; Halliwell and Guteridge, 2007).
- Effect of different colours on an individual's impression of the environmental parameters of thermal comfort, sound and light (Mahnke and Mahnke, 1996).

*Thermal stress* occurs when one is not able to regulate thermal balance or when one believes or perceives this isn't possible. The psychological effect of expectations and the perceived individual level of control seems important (Auliciems, 1981; Wilson and Hedge, 1987; Bluyssen *et al.*, 2011a). However, recent studies indicate that increased exposure to thermal neutral conditions might be related to increased adiposity (Marken Lichtenbelt *et al.*, 2009; Johnson *et al.*, 2011). Additionally, it has been shown that the thermal environment can affect sleep, specifically the REM (rapid eye movement) sleep (Muzet *et al.*, 1983; Libert *et al.*, 1988).

## 1.3.3 Moderators and mediators

How we evaluate and respond to our environment not only depends on the external stressors involved (physical and psychosocial), but also on personal factors and processes that occur over time (memory and learning), influenced by past events and episodes. They all determine the way external stressors are handled at the moment or over time.

### Memory and learning

Our memory plays an important role in the way we recognize, organize and make sense of the sensations we receive from environmental stimuli. Learning, a change in a response to a stimulus as a result of experience, influences not only our bodily responses to these stressors, but can even play tricks on us.

It is said that physiological effects may be conditioned following the same rules as for classical conditioning (Riether *et al.*, 2008). Classical conditioning occurs when a neutral stimulus is received just before an unconditioned stimulus, one that automatically elicits a behaviour. Some responses occur reflexively (unconditioned response), others are *conditioned*, or caused by associative learning. Pavlovian conditioning has been shown for fear conditioning with an olfactory, auditory or visual conditioned stimulus (CS) (Otto *et al.*, 2000), and has been seen in asthmatic patients suffering from skin sensitivities to house dust and grass pollen, who were exposed to these allergens by inhalation. After series of conditioning trials, they experienced allergic attacks after inhalation of the neutral solvent used to deliver the allergens (Riether *et al.*, 2008). It is hypothesized that individuals who suffer from IEI (idiopathic environmental intolerance) or MCS (multiple chemical sensitivity) have stronger implicit associations between odours and sickness, than healthy individuals (Bulsing *et al.*, 2009).

Some studies even indicate that the explanation for certain *unexplained symptoms* lies in the way we process information and that this processing may be influenced by our personality, expectations and the duration of a certain stress response (Matthews and Mackintosh, 1998; Brown, 2004; Pieper, 2008). It is hypothesized that unexplained symptoms (physical illness for which no adequate reason can be found, e.g. symptoms of somatic illness such as pain, fatigue, general malaise) can arise when chronic activation of stored representations in memory causes the selection of inappropriate information (Brown, 2004). According to Brown (2004) all people possess material about symptoms that could provide the basis for these so-called rogue representations. For example physical components of emotional states, such as those associated with anxiety, leave representations in the memory that could provide the basis for the later development of unexplained symptoms. *Self-focused attention* (as a feature of depression and anxiety) and negative affect but also boring environments seem to have an influence on this process. Both negative affect and the process of symptom misattribution trigger *illness worry and rumination*, named *preservative cognition* (the repeated or chronic activation of the cognitive representation of stress-related content) by Pieper (2008), in which the duration of the stress response is important.

## Personal factors

Personal factors are factors that are related to the person and that can influence the way that environment (the stressors) are handled (Table 1.3 and Figure 1.2). Personal factors that may have a relationship with the exposure and/or the response, include *demographic variables* (e.g. age, gender, ethnicity, social status, income and education), *states and traits* (personality at the moment (motivation and emotion) and over a longer time), *life style and health status* (life-style-related variables such as food patterns and physical activity, and health state variables such as allergies and obesity), and *genetics, events and exposures* (e.g. the risk for certain diseases and disorders, premature birth and low birth weight, substance abuse and attitudes).

**Table 1.3** *Several personal factors that can affect the way we perceive and respond to stressors at the moment and over time*

| States and traits | Demographic variables | Lifestyle and health status | Events and exposures |
|---|---|---|---|
| Negative and positive affect | Age | Food pattern (diet/nutritional status) | Exposures in womb and as child |
| Worry – nervousness – fear – anxiety (neuroticism) | Gender | Physical activity (exercise regime) | Previous smoking |
| Anger – hostility aggressiveness (psychoticism) | Ethnicity | Drugs (ab)use (smoking, coffee, alcohol, medications) | Expectations and worries Attitudes |
| Sadness – depression | Social status (education and income) | Obesity and diabetes | Episodes of depression and anxiety |
| Happiness – satisfaction – joy – ecstasy | | Allergies and asthma | Sleep deprivation and disruptions |
| Locus of control, coping skills, self-efficacy | | Pre-existing diseases (e.g cardiovascular diseases) | Traumatic experiences |
| Introversion – extraversion | | | Risk for certain diseases and disorders (genetics) |
| Boredom proneness – interest – intelligence | | | Daily activities |

Sources: among others, deFur *et al.*, 2007; Janssen and van Dongen, 2007; Pieper, 2008.

**Figure 1.2** *People*
Source: drawing by Sebastian Meertins (seven years old), September 2011.

## 1.4 Links with Parts II and III

From the synthesis of this Part I on the mechanisms of the human body, it can be concluded that in order to answer the question 'How and why do people respond to external stressors?', all potential stressors and factors that can have an influence on the balance of the human systems could potentially be relevant to include (see also Box 1.7). It can also be seen that all human systems, the basic and the control systems, are involved. Major health effects presented seem to be associated with more than one stress response mechanism and with more than one stressor (see Figure 1.3). Time, as in previous exposures, but also future anticipations, as well as in regards to duration of exposures, and as in combined exposures of stressors, need to be included!

While in Part II of this book – the inventory and discussion of different assessment techniques for indicators and parameters from the human-being, as well as the (physical- and psychosocial-)environment, point of view – points out the need for a different approach, Part III tries to specify that need or approach. It tries to give an answer on what we (still) need to be able to assess wellbeing of people indoors, based on an inventory of available performance indicators and risk analysis models for 'classifying' the health and comfort status of people in different scenarios.

---

### Box 1.7  The chicken or the egg?

In the global status of disease report on noncommunicable diseases 2010 (WHO, 2011a), cardiovascular diseases, cancers, diabetes and chronic lung diseases are said to be largely caused by four shared behavioural risk factors: tobacco use, physical inactivity, harmful use of alcohol and unhealthy diet. It could however be argued what was first: the chicken or the egg? Do psychosocial and physical stressors cause people to smoke more, move less, use more alcohol and eat unhealthy? Or is it the other way round?

---

**Figure 1.3**
*Possible associations between stressors, mechanisms and diseases and disorders*
Source: Bluyssen.

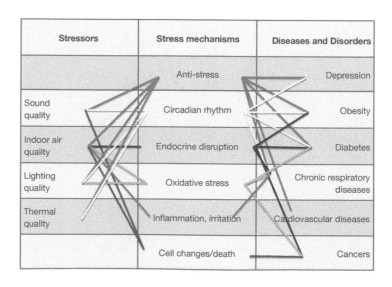

# 2
# Bodily processes

*In this chapter the systems that form the basis of our bodily processes (e.g. metabolism, digestion, circulation, excretion, respiration and motor activities) are described, as well as the mechanisms that regulate those processes and keep our body in balance: protection and defence by our immune system, hormonal regulation by our endocrine system, perception and behaviour by our nervous system, and last but not least the overall integration and control by the central nervous system. External stressors can influence all three control systems of the human body and can result in malfunctioning of the bodily processes, presented in both mental and physical effects. With that in mind this chapter provides the background knowledge required to understand the stress mechanisms that can lead to those effects.*

## 2.1 Introduction

The human body is an open system with an 'internal milieu' (Box 2.1), which on the one hand has to protect itself from the 'chaos' of the external environment, and on the other open itself to exchange heat, oxygen, food and waste products with the environment (Silbernagel and Despopoulos, 2008).

The cells of the human body are surrounded with fluid in the so-called extra-cellular space, which offers the conditions that are required to *survive* and *grow*. Through this fluid the exchange of metabolic products takes place with the blood *circulation* (via the interstitial fluid). For uptake and digestion of food, *metabolism* and *transport* through the body, the digestive system and the liver are mainly responsible. The lungs take care of the *exchange of gasses* (uptake of oxygen and exhaust of carbon dioxide), the liver and the kidneys of the *excretion* of waste and foreign matter and *osmoregulation* (water and salt balance), and the skin takes care of the exchange of heat with the external environment.

For all of the bodily processes to function properly, integration and regulation is required, which is established through the information transfer of the endocrine system (hormones), electrical signalling in the nervous system and transport processes between and in cells or over larger distances via blood and urine. Via hormones, the processes to regulate, for example, the sleep–awake rhythm and reproduction are controlled. The nervous system assures that the body activities and reaction to stimuli are regulated properly via *perception and behavioural* processes. To protect the human body from the external environment, we are equipped with *layered protection mechanisms*. The central nervous system (CNS) is responsible for the integration of all of the above.

---

### Box 2.1 Internal milieu

The organism is merely a living machine so constructed that, on the one hand, the outer environment is in free communication with the inner organic environment, and, on the other hand, the organic units have protective functions, to place in reserve the materials of life and uninterruptedly to maintain the humidity, warmth and other conditions essential to vital activity. Sickness and death are merely a dislocation of disturbance of the mechanism, which regulates the contact of vital stimulants with organic units. In a word, vital phenomena are the result of contact between the organic units of the body with the inner physiological environment.

(p. 76)

If we break up a living organism by isolating its different parts, it is only for the sake of ease in experimental analysis, and by no means in order to conceive them separately. Indeed when we wish to ascribe to a physiological quality its value and true significance, we must always refer to this whole, and draw our final conclusion only in relation to its effects in the whole.

(p. 89)

Only in the physico-chemical conditions of the inner environment can we find the causation of the external phenomena of life. The life of an organism is simply the resultant of all its inmost workings.

(p. 99)

(Claude Bernard, 1865)

---

More detailed information on the mechanisms and processes that take place in the human body can be found in Sections 2.2 to 2.7 (based on information acquired from Kapit *et al.* (2000) and Silbernagel and Despopoulos (2008) unless otherwise stated) (see Table 2.1 and Figure 2.1 on p. 20).

## 2.2 Metabolism

*Metabolism is the mechanism that allows our body to grow and reproduce, maintain its structure and respond to the environment. The chemical reactions that take place in our cells and tissues metabolize food into proteins, vitamins, minerals, carbohydrates and fats, which the body needs for energy, growth and repair.*

### 2.2.1 Cells and tissue

About two-thirds of a human's body weight comprises cells (Kapit *et al.*, 2000), in which chemical reactions take place that allow our body to grow and reproduce, maintain its structure and respond to the environment. Those chemical reactions are defined as metabolism and have two parts:

- *Catabolism*: in which food is broken down into small molecules to create chemical energy.
- *Anabolism*: in which energy is used to synthesize large molecules (e.g. the three basic classes of molecules: proteins, carbohydrates and fats (lipids)).

The two are linked through the reactions of phosphate transfer involving mainly ATP (adenosine triphosphate), the major carrier of energy for all forms of living matter (Encyclopaedia Brittanica, 1991), formed in the mitochondria of a cell (see Figure 2.2 on p. 21).

*Proteins* are large molecules made of amino acids forming a chain with a special chemical bond (peptide bonds). Tissues require amino acids for growth, repair and normal production of cellular proteins. The proteins, general and cell specific (such as antibodies from white blood cells (leukocytes) and haemoglobin (oxygen-transporting protein) found in red cells) are continuously formed on ribosomes and broken down in lysosomes (see Figure 2.2 on p. 21). Of the 20 amino acids

**Table 2.1** *Functions, systems, organs and processes/mechanisms*

| Function | System | Organs | Processes/mechanisms |
|---|---|---|---|
| Breakdown of food and absorption for use as energy (for internal and external activities) | Digestive system | Stomach, liver, teeth, tongue, pancreas, intestine, esophagus | Digestion (see Section 2.3) |
| | Cells and tissues | Cells | Metabolism (see Section 2.2) |
| Intake of oxygen and removal of carbon dioxide from body | Respiratory system | Lungs, nasal passages, bronchi, pharynx, trachea, diaphragm, bronchial tubes | Exchange of gases (see Section 2.3) |
| Transport of nutrients, metabolic wastes, water, salts and disease fighting cells | Circulatory system | Blood, blood vessels, heart, lymph | Circulation (see Section 2.3) |
| Protection and movement | Skeletal and muscular system | Bones, muscles | Motor activities (see Section 2.3) |
| Control of water and salt balance | Excretory system | Kidneys, bladder ureters, skin[1] | Excretion (see Section 2.3) |
| Reproduction | Reproductive system | Male and female reproductive systems | Hormonal cycles (see Box 3.16) |
| Protection of body from injury and bacteria | Immune system (incl. the lymphoid system) | Skin, bone marrow, thymus, spleen, lymph nodes and lymphoid tissue | Protection and defence (see Section 2.4) |
| Production of hormones and body regulation | Endocrine system | Pituitary gland, adrenal gland, thyroid gland, gonads | Hormonal regulation (see Section 2.5) |
| Control of body activities and reaction to stimuli | Peripheral nervous system | Nerves, sensory receptors and motor effectors | Perception and reaction (see Section 2.6) |
| Integration | Central nervous system | Brain centres and spinal cord | Integration and control (see Section 2.7) |

1 The skin can also be seen as a separate system, which protects the body from injury and bacteria, takes care of maintenance of tissue moisture, holds receptors for stimuli response, and performs body-heat regulation.

Source: Bluyssen.

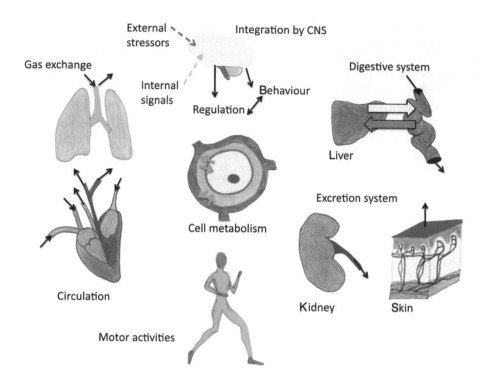

**Figure 2.1** *Maintenance of internal milieu: basic bodily processes supervised by the central nervous system (CNS) (including Figures 2.2, 2.5, 2.7 and 2.8)*
Source: Bluyssen.

proteins are made of, the body can only synthesize 12, starting with glucose or fatty acids. The other eight must come from the food we eat and drink. The sequence of the amino acids (the same can occur more than once) determines the protein and its properties. Over 100,000 proteins are thought to exist in the body.

*Fats* comprise of triglycerides (TG (triacylglycerols)), for example palmitic, stearic and oleic acids, which comprise of esters of glycerol (alcohol), and three fatty acids (FA). FA are long hydrocarbon chains with a single carboxylic acid group. In fat formation (esterification), the fatty acids are bound to glycerol and with lipolysis they are broken down. Omega-3 fatty acids (alpha-linolenic acid and its derivatives) and omega-6 fatty acids (linoleic acid and its derivatives) are so-called essential fatty acids, which are polyunsaturated fatty acids, because they are important to human health (Nicolle and Hallam, 2010). Body fats can be divided into fuel and structural types. The *fuel types* are stored in fat depots of the adipose tissue, comprising fat cells with large cytoplasmic stores of fat, which are continuously forming and degrading fats. The fuel types can be found in the abdominal cavity, within or around organs (muscle, heart) and under the skin. Brown fat is a form of a subcutaneous fat with many mitochondria. Those mitochondria oxidize the fat to produce mainly heat (not ATP) to protect against cold temperatures. The *structural types*, phospholipids in cell membranes and cholesterol, are not used for energy. Cholesterol is a steroid lipid with polar hydroxyl group and a non-polar hydrocarbon ring (see Box 2.2).

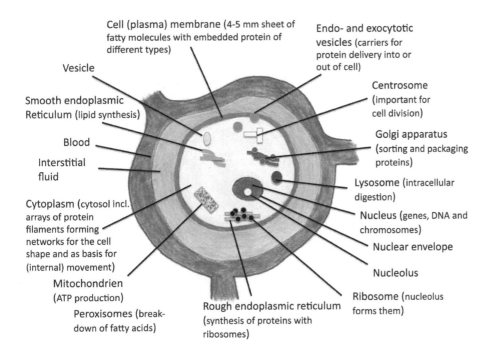

**Figure 2.2** *Cell structure: cells differ in size, shape and internal structure, but they do have several structures and organelles (mini-organs specialized to perform a certain function) in common*

Source: Bluyssen; based on Kapit *et al.*, 2000 and Silbernagel and Despopoulos, 2008.

## Box 2.2  Cholesterol and its relation to atherosclerosis and heart disease

Cholesterol may be formed from acetate in tissues (mainly the liver) or supplied by animal foods. Cholesterol has a function in the synthesis of steroid hormones in the gonads and adrenal cortex, the synthesis of vitamin D in the skin, and is a part of neural myelin tissue. In the liver, cholesterol serves as a precursor for bile salts (that help digest fat by emulsification – see Box 2.10 in Section 2.3.2) and in the skin. Cholesterol prevents the loss and permeation of water. Because cholesterol is only slightly soluble in water and insoluble in blood, it is transported in the circulatory system within lipoproteins. There are several types of lipoproteins within blood, for example, low-density lipoprotein (LDL) and high-density lipoprotein (HDL).

When the inner wall of an artery is damaged, platelets (cell fragments or cells without a nucleus) stick to the site of damage, stimulating fibrosis. Plasma cholesterol (in this case LDL cholesterol or 'bad' cholesterol) is deposited on these lesions together with calcium ions, forming hard, calcified cholesterol plaques. Plaques harden the arteries (arteriosclerosis) causing reduced elasticity and blood flow and increased blood pressure and clot forming. Principal causes of heart attacks and strokes are plaques in the heart or brain.

High concentrations of HDL cholesterol on the other hand seems to correlate with better health outcomes.

Source: based on Kapit *et al.*, 2000, plate 135.

---

## Box 2.3 pH

pH is the acidity or basicity of an aqueous solution. Pure water is neutral with a pH of 7. When pH < 7 the solution is acid, and when pH > 7 the solution is basic or alkaline. pH is defined as a negative decimal logarithm of the hydrogen-ion activity in a solution (Silbernagel and Despopoulos, 2008, p. 384):

$$pH = -\log_{10}(\alpha_{H_+}) \qquad\qquad [2.1]$$

Where $\alpha_H+$ is the activity coefficient of hydrogen ions in units of mol/l (molar concentration).

Activity has a sense of concentration, however activity is always less than the concentration and is defined as a concentration (mol/l) of an ion multiplied by an activity coefficient. The activity coefficient for diluted solutions is a real number between 0 and 1 (for concentrated solutions may be greater than 1) and depends on the many parameters of a solution, such as the nature of ion, ion force, temperature and so on. For a strong electrolyte, the activity of an ion approaches its concentration in diluted solutions.

---

*Carbohydrates* (e.g. starch, glycogen, sucrose) are straight-chain aldehydes or ketones with many hydroxyl groups that exist as straight chains or rings. They can be broken down into simple six-carbon sugars similar to glucose. The liver converts all digested carbohydrates into glucose. The liver releases glucose into the blood, while part of the excess glucose is stored within the liver as glycogen by glycogenesis (but this has a limit). The extra excess glucose entering the liver is converted into amino acids and proteins as well as fats (TGs) via the formation of glycerol and fatty acids.

Our food also contains minerals and vitamins. *Minerals* are solid inorganic chemicals such as magnesium, sodium, potassium and calcium, which can act as ions to maintain osmotic pressure and pH (see Box 2.3) but are also important for muscles and nerves. Most *vitamins*, which cannot be made by cells, function as co-enzymes after modification. A co-enzyme carries chemical groups between different reactions.

## 2.2.2 Mechanisms

### Cell division

Growth occurs mainly by increasing the number of cells by cell division rather than increasing the mass of individual cells. Old cells are replaced by new cells all the time. While brain cells may live for 60 years or more, on average, intestinal cells live for 36 hours, white blood cells for 2 days and red blood cells for 4 months. Each new cell needs to have an exact copy of the DNA present in the old cell. DNA is the hereditary material in humans and almost all other organisms. Almost every cell in a person's body has the same DNA, and most DNA is located in the cell nucleus (named nuclear DNA). A small amount can also be found in the mitochondria (named mitochondrial DNA or mtDNA). The information in DNA is stored as a code made up of four chemical bases: adenine (A), guanine (G), cytosine (C), and thymine (T) (see Figure 2.3).

Human DNA consists of about 3 billion bases; more than 99 per cent of those bases are the same in all people. The order, or sequence, of these bases determines the information available for building and maintaining an organism. Cell division takes place in three phases:

- *Interphase*: the cell increases in mass.
- *Mitosis*: DNA is replicated and moved.
- *Cytokinesis*: the cell divides.

## Protein synthesis

DNA contains detailed information on each protein that is synthesized, even though it never leaves the nucleus. Proteins control the enzymes (catalysts that 'turn on' a specific reaction simply by speeding it) and enzymes control the chemical reactions and cellular activities. A copy of the 'blueprint' of the DNA ladder (see Figure 2.3) is made with the help of RNA (ribonucleic acid) polymerase enzymes, resulting in RNA molecules such as messenger RNA (mRNA) and transfer RNA (tRNA). Both are transported into the cytoplasm where mRNA connects with particles called ribosomes (R) in the cytoplasm and tRNA collects unbounded amino acids that have been activated (energized) in preparation for use. Each tRNA, with a single amino acid, migrates to the ribosomes, where it binds with the R-mRNA complex, according to the matching of the codes in the mRNA with those in the tRNA, forming a polypeptide (protein) chain. The chain dissociates from the R when the synthesis is completed.

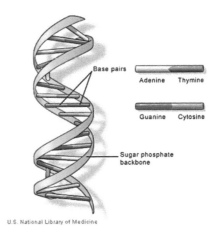

U.S. National Library of Medicine

**Figure 2.3** *DNA*

Source: 'What is DNA?', *Genetics Home Reference*, http://ghr.nlm.nih.gov/handbook/basics/dna, accessed 9 August 2011; public domain information.

## Energy supply in cell

Metabolic rate (MR) is defined as the rate of oxidation of foodstuff per unit of time. Metabolic rate measured under basal conditions (resting, fasting) is named basal metabolic rate (BMR). Oxidation of food generates heat to keep the body warm and increase the speed of the chemical reactions, and it generates ATP. Proteins, fats, and carbohydrates are important for ATP generation, which constitutes the energy supply in the cells. ATP is the central co-enzyme for transferring chemical energy between different chemical reactions (Box 2.4). ATP can be formed with or without oxygen ($O_2$): aerobically or anaerobically.

During the *aerobic metabolism*, glucose, glycerol and fatty and amino acids are recovered from proteins, fats and carbohydrates present in our food. Most of them are then broken down into acetate (a two-carbon fragment) that attaches to the co-enzyme A (CoA) forming the compound acetyl-CoA (see Figure 2.4) (Kapit *et al.*, 2000). During *anaerobic metabolism* or glycolysis, ATP is created from carbohydrates only, without using oxygen. Glucose is broken down into pyruvate (a three-carbon structure) forming lactic acid (also produced in muscles during exercising) and energy, which is captured in ATP.

## Box 2.4 ATP and energy

ATP transfers energy from food to the cellular machine:

$$ATP \Leftrightarrow ADP + P + energy \qquad\qquad [2.2]$$

Where:

ADP = adenosine diphosphate

P = phosphate.

ATP is composed of a ring of nitrogen containing ring compound, adenine, which is linked to three phosphates via ribose, a five-carbon sugar. Removing a phosphate will leave ADP.

Source: Kapit *et al.*, 2000, plate 5.

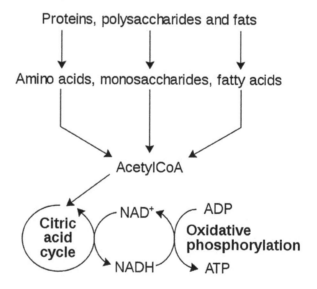

**Figure 2.4** *A simplified outline of the catabolism of proteins, carbohydrates and fats. Acetyl-CoA enters the so-called citric acid cycle, in which citric acid is transferred into six different acids. The energy released is used in the respiratory chain to produce NADH from $NAD^+$ as actyl-CoA is oxidized with $CO_2$ (carbon dioxide) as a waste product. During the re-oxidation of NADH the energy that is released is used to make ATP*

Source: Tim Vickers, 'A simplified outline of the catabolism of proteins, carbohydrates and fats', http://en.wikipedia.org/wiki/File:Catabolism_schematic.svg; public domain.

While the anaerobic energy pathway produces energy for short, high-intensive activities lasting no more than several minutes, during aerobic metabolism far more ATP is produced, used for long duration activity. The citric acid cycle and ATP synthesis take place within the mitochondria. Other functions (e.g. glycolysis) occur in the cytoplasm (see Figure 2.2).

## Transport between cells

Differences in concentrations, pressure and electrical charge between the two sides of the membrane of a cell are responsible for movements (respectively *diffusion*, *bulk flow* and *flow of ions*) through the membrane. Additionally, osmosis, the process in which water flows to the side of the membrane that has the most concentrated solute, creates an osmotic gradient (*osmotic flow* is from low to high and is responsible for the swelling and shrinking of tissue) (see Box 2.5).

Lipid soluble solutes (e.g. steroid hormones, fat-soluble vitamins, oxygen and carbon dioxide) dissolve in the lipid bilayer portions of the membrane and diffuse to the other side passively (down its concentration gradient). The more polar solutes, including ions, glucose and amino acids, move through special pathways provided by proteins in cell membranes, via channels, such as small solutes like $Na^+$ (sodium ion), or by facilitated diffusion. In facilitated diffusion, the solute binds to a protein carrier that goes back and forth in the membrane (such as glucose) until concentrations on both sides are the same.

When energy is needed for a facilitated diffusion transport (derived from the splitting of ATP), in the case of transport against a gradient, active transport takes place. This is called *primary active transport*. It is called *secondary active transport* when a passive solute transport is used to transport another solute. *The sodium-potassium pump* ($Na^+$–$K^+$ pump) is an active transport system that pumps $Na^+$ out of and $K^+$ into body cells. The energy this pump uses accounts for more than a third of the energy consumption of the body in rest. Other ATP-powered ion pumps are $Ca^{++}$ATPase (in all cell membranes and endoplasmic reticulum), $H^+$ATPase (maintains acid interior in lysosomes and endocytotic vesicles) and $H^+$-$K^+$ ATPase (regulates acidity in the stomach and kidney).

## Communication between cells

Cell communication takes places via signal molecules (e.g. hormones and neurotransmitters) to control vital processes including metabolism, movement, secretion and growth. Those signal molecules (or first messengers) can bind to the external surface of a receptor protein, causing a change in the receptor. This change activates the receptors site exposed to the cytoplasm, which then activates G-protein (GDP is exchanged for GTP in an analogue of the relationship between

---

### Box 2.5 Osmosis

Osmosis is diffusion of water due to an osmotic pressure difference.

When osmotic pressure is low (concentration of water molecules is lower in cell plasma than in interstitium), water is drawn into cells and the cells swell. This can lead to symptoms such as nausea, malaise, confusion, lethargy, seizures and coma.

When osmotic pressure is high (concentration of water molecules is higher in cell plasma than in interstitium), water is pushed out of cells and cells shrink. This can lead to symptoms such as lethargy, weakness, seizures and coma.

ATP and ADP) that splits into two activated parts (alpha and beta-gamma) activating effectors. These effectors catalyse the formation of many intracellular cAMP (cyclic adenosine monophosphate) molecules, so-called second messengers that carry the signal into the cytoplasm where it may diffuse to any site. Second messengers activate an enzyme named protein kinase A, which catalyses the phosphorylation of other enzymes or target proteins, causing a specific response (e.g. secretion, division). An enzyme that catalyses phosphorylation is named a kinase; an enzyme that catalyses dephosphorylation is called a phophatase. The process ends when the signal molecule leaves the receptor. Alpha and beta-gamma are reunited. cAMP levels decay.

G-proteins refer to a large family of regulatory molecules that bind to GTP. Some activate, some inhibit. Some *G-proteins act directly on membrane channels*, they do not need a second messenger. For example, binding of acetylcholine (neurotransmitter) by receptors on the heart cell membrane activates a G-protein that dissociates into the alpha and beta-gamma complex. The beta-gamma part migrates to the $K^+$ channels and opens them (result: slowing heart rate).

In many cells, for example, smooth muscle, liberation of $Ca^{++}$ into the cytosol from internal storage reservoirs is mediated by a *G-protein system* (including protein kinase C activated by $Ca^{++}$). In some cells, for example, nerve endings, $Ca^{++}$ enters the cytosol through $Ca^{++}$ channels that are opened (activated) by changes in the membrane potential (*ion channels*) (see Section 2.6.2). The receptor is thus not linked to G-proteins. Other receptors not linked to G-proteins include *intracellular receptors* for lipid-soluble hormones and receptors that are enzymes (*catalytic receptors*).

Maintaining GJIC (gap junctional intracellular communication) is crucial. The GJIC mechanisms link the extracellular endogenous communication factors, which represent thousands of entities (secreted factors, such as hormones, cytokines, chemokines, growth factors, nutrients, pressure and tension on cell membranes), specific extracellular matrices, and an almost infinite number of exogenous chemicals to a finite number of intracellular signalling pathways (Kang and Trosko, 2011).

### 2.2.3 Diseases and disorders

Several diseases have been related to the processes in and around the metabolism of our body, especially related to *excess of fats*, also named obesity (see Box 2.6). A cluster of metabolic risk factors (named metabolic syndrome by some researchers and insulin resistance syndrome by others) seem to relate to cardiovascular diseases (Grundy *et al.*, 2004):

- Obesity and abnormal body fat distribution: abdominal obesity (increased waist circumference) and atherogenic dysplidimia (e.g. low HDL and high LDL cholesterol – see Box 2.2) leading to hypertension (raised blood pressure) and increase in inflammatory cytokines (pro-inflammatory state) and fibrinogen (prothrombotic state).
- Insulin resistance or glucose intolerance: insulin is less effective to lower blood sugars resulting in high glucose levels, which can precede type 2 diabetes mellitus.

---

## Box 2.6 WHO on obesity

In 2008, 1.5 billion adults, 20 and older, were overweight. Of these over 200 million men and nearly 300 million women were obese. Sixty-five per cent of the world's population lives in countries where overweight and obesity kills more people than underweight. Nearly 43 million children under the age of five were overweight in 2010.

Overweight and obesity are defined as abnormal or excessive fat accumulation that may impair health. Body mass index (BMI) is a simple index of weight-for-height that is commonly used to classify overweight and obesity in adults. It is defined as a person's weight in kilograms divided by the square of his height in meters ($kg/m^2$). The WHO definitions are:

A BMI greater than or equal to 25 signals overweight.
A BMI greater than or equal to 30 signals obesity.

Raised BMI is a major risk factor for non-communicable diseases such as: cardiovascular diseases (mainly heart disease and stroke, which were the leading causes of death in 2008), diabetes, musculoskeletal disorders (especially osteoarthritis – a highly disabling degenerative disease of the joints) and some cancers (endometrial, breast and colon).

The risk for these non-communicable diseases increases with the increase in BMI. Childhood obesity is associated with a higher chance of obesity, premature death and disability in adulthood. But in addition to increased future risks, obese children experience breathing difficulties, increased risk of fractures, hypertension, early markers of cardiovascular disease, insulin resistance and psychological effects.

Source: WHO, 2011e.

---

- Independent factors that mediate specific components of the metabolic syndrome: e.g. genetics, blood pressure regulation, glucose level dependency on insulin-secretory capacity and insulin sensitivity.
- Other contributing factors such as age, but also pro-inflammatory state and endocrine factors, that have been related to metabolic syndrome.

The brain relies only on glucose for its energy needs. Depriving the brain of glucose can lead to serious damage of the brain cortex (Kapit *et al.*, 2000). The regulation of fat formation is described in Section 2.5.2.

On the other hand, a *deficiency of the polyunsaturated fatty acids* (PUFAs) has been associated with a whole range of diseases and disorders: skin problems, allergies, immune dysregulation, cardiovascular complications, depression, fatigue and even behavioural issues (Nicolle and Hallam, 2010). Adequate supplies of PUFAs play a key role in brain structure and function and cell signalling. Evidence suggests that PUFAs may also play a role in development disorders of learning and behaviour, as well as in some psychiatric conditions such as depression (Hanciles and Pimlott, 2010).

Uncontrollable cell division and growth, possibly leading to cancer, have been related to hyperactive *ras proteins* (a membrane bound protein similar to the alpha sub-unit of G-protein). If ras protein is activated and cannot deactivate itself, the cell behaves as if it is under continuous stimulation (Kapit *et al.*, 2000).

## 2.3  Basic processes

*Basic bodily processes comprise of digestion of food, exchange of metabolic products via the blood circulation, exchange of gases via the respiratory system, excretion of waste products and movement via the skeletal and muscular system. These processes are essential to survive and grow. For uptake and digestion of food, the digestive system and the liver are mainly responsible. The circulatory system, comprising the heart, arteries and veins, is the body's transport system. The respiratory system (e.g. the lungs) removes carbon dioxide and brings oxygen into the body required for metabolism, and for movement of the body via the skeletal and muscular system. The secretory and digestive system (e.g. liver, kidneys and skin) take care of the excretion of waste and foreign matter, the water and salt balance, and the exchange of heat with the external environment.*

### 2.3.1  Basic systems

Respiratory tract, cardiovascular system, skeleton and muscles, and digestive and excretory systems are the systems that are available to execute the basic bodily processes: respiration, circulation, movement, digestion and excretion.

### Respiratory tract

Besides food, our body requires oxygen ($O_2$) as major component for our metabolism. The respiratory system brings air (including oxygen) into the body and removes carbon dioxide (see Figure 2.5). When you breath in air through nose or mouth, the temperature and water-vapour content of the air is adjusted and large particles are removed. Then the air moves on to the pharynx (throat), the larynx, and enters the trachea, a long tube, which branches into two bronchial tubes, or primary bronchi, that go to the lungs. The primary bronchi branch off into even smaller bronchial tubes, or bronchioles, which end in the alveoli, or air sacs. Approximately 30 million alveoli in our lungs with a total of about $85m^2$ alveolar surface are available for gas exchange with blood (Kapit *et al.*, 2000). Oxygen passes through the walls of the air sacs and blood vessels and enters the blood stream. At the same time, carbon dioxide produced passes into the lungs and is exhaled.

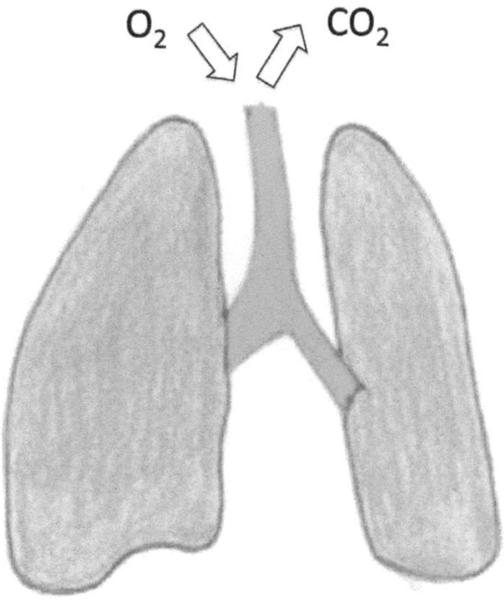

**Figure 2.5** *Gas exchange in the lungs*
Source: Bluyssen.

### Cardiovascular system

The circulatory system is the body's transport system: nutrients, oxygen and carbon dioxide are transported to and

## PULMONARY CIRCULATION

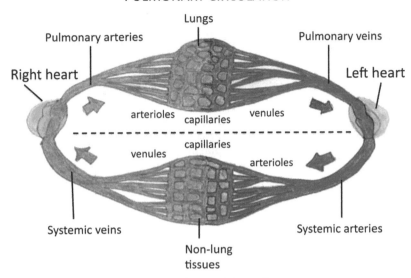

SYSTEMIC CIRCULATION

**Figure 2.6** *Schematic view of the cardiovascular system, including the heart and two circulations, as well as the different type of vessels*
Source: Bluyssen; based on Kapit *et al.*, 2000, plate 37.

from body tissues. The heart pumps the blood and the arteries and veins transport it. Oxygen-rich blood leaves the left side of the heart and enters the biggest artery called the aorta, which branches into smaller arteries and even smaller vessels that travel all over the body. Capillaries are the smallest blood vessels found in body tissue. They give nutrients and oxygen to the cells and take in carbon dioxide, water and waste. Veins then carry those waste products away from the cells to the right side of the heart, which pumps it to the lungs to pick up oxygen and eliminate waste carbon dioxide.

The cardiovascular system comprises the heart (left and right) and two circulations:

- *The pulmonary circulation*, which supplies the lungs with oxygenated blood via the pulmonary arteries from the right heart and collects deoxygenated blood via the pulmonary veins entering the left heart.
- *The systemic circulation*, which supplies the rest of the body tissues with the systemic arteries from the left heart and collects via the systemic veins entering the right heart.

Blood flows from aorta, to arteries, to arterioles and then enters the capillaries. In the capillaries, nutrients and oxygen are exchanged for carbon dioxide and waste products in the tissues. The walls of the capillaries are composed of very thin, porous endothelial cells, through which most solutes, except for proteins, can diffuse (based on concentration difference with tissue space – interstitial fluid). Exceptions are capillaries in:

- *Spleen, bone marrow and liver*: which have large gaps between cells through which also proteins can travel.
- *Kidneys, intestines and endocrine glands*: which have large surface window areas for permeation.
- *Brain*: which has the *blood–brain barrier*, a special blood capillary transport system that transports required nutrients such as glycose and amino acids. It is highly permeable for carbon dioxide, oxygen and water, but only slowly permeable for other substances.

Blood comprises plasma (about 55 per cent of the blood volume) and haematocrit (blood cells floating in the plasma). Plasma comprises 91 per cent water, 7 per cent proteins (e.g. fibrinogen, albumins and globulins) and 2 per cent electrolytes, nutrients and hormones. Haematocrit consists of mainly red blood cells and a small part of white blood cells (leukocytes) and platelets (thrombocytes). Blood cells are formed mainly in the bone marrow.

## Skeleton and muscles

The skeletal and muscular system is made of bones, ligaments, tendons and muscles. It shapes the body, protects organs and controls movement of the body (see Figure 2.7). Marrow, which is soft, fatty tissue that produces red blood cells, many white blood cells, and other immune system cells, is found inside bones.

*Bone* is a living vascular structure, composed of organic tissue (cells, fibres, extracellular matrix, vessels, nerves) and mineral (calcium hydroxyapatite), representing respectively about 35 and 65 per cent of a bone's weight (Kapit and Elson, 2002). Bone serves as a support structure, protection of organs and nerves, a site of attachment for skeletal muscle, ligaments, tendons and joint capsules, a source of calcium, and a significant site of blood cell development for the entire body.

*Bone marrow* comprises red and yellow marrow. The red marrow, which is the active bone marrow (primary source of blood cells), is found in bones of trunk, head (sternum, ribs, vertebrae and skull) and hipbones. Yellow bone marrow is fatty tissue that is not productive that can be found in the epiphyses and medullary cavities of long bones, and sponge bone of other bones.

A *skeletal muscle* is made of bundles of cylindrical striated cells called fibres, which can range from 5 to 100 μm in diameter, and several thousand times longer (from one bone to another). Along the length of a muscle cell, long fibrous cylinders (myofibrils), the contractile elements can be found. Each myofibril is composed of thick (actin) and thin (myosin) filaments. Filaments are ordered assemblies of protein molecules. The actin molecules are pear shaped, while the myosin molecules have long rod-shaped tails with globular heads that form cross bridges between thick and thin filaments.

Unlike skeletal muscles, smooth muscles responsible for movements of the viscera and blood vessels are involuntary and adapted for long sustained contraction.

**Figure 2.7** *Movement of the body through the skeletal and muscular system*
Source: Bluyssen.

## Digestive and excretory systems

While the digestive system ingests, digests and absorbs nutrients for further processing in the cells and organs of our body (Section 2.2), the excretory and urinary system eliminate waste from the body, in the form of faeces and urine (see Figure 2.8). Additionally, via the skin and its glands, water and chemicals can be secreted (see Box 2.7).

*The digestive system* ingests, digests and absorbs nutrients into the bloodstream and eliminates remaining wastes by mechanical and chemical digestive processes in the mouth, stomach, small and large intestines. In the mouth the salivary glands secrete saliva to support mechanical digestion and dissolving of food. Via the pharynx and esophagus, the food enters the stomach, where food is mixed with gastric juices, containing mucus, acid and enzymes, secreted by the stomach glands. In the small intestines the dissolved food is mixed with intestinal juice (alkaline) containing secretions of the pancreas and the liver. Across the lining of the small intestines, absorption of nutrients takes place. All water-insoluble material is taken to the liver for processing, and all fatty nutrients enter the lymph vessels. From the liver as well as from the lymphatic vessels, absorbed and digested nutrients are transported by blood to the body cells. In the large intestines (colon), remaining water and salts are absorbed, and waste products of digestion (faeces) are excreted through rectum and anus.

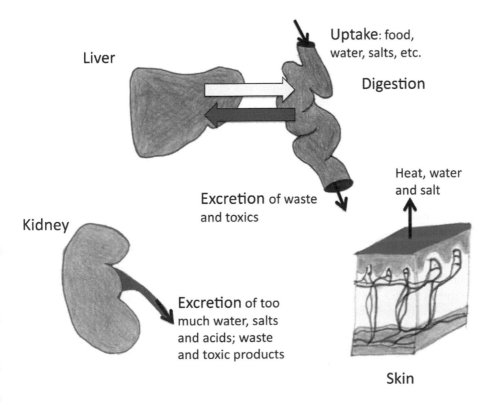

**Figure 2.8** *Digestive and excretory systems*
Source: Bluyssen; based on Silbernagel and Despopoulos, 2008, p. 3.

---

### Box 2.7  Types of glands

A gland is an organ that synthesises a substance for release of substances often into the bloodstream (*endocrine* gland – see Section 2.5.1) or into cavities inside the body or its outer surface (*exocrine* gland). The exocrine glands secrete their products through a duct or directly onto the apical surface, and can be divided into:

- *Apocrine glands* (genital and axillary areas): these secrete protein, cholesterol and steroids. A portion of the secreting cell's body is lost during secretion. Bacteria present in the axillae, scalp and feet breakdown the secretions and the breakdown products cause the particular smells for these areas: cheesy smell (methanethiol) from feet, the smell of unwashed hair (γ-decalactone) or the sweat smell of the axillae (mix of musky, urine and sour smell caused by androstenol, androstenone and isovaleric acid, respectively).
- *Holocrine glands* (e.g. sebaceous glands): the entire cell disintegrates to secrete its substances. The sebaceous glands can be found all over the human body (most in forehead, face and scalp) except for the palms and foot soles. They secrete lipid materials such as triglycerides and esters resulting in a slightly pleasant odour.
- *Merocrine glands* (*eccrine glands* e.g. sweat glands): cells secrete their substances by exocytosis. The sweat (thermoregulatory) glands secrete sweat containing c.99 per cent water and c.1 per cent electrolytes, amino acids, vitamins and miscellaneous compounds. Sweat glands are distributed throughout most of the skin surface, but are most numerous in palms and soles.

While the eccrine glands respond to physical activity, the sebaceous glands are controlled by hormones and the apocrine glands secrete continuously. The eccrine glands are controlled by sympathetic cholinergic (alkaline amino) nerves, which are controlled by a centre in the hypothalamus. From directly sensing the core temperature and from input from temperature receptors in the skin, the hypothalamus adjusts the sweat output, together with other thermoregulatory processes.

---

*The urinary system* eliminates wastes from the body, in the form of urine. The kidneys produce urine from blood that passes through (about 1,300 ml blood passes per minute, of which 1–2 ml leaves as urine). With this process, the kidney controls the water and salt level. The kidneys remove waste from the blood, which is combined with water to form urine. The urine then travels down two thin tubes called ureters to the bladder. When the bladder is full, urine is discharged through the uretha. Urine comprises water, salts, small amounts of acid (e.g. uric acid), and a variety of waste products, such as urea (contains nitrogen derived from metabolism of amino acids and proteins). The composition of urine as well as the volume can change to compensate for fluctuations of volume and composition of body fluids.

The largest part of the human body is *the skin* with a mean surface area of around 1.8 m². The skin comprises two pressure senses (for light and deep stimulation), two temperature sensitivity senses (warm and cold) and a pain sense (nociceptor) (a description can be found in Bluyssen, 2009a: Section 2.2.1). Besides these senses, at the skin three different systems for secreting chemicals to the skin surface can be found, namely eccrine, sebaceous and the apocrine glands (see also Box 2.7). With the sweat glands (eccrine glands) the amount of sweat that a body needs to secrete to help to keep the body in thermal balance can be controlled.

### 2.3.2 Mechanisms

The mechanisms of the basic bodily processes e.g. respiration, circulation, movement, digestion and excretion, except for heat regulation (see Section 2.7.2) are described.

### Respiration

#### Gas exchange

The gas-exchange area for oxygen and carbon dioxide begins in the alveoli and comprises air, blood and tissue (see Figure 2.9). The interface between blood and air, where the gas exchange takes place, is made of very thin cells. From the alveoli oxygen diffuses down its partial gradient (difference in partial pressure, see Box 2.8) into the blood, while carbon dioxide diffuses in the opposite direction (concentration of carbon dioxide is higher in blood than in air).

Oxygen can dissolve in the blood plasma but this amount is small (about 1 per cent). The other 99 per cent carried by the blood is combined with haemoglobin (Hb), an iron containing protein within the red blood cell, forming oxyhaemoglobin:

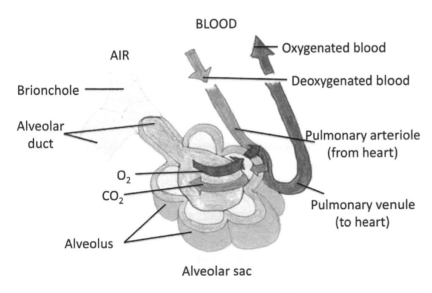

**Figure 2.9** *Gas exchange between blood and alveoli: $O_2$ and $CO_2$ traverse two cell layers in passing between alveolus and capillary via diffusion*
Source: Bluyssen; based on Kapit *et al.*, 2000, plate 48.

---

### Box 2.8 Partial pressure

Partial pressure is a measure of gas concentration; the more molecules of a gas present in the same space, the higher the pressure. The ideal ratio of partial pressures is the same as the ratio of molecules. The partial pressure of a gas is a measure of thermodynamic activity of the gas's molecules. Gases will always flow from a region of higher partial pressure to one of lower pressure. Gases dissolve, diffuse and react according to their partial pressure, and not according to their concentrations in gas mixtures or liquids. In a mixture of gases each component acts independently of the others and each molecule makes the same contribution.

Dalton's Law of partial pressures states: $P = P_{N_2} + PO_2$      [2.3]

In which:

$P_{N_2}$ = the partial pressure of $N_2$

$PO_2$ = the partial pressure of $O_2$.

The amount of oxygen required for human respiration is set by the partial pressure of oxygen. Air comprises approximately 20 per cent $O_2$ and 80 per cent $N_2$ (nitrogen). Atmospheric air has a pressure of 760 mm Hg (mercury) at sea level (height that Hg rises in a closed-ended tube). Thus $PO_2$ and $P_{N_2}$ yield respectively 152 (20 per cent of 760) and 608 (80 per cent of 760). In the lungs $PO_2$ amounts 105 mm Hg, while in the tissues $PO_2$ averages 40 mm Hg and in active muscles even as low as 20 mm Hg (Kapit et al., 2000).

---

$$O_2 + HHb \Leftrightarrow HbO_2^- + H^+ \qquad\qquad [2.4]$$

In which:

HHb = deoxyhaemoglobin

$HbO_2^-$ = oxyhaemoglobin

$H^+$ = hydrogen ion (excess in the blood plasma passing the alveoli is used by the reaction presented in Equation 2.5).

Red blood cells have no nucleus or cytoplasmic organelles. Instead they are packed with haemoglobin, the oxygen binding protein. The amount of haemoglobin in the blood determines the blood's capacity to carry oxygen. Each haemoglobin molecule can bind and transport up to four oxygen molecules.

$CO_2$ can dissolve in the blood plasma (about 9 per cent), but is also carried in different combined forms in the red cells. $CO_2$ reacts with water to form carbonic acid, which dissociates into $H^+$ and bicarbonate. 67 per cent of $CO_2$ is carried as bicarbonate ($HCO_3^-$) (Kapit et al., 2000). In the alveoli, $CO_2$ is released to the air (with the enzyme carbonic anhydrase as a catalyst):

$$H^+ + HCO_3 \Leftrightarrow H_2CO_3^- \Leftrightarrow H_2O + CO_2 \qquad\qquad [2.5]$$

In which:

$H_2O$ = water

$H_2CO_3$ = carbonic acid

$HCO_3^-$ = bicarbonate

$H^+$ = hydrogen ion (excess in the blood plasma passing the tissue cells is used by the reverse reaction presented in Equation 2.4).

Another fraction of $CO_2$ (24%) combines with some of the amino groups on polypeptide potions of Hb to form carbaminohaemoglobin (Kapit *et al.*, 2000).

The oxygenated blood is transported via circulation (see Circulation, below) to the organs and tissues where oxygen is required. There the $O_2$ is released and consumed, while $CO_2$ is produced (*internal respiration*) and carried in the blood plasma on its way to the lungs. The reversed reactions of the above take place.

### Removal of particles

The smaller the particles in the air that are breathed in, the further they can come into the respiratory tract. It is said that particles smaller that 200 μm can reach the throat, particles smaller that 10 μm can reach the larynx, particles smaller that 5 μm can reach the alveoli (Bluyssen, 2009a).

Most inhaled fibrous material is cleared by *the mucociliary escalator clearance mechanism*, a synchronized, rhythmic beat of the cilia in a layer of fluid on the epithelium directed outward, which results in ingestion. Other inhaled particles are retained in the lungs, were they accumulate. Some fibres, through uncertain routes, migrate to the pleura (body cavity that surrounds the lungs) and stay there. Fibre transportation from the lung also occurs via the haematogenous (blood) and lymphatic systems, with eventual accumulation of fibres in virtually every organ of the body.

The alveolar macrophage cells in the airway and the alveoli of the lungs can ingest and destroy bacteria and viruses and remove small particles (see also Section 2.4.1). They secrete chemicals that attract white blood cells to the site, which initiate an inflammatory response in the lung. Particles are removed into the lymphatic system of the lung and stored in adjacent lymph glands. Soluble particles (i.e. small lead particles emitted by an automobile exhaust) are removed into the bloodstream and excreted by the kidney (takes about 12 hours).

### Circulation

### Blood flow and pressure

Pressure difference is the basic mechanism why blood flows. The pressure drop over the arteries to the capillaries is proportional to the resistance (see Box 2.8). The greatest pressure drop occurs across the terminal arteries, the arterioles that enter the capillaries (see Figure 2.6). Smooth muscles wrapped around the arterioles control blood pressure and flow to specific tissues by increasing/decreasing the radius of the arterioles. These muscles are controlled by nerves and hormones.

Most tissues contain tiny lymph vessels (named lymph capillaries), which are open on only one end. They merge into larger vessels (collecting ducts), which in turn merge into larger vessels (lymph ducts) – and so on – until the largest ducts drain into the circulation via connections with large veins. This *lymphatic system* drains excess fluid and plasma proteins leaked into tissue spaces. Besides this important function, the lymphatic system is a major part of our defence system (see Section 2.4).

The veins serve as a low-resistance way to return blood to the heart. They also provide a low-pressure storage system for blood. Only 5 per cent of our blood fills the capillary bed (at any moment). At rest, the veins contain about two-thirds of our blood. Smooth muscle-cells in the walls of the veins control the volume (the diameter) of the veins. The blood vessel diameter controls the blood flow and the flow resistance of the vascular bed, one of the main indicators of blood pressure. The activity of smooth muscle in the blood muscle can be controlled chemically or via sympathetic stimulation. More information on blood pressure regulation is presented in Section 2.7.2.

### Blood cells and bone marrow

Blood cells are formed in the red marrow from the proliferation and differentiation of *stem cells*, which are permanently present. One line forms the red cells, another the white cells, and another the platelets. The process by which red blood cells (erythrocytes) are produced is named *erythropoiesis*. The stem cells that form red blood cells divide, forming forerunners of red cells (erythroblasts) with nuclei. Within two to three days the cells develop, fill up with haemoglobin, lose their nuclei and enter circulation.

The production of red blood cells is regulated by several factors such as the arterial oxygen pressure ($PO_2$). Low $PO_2$ (high altitudes) stimulates the kidneys to release a hormone, erythropoietin, which stimulates the bone marrow to increase the production. Old red blood cells (about four months) are destroyed by liver and spleen macrophages and metabolized into, among other things, iron and bilirubin. The iron is recycled by bone marrow, while bilirubin is removed by the liver (giving the yellow colour to urine).

### Muscle contraction

Movement or motor activities, such as beating of the heart, breathing air, blinking an eye, are all caused by muscular contraction. A muscle cell contracts because the myofibrils contract. In the relaxed muscle the cross bridges are detached from actin filaments. ATP supplies the energy for contraction, while $Ca^{++}$ is required for contraction (see also Box 2.9). If sufficient $Ca^{++}$ is present, attachment can occur. When there is not enough, the muscle relaxes. When nerve impulses activate muscles (see Section 2.6.2 nerve impulse), a muscle potential spreads over the muscle cell and reaches the interior via transverse tubules, a system of tiny tubes. ATP is supplied via three sources: the amount of ATP present in muscle cells (which is used after 20–25 seconds of intense activity), the glycolysis-lactic acid system (anaerobic metabolism) and the aerobic metabolism (see Section 2.2.2).

The heart is a hollow organ made from *cardiac muscle*, similar to skeletal muscles. The contractions (are always brief and entire heart participates) and duration (100 times longer) are however different. While the cardiac muscle exits itself, skeletal muscles will normally only contract when they receive an impulse.

---

**Box 2.9  Ca$^{++}$ as regulator of cellular processes**

Ca$^{++}$ regulates:

- Skeletal, cardiac and smooth muscle contraction.
- Amoeboid movement (a crawling-type of movement of cell cytoplasma).
- Exocytosis (process by which a cell directs the contents of secretory vesicles out of the cell membrane.
- Synaptic transmission (see Section 2.6.2).
- Enzyme activation.
- Cell cleavage (division of cells in the early embryo).

---

The heart pumps intermittently resulting in a minimum blood pressure (diastolic pressure) and a maximum blood pressure (systolic blood pressure), which lie respectively around 60–80 and 90–120 mm Hg for a normal young adult.

## Digestion

### Liver: formation of bile
The liver plays an important role in digestion, through its exocrine function, namely the formation of bile. Bile is secreted by liver cells (hepatocytes) and comprises about 97 per cent water, organic bile salts and bile pigments and inorganic salts (sodium chloride and sodium carbonate). Organic bile salts (also named bile acids), such as cholic acid and deoxycholic acid, are formed from cholesterol within the liver cells. Liver cells are the major producers of cholesterol. Bile pigments are derived mainly from bilirubin, a metabolite of heme in the haemoglobin. The yellowish water-soluble bilirubin glucuronide is derived from bilirubin and is secreted by the liver cells. Most of the bile pigments are excreted with the faeces, what is left is re-absorbed and excreted in the kidney. Bile salts, such as cholate and deoxycholate, facilitate fat digestion by emulsification (see Box 2.10 and Figure 2.10). They act as fat solubilizing agents. When emulsified, fats can be better digested by the water-soluble enzyme lipase from the pancreas.

### Kidney
Each kidney has about one million tubular nephrons that produce urine from a filtration of blood. Filtration occurs continuously through the tubules on its way to the collecting duct. During this process, nutrients and most of the fluids are withdrawn. However, the tubules re-absorb most of the fluid and some minerals. Re-absorption is mostly based on the Na$^+$–K$^+$ pump (see Section 2.2.2 and Box 2.11).

Cl$^-$ follows Na$^+$ (NaCl) and drags water along (osmosis see Box 2.5). K$^+$ is also re-absorbed. Glucose and amino acids are co-transported with Na$^+$, but also HCO$_3^-$. H$^+$ is counter transported in exchange of Na$^+$. Re-absorption capacity adjusts the filtration rate through negative feedback in each single nephron.

## Box 2.10  Emulsification

Fats and other lipids do not dissolve in water; instead they tend to congeal together into large masses. The same thing happens in a salad dressing, when the oil and vinegar (which is mostly water) are allowed to separate.

Bile consists of molecules that have a dual nature. Half of the molecule is attracted to water, and the other half is attracted to fats. The bile molecules therefore place themselves in between fat and water. In this way the fat droplets remain suspended in water rather than merging together. This process is called emulsification, and is similar to the way detergents remove grease from dirty dishes. In the digestive tract, emulsification allows lipase to gain access to the fat molecules and thus aids digestion.

Source: http://biologyinmotion/bile/index.html (accessed November 9, 2011).

**Figure 2.10**
*Two immiscible fluids on the left and two emulsified fluids on the right*
Source: Bluyssen.

     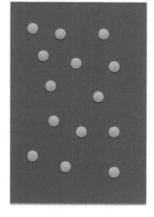

immiscible                    emulsion

## Box 2.11  Sodium (Na$^+$) and potassium (K$^+$)

Potassium inside cells (most of K$^+$ lies within cells, only about 2.5 per cent in extracellular fluid) is important for:

- Optimal growth and DNA and protein synthesis.
- The performance of many enzyme systems.
- Maintaining cell volume, pH and membrane potential.

Sodium is important for:

- Excitability functions of all cell membranes, especially of nerve and muscle tissue.
- Regulating plasma and extracellular water volume and blood pressure.

The kidneys control $HCO_3^-$ and the lungs control $CO_2$ (see also equation 2.5, above). Together they maintain a viable $H^+$ concentration in the body fluids (or the acid–base balance). Kidneys extract $H^+$ and re-absorb $HCO_3^-$. The kidneys also control the water–salt balance (see Section 2.5.2) and the volume of the extracellular fluid (and thus plasma volume) through the total amount of NaCl (not concentration) in the extracellular spaces, which determines osmotic pressure.

## 2.3.3 Diseases and disorders

### Respiratory tract

The respiratory tract can suffer from many diseases and disorders, most of them caused by infection (bacterial, viral or fungal), inflammation (allergic reaction), damage or tumours. In most cases they are caused by an external pollutant, causing an immediate or long-term effect. It is not always clear whether the cause is an infection or an allergic reaction, because the symptoms might be similar (wheezing, coughing with sputum, shortness of breath (dyspnoea) or chest pain). *Asthma* (see Box 2.12) and *bronchitis* are examples of this (see Section 2.6.3 in Bluyssen, 2009a). Next to that, there are a number of pollutants of which the health effect has been given a specific name (see Bluyssen, 2009a, Section 2.6.3).

An important disease of our time is COPD (*chronic obstructive pulmonary disease*), caused by noxious particles or gases, which trigger an abnormal inflammatory response in the lungs (WHO, 2011d). COPD is a chronic disease of the lungs in which the airways become narrowed, which leads to a

---

### Box 2.12 WHO on asthma

The WHO estimates that some 235 million people currently suffer from asthma. It is the most common chronic disease among children. The fundamental causes of asthma are not completely understood. The strongest risk factors for developing asthma are a combination of genetic predisposition and environmental exposure to inhaled substances and particles that may provoke allergic reactions or irritate the airways, such as:

- Indoor allergens (e.g., house dust mites in bedding, carpets and stuffed furniture, pollution and pet dander).
- Outdoor allergens (such as pollens and moulds).
- Tobacco smoke.
- Chemical irritants in the workplace.
- Air pollution.

Other triggers can include cold air, extreme emotional arousal, such as anger or fear, and physical exercise. Even certain medications can trigger asthma: aspirin and other non-steroid anti-inflammatory drugs and beta-blockers (which are used to treat high blood pressure, heart conditions and migraine).

Source: WHO, 2011c.

limitation of the flow of air to and from the lungs causing shortness of breath. The limitation of airflow is poorly reversible, usually gets worse over time and is not curable but is preventable. It is frequently diagnosed in people aged 40 or older because it develops slowly. The primary cause is tobacco smoke; other risk factors include indoor and outdoor air pollution, occupational dusts and chemicals, and frequent lower respiratory infections during childhood. More than 3 million died of COPD in 2005 (equal to 5 per cent of all deaths globally that year) and it is predicted that in the next 10 years this death rate will increase by 30 per cent if nothing is undertaken.

Furthermore, the gas exchange itself can also cause problems, such as hypoxia and hyperoxia. *Hyperoxia*, an excess of $O_2$ in body tissues, occurs when elevated concentrations of oxygen are inhaled (can happen for example during scuba diving). *Hypoxia*, a deficiency of $O_2$ in the body tissues, can occur when:

- The arterial $PO_2$ is low (e.g. at high altitudes), named hypoxic hypoxia.
- Hb is low, named anemic hypoxia.
- Circulation is failing, named stagnant hypoxia.
- Utilization in the cells is poor, named histotoxic hypoxia.

Severe hypoxia leads to unconsciousness in about 20 seconds and death in 4–5 minutes. Less severe hypoxia can lead to impaired judgement, drowsiness, disorientation, headache, nausea, vomiting and rapid heart beat. Hyperoxia may lead to effects on the central nervous system (convulsions, unconsciousness) and during prolonged exposure to effects on pulmonary organs (lungs) (difficulty in breathing and pain within chest) and ocular organs (alterations to the eyes) (Clark and Thom, 2003).

## Cardiovascular system

*Cardiovascular diseases* (CVDs) are the number one cause of death globally (see Box 2.13). Cardiovascular diseases are a group of disorders of the heart and blood vessels and include (WHO, 2011b):

- *Coronary heart disease*: disease of the blood vessels supplying the heart muscle.
- *Cerebrovascular disease*: disease of the blood vessels supplying the brain.
- *Peripheral arterial disease*: disease of blood vessels supplying the arms and legs.
- *Rheumatic heart disease*: damage to the heart muscle and heart valves from rheumatic fever, caused by streptococcal bacteria.
- *Congenital heart disease*: malformations of heart structure existing at birth.
- *Deep vein thrombosis and pulmonary embolism*: blood clots in the leg veins, which can dislodge and move to the heart and lungs.

Capillary blood pressure normally prevents flow of fluids. Osmotic pressure is caused by unequal amounts of proteins and can cause fluid and protein to go through anyway. Lymphatic vessels drain excess fluid in the tissue spaces together with plasma proteins that leak out of capillaries, and return them to the circulation via the large veins. *Oedema*, accumulation of fluid in tissue spaces,

---

### Box 2.13  WHO on cardiovascular diseases

In 2008, 17.3 million people died from CVDs, representing 30 per cent of all global deaths. Of these deaths, an estimated 7.3 million were due to coronary heart disease and 6.2 million were due to stroke. By 2030, almost 23.6 million people will die from CVDs, mainly from heart disease and stroke, projected to remain the single leading causes of death.

The most important behavioural risk factors of heart disease and stroke are unhealthy diet, physical inactivity, tobacco use and harmful use of alcohol. Behavioural risk factors are responsible for about 80 per cent of coronary heart disease and cerebrovascular disease. The effects of unhealthy diet and physical inactivity may show up in individuals as raised blood pressure, raised blood glucose, raised blood lipids, and overweight and obesity; these are called 'intermediate risk factors' or metabolic risk factors. There are also a number of underlying determinants of CVDs, or 'the causes of the causes', which are a reflection of the major forces driving social, economic and cultural change – globalization, urbanization and population ageing. Other determinants of CVDs include poverty, stress and hereditary factors.

Source: WHO, 2011b.

---

is a disorder that can occur when the fluid balance is upset. The causes can be several:

- Increase in capillary blood pressure (e.g. from dilation of arterioles or venous congestion).
- Decreased plasma oncotic pressure (e.g. from starvation; oncotic pressure is osmotic pressure exerted by the plasma protein) by disturbed permeability of capillary walls to protein.
- Increased permeability of capillary walls to protein (in response to allergies, inflammation and burns).
- Disturbance of lymph drainage (e.g. from obstruction of lymph vessels).

When chemical vasodilation control opposes sympathetic vasoconstriction control (e.g. blood pressure failing while organ had an inadequate blood supply) a problem can arise. Vasodilation most often dominates. Some people experience dizziness, impaired vision and buzzing in ears when standing up after lying down for a long time. These are signs of inadequate cerebral circulation that arises from the drop in blood pressure following a sudden change to upright position. Similar reactions may occur with healthy people, when blood vessels in skin or muscles are dilated because of heat or exercise. Regulatory responses may fail because of intense demands of heat regulation and metabolism priority.

*Hypertension* is a chronic elevation of blood pressure of which the causes are not always known. Blood pressure is regulated but not at a healthy level. Blockage of arteries can occur via damage of blood vessels, causing heart attacks and strokes.

## Skeleton and muscles

*Repetitive strain injury* (RSI), cumulative trauma disorders (CTDs) or repetitive stress injury are injuries of the musculoskeletal system – including the joints, muscles, tendons, ligaments, nerves and blood vessels – that may develop from the accumulation of repeated small injuries or stresses to our musculoskeletal system. It is a response to excessive repeated demands on our body without enough time to recover before adding more stress (Healthpages.org, 2011). Risk factors for RSI are, among others: motions using force, sustained awkward positions, holding the same body position without moving or resting for long periods, and repetitive motions. Psychosocial factors also seem to play a role (not enough time to recover, psychological stress and insufficient social support) (Health Council of The Netherlands, 2000).

*Musculoskeletal disorders* (e.g. pain in neck, shoulders and low back) are intertwined with mental disorders, with mechanical (posture, manual material handling, muscle load, peak load, movement pattern, physical demands, mechanical variation, etc.), as well as psychosocial risk factors or stressors (despair, mental variation, effort–reward imbalance, psychological and emotional demands, mental energy, role overload, etc.) (Westgaard and Winkel, 2011).

*Tendonitis*, an inflammation in the tendon (e.g. tendons of the wrists, elbows and shoulders), is a common CTD as are nerve disorders such as carpal tunnel syndrome (the nerve that passes the carpal tunnel in the wrist) and cubital tunnel syndrome (entrapment of nerve over the cubital tunnel at the elbow) (Tulder *et al.*, 2007).

In crippling diseases such as *multiple sclerosis* and *Guillian-Barré*, the white, fatty myelin sheaths (see Figure 2.19) around axons (broken at intervals or nodes) are destroyed. Impulse conduction is the same as for axons without myelin, but faster because impulses jump from node to node.

*Anemias* (anemia means lack of blood) are diseases associated with reduced content of blood haemoglobine, which decreases the blood's capacity to transport oxygen to the tissues. Dietary deficiencies of folic acid and iron may cause anemia, but also a vitamin $B_{12}$ deficiency (required for erythropoiesis).

## Digestive and excretory systems

*Acidosis* is an increased acidity in the blood (plasma more acid than normal ($H_2CO_3$ or $CO_2$)). It can be caused by an increased production of metabolic acids or disturbances in the ability to excrete acid via the kidneys (metabolic acidosis) (the increase of $H^+$ stimulates breathing and can cause hyperventilation) and by a build-up of carbon dioxide in the blood due to hypoventilation (respiratory acidosis). A high level of uric acid may lead to *gout*, a medical condition characterized by recurrent attacks of acute inflammatory arthritis (red, tender, hot, swollen joint). The elevated levels of uric acid in the blood crystallize and deposit in joints, tendons and surrounding tissues.

*Alkalosis* is a decreased acidity in the blood (plasma less acid than normal ($HCO_3^-$) caused by hyperventilation (respiratory alkalosis) resulting in a loss of carbon dioxide, or caused by prolonged vomiting (metabolic alkalosis). The decrease of $H^+$ stimulates retention of $CO_2$ and $H_2CO_3$.

Most common digestive disorders are *diarrhoea* and *constipation*, caused by respectively increased and decreased intestinal motility. For diarrhoea, certain food (e.g. prunes), microbial infections or anxiety can be the reason; for constipation lack of dietary fibre, can be the reason. An imbalance of the gastro-intestinal tract can contribute to the development of many chronic diseases, including inflammatory bowl disease and rheumatic disease (Trueman and Bold, 2010).

*Vomiting* is a physiological defence reflex in which undesirable food is expelled out of the stomach. Chemoreceptors and stretch receptors in the stomach are activated respectively by microbial toxins and excessive food. These sensory signals are sent to a vomiting centre in the brain medulla. *Ulcers* are wounds occurring in the inner lining of the stomach and small intestine, caused by corrosive and noxious effects of acid on the gut wall. Bacteria that disrupt the protective barrier of the stomach against acid seem to be a major cause of ulcers. Other causes are excessive use of alcohol or anti-inflammatory drugs, such as aspirin and ibuprofen, and prolonged and excessive acid production as a result of increased vagus nerve activity or excessive gastrin secretion.

*Jaundice* is characterized by a yellowish colour of the skin and eyes caused by deposition of bilirubin and related bile pigments in the capillaries and tissue spaces. It can occur because of excessive haemolysis of the red cells, producing a large amount of bilirubin, or, it can be caused by gallstones that obstruct the bile from leaving the liver. The bile flows back into the liver and leaks into the blood. The majority of *gallstones* comprise of cholesterol (and some calcium bilirubinate), which are caused by excess amounts of the water-insoluble cholesterol (part of bile). The gallbladder stores bile for release after meals.

Possible *diseases and disorders of the skin* can be caused by an infection (bacterial, viral or fungal), inflammation (allergic reaction), radiation-related disorders (such as sunburn) and psychological reasons (e.g. anxiety). Itching, rashes, hives, blisters and redness may be the symptoms. *Sebaceous glands* are involved in skin problems such as acne. In the skin pores, sebum and keratin can create a hyperkeratotic plug. *Sweat glands* produce a clear, odourless substance. The odour from sweat is caused by bacterial activity on the secretions of the apocrine glands.

## 2.4 Protection and defence

*The immune system is our body's defence against (potential) disease (irritation, allergy, infection, toxicity) through the lymphoid system. It comprises a collection of layered protection mechanisms of increasing specificity against disease. It starts with physical barriers (mechanical, chemical and biological) to prevent threatening substances from entering your body. If substances do get by, the natural or innate immune system (inflammatory system) attacks. And finally if that fails the acquired or adaptive immune system starts to work. The lymphoid system, comprising the primary organs (bone marrow and thymus), the secondary organs (spleen, lymph nodes and lymphoid tissue) and the white blood cells, helps the body in this defence. The lymphoid tissues and organs are collections of lymphocytes and related cells, supported by fibres and cells. Lymphocytes are among the principal cells of the immune system. From their generative organs (bone marrow and thymus), they are transported in the lymph vessels and lymphoid tissues and organs.*

### 2.4.1 Immune system

The immune system is our body's defence system against infections and disease (Figure 2.11). *The lymphoid system*, comprising the primary organs (bone marrow and the thymus), the secondary organs (spleen, lymph nodes and lymphoid tissue) and the white blood cells, helps the body in this defence (see Figure 2.6 in Bluyssen, 2009a).

### White blood cells

White blood cells (leukocytes) originate in the bone marrow and comprise of two types of cells, the granulocytes (with cytoplasmic granules: neutrophils, eosinophils, basophils) and the agranulocytes (without granules: monocytes, macrophages, *lymphocytes*). White blood cells can be divided into two categories:

- Those that participate in inborn, immune responses to infections and inflammations caused by tissue injuring.
- Those that take part in acquired immune response.

Lymphocytes are the main white blood cells for the second category. They are among the principal cells of the immune system (see Box 2.14) and are transported from their generative organs (bone marrow and thymus) in the lymphatic vessels (*lymph* is plasma including white blood cells) and lymphoid tissues and organs. All other white blood cells take part in the first.

**Figure 2.11** *Monsters: free interpretation of little animals that cause infections and disease*
Source: Drawings by Anthony (11) and Sebastian Meertins (7) on November 23, 2011.

## Box 2.14  Lymphocytes and other substances in the immune system

*Lymphocytes*:

- *B (bone marrow)-lymphocytes*: formed in bone marrow (some of lymphocytes become B-lymphocytes; the larger ones enter circulation to functions as *natural killer (NK) cells*; other partly differentiated migrate via blood to the thymus).
- *T (thymus)-lymphocytes*: differentiated B-lymphocytes migrated via the blood to the thymus where they become T-cells and differentiate further and re-enter circulation and migrate to secondary lymphoid organs where antigens activate lymphatic operation.
- *Helper T-lymphocytes*: bind and interact with B-lymphocytes and macrophages, releasing cytokines (lymphokines and interleukins) to regulate their functions as well as those of the T-lymphocytes. Cytokines can also be released from injured tissues and control the migration of white blood cells in the natural and acquired immunity.
- *Cytotoxic T-lymphocytes*: bind to and destroy infected cells and form memory cells.
- *Suppressor T-lymphocytes*: inhibit activity of other T- and B-lymphocytes.
- *Memory T-lymphocytes*: antigen stimulated lymphocytes circulate in the blood and lymph for years. When the organism is re-exposed to the antigen, the immune response can take place faster and stronger.

*Antibodies*: protein molecules (immune-globulins, Ig) are Y-shaped molecules that do not destroy pathogens themselves – they provide markers for the effector cells such as T-cells and NK cells to complete the elimination:

- *IgA type*: secretory immune-globulins released into secretions of gastrointestinal and respiratory mucosa and milk.
- *IgE*: participate in allergic reactions.
- *IgD*: exist on the surface of B-lymphocytes, recognize antigens.
- *IgG and IgM*: fight bacterial and viral infections.

The *complement system* is a biochemical cascade that attacks the surfaces of foreign cells. It contains over 20 different proteins and is named for its ability to 'complement' the killing of pathogens by antibodies. Complement is the major humoral component of the innate immune response.

Recognition receptors of the innate immune system: the specificity of each receptor is genetically predetermined; they recognize elements and are currently identified in four divisions:

- *Toll-like receptors (TLRs)*: recognize conserved molecular patterns from bacteria, viruses, protozoa and fungi.
- *C-type lectin receptors*: recognize mainly fungi, which when activated trigger phagocytosis.
- *Rig-1-like receptors*: sense viral DNA and RNA or bacterial products.
- *Intracellular nucleotide receptors*: sense endogenous products such as cytosols released by dying cells.

The TLRs, found on most structural, immune and inflammatory cells (especially in the gastro-intestinal tract), capture microbes and present them, pre-digested, to either naïve or activated T-cells (antigen presentation).

Sources: Kapit and Elson, 2002; Mygind, 1986; Ash, 2010.

### Thymus

The thymus is a primary lymphatic organ in the chest cavity. The thymus is largest and most active during the neonantal and pre-adolescent periods. It helps to mature T-lymphocytes (B-lymphocytes become T-lymphocytes) and secretes thymosin hormone. The T-lymphocytes are mostly produced during early life. Thymosin promotes maturation of T-cells in the thymus and periphery. Thymosin secretion declines after middle age and may cause reduced cell-mediated immunity (see Section 2.4.2) in the aged.

### Spleen

Besides removing old red blood cells, the spleen metabolizes haemoglobin, synthesizes antibodies and removes anti-body coated bacteria. It can be seen as a large lymph node, which is active in immune response through humoral and cell-mediated pathways (see Section 2.4.2).

### Lymph nodes

Lymph nodes are distributed throughout the body, including the armpit and stomach/gut and are linked by lymphatic vessels. They act as filters or traps for foreign material and are a storage place for white blood cells (i.e. lymphocytes and macrophages).

## 2.4.2 Mechanisms

The defence system comprises of a collection of layered protection mechanisms of increasing specificity against disease. It starts with physical barriers (mechanical, chemical and biological) to prevent threatening substances from entering your body. If substances do get by, the natural or innate immune system (inflammatory system) goes on attack. And finally if that fails the acquired or adaptive immune system is put to work.

### Layer 1: physical barriers

*Physical barriers*, such as the skin, prevent pathogens (e.g. bacteria and viruses) from entering the body. Other systems act to protect body openings such as the lungs and gastro-intestinal tract. In the lungs, coughing and sneezing eject pathogens and other irritants from the respiratory tract. Tears and urine are also a way to mechanically expel pathogens, while mucus secreted by the respiratory and gastrointestinal tract serves to trap and entangle microorganisms. Chemical barriers, such as antimicrobial peptides and enzymes in saliva, tears, and breast milk, also protect against infection. Biological barriers can be found in the gastrointestinal tract. The gastro-intestinal but also the blood–brain barrier provide a complex immune protection and communicate between themselves intensively (Müller and Yeoh, 2010).

### Layer 2: the innate immune system

If a pathogen breaks through the barriers, *the innate immune system* provides an immediate and short-lasting response. Although it has a limited memory, the innate immune system seems to be the most significant driver of health and disease (Ash, 2010). The innate immune system:

- Recruits immune cells to sites of infection and inflammation for example via the production of cytokines, chemical mediators, secreted by inflammatory leukocytes and some non-leukocytic cells.
- Activates the complement protein cascade to identify bacteria, activate cells and to promote clearance of dead cells or antibody complexes.
- Identifies and removes foreign substances present in organs, tissues, the blood and lymph, by specialized white blood cells.
- Activates the adaptive immune system through antigen presentation (see Box 2.14).

Upon an injury of the epithelium (the 'skin' of an organ), the following is happening:

- Microbes enter the body, release toxins (antigens) and create local infection.
- The mast cells in the tissue release granules containing *heparin* (prevents blood coagulation) and histamine (causes vasodilatation and increased permeability of local blood vessels to blood proteins and blood cells) within the tissue space; basophiles (type of white blood cells) do the same in blood.
- Blood proteins and fluids leak into the injured site causing oedema (swelling).
- Gradually the fluid in the swelling clots traps the microbes and prevents further penetration.

*First line of defence*: tissue macrophages attack microbes and destroy them by phagocytosis (leading to digestion by lysosomal enzymes) (see Figure 2.12).

*Second line of defence*: if infection persists, *neutrophils* (type of white blood cells) migrate to the injury site and phagocytize the microbes.

*Third line of defence*: if still not enough, agranular monocytes (type of white blood cells) move to the site, grow, and begin to phagocytize microbes and the dead neutrophils.

## Layer 3: adaptive immune system

If pathogens successfully avoid the innate response, a third layer of protection, *the adaptive immune system*, is activated. Afterwards, the acquired immunity is adapted and/or improved. When a new attack is made by the same pathogen, the adaptive immune system should respond faster and stronger. This system works slow and specific, but is capable of life long memory. The adaptive immune system comprises two types of acquired immunity: humoral or anti-body mediated responses, carried out by B-lymphocytes, and the cell-mediated responses, carried out by T-lymphocytes.

### Humoral or active immunity

The humoral immunity always requires the production of antibodies (humour comes from body fluid, were the components are found) (Ash, 2010). Anti-body mediated responses are carried out by *B-lymphocytes*, which recognize the

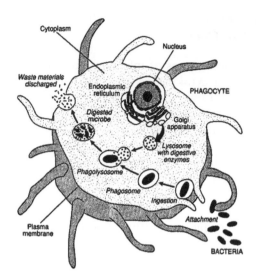

**Figure 2.12** *Phagocytosis is a nonspecific defense mechanism in which various phagocytes engulf and destroy the microorganisms of disease. Neutrophils and monocytes are important phagocytes. In the tissues, the monocytes are transformed into phagocytic cells called macrophages, which move through the tissues of the body performing phagocytosis and destroying parasites. The process of phagocytosis begins with attachment and ingestion of microbial particles into a bubble-like organelle called a phagosome. Inside the phagocyte, the phagosome joins with a lysosome, which contributes enzymes, resulting in a phagolysosome*

Source: Phagocytosis, *Cliffnotes.com*, 3 January 2013, www.cliffsnotes.com/study_guide/Phagocytosis.topic ArticleId-8524,articleId-8471.html (accessed 18 September 2011).

presence of certain types of antigens (genetically). Upon activation by antigen they develop and form memory and plasma cells, which both secrete antibodies. These antibodies attach to an antigen and facilitate its phagocytosis (elimination) by phagocytes. The formation and proliferation of plasma cells are controlled by the release of cytokines from helper T-lymphocytes. Antibody production takes days to weeks. Antibodies can also activate the complement system (see Box 2.14). Passive immunity refers to antibody transfer across placenta and via milk, artificial immunity to vaccination of memory cells.

### Cellular immunity

Cell-mediated responses are carried out by *T-lymphocytes* (the cytotoxic). The cell-mediated immunity involves many cell types, inducing macrophages, T-helper cells, cytoxic cells, delayed hypersensitivity T-cells, natural killer cells, granulocytes and Treg cells. Antibodies play a minor role in some responses (Ash, 2010). Upon recognition of the infected cell (host cell), cytotoxic T-lymphocytes bind to antigens on the host cell, and release cytoplasmic granules that contain perforin, which makes the host cell swell and die. Microbes are released and phagocytized by macrophages. Abnormal body cells (tumour and cancer) produce endogenous antigens and are also recognized and attacked in the same way. Recognition is based on the differentiation between 'self' (specific protein complex on cell membrane – genetics) and 'non-self' (with antigen), which is acquired during their stay in the thymus (Kapit *et al.*, 2000).

## 2.4.3  Diseases and disorders

*Autoimmune diseases* have been related to environmental toxins such as dioxins (mostly man-made family of 210 chlorinated compounds, a by-product of any combustion process). Autoimmune diseases arise from an overactive immune response of the body against substances and tissues normally present in the body. The immune system mistakes some part of the body as a pathogen and attacks it.

*Allergy and infectious diseases* are other reactions of the human body. Allergy is the reaction of the body to an in itself unharmful substance, while with an infectious disease it's the reaction of the body to a substance that can cause harm (Mygind, 1986). In both, inflammation-like reactions (e.g. swelling, pain, redness and/or heat) occur and both are hypersensitivity reactions of the immune system, which can be divided into four classes based on the mechanisms involved and the time course of the hypersensitive reaction:

- Type I hypersensitivity is an immediate reaction, often associated with allergy. Symptoms can range from mild discomfort to death.
- Type II hypersensitivity occurs when antibodies bind to antigens on the patient's own cells, marking them for destruction. This is also called antibody-dependent hypersensitivity.
- Type III hypersensitivity reactions are triggered by immune complexes (aggregations of antigens, complement proteins and antibodies).
- Type IV hypersensitivity, also known as cell-mediated or delayed type hypersensitivity, usually takes between two and three days to develop. Type IV reactions are involved in many autoimmune and infectious diseases.

Inflammation should not be confused with infection; inflammation is the body's response to the cause of the infection. One can recognize some inflammatory conditions because their names end in 'itis' – arthritis, appendicitis, gastritis, laryngitis, pancreatitis, dermatitis, meningitis (inflammation of the membranes that cover the central nervous system), peritonitis (inflammation of the membrane that covers all the abdominal organs and inside walls of the abdomen). Two inflammatory conditions that don't end in 'itis' are asthma (usually caused by an allergy to an ingested substance) and pneumonia, an inflammatory reaction to the invading microorganisms in the lungs (Nadler, 2007).

The main response of the immune system to *tumours* is to destroy the abnormal cells. However, tumour cells are often not detected, some tumour cells also release products that inhibit the immune response. And, immunological tolerance may develop against tumour antigens, so the immune system no longer attacks the tumour cells. Paradoxically, sometimes tumour growth is promoted (see Section 3.6.2).

*Fever* is an increase in core temperature of one to several degrees and is caused by release of cytokines (e.g. interleukins) from white blood cells in response to infections. These cytokines stimulate hypothalamic neurons to raise their set-point. As a consequence of this hot state, bacteria cannot grow and toxins are deactivated.

## 2.5 Hormonal regulation

*The endocrine system receives and sends information via blood vessels to endocrine glands that produce specific hormones to regulate a number of functions of the human body (such as metabolism, mood and growth). The effects take much longer and last longer than the nervous system information channel. Endocrine glands located all over the human body release specific chemical messengers (hormones) into the bloodstream. These hormones regulate the many and varied functions of an organism through binding with the appropriate receptors (e.g. mood, growth and development, tissue function, and metabolism), as well as sending messages and acting on them. The endocrine system comprises of endocrine glands (the pituitary gland, thyroid gland, the parathyroid glands, adrenal glands, thymus gland, pineal body, pancreas, ovaries and testes) and of organs with partial endocrine function, and additionally local produced hormones are involved.*

### 2.5.1 The endocrine system

The endocrine system is an information signal system mainly using blood vessels as information channels. This in general means the effects take much longer and last longer than the signals sent via the nervous system (see Section 2.6). Endocrine glands located all over the human body release specific chemical messengers (hormones) into the bloodstream (see Figure 2.4 in Bluyssen, 2009a). These hormones regulate the many and varied functions of an organism through binding with the appropriate receptors (e.g. mood, growth and development, tissue function and metabolism), as well as sending messages and acting on them (Figure 2.13).

**Figure 2.13** *The hormones produced by the endocrine system regulate the many and varied functions of an organism through binding with the appropriate receptors (e.g. mood, growth and development, tissue function and metabolism), as well as sending messages and acting on them*
Source: Bluyssen.

## Hormones

Two groups of secreted hormones can be distinguished: the *slow-acting* but long-lasting (i.e. steroid hormones of the adrenal cortex and gonads, amine hormones from the thyroid gland, and hormones derived from vitamin $D_3$), and the *fast-acting* and not long-lasting hormones (e.g. peptide hormones of the hypothalamus and the pituitary gland, and catecholamine hormones of the adrenal medulla). The long-lasting in general bind with nuclear receptors and via the DNA and messenger RNA initiate the synthesis of new proteins (takes hours to days), or, they first connect to another hormone. And the fast acting binds with plasma membrane receptors or G-proteins (see Section 2.2.2) releasing intercellular messengers (cAMP or $Ca^{++}$) that activate cellular enzymes (within seconds to minutes). Some rapidly acting peptide hormones (e.g. insulin and growth hormone) act via membrane enzyme receptors, their action doesn't involve G-proteins or second messengers. The receptor has an intracellular domain that can act as an enzyme.

There are three main chemical groups of hormones: amines, peptides and steroids, of which the first two show overlap with the neurotransmitters found in the CNS. *Locally produced hormones*, which act on the same cell as it is released from or cells in the tissue environment, are the *prostaglandins* and related substances (thromboxane and leukotines). Serotonin and histamine also act sometimes as local hormones (e.g. in blood or stomach mucosa). They may induce vasodilatation and bronchiole dilation by causing relaxation of the bronchiole smooth muscle, an effect with therapeutic value in the respiratory disorders of asthma. Some are produced during inflammatory responses, for example in arrhythmic disorders of joints (Kapit *et al.*, 2000). The release of prostagladins in the hypothalamus raises body temperature, and when excessive this leads to fever. Prostagaldins also inhibit stomach acid secretion (aspirin is thought to inhibit this, leading to ulcers). Certain hormones activate receptor mechanisms locally that cause the release of prostagladins, which amplify, antagonize or just mimic the effect of the systemic hormone.

## Glands

The endocrine system comprises of endocrine glands (the pituitary gland, thyroid gland, the parathyroid glands, adrenal glands, thymus gland, pineal body, pancreas, ovaries and testes) and of organs with partial endocrine function, and additionally local produced hormones are involved (see Figure 2.4 in Bluyssen, 2009a).

The *pituitary gland* is vital to the complex functioning of the whole hormonal system. It is the main link between the nervous and the endocrine system. The pituitary gland (hypophysis) is divided in three lobes of which the anterior lobe is the true gland. Its secreted growth hormone promotes tissue and body growth by stimulating cells to multiply and make more proteins. In the *posterior lobe* (in fact an extension of the hypothalamus) the hormone ADH (antiduretic hormone or vasopressin), involved in maintaining the water balance, is stored (see water and salt control, Section 2.5.2). And the intermediate lobe comprises only a few cells with no known functions (Kapit *et al.*, 2000; Parker, 2003).

The *adrenal glands*, divided into two layers: the adrenal cortex and the adrenal medulla, produce several of the fast-acting hormones to help the body cope with all kinds of stress (mentally and physically). The *adrenal medulla* is part of the sympathetic nervous system and is in fact a modified synaptic ganglion. The adrenal medulla secretes catecholamines (about 80% epinephrine and 20% norepinephrine), regulated via the symphatic nervous system, that are synthesized from the amino acid tyrosine via several chemical reactions:

$$\text{tyrosine} \Rightarrow \text{dopa} \Rightarrow \text{dopamine} \Rightarrow \text{norepinephrine} \Rightarrow \text{epinephrine}$$

Catecholamines help prepare the body to stressful situations such as stress, emotions and exercise (see Section 3.2). Dopamine and endorphin (may have anti-pain effects) are also secreted.

The *adrenal cortex* has three zones, each secreting a different type of steroid hormone (corticosteroids):

- *Aldosterone*, involved in regulation of plasma salts (sodium and potassium), blood pressure and blood volume.
- *Glucocorticoids* (mainly cortisol), which regulate metabolism of glucose, especially under stress.
- *Sex steroids*, mainly androgens and some estrogen and progesteron.

The *thyroid gland* is located in the neck and secretes two hormones: thyroxine (T4, tetra-iodothryonine) and tri-iodothyronine (T3). The thyroid hormones regulate metabolic rate (see Section 2.5.2). The *parathyroids* are four small glands embedded in the thyroid tissue, which excrete the hormone parathormone, to regulate the levels of calcium in the blood (see Section 2.5.2). This is important for strong bones and teeth and healthy nerves.

Hormones from the *pancreas* control the glucose level in blood and cells, the body's energy source (Kapit *et al.*, 2000) (see Section 2.5.2). The *pineal gland* has links with the hypothalamus and the pituitary gland, and it has links with the nerves that carry signals from the eyes to the brain and with the brain's biological clock. It produces melatonin, the sleep hormone, which makes one sleepy (see Section 3.3.2). The *thymus* works as a lymph gland of the lymphatic system and it has a major role in the immune system (Parker, 2003). One of its hormonal roles is to process cells from the bone marrow to make specialized cells of the immune system so that they can fight infection (see Section 2.4.1).

In Table 2.2 main hormones of different glands and their effects are presented.

## 2.5.2 Control mechanisms

Three control mechanisms exist to regulate the excretion of hormones:

1  *Simple hormonal regulation*, in which the blood hormone level and physiological parameter regulated, interact according to pre-determined and set limits to maintain hormone secretion. Only one gland is involved. Examples of simple hormonal regulation are the regulation of plasma calcium by parathormone (Figure 2.14) and the regulation of blood sugar by insulin and glycagon from pancreatic islets (Figure 2.15).

**Table 2.2** *Hormones and their effect(s) for the main glands*

| Hormone | Effect |
|---|---|
| ***Pituitary gland – anterior lobe*** | |
| Growth hormone | Promotes tissue and body growth; increase metabolic rate |
| Trophin or tropin hormones | Induces the release of other hormones |
| Beta-endorphin | Pain inhibition |
| Melanocyte stimulating hormone | Affects melanocytes in skin which produces melanin (makes skin coloured) |
| Prolactin | Involved in breast milk flow and child birth |
| ***Pituitary gland – posterior lobe*** | |
| Antidiuretic hormone (ADH) | Regulates overall water balance of body |
| Aasopressin | Moderates vasoconstriction |
| Oxytocin | Involved in breast milk flow and child birth; involved in trust between people and circadian homeostasis (body temperature, activity level, wakefulness) |
| ***Pineal body*** | |
| Melatonin (sleep hormone) | Antioxidant and causes drowsiness |
| ***Thyroid gland*** | |
| Thyroxin (tetra-iodothyronin) and tri-iodothyronin) | Increases metabolic rate and influences cardiovascular functions |
| Calcitonin | Decreases plasma calcium levels (opposite of PTH), stimulates absorption of calcium and its deposition into bone (important for growing children) |
| ***Parathyroid glands*** | |
| Parathyroid hormone (PTH) | Acts on bone and kidney to elevate plasma calcium level |
| ***Thymus*** | |
| Hormones and hormone-like substances known as 'factors': e.g. thymosin | Effects on both immune (involved in production of T-lymphocytes and antibodies produced by B-lymphocytes) and reproductive system |
| ***Pancreatic islets*** | |
| Insulin and glucagon | Regulates carbohydrate metabolism in tissues and ensures maintenance of optimal plasma glucose levels |
| Somatostatin | Regulates insulin and glucagon secretion locally; may inhibit growth hormone from the pituitary |
| ***Adrenal glands – cortex*** | |
| Sex steroids | In females: stimulate formation of red blood cells, have mainly anabolic effects; can be converted to testosterone (much less potent than that from testis) |
| Aldosterone (a mineralcorticoid) | Involved in regulation of plasma salts (sodium and potassium), blood pressure and blood volume |
| Cortisol (glucocorticoid) | Regulates metabolism of glucose, especially in times of stress |
| ***Adrenal glands – medulla*** | |
| Catecholamines: epinephrine and norepinephrine | Prepares for stressful situations, such as fight-flight |
| Dopamine | Increases heart rate and blood pressure |
| Endorphin | Anti-stress analgesic (anti-pain) effects |

Source: Parker, 2003; Kapit *et al.*, 2000.

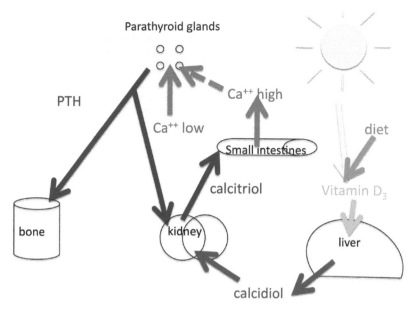

**Figure 2.14** *Hormonal regulation of plasma calcium: in response to a decrease in the level of calcium ions (Ca⁺⁺) (monitored by specific membrane calcium receptors), the parathyroid glands secrete parathormone (PTH) into the circulation. PTH acts directly on bone (calcium is mobilized to increase plasma calcium) and the kidneys (calcium is retained) and indirectly on intestinal mucosa to raise plasma calcium (which reduces again PTH secretion). In the small intestine absorption of calcium is increased by the formation of the active form of vitamin D in the kidney. Active vitamin (calcitriol) is formed by converting vitamin $D_3$ in the liver, stimulated by PTH. Vitamin $D_3$ can be obtained in food or produced from cholesterol derivates in the skin in the presence of ultraviolet radiation in sunlight*

Source: Bluyssen based on Kapit *et al.*, 2000, plate 120.

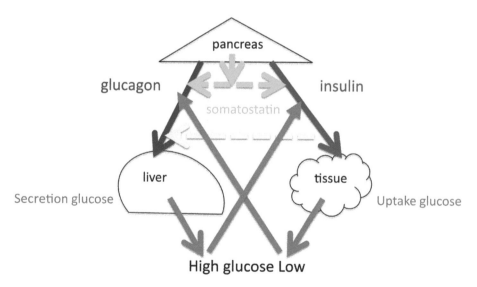

**Figure 2.15** *Hormonal regulation of glucose blood level: insulin is released by the endocrine cells of the pancreas into the blood when blood glucose (sugar) is high. Insulin decreases blood sugar by promoting uptake of glucose by tissues. Low blood sugar stimulates glucagon release by endocrine cells of the pancreas into the blood, raising blood sugar. Somatostatin, also secreted by the pancreas but locally, regulates insulin and glucagon secretion. Glucagon stimulates insulin secretion while insulin inhibits glucagon secretion*

Source: Bluyssen; based on Kapit *et al.*, 2000, plates 122 and 123.

2   *Complex hormonal regulation*: in which the activity of one gland is
    controlled by hormones of another. An example of complex hormonal
    regulation is water and salt control (see Figure 2.16).
3   *Complex neurohormonal regulation*: in which interaction between brain
    (hypothalamic hormones) and endocrine system (production of hormones
    by anterior pituitary gland) takes place (see Box 2.15). Examples of com-
    plex neurohormonal regulation are the regulation of the fat metabolism
    (Figure 2.17) and the regulation of metabolic rate (Figure 2.18).

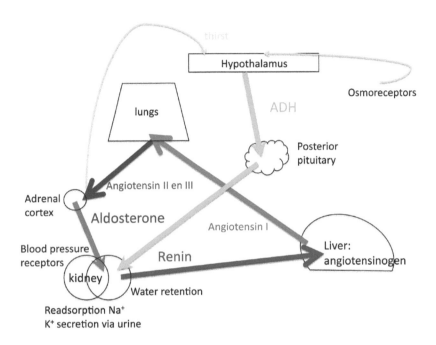

**Figure 2.16** *Water and salt control: Renin (an enzyme) is released by the kidney when
extracellular volume is depleted (reduce blood pressure measured by low pressure volume
receptors in kidney). Renin acts on angiotensinogen (plasma protein produced by liver) and
splits off angiotensin I. Angiotensin I is converted into angiotensin II when it passes (in blood)
the lungs, and is split into angiotensin III. Angiotensin II and III both stimulate secretion of the
hormone aldosterone by the adrenal cortex and both stimulate thirst. Aldosterone stimulates
$Na^+$ readsorption in the kidney. ($Cl^-$ follows $Na^+$, therefore net results: readsorption of NaCl
by kidney tubules.) The total amount of NaCl (not concentration) in the extracellular spaces,
determines the osmotic pressure (see Box 2.5). The hormone ADH (antidiuretic hormone
or vasopressin), produced in neural cells of the hypothalamus (upon impulses from neural
osmoreceptors) and stored in the posterior pituitary, tunes the relative proportions of NaCl
and water gained (by kidney). Sodium readsorption also increases readsorption of water and
indirectly decreases potassium levels by promoting potassium secretion in tubules of kidney*

Source: Bluyssen; based on Kapit *et al.*, 2000, plate 126.

## Box 2.15  Hormones in complex neural hormonal regulation

The *anterior pituitary gland* secrets tropin or trophin hormones. Hormones of the *hypothalamus* regulate the pituitary and those of the pituitary regulate hormones of the target glands. Hormones of the target glands regulate secretion of the respective hypophysiotropin hormones and the pituitary tropins (via negative and positive feedback effects):

- The thyroid-releasing hormone (TRH) regulates thyroid-stimulating hormone (TSH); TSH promotes thyroid secretion of thyroxine (T4) and tri-iodothyronine (T3) (T4 is converted to T3) (see Figure 2.18).
- Corticotropin-releasing hormone (CRH) regulates adrenocortico tropin hormone (ACTH); ACTH promotes secretion of cortisol by the adrenal cortex (see Section 3.2.2).
- Gonadotropin-releasing hormone (GnRH) regulates FSH (follicle stimulating hormone) and LH (luteinizing hormone); FSH and LH promote secretion of sex steroids (estrogen, progesterone and testosteron).
- Growth hormone-releasing hormone (GRH) stimulates the release of growth hormone (GH); GH promotes the conversion of stored fat to fatty acids and glycerol in the adipose tissue. Via the release of somatomedins by the liver, the uptake of amino acids in muscle tissue is promoted and the uptake of glucose is inhibited. (See Figure 2.17.)
- Somatostatin (GIH) inhibits the release of GH (see Figure 2.17).
- Prolactin-releasing hormone (PRH) stimulates release of prolactin (involved in breast milk flow and child birth).
- Prolactin-inhibiting hormone (PIH) inhibits release of prolactin.

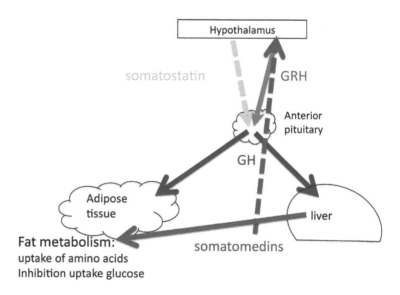

**Figure 2.17** *Regulation of fat metabolism: the growth hormone (GH) promotes tissue and body growth, but also influences metabolism. GH promotes the conversion of stored fat to fatty acids and glycerol in the adipose tissue. Via the release of somatomedins by the liver, the uptake of amino acids in muscle tissue is promoted and the uptake of glucose is inhibited. Secretion of GH by the anterior pituitary is stimulated by a hypothalamic-releasing hormone and inhibited by somatostatin, which can be elevated during stress (e.g. sustained exercise or fasting). Increased blood levels of somatomedins, fatty acids and glucose cause the hypothalamus to reduce GH secretion*
Source: Bluyssen; based on Kapit et al., 2000, plate 118.

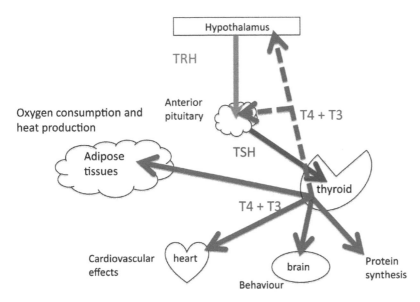

**Figure 2.18** *Regulation of metabolic rate: the thyroid-releasing hormone (TRH) regulates thyroid-stimulating hormone (TSH); TSH promotes thyroid secretion of thyroxine (T4) and tri-iodothyronine (T3) (T4 is converted to T3). T4 increases metabolic rate and heat production by increasing oxygen consumption and heat production in many body tissues. They promote protein synthesis, increase heart rate, heart contractility and blood pressure, and promote the effects of catecholamines on their targets (heart, adipose tissue, brain)*

Source: Bluyssen; based on Kapit *et al.*, 2000, plate 119.

## 2.5.3 Diseases and disorders

There are numerous diseases and disorders related to the production of hormones such as insulin and thyroxin.

There are two types of *diabetes mellitus*: type I insulin dependent diabetes and type II non-insulin dependent diabetes melitus. With type I insulin dependent diabetes one suffers from an insulin deficiency (one does not produce insulin), and it occurs in most cases with juveniles. It is characterized by reduced uptake of glucose into the muscle and fat tissue (increased breakdown of fat and protein reserves) and increased release of glucose into blood, leading to high blood sugar levels (hyperglycemia), frequent urinating and excess water in urine. In type II, one suffers from insulin resistance and it occurs most often with people above 40. Blood insulin levels are higher than normal, caused by a decreased number of insulin receptors in muscle and fat cells in response to excessive and long-term insulin production, and reduced membrane glucose transportation. Insulin production slows down.

Lipid accumulation in skeletal muscle is associated with the development of insulin resistance. Increased consumption of carbohydrates and fats contributes to intracellular oxidative stress and reduces cellular capacity to neutralize reactive oxygen species (ROS) and other free radicals (see Section 3.5) (Culp, 2010).

Disorders of the thyroid function can comprise an active thyroid (*hyper-thyroidism*) or a slow thyroid (*hypothyroidism*). An active thyroid leads to an increased metabolic rate, sweating, poor heat tolerance, increased appetite, nervous behaviour and reduced weight. A slow thyroid results in sluggishness, reduced metabolic rate and poor tolerance to cold. A *goiter*, a swollen thyroid gland, occurs when the thyroid gland doesn't have enough iodine to make thyroxine. The pituitary releases more TSH to make the thyroid grow and produce more hormones.

When the pituitary gland doesn't produce enough ADH, a disorder named *Diabetes Insipidus* can occur. Constant thirst and production of a lot of very weak or dilute urine are the symptoms.

Overproduction of steroid hormones by the adrenal cortex can cause *Cushing's disease*. Body, face and shoulders become rounder and fatter, while their muscles shrink and weaken. Bones weaken and skin develops steaks, spots and bruises. Tiredness and depression can be related as well.

*Addison's disease* is caused by a lack of cortisol, resulting in disruption of blood glucose supply. Symptoms are weakness and lack of energy, weight loss and loss of appetite, and darkened skin.

## 2.6 Perception and reaction

*The nervous system receives and sends information via nerves to brain and vice versa (odour, sound, colour, taste, pain). Responsibilities of the nervous system are sensory and motor functions, instinctive and learned behaviour, and regulation of activities of the internal organs and systems. The nervous system can be divided into two parts: the central nervous system (CNS) and the peripheral nervous system (PNS). The PNS comprises nerves, sensory receptors and motor effectors, while the CNS comprises the lower and higher centres in the brain and the spinal cord. The PNS has an autonomic part, which regulates involuntary action, and a somatic part (voluntary).*

### 2.6.1 Peripheral nervous system

Through sensory receptors, via nerves (spinal cord and cranial), our senses can provide information to the brain, which is processed and used to send messages with prescribed actions to the relevant parts of the human body. The nervous system is responsible for sensory and motor functions, for instinctive and learned behaviours, and for regulating activities of internal organs and systems. While the central nervous system (CNS) takes care of processing information perceived, integrating and producing the appropriate motor commands (see Section 2.7), the peripheral nervous system (PNS) is responsible for perception and behaviour (responses). In Box 2.16 an overview is presented of the basic components of our PNS.

The PNS has an autonomic part, which regulates involuntary action, such as heartbeat and digestion, and a *somatic* part (*voluntary*), which mainly responds to external stimuli by an external action (behaviour, e.g. flight). The somatic system comprises sensory nerves that connect the sensory receptors to the spinal cord and the brain (CNS), and motor nerves that connect the CNS to the voluntary skeletal muscles. The *autonomic nervous system* (ANS) (vegetative or visceral nervous system) deals with the visceral effects (glands and smooth muscle).

---

## Box 2.16  Basic components of peripheral nervous system

*Sensory receptors*: Receptor cells located at a receiving membrane or surface in the body, which are specifically sensitive to one class of stimuli. They connect with a secondary, on-going nerve cell, that carries the nerve impulse (action potential) along (see Table 2.3).

*Neurons or nerve cells*: These transmit 'messages' or impulses throughout the nervous system. A typical neuron comprises of *dendrites* (receiving stimuli), cell body and axon (conducting impulses) or *nerve fibre*.

*Nerve fibres*: These can be classified, based on size and conduction velocity, in three general types A, B and C, with four subtypes for A (Aα, Aβ, Aγ and AΔ). Type A are the largest and fastest and type C the slowest and smallest.

*Nerves*: These can be afferent (towards the CNS) or efferent (away from the CNS). Nerves comprise of thousands of nerve fibres that form fascicles (see Figure 2.19). Twelve pairs of cranial nerves emerge directly from different brain sites and 31 pairs of spinal nerves are associated with the spinal cord. The spinal nerves are divided into 8 cervical (neck), 12 thoracic (chest), 5 lumbar (loin, lower back), and 5 sacral (sacrum bone). And there is 1 coccygeal nerve.

*Ganglia*: These are intermediary connections between different neurological structures in the body, such as the peripheral and central nervous system. The basal ganglia in the brain is a group of nuclei interconnected with the cerebral cortex thalamus and brainstem, and is associated with a variety of functions: motor control, cognition, emotions and learning. In the autonomic nervous system, fibres from the central nervous system to the ganglia are named preganglionic fibres, while those from the ganglia to the effector organ are called postganglionic fibres. Dorsal root ganglia (also named spinal ganglia) contain the cell bodies of sensory (afferent nerves).

*Synapse*: This is a site where nerve impulses are transmitted from nerve cell to nerve cell, from nerve to muscle or from nerve to gland.

*Neurotransmitter*: This is released during a synaptic transmission of impulse. Neurotransmitters activate (bind to) postsynaptic receptors (proteins) which then causes activation or inhibition processes.

## Perception and reaction

Sensory inputs (stimuli) enter through the eyes (vision), ears (audition), mouth (taste), nose (olfaction) and body sensors (touch and internal organ configurations). The sensory cells connect with secondary, ingoing nerve cells that carry the nerve impulse (action potential) along (some are very short such as in the eye, and some are very long such as in the skin). From such *afferent* nerves, higher order neurons make complex connections with pathways of the brainstem and deeper parts of the brain (e.g. the thalamus) that eventually end in specific areas in the cerebral cortex of the brain. Vision goes to the occipital lobe in the back, smell to the frontal lobe (visual cortex: for early vision), audition to the temporal lobes on the sides, and touch (tactual function) to the parietal lobes towards the top of the brain. Associations between stimuli are developed by memory in various cortical sites. Conscious and unconscious actions are then coded and sent back

**Table 2.3** *Types of sensory receptors found in or on the human body and senses (besides these sensory receptors the human body also has motor and integration receptors). A basic description of the characteristics and the mechanisms of the human senses (the eyes, ears, skin, nose) is presented in Chapter 2 of the Indoor Environment Handbook (Bluyssen, 2009a)*

| Receptor | Where | Sensitive to/detect stimuli | Function |
|---|---|---|---|
| Mechanoreceptor | Skin, muscles, joints and visceral organs | Mechanical deformation by indentation, stretch and hair movement | Touch, muscle length, tendon and limb position<br>Hearing and balance<br>Blood pressure |
| Chemoreceptor | Olfactory bulb (nose)<br>Taste buds in mouth<br>Internally: e.g. blood, digestive tract | Chemicals | Smell (odour)<br>Taste<br>Blood levels (oxygen, carbon dioxide, glucose, osmopolarity)<br>Hunger, thirst |
| Thermoreceptor | Free nerve endings in skin | Warmth and cold | Internal temperature (hypothalamus) |
| Nociceptor | Free nerve endings in skin and visceral tissues | Noxious stimuli | Pain |
| Photoreceptor | Retina in eye | Light energy | Vision |
| Kinaesthetic (motion) sense | Muscles, tendons and joints | Position of legs and arms and the perception of the active or passive movement of a limb | Regulation of reflex and voluntary movement |
| Sense of balance or equilibrium | Inner ear | Rotational and linear acceleration (changes of body's position in relation to gravity) | Ensure balance and equilibrium |

Source: adapted from Kapit *et al.*, 2000, plate 89.

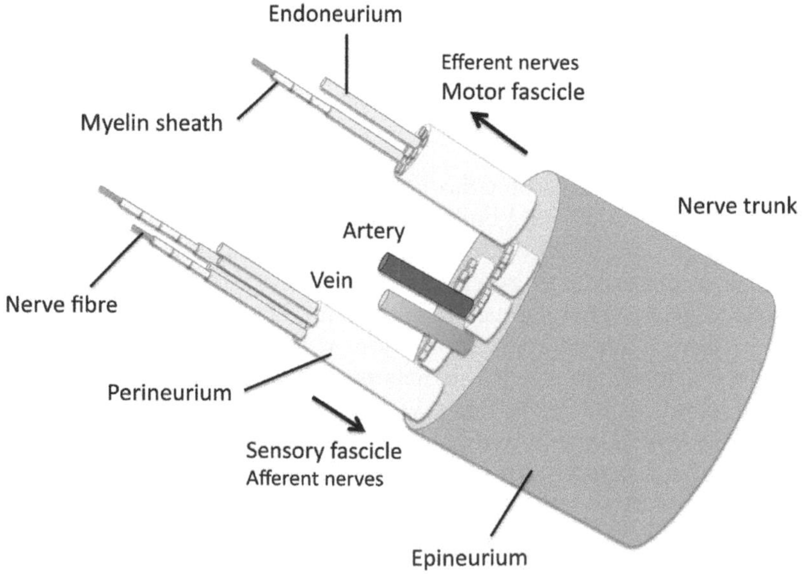

**Figure 2.19**
*A peripheral nerve: comprises thousands of nerve fibres that form fascicles. Each fascicle has a distinct function (sensory or motor) or target. Each fibre in the fascicle is the axon of a sensory, motor or autonomic neuron. The whole of trunk with fascicles may therefore be sensory, motor or mixed, and can also contain blood vessels. A peripheral nerve can be part of spinal, cranial or autonomic groups. Each nerve fibre inside a fascicle has a sheath (endoneurium), each fascicle is surrounded with a coat (perineurium) and each trunk has a coat (epineurium)*

Source: Bluyssen; based on Kapit *et al.*, 2000, plate 86.

Endoneurium

Efferent nerves
Motor fascicle

Myelin sheath

Nerve trunk

Artery

Vein

Nerve fibre

Perineurium

Sensory fascicle
Afferent nerves

Epineurium

to the relevant parts of the human body via *efferent* nerves, where they are executed (see also Section 2.7.2).

In the autonomic system, visceral sensory fibres (input) and the sympathetic and parasympathetic fibres (motor output) connect the visceral organs and effectors to the sympathetic and parasympathetic ganglia as well as to the CNS. Sympathetic motor output runs via the thoraric and lumbar spinal nerves (also targets in the head, e.g. iris of the eyes). The usual *postganglionic* fibres, driven by preganglionic sympathethic neurons in the spinal cord, innervate practically all visceral and peripheral organs in the body, in particular the blood vessels. The parasympathetic nerves, basically *preganglionic*, are associated with four cranial nerves (III, VII, IX, X) and with the sacral spinal nerves. The short postganglionic neuron emerges from a peripheral ganglia located near or in the target organ.

## 2.6.2 Mechanisms

### Nerve impulse

Nerve impulses are action potentials. At rest the interior of the axon has a negative charge compared to the outside. When the axon is stimulated, the membrane potential reverses momentarily along the axon. $Na^+$ permeability is increased and $Na^+$ moves in via $Na^+$ channels (depolarization). A moment later $K^+$ permeability is very high and $K^+$ moves out via $K^+$ channels (repolarization). After several seconds the rest status (polarization) is back again. This process moves along the entire length of the axon. The discrepancy in charge between excited and unexcited regions of the axon will cause the charge to move along the axon. A stimulus is only effective when it depolarizes the membrane. The strength of a stimulus to just excite is called the threshold. Stimuli below the threshold do not work. But all action potentials are the same, no matter how large the stimulus: the response is all-or-none. In Figure 2.20 an example is shown for a Purkinje cell in the cardiac muscle.

### Synaptic transmission

When nerve impulses are transmitted from nerve cell to nerve cell, from nerve to muscle or from nerve to gland, this is called synaptic transmission. When an impulse arrives at the synaptic terminal it opens, the vesicles with presynaptic cell membranes fuse, release neurotransmitters, which diffuse across the synaptic cleft to the postsynaptic membrane (see Figure 2.3 in Bluyssen, 2009a).

In some cases, transmission is electrical (ion flow), but in most cases, transmission is chemical: the action potential opens $Ca^{++}$ channels and $Ca^{++}$ moves inwards the synaptic terminal, which promotes the release of a neurotransmitter. Neurotransmitters activate (bind to) postsynaptic receptors (proteins). Some of those receptors open ion channels directly, others activate second messengers/ enzymes. Some ion channels excite (depolarization), others inhibit (hyperpolarization), depending on the ions that pass through the channel. When $Na^+$ and $K^+$ are moving it concerns an excitatory channel; when $Cl^-$ and $K^+$ are moving it concerns an inhibitory channel.

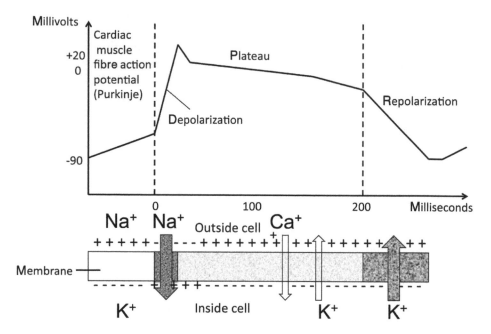

**Figure 2.20** *Action potential of the heart: a rise in potential (necessary for contraction) is due to the opening of Na+ channels which allow Na+ ions to enter the cell, followed by inactivation of Na+ channels and opening of K+ channels (which is delayed in cardiac muscle causing the contraction to take longer). The K+ that is leaking out before the opening of the K+ channels is compensated by Ca++ through Ca++ channels. After K+ channels open, Ca++ channels close. The potential falls to minimum and the cycle starts again*
Source: Bluyssen; based on Kapit *et al.*, 2000, plate 32.

The action of a neurotransmitter does not persist because it is continuously removed from the synaptic cleft either by enzymatic attack or by re-uptake by nerve terminals. A persistent response of the postsynaptic cleft can be obtained only by delivery of an equally persistent amount of nerve impulses to the synapse.

### Inhibition or excitation

Different synapses use different neurotransmitters: synapses at neuromuscular junctions use acetylcholine and are always excitatory. Those in visceral organs (i.e. autonomic synapses) use norepinephrine or acetylcholine and can be either excitatory or inhibitory. Autonomic nerves first make synaptic connections with other neurons (at ganglia), which then relay the impulses to the organs. Both parts use the same neurotransmitter (acetylochine) towards the ganglia, but from the ganglia on parasympathetic transmission uses acetylcholine, while sympathetic transmission uses norepinephrine.

### Sympathetic versus parasympathetic actions

The ANS is divided into two parts: the sympathetic and the parasympathetic nervous system, whose actions are antagonistic – running or fighting versus resting or recovering. Many organs are supplied by nerves from each part (see Table 2.4), as for example the beating of the heart. Alterations of heartbeat (e.g. heart rate and contraction force) are mainly caused by the sympathetic and parasympathetic nerves on the heart:

- The sympathetic nerves liberate *norepinephrine*, increasing $Ca^{++}$ permeability and $Na^+$ flux, which stimulates the heart, increasing rate and force of contraction.
- The parasympathetic nerves liberate *acetylcholine*, increasing $K^+$ permeability, which slows the heart rate.

## Reflexes

Reflexes are programmed, predictable motor responses to certain sensory stimuli. Some reflexes are defensive, others help to maintain balance and posture, and some help regulate the internal environment by affecting exocrine glands, heart and visceral smooth muscles (autonomic reflexes). Somatic reflexes involve body's skeletal muscles and motor behaviour, such as the well-known knee-jerk reflex.

**Table 2.4** *Autonomic effects of some selected organs*

| Organ | Effect of sympathetic stimulation | Effect of parasympathetic stimulation |
|---|---|---|
| *Heart* | | |
| • Muscle | Increase rate and contraction force | Decrease rate and contraction force |
| • Coronaries | Dilation; constriction | Dilation |
| *Systemic arterioles* | | |
| • Abdominal | Constriction | – |
| • Muscle | Constriction; dilation | – |
| • Skin | Constriction | – |
| *Lungs* | | |
| • Bronchi | Dilation | Constriction |
| • Blood vessels | Constriction (slightly) | |
| *Adrenal medullary secretion* | Increase | – |
| *Liver* | Release glucose | Small glycogen synthesis |
| *Sweat glands* | Copious sweating | – |
| *Glands (nasal, lacrimal, salivary, gastric)* | Vasoconstriction and some secretion | Copious secretion |
| *Kidney* | Decrease urine and renin secretion | – |
| *Basal metabolism* | Increase | None |
| *Eye* | | |
| • Pupil | Dilation | Constriction |
| • Ciliary muscle | Slight relaxation | Constriction |

Source: adapted from Kapit *et al.*, 2000, plate 29.

### 2.6.3 Diseases and disorders

Besides from pain and tiredness, several diseases and disorders have been related to the different senses of our body (see Table 2.5). Other than those, diseases related to the peripheral system are mostly a consequence of a dysfunction or degradation of the control systems in the CNS. As an example Parkinson disease is related to the degeneration of dopamine releasing neurons, Huntington's disease is related to the loss of GABA (gamma-aminobutyric acid)-releasing neurons.

*Pain or nociception* involves sensation as well as feelings and emotions. Pain usually has a strong negative affective component, which is believed to modulate pain perception (Janssen, 2002). Pain can involve a fast, short-lasting experience (sharp localized sensation) or a dull sensation or throbbing pain, which is long-lasting and diffuse. Fast pain is experienced via type Aδ fibres and dull pain via type C sensors. Fast pain fibres project directly to the thalamus and up to the sensory cortex. Slow pain fibres make a major input into the brain stem reticular formation, which mediates the central inhibition and arousal effects of this pain. These fibres terminate largely in the thalamus, with further input to the limbic system, in particular the cingulate gyrus, where the emotional, hurting component of pain is processed. Descending fibres release neurotransmitter serotonin that excites certain inhibitory interneurons so a peptide neurotransmitter, enkephalin (an endorphin), is released, which suppresses relay of pain signals by afferent type C-fibres through binding of the opiate receptor molecules present. Morphine and other opiate analgesics (pain killers) bind to the same receptors. A peptide called nociceptin, which resembles dynorphin, has been found to have the opposite effect.

**Table 2.5** *Diseases and disorders of the human senses in the indoor environment*

| Level | Skin | Eyes | Ears | Nose | Respiratory tract |
|---|---|---|---|---|---|
| Discomfort | Warm, cold, sweat, draught | Too much/too little light, blinding, glare, reflections | Disturbance, hearing/ understanding problems | Smell, irritation | Cough, shortness of breath |
| Allergic or irritant reaction | Contact dermatitis, dry, itchy, red skin | Redness, itching, dry feeling | | Blocked or runny nose, sneezing | Asthma, bronchitis, hypersensitivity reactions |
| Infectious diseases | Infection (bacterial, viral or fungal) | Rare: dry eyes syndrome | Inflammation of the inner ear | Blocked/runny nose, temporary loss of smell | Infection (bacterial, viral or fungal) |
| Toxic chronic effects | Radiation-related disorders (such as sunburn) | Damage to the eye by UV light, cataract formation | Severe and permanent loss of hearing | Permanent loss of smell | Damage and/or tumours |

Source: Bluyssen, 2009a, Table 1.1, p. 5.

## 2.7  Integration and control

### 2.7.1  Central nervous system

The central nervous system (CNS) consists of the brain and spinal cord. The CNS performs actions with the sensory, motor and association (integrative) centres in the brain. The lower centres of the CNS (spinal cord) are in direct contact with the PNS via sensory and motor nerves.

### The brain

The brain may be divided into a lower brain stem and a higher cerebrum (forebrain) (see Figure 2.21 and Box 2.18). The brain is divided in two halves, left and right. Neural integration involves binding the two functions of the two sides of the brain (horizontal or bilateral integration) and vertical integration involves binding the activity of at least the brainstem, the limbic area and the cortex (Siegel, 2011).

---

### Box 2.17  What is the human mind?

The human mind is a relational and embodied process that regulates the flow of energy and information.

Source: Siegel, 2011, p. 52.

---

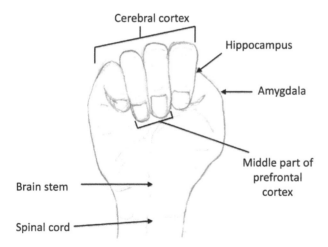

**Figure 2.21** *Hand model of half of the brain: if you put your thumb in the middle of your palm and you curl your fingers over the top, your wrist represents the spinal cord, the inner brainstem is represented in the palm, while behind your thumb you will find the location of the limbic area. Your curled fingers represent the cortex*
Source: Bluyssen, based on Siegel, 2011, p. 15.

# Box 2.18 Parts of the brain

The *forebrain*, comprising:

- The *lower diencephalon* structure, with:
  - The *hypothalamus*: for regulating the internal environment (e.g. body temperature, blood sugar, hunger and satiety) and sexual behaviour, the diurnal cycles by its biological clock and activities of the endocrine system and hormones.
  - The *thalamus*: for integrating sensory signals and relaying them to the cerebral cortex, participating in motor control and in regulating cortical excitation and attention.

- The *higher telencephalon*, comprising two nearly symmetrical hemispheres, connected by the corpus callosum, with:
  - The *cerebral cortex*: network of nerve cells with:
    - The *frontal lobe*: posterior areas are specialized in motor functions; anterior areas are involved in learning, planning, speech and other psychological functions.
    - The *occipital lobe*: visual functions.
    - The *parietal lobe*: somatic sensory function (e.g. skin senses) and related association and integrative roles; areas for cognitive and intellectual processes.
    - The *temporal lobe*: hearing centre and related association areas, including speech centres; other areas are important in memory; anterior and basal areas are involved in the sense of smell and functions related to the limbic system.
    - The *insular lobe*: not visible externally.
  - The *basal ganglia*: work with the motor areas of the cortex and cerebellum to plan and coordinate gross voluntary movements.
  - The *limbic system*: centre of emotional behaviour (oldest part of cortex), comprising:
    - The *hippocampus*: memory acquisition and recall; formation of long-term memory.
    - The *amygdala*: important for processing and memory of emotional reactions.
    - The *cingulate cortex* (or limbic lobe): affective significance.
    - The *septal nucleus*: pleasure and reward.
    - *Fornix*: carries signals from hippocampus to the mammillary bodies and septal nuclei.
    - *Mammillary bodies*: important for the formation of memory.
  - Sometimes also said to be part of limbic system:
    - *Nucleus accumbens:* for reward, pleasure and addiction.
    - *Orbitofrontal cortex:* interacts with limbic system in support of higher-order functions such as association, integration and regulation of central autonomic processes, mood and affect, and those motor functions that are under limbic control.

The *brainstem*, comprising:

- The *midbrain*: with somatic motor centres involved in regulation of walking, posture and reflexes for head and eye movements.
- The *pons*: with the inhibition control centres for respiration; it interacts with the cerebellum.
- The *medulla*: the lowest of the brain's areas with the centres for respiration, cardiovascular and digestive functions.
- The *cerebellum*: involved in movement coordination.

Source: Kapit *et al.*, 2000; Kapit and Elson, 2002; Michael-Titus *et al.*, 2010.

The brain stem or the *vegetative/reflexive brain* consists of medulla, pons and midbrain (see Figure 2.1 in Bluyssen, 2009a). The brain stem is concerned with controlling vital bodily functions (respiration, digestion, circulation), the basic processes. The brain stem directly controls our state of arousal, determining, for example, if we are hungry or satisfied, asleep or awake. In cooperation with your limbic system, the brain stem also determines whether we respond to threats (fight–flight–freeze) and the brain is important for satisfying your basic needs (drivers) for food, shelter, reproduction and safety (Siegel, 2011).

The forebrain consists of a lower diencephalon (hypothalamus and the thalamus) and a higher telencephalon, including the cerebral cortex (also named the *adaptive/skilled brain*), which is concerned with complex perception and execution of skilled sensory and motor functions (e.g. hand movements, speech) as well as higher mental functions (e.g. learning, thoughts, introspection, planning). The forebrain also comprises the *basal ganglia*, a complex of motor structures working together with the motor areas of the cortex and cerebellum, and the *limbic system* (see Figures 2.2 and 9.2 in Bluyssen, 2009a), a circuit comprising different structures. Together with the hypothalamus, the limbic system (e.g. hippocampus, amygdala, cingulated cortex, septal nuclei, formix, mammillary body) controls the expression of instinctive behaviour, emotions, and drives, and is also named the *intermediate brain* (see also Section 3.7).

Signals for somatic motor reactions (smiling) are sent to brainstem motor centres; for visceral motor effects (heart rate) they are sent to the autonomic nervous centres; for neuro-hormonal effects, they go to the pituitary/endocrine system. Feelings are integrated at the higher cortical levels.

Besides specialized areas for sensory and motor functions, the greater part of the cerebral cortex contains association areas involved in integrative functions such as speech and language, and planning. The left and right hemispheres of the brain are fairly symmetrical for sensory, motor and association areas. But for certain integrative functions they show asymmetry: the left hemisphere is specialized in verbal and analytical tasks, such as speech, hand control (right-handedness) and logical and analytical functions. The right hemisphere is specialized in spatial and holistic tasks, such as perception and discrimination of musical tones and speech intonations, emotional responses, and in appreciating humour and metaphor.

## Cortex

The outer layer of the brain is the cortex. The frontal part of the cortex gives us insight into the inner world, it allows us to think about thinking, to imagine, to recombine facts and experiences, to create. The posterior cortex (hand model Figure 2.21: from your second knuckle to the back of your head) includes the occipital, parietal and temporal lobes and generates our perceptions of the outer world through the five senses and keeps track of location and movement of our physical body through touch and motion perception. In the prefrontal cortex (hand model Figure 2.21: area from first knuckle to fingertips), we create representations of concepts such as time, a sense of self and moral judgements. The middle prefrontal cortex links the cortex, the limbic system and brainstem, and it links signals from all those areas to signals we send and receive in our social world (Siegel, 2011).

## The limbic system

The limbic system works closely with the brainstem to create our basic drives and emotions. The limbic system evaluates our current situation ('Is this good or is this bad?') and the limbic system is crucial for how we form relationships and become emotionally attached to one another. Through the hypothalamus, the limbic system plays an important regulatory role. The limbic system also helps to create several forms of memory through the hippocampus and the amygdala, two clusters of neurons located to either side of the central hypothalamus and pituitary (Siegel, 2011) (see Section 3.7.1).

The hippocampus receives information from all association areas of the brain and sends information back to them. In addition, the hippocampus has two-way connections with many regions in the interior of the cerebral hemisphere. One reason that very few people remember events that occurred during infancy may be the immaturity of the hippocampus. It is not until a child is two to three years old that most of the brain structures are developed (Martin *et al.*, 2010).

## The hypothalamus

The hypothalamus, the head ganglion of the sympathetic nervous system, is the major brain centre for regulation of internal body functions and integration of activities. The hypothalamus receives input from:

- *Major sensory systems*: information about environmental conditions, smell, information from digestive system, blood vessels and so on.
- *The limbic system*: information on the state of an individual's drives (hunger, thirst, sex) and emotions.
- *Hormones and other blood-substances* (such as sodium ions and glucose): information on salt, water and energy situation.

Output goes to:

- *The limbic system* to interact with the structures controlling emotions and drives.
- *The midbrain motor centres* for controlling somatic motor responses during emotional behaviours.
- *Sympathetic and parasympathetic autonomic centres* in the medulla and spinal cord for controlling visceral organs.
- *Pituitary gland* to control water, salt, metabolic and hormonal parameters.

The hypothalamus controls the three control systems:

- *The autonomic nervous system*: helps integrate autonomic responses with activities of other brain areas, with the individual emotional state and with environmental conditions conveyed via the senses; produces sympathetic responses (increase in cardiac activity, peripheral vasoconstriction, vasodilation in the skeletal muscles); integrates autonomic and visceral responses to emotional stress (e.g. fear) in a similar way.
- *Endocrine system and hormones*: cyclicity and pulsatile release pattern of pituitary hormones (see Section 2.5.1).

- *The immune system functions partly*: adrenal steroids and cytokines interact through the hypothalamus. Mild stress stimulates the immune system and severe stress depresses immunity and promotes certain diseases (see Section 3.2). Cytokines from white blood cells may be involved in temperature regulation and sleep changes during illness and infections.

And the hypothalamus controls:

- *Body temperature*: the hypothalamus contains a thermostat that regulates the body temperature at a set point through the integration of diverse sympathetic responses to cold (cutaneous vasoconstriction, secretion of epinephrine and thyroxine, and consequent increase in metabolic rate).
- *Food intake, salt and water balance*: feeding behaviour, energy balance and possibly body weight.
- *Diurnal cycles and sleep*: entrained by light input in the eyes. Pineal gland (release of melatonin) interacts.
- Sex and sexual behaviour.

### The spinal cord

The spinal cord runs through the vertebral column, from the neck to the lower back. It receives sensory messages from all body parts (except the head) and sends motor fibres to voluntary muscles for movements of the limbs, trunk and neck, as well as the involuntary muscles and glands of visceral organs. In the spinal cord, the anterior (ventral) structures carry out motor functions while posterior (dorsal) structures perform sensory functions. The cord's middle region is concerned with association functions, connecting sensory with motor areas and the right half with the left half.

### Synapses and neurotransmitters

Synapses of the CNS are responsible for the integrative functions of the CNS. Billions of neurons (the excitable cells of the nervous tissue), generate and conduct action potentials (pulses). Trillions of synapses and neural circuits make the release of so-called neurotransmitters possible (see Table 2.2 in Bluyssen, 2009a).

A CNS neuron may receive thousands of synapses from other neurons and make hundreds of synapses upon other neurons. They can inhibit (suppress) or excite a neuron, depending on the level that is reached (by spatial or temporal summation of the synaptic interaction). Synapses can be fast responding, which is good for relay but not for integration of responses, and synapses can be slow responding (long time delay between arrival of impulse and response). The latter often involve neuromodulators and neuropeptides and many work in arousal, attention and neural plasticity (learning and memory – see Section 3.7).

### 2.7.2 Mechanisms

The mechanisms of several bodily functions (e.g. respiration, digestion and circulation, regulation of body temperature, diurnal cycles and sleep, perception and reactions) are presented hereafter.

## Respiration

Breathing (*external respiration*) is controlled by two separate neural systems: the voluntary neural control system in the lower brain respiratory centres (pons and the medulla) and the automatic neural control system in the cerebral cortex. The latter is initiated by low $PO_2$, low pH (see Box. 2.3) or high $PCO_2$ in plasma. $PCO_2$ is thus the most important indicator. An increase of $PCO_2$ will increase ventilation (breathing in), which normally only occurs with exercise. While breathing is an automatic and rhythmic activity produced by networks of neurons in the medulla, stretch receptors prevent too much inhalation and chemoreceptors in arteries and in ventral surface of the medulla respond to changes in partial pressure of $O_2$ and $CO_2$, or pH. When arterial $PO_2$ drops very low (e.g. at very high altitudes), ventilation increases. This is caused by $O_2$-sensitive receptors in the arteries (aortic and carotid bodies), which are stimulated by a drop in $PO_2$, and causes them to secrete a neurotransmitter (dopamine). Dopamine stimulates sensory nerves, increasing impulses to send to the respiration centres.

Receptors in the respiratory muscles and in the lung can also affect breathing patterns. The most important function of these receptors (irritant receptors) is to protect the lung against noxious material. When stimulated by rapid lung inflations and by chemicals such as histamine and prostaglandins, they constrict the airways and cause rapid shallow breathing, which inhibits the penetration of pollutants. Stimulation can also cause coughing.

## Digestion

Digestion is controlled by the autonomic nervous system, mainly the motor and sensory fibres of the parasympathetic vagus nerve and via hormonal regulation (see Box 2.19).

Three phases can be distinguished:

- *Cephalic phase*: regulated by higher brain centres (medulla oblongata) before the food is ingested and while it is still in the mouth. Odours and thoughts of food evoke salivary secretion and gastric juice production. The gut lumen is prepared to receive food.
- *Gastric phase*: food enters the stomach. Mechanical stretch receptors sense increase in bulk and chemo-receptors detect presence of peptides. The brain medullary digestive centres and the local neurons are signalled, which causes an increase in stomach secretion and motility.
- *Intestinal phase*: arrival of chyme in small intestines. At first gastric secretion and motility are increased, but as the chyme fills the small intestine inhibitory signals (hormonal) decrease stomach activity.

Many neurotransmitters regulate the secretion of digestive enzymes and electrolytes in the gut. This is also named the *gut–brain connection*. For example histamine is a potent stimulator of hydrochloric acid secretion and acetylcholine elicits salivary, gastric and pancreatic enzymes. The greatest amount of serotonin is believed to be in the gut, not in the brain (Puri and Lynam, 2010).

## Box 2.19 Gastrointestinal hormones

- *Gastrin*, secreted by G-cells in the lateral walls of the stomach glands, when stimulated directly or indirectly by food peptides (via chemo- and stretch-receptor cells which signal g-cells). Gastrin stimulates stomach glands and smooth muscle to enhance gastric secretion (acid) (indirectly via release of histamine that link to histamine $H_2$ receptors, which stimulate secretion) and motility.
- *Secretin*, secreted by musosal walls of small intestine upon presence of acidic chyme:
  - Stimulates release of pancreatic bicarbonate produced by pancreatic duct cells, which neutralizes acid.
  - Inhibits gastric activity of smooth muscle cells and glands, which slows down the delivery of chyme.
- *GIP (glucose dependent insulinotopic peptide)* stimulates release of insulin in response to glucose in the small intestine.
- *Choleocystokinin*, secreted by mucosal walls of small intestine, when chyme arrives. It stimulates:
  - Contraction of gallbladder, which releases stored bile. Bile neutralizes acid and emulsifies fat.
  - Production and release of pancreatic enzymes.
- *Motilin*, also secreted by mucosal walls of intestine, acts on smooth muscles of intestinal walls to enhance intestinal contractions and movements, resulting in stomach emptying.

Source: Kapit *et al.*, 2000.

## Circulation

Regulation of blood pressure is performed by a number of mechanisms classified as short-term, intermediate and long-term regulators. Short-term regulators, such as the *baroreceptors* (arterial presso-receptors), work within seconds. Baroreceptors in the walls of the aortic arch (blood supply to systemic circulation) and the internal carotid arteries (blood supply to brain) are sensitive to stretch (actual and rate of change). When arterial pressure drops, less stretch occurs, causing fewer impulses from the baroreceptors. The medulla cardiovascular centres respond by exciting sympathetic and inhibiting parasympathetic nerves (autonomic control). In this way the arterial pressure doesn't decrease when dilation occurs.

Intermediate regulators start in minutes but it takes hours before they are fully effective. These include:

- Trans-capillary volume shifts (movement of fluid to and from capillary) in response to changes in blood pressure.
- Vascular stress relaxation of venous system (slow expansion or constriction of blood vessels in response to increased or decreased stretch).

- Secretion of renin in response to decreased blood flow to kidney (renin produces angiotensins from plasma proteins that cause intense vasoconstriction).

Long-term regulation is accomplished by the kidney, which regulates body fluid volume (via thirst-intake and excretion). When arterial pressure rises, the kidney excretes more urine (see Section 2.5.2).

## Regulation of body temperature

The thermoregulatory adjustments of the body are controlled by a hypothalamic thermostatic-like centre, which responds to changes in skin and blood temperature (see Figure 2.22). Two parts can be distinguished:

1  The anterior hypothalamus, which takes care of the regulation of the temperature when the body is overheated (i.e. vascular dilatation, sweating).

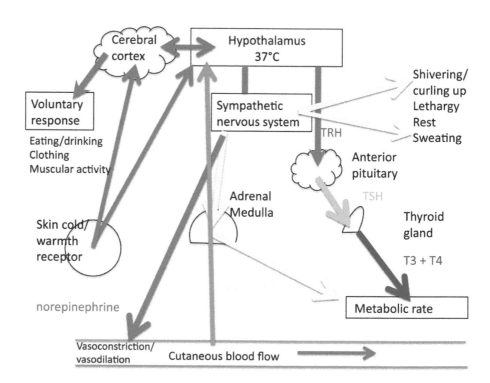

**Figure 2.22** *Regulation of body temperature: a temperature change is detected via a signal to the hypothalamus of skin cold and warmth receptors as well as the temperature of blood. In the case of cold exposure, the hypothalamus initiates responses that promote heat gain (voluntary: eating, extra clothing, muscular activity; sympathetic: increase metabolic rate, shivering) while inhibiting heat loss (voluntary: curling up; sympathetic: vasoconstriction skin, decrease blood flow in skin). And in the case of heat exposure, the hypothalamus initiates responses to increase heat loss (voluntary: wearing of loose clothing, drinking; sympathetic: sweating, lethargy, rest) and to decrease heat production (sympathetic: decrease metabolic rate). Prolonged exposure to a cold climate increases the basal secretion rate of thyroid hormones, which increases heat production*

Source: Bluyssen, based on Kapit *et al.*, 2000, plate 141.

2   The posterior hypothalamus, which defends the body against cold (i.e.
    vascular narrowing, shivering).

To keep the human body in thermal balance, the heat transmission should
be equal to the heat production. This temperature regulation is influenced
by the core temperature, the temperature of the central parts, or vital organs
of the human body (blood temperature), and the skin temperature. The human
being has an average core temperature of 37°C, which can easily increase during
sport or physical labour to 38–9°C. After ending the activity, the temperature
will go back to around 37°C. This core temperature of about 37°C is required
for certain processes in the body (beating of the heart, breathing, functioning
of the intestines). For a person in thermal balance, the skin temperature lies at
around 34°C.

Heat exchange takes place mainly via the largest part of the human body,
the skin. Table 2.6 presents percentages of heat loss for different environmental
temperatures when naked and at rest (Silbernagel and Despopoulos, 2008).
Physical mechanisms of heat exchange include radiation (e.g. from the sun),
convection (e.g. from a heater or an air conditioning system in a room) and
conduction (e.g. direct contact with something warm or cold).

Heat can be lost through the skin in three ways:

*   *Direct exchange of heat* between blood circulating in the skin and the outside
    (*radiation*).
*   *Evaporation* of water from the skin surface.
*   *Conduction and convection* of heat: heat from blood is conducted to the air
    layer on the skin and via convection the heat is transported away from the
    skin. Warm clothes or body hair (when it is cold, hairs stand up) form an
    insulating layer.

For direct exchange of heat, the skin contains special blood capillaries
that do not exchange nutrients with the skin cells, but only exchange heat with
the external environment. Blood flows in these vessels whenever they are open.
When it is cold outside, these vessels close up (cutaneous *vasoconstriction*),
when it is warm they open (cutaneous *vasodilation*). At its peak, the blood
flow in the skin can be as much as 10 per cent of the total cardiac output (Kapit
*et al.*, 2000).

Evaporation of water from the skin can occur by *insensible perspiration* and
by *sweating*. With insensible perspiration (water loss at low temperatures) no

**Table 2.6** *Percentages of heat loss for different environmental temperatures when naked and at rest*

| Environmental temperature [°C] → Ways of heat loss ↓ | 20 | 30 | 36 |
| --- | --- | --- | --- |
| Evaporation [%] | 13 | 27 | 100 |
| Conduction and convection [%] | 26 | 27 | 0 |
| Radiation [%] | 61 | 46 | 0 |

Source: adapted from Silbernagel and Despopoulos, 2008, p. 225.

sweat drops are formed; the water diffuses through the skin cells and pores and evaporates. It accounts for the loss of more than 0.5 l of water per day (Kapit et al., 2000). Sweating on the other hand occurs when the internal body temperature increases to above the internal threshold (which lies at around 37°C).

## Diurnal cycles and sleep

The biological clock can be found in a small brain area, the suprachiasmatic nucleus (SCN), which is part of the hypothalamus. The biological clock has an important role in several activities of the human being, such as the sleep–awake rhythm, heart beat rhythm, functioning of organs and brain. The hormone melatonin, of which the production of melatonin is controlled by the SCN, has a major role in regulating these circadian rhythms of the body (see Section 3.3).

In the brain cortex on-going waves of electrical activity, so-called spontaneous EEG (electro-encephalogram) waves, can be seen, even in absence of all sensory stimuli. This EEG is slow and synchronous (alpha waves) during rest and relaxation, while during alertness and concentration the EEG is fast, desynchronized with a low amplitude (beta waves). During sleep the EEG changes in different stages: four stages of slow wave sleep and REM (rapid eye movement) sleep (Kapit et al., 2000; Michael-Titus et al., 2010). In stage 1 one feels drowsy and the EEG is slow (alpha). From stage 2 to 3 one moves from a light sleep to an intermediate sleep, with a faster EEG and lower amplitudes (theta waves). In stage 4 deep sleep is reached with very large and slow waves (delta waves). The REM sleep then shows a fast EEG (beta waves, like wakefulness) during which the body is limp, the wakening threshold is very high, teeth grind and eyes move rapidly under closed lids. Sleepwalking and bed-wetting occur during stage 4, while dreaming (one of the important ways memory and emotion is integrated (Siegel, 2011)) and increased heart and respiratory activity occurs during REM. Each cycle of sleep stages takes about 1.5 to 2 hours, while the duration of the different stages in each cycle are not constant in time and with age, i.e. duration of REM increases as night progresses; infants spend more sleep in REM than adults; and elderly have less total night sleep, even less REM and no stage 4.

## Perception and reaction

Sensory stimuli excite peripheral sensory receptors and evoke nerve impulses (see Figure 2.23).

Impulses are transmitted via the afferent sensory nerves to the lower centres of the CNS, where they are analysed and integrated. Complex stimuli are transmitted to higher centres before they are integrated. Final motor commands are sent out to *the peripheral effectors* (i.e. skeletal muscles of the somatic nervous system, via lower motor neurons and peripheral motor efferent nerves, which generate bodily movements) or *the autonomic efferent nerve fibre* (via the older structures in the brain (limbic system, hypothalamus and medulla oblongata), which regulate the activity of blood vessels, the heart and digestive and other systems).

Conversely, some hormones are regulated by the immune system, such as the thyroid hormone activity.

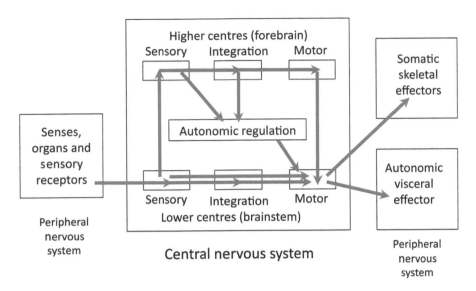

**Figure 2.23**
*Perception and reception pathways: impulses are transmitted via the afferent sensory nerves to the lower centres of the CNS, where they are analysed, integrated and, if required, reflex responses are produced by activation of lower integration and motor systems. For complex stimuli, sensory signals are transmitted, via central pathways to higher sensory structures, where they are analysed, integrated and transmitted to the higher motor centres, via the central motor pathways to the lower motor centres. Final motor commands are sent out to skeletal muscles of the somatic nervous system, or the autonomic visceral effectors*
Source: Bluyssen, based on Kapit *et al.*, 2000, plate 82.

## 2.7.3 Diseases and disorders

Deficiency or surplus of certain biogenic amines can cause disorders. For example a deficiency of norepinephrine can lead to *depression*, while a surplus of dopamine can lead to *addiction* (see Section 3.8.4). Neurotransmission may be compromised by a deficiency of neurotransmitters (unlikely to occur except under extreme food deprivation), a deficiency of co-enzymes and co-factors, a deficiency of PUFAs in the cell membrane in which the neurotransmitter receptors are located, or errors in inactivation of neurotransmitters (Puri and Lynam, 2010).

Blood flow changes in the brain are associated with certain brain diseases. *Schizophrenia*, *depression*, *Alzheimer's disease*, involving reduced cognitive and memory capacities, are associated with reduced blood flow (metabolic activity). *Epilepsy*, involving convulsions, creates excessive electrical activity and increased blood flow and metabolic activity.

*Insomnia* is a disturbance of normal nocturnal sleep patterns that affects daily activities (Michael-Titus *et al.*, 2010). Insomnia can be transient (caused by an acute disturbance or environmental stress; lasts for less than a week; jetlag), short-term (often associated with stressors such as illness or emotional upset; lasts 3–4 weeks) or chronic. Chronic insomnia can be associated with a chronic medical condition or a major psychiatric disease (e.g. depression – see Box 2.20) and lasts longer than 3 weeks.

The inability to form new memories is called anterogade amnesia. The impairment in the ability to retrieve memories from before the brain injury is called retrogade amnesia. Brain damage can be caused by: the effects of long-term alcoholism, severe malnutrition, stroke, head trauma or surgery (Martin *et al.*, 2010).

---

### Box 2.20  WHO on depression

Depression is a common mental disorder, characterized by sadness, loss of interest or pleasure, feelings of guilt or low self-worth, disturbed sleep or appetite, low energy and poor concentration. These problems can become chronic or recurrent and lead to substantial impairments in an individual's ability to take care of his or her everyday responsibilities. At its worst, depression can lead to suicide, a tragic fatality associated with the loss of about 850,000 lives every year.

Depression is the leading cause of disability as measured by YLDs (years lived with disability) and the fourth leading contributor to the global burden of disease (in DALYs = disability adjusted life years) in 2000. By the year 2020, depression is projected to reach second place of the ranking of DALYs calculated for all ages, both sexes. Today, depression is already the second cause of DALYs in the age category 15–44 years for both sexes combined. Depression occurs in persons of all genders, ages and backgrounds.

Depression is common, affecting about 121 million people worldwide. Depression is among the leading causes of disability worldwide. Depression can be reliably diagnosed and treated in primary care. Fewer than 25 per cent of those affected have access to effective treatments.

Source: WHO, 2011g.

# 3

# Stress response mechanisms

*External stressors (mainly physical) cause an imbalance in one of the bodily systems characterized by changes in production of hormones and cytokines and other physiological processes to restore the balance. Stressors causing a response may in the long-term produce illness by weakening the body's ability to defend against external challenges. Thus over or underproduction of certain chemicals by our body, caused (in)directly by over- or under-stimulation by external stressors, may deteriorate or damage our systems. The possible responses by the bodily systems when triggered by physical and/or psychological stressors and the processes involved are presented. How we evaluate and respond to our environment does not only depend on the external stressors involved, but also on several factors such as personal factors, other factors (such as time of day and season), and past events and episodes, which can influence the way external stressors are handled at the moment or over time. Those factors of influence are described and discussed.*

## 3.1 Introduction

Considering that diseases and disorders of the human body could be caused by or related to an 'unhealthy indoor environment', in the *Indoor Environment Handbook* the following division of diseases and disorders was introduced (Bluyssen, 2009a, p. 229):

> Diseases and disorders that are stress induced by external stress factors and which are 'handled' by the cooperation between the nervous system and the endocrine system, but can be influenced by the status of the immune system. For example:
>
> * Direct noticeable comfort-related complaints by the human senses: such as smell, noise, heat, cold and draught.
> * Systemic effects such as tiredness and bad concentration.
> * Psychological effects such as not being in control, depression and anxiety.
>
> Diseases and disorders that are induced by external noxious effects and are 'handled' by the cooperation between the immune system and the endocrine system, but the handling can be influenced by the nervous system. For example:
>
> * Irritation, allergic and hyper reactive effects such as irritation of mucous membranes of skin and respiratory tract, asthma, rashes on skin caused by

allergic reactions to certain pollutants, sun burn, hair loss and damage of eyes caused by too bright light.
- Infectious diseases such as Legionnaires' disease.
- Toxic chronic effects, slowly increasing or appearing, such as cancer.

In each of these categories, a stressor or stress triggers a mechanism or several mechanisms and causes an effect (or multiple effects) immediately (within seconds) or in the medium-term (within minutes to hours) or long-term (days to years). When exposed to this stressor, stress or stress situation continuously (chronically) or repeatedly, this may lead to an imbalance of the bodily systems as described in Section 1.1, or an allostatic state (a chronic imbalance in the regulatory system, reflecting excessive production of some mediators and inadequate production of others (McEwen, 1998)). However, stress can also be experienced as positive stress.

Stress can be defined as '*a physical, chemical or emotional factor that causes bodily or mental tension and may be a factor in disease causation*' (see Box 3.1). Interactions may occur between stressors in complex, real-life exposure as well as between various body responses to exposure(s). Our senses perceive individually, but interpretation occurs together. To truly evaluate the effect of an indoor environmental situation, therefore, all routes of exposure (both physiological and psychological) and all interactions between and in the human systems are) in principle worth considering.

---

## Box 3.1 Definition of stress

1    constraining force or influence: as

    a    a force exerted when one body or body part presses on, pulls on, pushes against, or tends to compress or twist another body or body part, especially: the intensity of this mutual force commonly expressed in kilogram per square centimetre.

    b    the deformation caused to a body by such force.

    c    a physical, chemical or emotional factor that causes bodily or mental tension and may be a factor in disease causation.

    d    a state resulting from a stress; especially: one of bodily or mental tension resulting from factor that tend to alter an existent equilibrium.

    e    strain, pressure.

2    emphasis, weight.

3    archaic: intense effort or exertion.

4    intensity of utterance given to a speech sound, syllable, or word producing relative loudness.

5    a    relative force or prominence of sound in verse.

    b    a syllable having relative force or prominence.

6    accents.

Source: Merriam-Webster Dictionary, 2011.

In the indoor environment physical and psychosocial factors can both lead to *changes in production of hormones and cytokines causing changes in bodily and regulatory processes*. Several stress response mechanisms have been brought forward that might explain why people can get sick or feel unwell in an indoor environment caused by indoor environmental stressors (physical, chemical and psychosocial) or why people feel and perform better in certain situations compared to others.

Mechanisms starting with the endocrine system but also having an effect on the nervous system, the immune system and other bodily systems:

- *The HPA (Hypothalamus-pituitary-adrenal) axis*: in response to various stresses, psychosocial and physical, an increase of secretion of anti-stress hormones can occur (Section 3.2).
- *Disturbance of sleep–awake rhythm*: improper lighting, noise pollution during night, and even thermal discomfort during night can lead to sleep disturbances (Section 3.3).
- *Endocrine disruption*: in the indoor environment, inhalation of air and ingestion of house dust have been considered important pathways for endocrine-disrupting chemicals (EDCs) (Section 3.4).

Mechanisms mainly starting with the immune system but also having an effect on the endocrine system, the nervous system and other bodily systems:

- *Oxidative stress*: can be caused by air pollution (lungs, eyes and brain, all organs), light (skin and eyes) and noise (ears) when there is an excess of free radicals (stolen electrons) over antioxidant defences (Section 3.5).
- *Inflammation*: the body's response to the cause of the infection, an allergic reaction, can be directly caused by air pollution and indirectly by other mechanisms such as oxidative stress (see Section 2.4).
- *Cell death and cell changes*: by physical, chemical or radioactive agents (Section 3.6).

In addition, these stress responses may be altered by genetic factors or early developmental influences or may be affected by life style (McEwen, 2004):

- *Time and events*: (un)conscious learning leading to behavioural conditioning and unexpected physiological effects (Section 3.7).
- *Personal factors*: from genetics to life style (Section 3.8).

## 3.2 Anti-stress responses

*External stressors (physical or psychosocial) causing stress can in the short-term cause the adrenal medulla to produce epinephrine (adrenaline) and prepare the body for action (fight-or-flight) (through the sympathetic nervous system by producing norepinephrine). If the stressor is limited in time and perceived intensity, in due time our parasympathetic nervous system makes certain the balance is restored. With prolonged stress, production of corticotrophin may increase causing an increased production of anti-stress glycocorticoids (e.g. cortisol) via the HPA (hypothalamus–pituitary–adrenal) axis.*

### 3.2.1 Mechanisms

During *normal daily* (minor) *stress*-causing events, the sympathetic nervous system is activated and can deal with the challenges without stress. The HPA axis is not activated. And although you may feel tense, irritable and tired after meeting the challenge, your health isn't disturbed or damaged. Stressors causing stress (i.e. emotions, exercise, starvation) in the short-term make the brain (the hypothalamus) signal (via the sympathetic nervous system) to the adrenal medulla to produce epinephrine (adrenaline) and to prepare the body for action. Preganglionic sympathetic nerves are stimulated and release acetylcholine in sympathetic ganglia, stimulating the postganglic nerves. They innervate visceral organs and skin, releasing norepinephrine (noradrenaline) to their targets.

With *acute or early chronic stress* the sympathetic nervous system is engaged and cortisol levels are elevated, which can lead to restlessness, irritability and exhaustion. Although you can experience sleep disturbance, pain and gastro-intestinal and infectious disorders, your body can recover from this allostatic load if the stress response is efficiently turned off. In the long-term, the brain signals via an increase of corticotropin to the pituitary gland, which signals to the adrenal cortex to produce anti-stress glycocorticoids (e.g. cortisol) (the HPA axis; see Figure 3.1), increasing heart rate, ventilation, myocardial contraction force, arterial vasodilation to working muscles, vasoconstriction to nonworking muscles, and dilating pupils and bronchi. From Figure 3.1 it can be seen that many systems and organs are involved in the response to stress. One can imagine that if one of the activated processes is not shut off after stress is terminated or is deregulated, a problem can occur. Or when compensatory increases by other systems are triggered, for example the secretion of inflammatory cytokines (which are counter-regulated by cortisol) if cortisol secretion does not increase in response to stress (McEwen, 1998).

When stress is long-term and persistent, *chronic or traumatic stress* causes the two stress response systems (HPA axis and sympathetic nervous system) either to work together (at chronically high or chronically low levels), or operate independently. Your stress response can then become out of balance with, for example, low cortisol levels but high activity of the sympathetic nervous system. This imbalance in hormone production, also named an unbalance in the brain–body connection by Hellhammer and Hellhammer (2008), has said to be caused by our allostatic load, or the wear and tear of the systems that results from chronic over or under activity of allostatic systems (McEwen, 1998) (see Section 1.1). The autonomic nervous system, the HPA axis, and the cardio-vascular, metabolic and immune systems protect the body by responding to internal and external stress. Thus, over- or underproduction of certain chemicals by our body, caused (in)directly by external stressors or by lacking the right amount of stimulation by external stressors, deteriorate or damage our systems.

But also an imbalance in the sympathetic and parasympathetic nervous system activity can cause problems. Frequent sympathetic nervous system dominance under repeated acute stress may interfere with growth and repair, especially important for children's development (Clougherty and Kubzansky, 2009). Excessive cortisol may have detrimental and harmful effects on the immune system. On the other hand high short doses may have therapeutic effects against inflammations produced by wounds, allergies or rheumatoid arthritis (joint disease) (Kapit *et al.*, 2000).

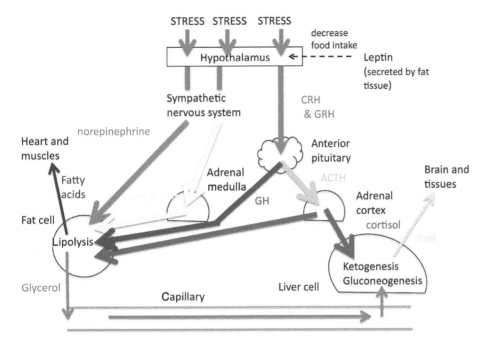

**Figure 3.1** *HPA mechanism and fat-glucose metabolism: hypoglycemia, due to strenuous physical activity or starvation, activates the hypothalamus to release norepinephrine (NE) from sympathetic fibres and epinephrine (E) from the adrenal medulla. NE and E act on fat cells, stimulating hormone sensitive lipase and increasing lipolysis. Prolonged stress stimulates release of growth hormone (GH) and ACTH from the pituitary. ACTH releases cortisol from the adrenal cortex. Cortisol increases blood glucose supply. In the presence of cortisol, GH increases lipolysis in fat cells, mobilizing fatty acids (FA) and glycerol (G). FA are used for fuel by the heart and muscles. In the liver FA are used for fuel for ketogenesis. Cortisol enhances this action. In prolonged starvation, keton bodies are used by the brain and tissues as fuel. In the liver mobilized G is converted into glucose*

Source: Bluyssen, based on Kapit *et al.*, 2000, plates 125–7.

'Stress' is a state of activation of physical and psychological readiness to act in order to help an organism survive external threats (Gee and Payne-Sturges, 2004). Stressors, the stressors that produce stress, can cause illness by weakening the body's ability to defend against external challenges. Two factors are said to determine an individual response to stress: how one appraises the situation, and their general state of physical health. Physical and psychosocial stressors may interact with one another. Some evidence suggests that stress may alter permeability of bodily membranes to chemical exposures, such that stress may alter systemic transport and chemical uptake into organs including the brain (Clougherty and Kubzansky, 2009).

Moreover, recently even a gene has been found that ties stress to obesity and diabetes (Kuperman *et al.*, 2010). Under the influence of stress, this gene starts to produce a hypothalamic protein known as urocortin-3 (Ucn3) in certain brain cells, that plays a role in regulating the body's stress response. Additionally, Mucha *et al.* (2011) identified a gene in the brain that produces the protein lipocalin-2 (Lcn2) in response to stress. This Lcn2 may lead to stress-related psychiatric diseases.

### 3.2.2  Stress hormones and effects

#### Stress hormones

The effects of chronic stress have been related to a chronic imbalance in the hormones released during stress. It has been proven, although simplified because there are other hormones and reactions involved (McClellan and Hamilton, 2010), that *cortisol* plays an important role in this:

- Too much cortisol is harmful: it degrades the immune system (atrophy of lymphatic nodes and reduction on white blood cells) and it causes hypertension and vascular disorders.
- High cortisol levels and the sympathetic nervous system in overdrive contribute to changes in carbohydrate and fat metabolism (*obesity*) and can lead to anxiety, depression and heart disease.
- Low cortisol production (leads to overactive immune actions, e.g. the secretion of inflammatory cytokines, which are counter-regulated by cortisol) and an overactive sympathetic system, which can lead to fatigue, allergies, asthma and increased weight.

Cortisol has significant influence on virtually all body functions, including brain activity, metabolism and immune function. Cortisol affects fat metabolism (obesity) via its direct influence on gluconeogenesis (the breakdown of protein and fat to provide metabolites that can be converted to glucose in the liver) as well as the activity of the immune system. Cortisol counteracts insulin, which is secreted after a carbohydrate meal. Insulin is the main hormone promoting fat formation (lipogenesis) in adipose tissue and liver (see Section 2.5.2). Cortisol is able to counteract certain components of the inflammatory response to tissue injury (i.e. by inhibiting the production or action of some of the chemical mediators of inflammation such as histamine, prostaglandins and leukotines). And, although cortisol is required for normal B-lymphocyte function, higher levels of cortisol suppress the immune response by decreasing both the number and the effectiveness of T- and B-lymphocytes (Hinson, Raven and Chew, 2010). Too much or too little cortisol can affect thyroid function: elevated levels suppress the release of TSH, which is important for the production of T4 and T3 (see Section 2.5.2) (Nodder, 2010).

But there are also other hormones involved in the response to stress (McClellan and Hamilton, 2010). Next to norepinephrine and epinephrine, dopamine and endorphin are secreted by the adrenal medulla. But also neurotransmitters such as serotonin and oxytocin, a brain hormone (neuropeptide), play a role (see Box 3.2).

It has also been proposed that different emotions can cause different neuroendocrine patterns (Henry, 1986):

- *Anger* in fight effort persistence with an increase in blood pressure, pulse and norepinephrine, and a slight increase in epinephrine.
- *Fear* in flight effort with a slight increase in blood pressure, pulse, norepinephrine and cortisol, and an increase in epinephrine.
- *Depression* in loss of control, subordination with an increase in ACTH, cortisol and endorphins, and a slight decrease in pulse.

---

### Box 3.2 Dopamine, endorphin, serotonin and oxytocin

*Dopamine*: 'The learning neurotransmitter' of the brain is involved in many complex pathways, including reward, long-term memory, motivation, drug disorders, sleep and motor systems. Norepinephrine and dopamine are converted from amino acid tyrosine ($\Rightarrow$ dopa $\Rightarrow$ dopamine $\Rightarrow$ norepinephrine).

*Endorphin*: A naturally occurring opiate-like neurotransmitter in the brain, decreases pain and enhances feelings of wellbeing.

*Serotonin*: Calms the stress response in the brain and influences appetite, mood, anger, fear and aggression. Serotonin appears to suppress the transmission of anxiety signals to the prefrontal cortex (Sterling, 2004). Serotonin is derived from tryptophan. Norepinephrine and serotonin are important in regulating moods, pleasure and brain excitability.

*Oxytocin*: Capable of buffering the fight-or-flight response, and is part of an affiliate system that signals the need for social connection ('tend and befriend'). Elevated levels of oxytocin may be a biological marker indicating inadequate levels of positive social affiliations (Taylor, S.E., 2006).

---

### Stress types

Hellhammer and Hellhammer (2010) developed a tool with 20 neuro-patterns that characterize a patient's stress response, or diagnose where the body's balance gets disrupted. Three principles underlie these neuropatterns:

- *Ergotropy*: a state of arousal in which the brain and sympathetic nervous system are activated that allows you to adapt to physical or mental stressors.
- *Trophotropy*: the opposite of ergotropy, in which the parasympathetic functions predominate. Sleep, relaxation, regeneration and recovery characterize this state.
- *Glandotropy*: the response of the HPA-axis to stress. This refers to the over- or under activity of cortisol.

McClellan and Hamilton (2010) indentified four stress types (mainly for women) by looking at how the HPA-axis and autonomic nervous system (your sympathetic (fight-or-flight) and parasympathetic systems (rest-and-restore)) interact under chronic stress (see Figure 3.2 and Table 3.1).

### 3.2.3 Psychosocial stressors

Stressors that can cause the HPA-axis to be activated comprise psychosocial stressors, but also several physical stressors that cause discomfort or annoyance. There is a growing interest in separating the health effects of social and physical environmental exposures and in exploring potential synergies among them (Clougherty and Kubzansky, 2009). It has been hypothesized that psychosocial stressors may make individuals more vulnerable or susceptible to physical stressors

(Gee and Payne-Sturges, 2004; Morello-Frosch and Shenassa, 2006). Stressors may directly lead to health disparities and stressors may amplify the effects of toxicants. Evidence suggests that stress may influence the internal dose of a given toxicant. It has also been suggested that one's level of experienced psychosocial stress is embodied in the cumulative amount of undesirable change or readjustment brought about by events occurring in one's life (see Section 3.7). A large number of studies have reported the significance of 'life stress' (as measured by life events via a checklist, see Section 6.5.4) (Cohen *et al.*, 1995). Lack of social recognition, high levels of job stress and high levels of perceived stress are associated with high levels of morning cortisol (Steptoe *et al.*, 2000; Wüst *et al.*, 2000).

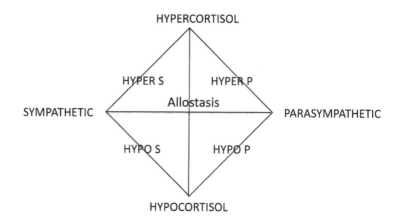

**Figure 3.2** *Stress types (for women mainly). Hyper and hypo refer to the HPA axis, P present parasympathetic and S sympathetic*
Source: Adapted from McClellan and Hamilton, 2010, p. 67.

**Table 3.1** *Description and symptoms of the four stress types*

| | |
|---|---|
| *HyperS*: High cortisol production and an overactive sympathetic nervous system, which makes one feel anxious and agitated. Cortisol stimulates wakefulness, decreases REM sleep and increases time spent awake during night. | *HyperP*: High cortisol and a charged parasympathetic nervous system (can be depleted from its major neurotransmitter norepinephrine). |
| *Symptoms*: i.e. headaches, anxiety, fear, irritability, restlessness, increased sweating, nervous stomach, trouble sleeping, lean but tend to gain weight in abdomen, high blood pressure, depression and heart disease. | *Symptoms*: i.e. extreme exhaustion, irritability, overemotional need for sleep, exercise intolerance, ulcers, lack of motivation and burnout. |
| *HypoS*: Low cortisol production (leads to overactive immune reactions) and an overactive sympathetic nervous system. | *HypoP*: Low cortisol levels and dominant parasympathetic nervous systems. One is withdrawn. |
| *Symptoms*: i.e. increased weight (hips and thighs), increase susceptibility to inflammation, fatigue/exhaustion, inactivity/lack of motivation, increased sleep, gastrointestinal complaints, allergies and asthma. | *Symptoms*: i.e. boredom, fatigue/exhaustion, decreased sweating, slow heart rate, asthma and inflammatory disorders. |

Source: McClellan and Hamilton, 2010.

*Psychosocial stressors* can be categorized into discrete stressful events (or stressors) and chronic stressors.

## Discrete stressful events

Discrete stressful events are stressors with an easily specifiable beginning and ending, comprising major and minor life events. *Major life events* can be a positive event, such as a marriage or a birth of a child, or negative, i.e. job loss or death of family member (Cohen *et al.*, 1995). At work sexual harassment early in the career can have long-term effects on depressive symptoms in adulthood (Houle *et al.*, 2011). *Minor daily events* that can cause psychosocial stress are an oral speech, catching the bus on time, and such like (Pieper, 2008). Previous exposures and events, such as smoking history and episodes of depression and anxiety, can also be important stressors (see Section 3.7).

## Chronic stressors

Chronic stressors can be characterized by their continuous presence or high frequency, for example, job strain, marital stress, perceived racism or having no privacy (deFur *et al.*, 2007; Pieper, 2008). Psychosocial chronic stressors comprise stressors in the social environment (community) and individual stressors (Gee and Payne-Sturges, 2004).

*Community stressors*, which are often considered to be factors of influence on the handling of stress (see Section 3.8), can be divided into stressors that occur (Gee and Payne-Sturges, 2004; deFur *et al.*, 2007):

- *Locally*: such as crowding (neighbourhood), social disorganization, neighbourhood quality, safety (crime and violence) and poverty.
- *Nationally*: such as fear and economic deprivation, social support, racial segregation and income inequality.
- *Worldwide*: such as fear for climate- and energy-related problems, fear for epidemic diseases and economic depression.

*Individual stressors* can be divided into stressors (Gee and Payne-Sturges, 2004; deFur *et al.*, 2007):

- *At work*: such as working hours, high demands and low control, flexible versus same working place and working pattern.
- *At home*: such as financial stress, access to health care, family turmoil, marital problems, cold-harsh parenting and separation from family.
- *During commuting conditions*: such as travel time, queuing and transportation means.

Long working hours have been found to be associated with cardiovascular and immunologic reactions, reduced sleep duration, unhealthy life style and adverse health outcomes, such as cardiovascular disease, diabetes, subjective health complaints, fatigue and depression (Virtanen *et al.*, 2009). Long working hours may have a negative effect on cognitive performance in middle age. Runeson *et al.* (2006) found psychosocial work environment (high demands in combination with low control) to be associated with SBS (Sick Building Syndrome)-related symptoms. (While low control means not having enough influence over the way the work should be performed, on a day-to-day basis, low support means not having sufficient support from co-workers and superiors.) An unbalanced psychosocial work situation increases the risk of cardiovascular disease and higher reports of psychosomatic disorders, particularly a combination of high demands, lack of control and lack of social support from superiors or colleagues (Wahlstedt and Edling, 1994).

In the indoor environment controllability (control of thermal, visual, acoustical and indoor air environment (ventilation)) seems to show a relation with satisfaction of the indoor environment as well as the self-reported building-related symptoms: the more control, the more satisfied and the lesser symptoms (Hedge *et al.*, 1992; Clements-Croome, 2002). The degree of control over the indoor environment, as perceived by the occupants of a building, seem more important for the prevalence of adverse symptoms and building-related symptoms than the ventilation mode per se (Toftum, 2010) (see also Section 9.4.2).

The *need to belong*, the desire for interpersonal attachment or *social engagement*, is a fundamental human motivation. Several studies have shown that social engagement can have an impact on subjective and objective health and mortality (Baumeister and Leary, 1995; Bennet, 2005). Social engagement was found to be an important predictor of health status. Deficits in belongingness at work, at home or anywhere else (such as perceived threats to social bonds or frustration when desperate attempts to establish or maintain relationships with other people fail) apparently lead to a variety of ill effects. Both psychological and physical health problems are more common among people who lack social attachments (Baumeister and Leary, 1995).

### 3.2.4 Physical stressors

Physical stressors that can cause annoyance or discomfort comprise of environmental stimuli that can be perceived by one of our sensory receptors, such as a sound (noise), chemical compounds (air quality), light (visual discomfort) and heat, cold or draught perception (thermal discomfort). In the nineteenth century, a distinction was made between the 'higher' senses, vision and audition (the senses of the intellect) and the 'lower' senses, touch, taste and smell (the senses of the body) (Köster, 2002). For noise exposure it has been well established that noise can activate the HPA axis in the same way as psychological stress does (Maschke *et al.*, 2000; Babisch, 2002): via the emotional and cognitive perception of the sound via the cortical and sub-cortical structures including the limbic region (Spreng, 2000). Bad odours, indoor environmental conditions that cause visual discomfort (i.e. eyestrain) or thermal discomfort can also cause annoyance. Although not proven in studies (that the author knows of), it seems likely that also those types of annoyance can cause activation of the HPA axis.

### Noise

Traffic noise exposure has been associated with changes in stress hormone levels and with cardiovascular changes (e.g. Babisch *et al.*, 2001; Babisch, 2006), and has been related to the parasympathetic and sympathetic balance (Graham *et al.*, 2009). Babisch proposed a reaction scheme in which he presents the cause–effect chain of exposure to noise. An adapted version of this schema is presented in Figure 3.3. From this scheme can be seen that noise causes direct and indirect stress reactions (HPA axis). Annoyance is an important aspect in this mechanism (see Box 3.3). Thus, noise effects do not only occur at high sound levels, but also at relatively low environmental sound levels, when certain activities such as concentration, relaxation or sleep are disturbed. In office buildings major indoor sources of noise are HVAC (heating, ventilating and air conditioning) systems and people (colleagues). Control over noise in an office environment has been significantly (negatively) related with discomfort (Bluyssen *et al.*, 2011a). In homes, major sources are noise from neighbours and noise from outside such as noise from wind turbines and traffic. Noise sensitivity seems to play an important role on how these types of noise are perceived (Miedema and Vos, 2003).

Noise exposure during the night also has been related to the parasympathetic and sympathetic balance (Graham *et al.*, 2009). During sleep overall parasympathetic nervous system activity is increased compared to PNS activity during wakefulness, whereas sympathetic nervous system activity is decreased during sleep depending on the sleep stage (see Section 2.7.2). Thus, parasympathetic nervous system activity is mostly influenced by the circadian system, whereas sympathetic nervous system activity is mostly influenced by the sleep system (Burgess *et al.*, 1997). Raised cardiac sympathetic tone leads to increased myocardial contractility and heart rate as well as increased blood pressure and vascular resistance (see Section 2.3.2). Increased cardiac parasympathetic tone leads to decreased heart rate and consequently a reduction in blood pressure. Protection of the cardiovascular system seems to be one of the functions of sleep (Carter *et al.*, 2002). Therefore, when sympathetic activity is increased during sleep, the protective function of sleep can be disturbed. Alterations in autonomic nervous system activity during sleep could affect several aspects of sleep (Muzet, 2007).

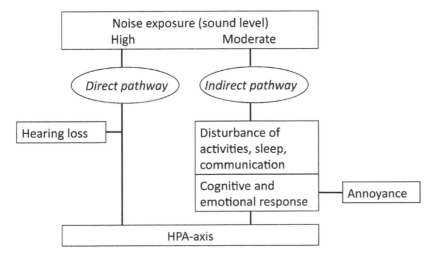

**Figure 3.3** *Noise effects reaction schema*
Source: adapted from Babisch, 2002.

---

## Box 3.3 Annoyance, discomfort and intensity

The different sensory modalities (light, noise, odour, heat/cold) can be measured by using the attributes (Bluyssen, 2009a):

- *Detection*: for example the limit value for absolute detection.
- *Intensity*: for example odour intensity, sensory irritation intensity, light intensity, noise loudness level.
- *Quality*: a value judgement such as annoyance, discomfort or acceptability.

According to the Merriam-Webster – Learner's Dictionary (www.learnersdiction ary.com):

- *Annoyance* is defined as:
    - Slight anger; the feeling of being annoyed.
    - Something that causes feelings of slight anger or irritation; a source of annoyance.

- *Discomfort* is defined as:
    - An uncomfortable or painful feeling in the body.
    - A feeling of being somewhat worried, unhappy, etc.

- *Intensity* is defined as:
    - The quality or state of being intense; extreme strength or force.
    - The degree or amount of strength or force that something has.

According to Stevens' power law the sensation of a sensory stimulus intensity or perceived intensity increases as a power function of the intensity of the stimulus (Stevens, 1957):

Perceived intensity = $b$ Stimulus intensity$^a$       [3.1]

Where: $a,b$ = constants (can vary per modality, per person and stimulus used).

---

Night-time traffic noise has been found to affect the cardiac (para)sympathetic tone during sleep without awaking them (Hofman *et al.*, 1995; Carter *et al.*, 2002). In a field study performed by Graham *et al.* (2009) in The Netherlands, a potential influence of road and rail traffic noise on parasympathetic tone during sleep, specifically during the second half of the sleeping period, was found. No evidence of an effect of traffic noise on cardiac sympathetic tone was found.

## Odour annoyance

Indoor air pollutants can be odorous and lead to annoyance or pleasure. For odour to be perceived, a molecule is in general volatilized from its source, inhaled into the nasal cavity and dissolved in the protective mucous layer (epithelium).

The nose comprises two nostrils with smelling organs, the sense of smell (for the odorous aspect) and the common chemical sense (for the irritant aspect), from which via the olfactory nerve and the trigeminal nerve impulses are sent to the olfactory lobe of the brain (see also Box 3.4). The mechanism is described in Section 2.4.2 of Bluyssen (2009a). The brain interprets the incoming signals by associating them with a previous olfactory experience. This is how the nose distinguishes between perceived air qualities (Geldard, 1972). The biological principles for receptor activation of odour and irritation are fairly well understood, while for the information processes at higher centres of the brain this is less clear (Berglund *et al.*, 1999b). How these basic processes relate to the more complex psychological responses to odorant/irritant stimulation, such as perceived air quality, annoyance and symptom reporting, is uncertain.

Olfaction was the first sense to evolve. It was the sense that provided information for communication for all of the most critical aspects of behaviour for surviving and thriving (e.g. finding food, knowing one's children, sniffing sources of danger) (Herz, 2002). The olfactory signal terminates in or near the amygdala and therefore odours are strongly linked to memories and can evoke emotions (see Box 3.5 and Figure 3.4).

The orbitofrontal cortex and amygdala have been shown to play major roles in stimulus-reinforced associative learning and olfactory processing has been shown to activate orbitofrontal cortex and to some extent the amygdala (Herz, 2002) (for description of the brain, see Section 2.7.1). The amygdala participates in the hedonic or emotional processing of olfactory stimuli, in particular the hedonic valence of the odorant influences amygdala activity, suggesting greater amygdala involvement in negative than positive emotions (Zald, 1997).

Exposure to odours can alter moods, change attitudes, influence perceptions of health, affect task performance and evoke memories. In general, pleasant

---

### Box 3.4 The sense of smell is . . .

- *A nominal sense*: detection, recognition and quality discrimination is much more important than verbal identification or intensity gradient. The sense of smell provides simple 'nominal' data about the presence of qualitatively different odours in our surroundings.
- *A 'near' sense*: people differ much more in the way they perceive odours than in the way they perceive visual objects. These large interpersonal differences in olfaction (e.g. threshold of sensitivities can vary by a factor of 1,000) have the origin not only in difference in interpretation, as in vision, but also in the sensory basis of the perception.
- *A 'hidden' sense*: smells can influence our moods, the time we spend in various locations, our perception of other people and performance of vigilance and even mathematical tasks (even when the subjects are unaware of the presence of the odour).
- *An associative and emotional sense*: there are no inborn preferences in olfaction. Emotions evoked by a given odour can be very different from one individual to another, depending on the different emotional encounters with that odour in each person's personal history.
- *Special with respect to memory*: olfactory memory is episodic and semantic in nature. We can remember where and when we encountered an odour before, but in most cases we cannot call up the name of the odour. Olfactory memory is in general better in women than in men.

Source: Köster, 2002.

## Box 3.5  Affective states, mood or emotions

Two prominent traditions in emotion research can be distinguished (Egloff *et al.*, 2003). In the first emotions are treated in terms of specific and distinct emotions: discrete emotions each reflect a specific set of eliciting stimuli and trigger a characteristic range of adaptive behaviour. An example are the 12 emotions (9 negative: fear, anger, disgust, contempt, inward hostility, sadness, guilt, shame, shyness; 2 positive: interest, enjoyment) in the DES (differential emotions scale) of Izard *et al.*, 1974.

The second point of view emphasizes negative affect (e.g. anxiety, depression) as a cumulative indicator of mental distress and psychological stress. The tripartite model of Clark and Watson (1991) suggests that anxiety and depression share a nonspecific component of generalized distress: called 'negative affect'. The affective experience is characterized as an ordering of affective states on the circumference of a cycle. Examples are the circumplex model of affect (Russel, 1980) and the two-dimensional structure of affect (Watson and Tellegen, 1985) (see Figure 3.4).

Affect is defined as an evaluation/appraisal of an object, person or event as good or bad, favourable or unfavourable, desirable or undesirable (Von Kempski, 2003). The construct mood (affective state) can be represented by two orthogonal factors – positive affect (PA) and negative affect (NA). The NA factor subsumes a broad range of aversive mood states including anger, disgust, scorn, guilt, fearfulness and depression. PA (positive affect) reflects one's level of energy, excitement and enthusiasm (Watson and Pennebaker, 1989). Later, positive and negative affect were respectively transferred to positive activation (PA) and negative activation (NA) in the PANAS (Positive and Negative Activation Scale (Watson *et al.*, 1999). Egloff *et al.*, 2003 demonstrated that the PANAS comprises three more specific affects: joy, interest and activation. Both mood factors (PA and NA) can be measured either as a state (i.e. transient fluctuations in mood) or as a trait (i.e. stable individual differences in general affective level). Persons with high negative affectivity experience discomfort under all circumstances, even in the absence of a clear agent that may cause discomfort (Watson and Pennebaker, 1989).

smelling odours induce positive moods, attitudes and behavioural changes, and unpleasant odours induce negative moods, attitudes and behaviours (Herz, 2002). Even when olfactory stimuli can no longer be consciously perceived (adaptation: loss of sensitivity as a result of prolonged stimulation) or are no longer attended to (habituation: reduction of attention and responsiveness to monotonous stimuli), they continue to exert influences on behaviour and mood (Köster, 2002).

### Visual discomfort

Lighting conditions that cause visual discomfort are likely to lead to eyestrain (Boyce, 2003). Symptoms of eyestrain are irritation of the eyes, evident as inflammation of the eyes and lids; breakdown of vision, evident as blurring or double vision; and referred effects, usually in the form of headaches, indigestion, giddiness and so on. Lighting conditions, which have been shown to produce

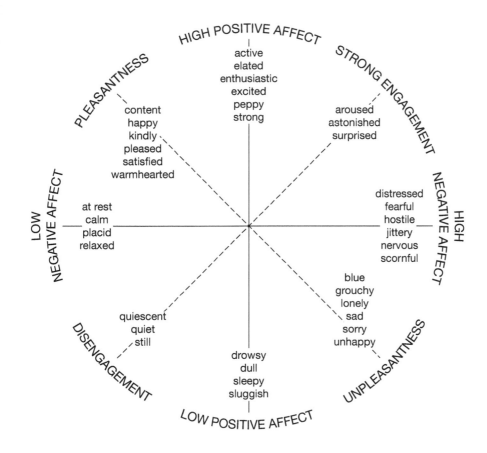

**Figure 3.4** *The two-dimensional structure of affect*
Source: Watson and Tellegen, 1985.

visual discomfort are inadequate illuminance for the task, excessive luminance ratios between different elements of a task, and lamp flicker, even when it is not visible.

The fluctuations in light output from a light source, specifically low frequencies (15 Hz to 50 Hz) seem to cause more headaches among office workers who get a headache now and then (Boyce, 2003). Reflection, blinding, too little light, bad colouring of the light can lead to tiredness of the eye or adaptation problems, decreased alertness and concentration problems. And neurophysiological responses suggest mechanisms for the effect of flicker (100 Hz modulation) on performance (Winterbottom and Wilkins, 2009).

Glare occurs when a part of what can be seen is much brighter than the rest. Glare can be of two types (Winterbottom and Wilkins, 2009):

- *Disability glare*: a decrease in visual performance resulting from a decrease in contrast due to light scattered within the eye.
- *Discomfort glare*: resulting in eyestrain and headaches.

Excessive lighting or excessive illumination may also result in increased discomfort. There is some evidence for increased discomfort at illuminance above 1,000 lux (Rea, 1982).

Healthy and comfortable lighting, however, depends on the eye task, the time of day, the weather and individual needs. The effect of lighting on performance and wellbeing depends on illuminance, lighting period, timing and spectral distribution. Additionally, psychological effects of light (colour, illuminance) can be different for different people. Because change in colour and illuminance is a quite normal process (natural light is changing all the time), it could well be that dynamic lighting has a positive influence on the wellbeing of a person (Boyce, 2003). Different colours can directly affect an individual's impression of (Mahnke and Mahnke, 1996):

- *Temperature*: light blue cooler; red/orange warmer.
- *Sounds*: shrill, high pitched may be offset by olive green.
- *Size of objects*: dark colours heavier; less saturated colours less dense.
- *Size of spaces*: light or pale colours recede and so increase perception of size; dark or saturated hues protrude and decrease apparent size.

When selecting colours, the nature of the task is relevant.

### Thermal discomfort

Thermal comfort is strongly related to the regulation of body temperature (described in Section 2.7.2). Thermal stress or thermal discomfort occurs when one is not able to regulate its thermal balance or when one believes or perceives it isn't possible. The psychological effect of expectations seems an important part of thermal comfort (Auliciems, 1981; Humphreys and Nicol, 1998). Thermal discomfort affects our state of wellbeing and shows a strong relation with productivity of office workers (Clements-Croome, 2002) (see also Section 8.3.2). But thermal comfort is more than a physiological experience (de Dear, 2004). The perceived individual level of control on our thermal comfort (Wilson and Hedge, 1987; Bluyssen *et al.*, 2011a) (see also Section 3.8.4), but also the unconsciously experienced thermal boredom (McIntyre, 1980), adaptation and motivation (Lan *et al.*, 2011), have been pointed out as important components. Additionally, the HRV (heart rate variation) index (see Section 6.3.1) has been shown to be closely related to thermal comfort sensations (Yao *et al.*, 2009).

Moreover, studies recently have indicated that thermal neutral conditions do not have to be necessarily healthy. A causal relation may exist between the time spent indoors in the thermal-neutral zone and an increased adiposity (Marken Lichtenbelt *et al.*, 2009; Johnson *et al.*, 2011) (see Box 3.6). Mild temperature challenges could thus perhaps increase the energy spend and subsequently reduce obesity.

## 3.3 Disturbance of sleep–awake rhythm

*Improper lighting, noise pollution during night and even thermal discomfort during night can lead to sleep disturbances. Under the influence of light, the hypothalamus signals to the pineal body to produce melatonin via the RHP (Retino-hypothalamic-pineal) axis, a hormone that makes us want to sleep. If exposed to light during*

---

**Box 3.6  White and brown adipose tissue (fat)**

Studies in animals have indicated that brown adipose tissue is important in the regulation of body weight (Johnson *et al.*, 2011). The primary function of brown fat is to generate body heat. The other fat tissue is white fat, which is used as a store of energy (see Section 2.2.2). Studies of the role of brown adipose tissue in human thermogenesis suggest that increased time spent in conditions of thermal comfort can lead to a loss of brown adipose tissue and reduced thermogenic capacity. Marken Lichtenbelt *et al.* (2009) studied the brown-adipose tissue activity in 24 healthy men – 10 of whom were lean (BMI < 25) and 14 of whom were overweight or obese (BMI ≥ 25), under thermal neutral conditions (22°C) and during mild cold exposure (16°C). Brown adipose tissue activity was observed in 23 of the 24 subjects during cold exposure, but not under thermo-neutral conditions. This activity was significantly lower in the overweight or obese subjects.

---

*the night, the production of the antioxidant melatonin is immediately stopped, alertness and core body temperature is increased and sleep is distorted.*

## 3.3.1 Mechanism

A circadian rhythm (day–night rhythm) is an endogenously (non-reliant on environmental cues) driven cycle in biochemical, physiological or behavioural processes. Examples of circadian rhythms include oscillations in core body temperature, hormone secretion, sleep and alertness (Rea *et al.*, 2008). There is a decrease in the hypothalamic set temperature during sleep, as well as a slowing down of vegetative functions such as respiration and circulation. Even gastric emptying is controlled by diurnal rhythms, with faster emptying occurring in the morning than in the afternoon (Trueman and Bold, 2010). Circadian oscillations also exist at a cellular level, including cell mitosis and DNA damage response.

These oscillations are a result of a small group of clock genes inside the cell nuclei creating interlocked transcriptional and post-translational feedback loops. The timing of these clock genes is generally orchestrated by a master biological clock located in the supra-chiasmatic nuclei (SCN) of the hypothalamus of the brain (Rea *et al.*, 2008). The SCN sends neural signals to many parts of the human system to coordinate physiological and behavioural activities.

Undisturbed sleep is important for human functioning. Insufficient sleep is associated with feelings of sleepiness and fatigue during the daytime, with decreased cognitive performance and with changes in physiological parameters, such as metabolic and endocrine function, altered cardiac autonomic nervous system activity and reduced acute immune system responses during stress (Kluizenaar *et al.*, 2009). Irwin *et al.* (1999) showed that sleep onset is associated with changes in levels of circulating catecholamines (norepinephrine and epinephrine). Wright *et al.* (2007) monitored sleep (including time awake after sleep onset (WASO)), the night before a laboratory stressor was introduced. Their findings raised the possibility that inadequate lymphocyte mobilization and poor recovery from acute stress may be one pathway by which sleep could impact health. Sleep and immune function seem to be linked (Bryan *et al.*, 2004): infections cause increased sleep (infections elicit immune responses (cytokines) that alter the expression of endogenous immunomodulatory substances, which in turn affect sleep), and sleep duration seems to be related to mortality. Partial sleep loss is linked to obesity and diabetes and it disrupts key components of appetite regulation and changes your carbohydrate metabolism. Leptin, an appetite suppressant, is produced during sleep. With less leptin produced, one eats more. Melatonin, produced at night (darkness), is also related to weight regulation and energy balance. The prolonged production of melatonin during winter months is associated with weight loss (McClellan and Hamilton, 2010).

### 3.3.2 Improper lighting

#### Mechanism

The anatomical linkage between the eye and the pineal gland is called the retino-hypothalamic-pineal (RHP) axis. The pineal gland synthesizes and secretes the hormone melatonin during the dark phase of the 24-hour light–dark cycle. Melatonin carries a message of time as determined by SCN, and is transported in the bloodstream. The period of oscillation of the RHP-axis is however longer than 24 hours. The presence of time cues, particularly a light–dark cycle, is necessary to prevent the circadian system from free running (Boyce, 2003). Because the cycle length, or period, of the circadian rhythm is near, but in most organisms, not exactly 24 hours, they must be synchronized or entrained to the 24-hour day on a regular basis. In most organisms, this process of entrainment occurs through regular exposure to light and darkness (Duffy and Czeisler, 2009). In the absence of light, and other cues, the internal oscillator located in the SCN continues to operate but with a period longer than 24 hours. External stimuli are necessary to entrain the internal oscillator to a 24-hour period and to adjust for the seasons. The light–dark cycle is one of the most potent of these external stimuli used for entrainment. The light–dark cycle also signals the passing of the seasons. Consistent electrical light in the evening after the sun has set, removes seasonal variation (Boyce, 2003).

Thus, the role of circadian systems is to establish an internal replication of external night and day. This internal representation is not just a passive response to external conditions but rather is predictive of external conditions to come. To maintain synchrony with the external world, the light–dark pattern incident on the retina resets the timing of the SCN. If the period of light–dark pattern is too long or too short, or if the light and dark exposures become aperiodic, the master clock can lose control of the timing of peripheral circadian clocks (Rea *et al.*, 2008). Currently, little is known about individual differences in circadian sensitivity to light, or about how polymorphisms in so-called 'clock genes' (or other genes) affect sensitivity to light (Duffy and Czeisler, 2009). Bright light given early in the night tends to delay the circadian cycle, but bright light given late in the night tends to advance the phase of the circadian cycle. But how much light is necessary to influence the circadian system is not clear (Boyce, 2003).

Exposure to light seems to be the principal external stimulus to the human circadian system. Some studies have shown that social cues, night-time activity and fitness training may also have an impact (Boyce, 2003). Repeated observations of blind people show that the timing of their circadian rhythms slowly drifts later week by week, indicating that they are not entrained. However, there are blind people with no conscious light perception that do retain circadian photoreception, and thus are able to entrain using light cues from the environment (Duffy and Wright, 2005). Based mainly on EEG (electro-encephalogram) and ECG (electrocardiogram) data, it was shown that light of sufficient irradiance increases alertness at night. Present results indicate that this effect is not mediated only by the circadian system, implicating other mechanisms through which light can also increase night-time alertness (Figuerio et al., 2009).

Exposure to light is essential for the visual system to operate and desirable for entraining the circadian system. Four kinds of photoreceptors, three cone types and one rod type, have been identified for the visual system (Boyce, 2003). Additionally, another type of light-sensitive cell has been discovered (Brainard et al., 2001) that serves as a photoreceptor for circadian and other non-image-forming responses. These cells transmit the biological stimulating part of light to the brain and are not used for the visual function. These specialized retinal ganglion cells are distributed throughout the retina, project to the SCN, are photosensitive, and contain melanopsin as their photo-pigment. While the photosensitive retinal ganglion cells can mediate circadian responses to light, there is also evidence that rod and cone photoreceptors can play a role in circadian responses to light. The relative contribution of different photoreceptors to circadian receptors is not well understood. It is likely that the intensity, spectral distribution and temporal pattern of light can all affect the relative contribution of different photoreceptors to circadian responses (Duffy and Czeisler, 2009). Nevertheless, it seems that the spectral sensitivity is highest at about 464 nm (Brainard et al., 2001), and therefore light at 464 nm can be used to stimulate our biological rhythm (clock) at a much lower illumination than other colours (Cajochen et al., 2000).

Under the influence of the transmitted light (specifically the blue-green part) the hypothalamus signals to the pineal body to produce melatonin, a hormone that makes us want to sleep (Figure 3.5). If exposed to light during the night, the production of melatonin is immediately stopped (Brainard et al., 2001; Boyce, 2003).

## Melatonin

Melatonin, synthesized from tryptophan (to 5-hydroxytrytophan to serotonin to N-acetylserotonin to melatonin), is produced at night by the pineal gland. Light falling on the retina, in particular short waves, suppresses the production of melatonin. The maximum concentration of melatonin produced during the night (in darkness) can vary between individuals (Waldhauser and Dietzel, 1985).

Melatonin is also an antioxidant, like vitamin C, that stimulates the recovery of cell functions and of general resistance (Hinson, Raven and Chew, 2010). Melatonin reduces oxidative stress (see Section 3.5) by several means. It is an effective hunter of both the highly toxic hydroxyl radical, produced by the three-electron reduction of oxygen, and the peroxyl radical, which is generated during the oxidation of unsaturated lipids and which is sufficiently toxic to propagate lipid peroxidation. Additionally, melatonin may stimulate some important antioxidative enzymes, i.e. superoxide dismutase, glutathione peroxidase and glutathione reductase (Reiter et al., 1997).

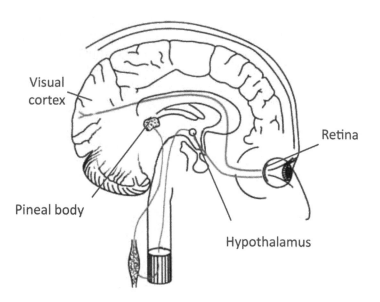

Visual cortex

Retina

Pineal body

Hypothalamus

**Figure 3.5** *Pathway of light signal to pineal body: the signal from the retina as a result of light is relayed through the hypothalamus down the spinal column, returning via the superior cervical ganglion to supply the pineal gland*
Source: Bluyssen; based on Hinson, Raven and Chew, 2010.

## Effects

The immediate effect of light exposure at night is the suppression of melatonin synthesis and the consequent increase in alertness (Cajochen *et al.*, 2000) and core body temperature. Alertness is a construct associated with high levels of environmental awareness, and is associated with self-reported high levels of wakefulness and low levels of fatigue, short response times, fast and more accurate tests of mental capacity, low power densities in alpha frequency range and high power densities in the beta frequency range in EEG (Figuerio *et al.*, 2009). People will experience difficulty in performing many sorts of tasks if they are asked to do it at a time when their circadian system is telling them to sleep, i.e. in its night phase (Boyce, 2003). The non-imaging forming (NIF) effects influence mood, concentration, alertness, sleep and reaction time.

The mechanisms underlying the alterations in entrainment for circadian rhythm sleep disorders are poorly understood (Duffy and Wright, 2005). But lack of synchrony between the master clock and the peripheral clocks can lead to asynchronics within cells (e.g. cell cycle) and between organ systems (e.g. liver and pancreas) (Rea *et al.*, 2008). Consequences of an inappropriate phase relationship between biological and environmental time include impaired alertness, memory and performance, disturbed endocrine functions, and upset gastrointestinal function (Klerman, 2005). Research on biological lighting demands has shown that improper lighting can lead to sleep distortions, trends of (winter) depression and loss of concentration (Duffy and Wright, 2005).

Moreover, the WHO has identified rotating shift work as a probable cause of cancer (Straif *et al.*, 2007). But also other disorders have been associated with rotating shift work, such as obesity and diabetes, suggesting a role for circadian disruption in the development and progression of diseases (Stevens and Rea, 2001; Stevens *et al.*, 2007).

Furthermore, *seasonal affective disorder* (SAD) has been linked to disturbances in the circadian system and regulation of the hormone serotonin. SAD is a sub-type of major depression that is identified by a regular relationship between the onset of depression and the time of year and full remission of depression at specific times of the year repeated over the last two years. There are two forms (Boyce, 2003):

- *Winter SAD*: increased feelings of depression and a reduced interest in all or most activities together with atypical symptoms such as increased sleep, increased irritability, and increased appetite with carbohydrate cravings and consequent weight gain.
- *Summer SAD*: increased feelings of depression and a reduced interest in all or most activities together with a decrease of sleep, poor appetite and weight loss.

### 3.3.3 Noise

Road traffic noise has been identified as a major cause of sleep disturbance (Berglund *et al.*, 1999a; Muzet, 2007). Effects of night-time road traffic noise exposure on aspects of sleep have been found in both laboratory studies and in field studies with subjects exposed to habitual noise in their home situation (HCN, 2004). These effects comprised awakenings or sleep stage changes, autonomic responses, body movements and self-reported noise-induced awakenings, difficulty falling asleep and reduced sleep quality (overview from Kluizenaar *et al.*, 2009). Long-term effects of road traffic noise have been found on self-reported noise-related sleep disturbance and general sleep quality (Kluizenaar *et al.*, 2009).

### 3.3.4 Thermal comfort

Some studies have shown that the thermal environment can affect the duration of the REM sleep (Muzet *et al.*, 1983). It seems that the effects of cold temperature on sleep stage distribution are more important in brief than prolonged exposure. And a cold environment seems to be more disruptive to sleep than a warm environment (Haskell *et al.*, 1981). With respect to prolonged heat exposure, no improvement in the sleep pattern from one night to another (no adaptation) during prolonged exposure to a hot condition (35°C) was seen, while an adaptive thermoregulatory process similar to that described in awake subjects was observed (Libert *et al.*, 1988). The latter seems to relate to a redistribution of the local sweating rates towards the limbs that seems to occur during prolonged heat conditions (Höfler, 1968). The various regions of the body surface (trunk, head, arms and legs) may therefore participate to a different extent in the increasing sweat production when exposed to heat.

## 3.4 Endocrine disruption

*Endocrine-disrupting chemicals (EDCs), such as phthalates and flame retardants, may alter endocrine function by affecting the availability of a hormone at a target issue or the cellular response to a hormone. Besides those neuroendocrine systems, the cardiovascular and the respiratory system are also a target of environmental chemicals that interfere with intracellular signalling of hormonal and inflammatory pathways.*

## 3.4.1 Mechanism

In the mid-1990s concerns about the potential for environmental chemicals, drugs and other stressors to alter endocrine physiology emerged and the term 'endocrine disruptors' was born (Coldborn *et al.*, 1997; Marty *et al.*, 2011).

> An endocrine-disrupting chemical (EDC) is a compound, either natural or synthetic, which through environmental or inappropriate developmental exposures alerts the hormonal and homeostatic systems that enable the organism to communicate with and respond with its environment.
>
> (Diamanti-Kandarakis *et al.*, 2009; 'Key points').

EDCs may alter endocrine function by affecting (Diamanti-Kandarakis *et al.*, 2009; Marty *et al.*, 2011):

- *The availability of a hormone at a target issue*: for example enzymatic pathways involved in steroid biosynthesis and/or metabolism, transport and Hypothalamic-pituitary-Gonadal (HPG)/Hypothalamic-pituitary-Thyroid (HPT)-axis function.
- *The cellular response to a hormone*: for example receptor binding such as nuclear receptors, nonnuclear steroid hormone receptors, non-steroid receptors (e.g. neurotransmitter receptors such as serotonin receptor, dopamine receptor, norepinephrine receptor) and orphan receptors.

Along with endocrine effects, EDCs can have neurobiological and neurotoxic effects. Of the neuroendocrine systems, the reproductive HPG axis is best studied. The evidence for adverse reproductive outcomes (infertility, cancers, malformations) is strong. But there is growing evidence for effects on other endocrine systems, including thyroid, neuroendocrine, obesity and metabolism, and insulin and glucose regulation (see Section 2.5). EDCs may also disrupt the HPT and the HPA axis (Diamanti-Kandarakis *et al.*, 2009). Besides those neuroendocrine systems, the cardiovascular and the respiratory system are also a target of environmental chemicals that interfere with intracellular signalling of hormonal and inflammatory pathways (Newbold *et al.*, 2007).

A large number of industrial chemicals have been shown to reduce circulating levels of the thyroid hormone. For example chemicals that interfere with the uptake of iodide into the thyrocyte, such as perchlorate. And several chemicals, such as PCBs (polychlorinated biphenyls), can bind to the thyroid receptors (Diamanti-Kandarakis *et al.*, 2009).

EDCs may act directly upon the glucocorticoid or mineralocorticoid receptors or on steroidogenic pathways. EDCs including PCB, dioxin, lindane and others can affect the synthesis of adrenal steroids (Harvey *et al.*, 2007). EDCs may act upon nuclear hormone receptors that are expressed in hypothalamic or pituitary cells, thereby exerting feedback effects. And they may exert actions via membrane steroid receptors. Numerous neurotransmitter systems such as dopamine, norepinephrine, serotonin, glutamate and others are sensitive to endocrine disruption, which might explain the possible effects of EDCs on cognition, learning, memory and other non-reproductive behaviours (Diamanti-Kandarakis *et al.*, 2009).

## 3.4.2 EDCs and their sources

Food and water contamination has been a major focus for endocrine disruptors. Diet is probably the most important source of EDCs for the general population (Brouwers *et al.*, 2009). In the indoor environment, inhalation of air (Adibi *et al.*, 2008) and ingestion of house dust (Bornehag *et al.*, 2005a) have been considered important pathways for phthalates and flame retardants. Especially for very young children, the oral and dermal uptake from house dust might be relevant for risk assessment (Wensing *et al.*, 2006). Occupational exposure such as working with certain industrial chemicals (Pesticides, PAHs, etc.) can be relevant as well (Brouwers *et al.*, 2009). But for the general population, the main exposure route of phthalates and flame retardants are via the food chain: for example, it has been estimated to be more than 95 per cent (Wensing *et al.*, 2006; Fromme *et al.*, 2009). More recently, it was argued that exposure to semi-volatile organic compounds (SVOCs) via dermal pathways can be important to consider as well (Weschler and Nazaroff, 2012). In Box 3.7 an overview of endocrine-disrupting chemicals found in the environment (indoors and outdoors) is presented.

A group of pollutants that has received a lot of attention in the indoor environment are the phthalates, which are SVOCs and are used as plasticizers

---

### Box 3.7 Endocrine-disrupting chemicals (EDC) in the indoor environment

- *Phthalates* (di(2-ethylhexyl)phthalate (DEHP); n-butyl benzyl phthalate (BBzP); di-n-butyl phthalate (DnBP); diisobutyl phthalate (DIBP); diethyl phthalate (DEP)), as used in vinyl, plastics, building materials, toys, cleaning products, cosmetics, hygienic products and fragrances.
- *Bisphenol A* is used in polycarbonate (hard) plastic in numerous common consumer products such as epoxy, building components, electronic equipment as well as protective coatings on food containers and baby bottles.
- *4-nonylphenol* derived from detergents, pesticide formulations, polystyrene plastics, carpet and dry cleaning products, paints and textiles.
- *Parabens* used to preserve products such as lotions and sunscreens.
- *Pesticides* (methodxychlor, chlorpyrifos, dichlorodihenyltrichloroethane (DTT)) used to control insects, weeds and other pests in agriculture, gardening and the built environment.
- *Polybrominated diphenyl ethers* (PBDEs) in consumer products for improving fire resistance (e.g. the pentaBDE preparation is primarily used in polyurethane for applications such as carpet padding and furniture upholstery; octaBDE and decaBDE are used in hard plastics that house electrical appliances such as TV sets and computers) (Sjödin *et al.*, 2008).
- *Perflurorinated chemicals* (PFCs) in textiles, food packages, etc.
- *Polychlorinated biphenyls* (PCBs) in electrical equipment, caulking, paints and surface coatings.

Source: Rudel and Perovich, 2009; Bornehag, 2009.

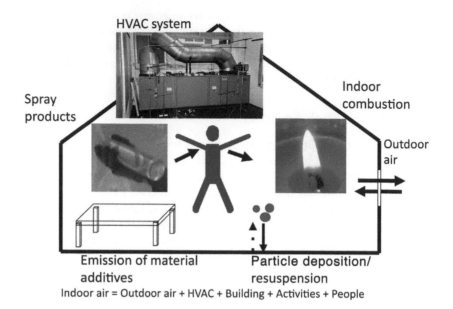

**Figure 3.6** *SVOCs in the indoor environment*
Source: Bluyssen.

in PVC (polyvinylchloride) plastics. Because of their chemical properties, exposure to phthalates does not result in bioaccumulation (Heudorf *et al.*, 2007).

Regarding polybrominated diphenyl ethers (PBDEs), it has been hypothesized that past and episodic current higher intakes are more important determinants of body burden than continuous background exposure (Roosens *et al.*, 2009). It has also been found that for PBDEs the body burden in North Americans is much higher than in Europeans, most likely due to the much higher indoor (dust) exposure experienced by the US population (Sjödin *et al.*, 2008). PBDEs burdens in the environment are expected to continue to increase despite the changes in their regulatory measures: penta and octa mixtures are now prohibited or voluntarily being phased out in many countries (Vonderheide *et al.*, 2008).

Many SVOCs are high production volume chemicals used in plastics, detergents, furniture, building material components, such as insulation and caulking, and other household and consumer products (Rudel and Perovich, 2009). SVOCs persist for a long time, even after the source has been removed (Weschler and Nazaroff, 2008). SVOCs can enter indoor environments along several pathways, including infiltration of outdoor air, indoor combustion, spray products and material additives (Xu and Zhang, 2011) (see Figure 3.6). Infants usually have a higher intake than adults, possible due to mouthing plastics and ingestion of indoor dust.

### 3.4.3 Effects

Exposure to EDCs may play a role in several diseases such as reproduction disorders including malformation of genitals, asthma and allergy, neurodevelopment disorders (e.g. autism), overweight/obesity and diabetes (Walker and Gore, 2007; Bornehag, 2009).

*Obesity* (see Box 2.6) has reached epidemic proportions in the United States, with more than 20 per cent of adults defined as clinically obese and an additional 30 per cent defined as overweight. An emerging hypothesis proposes that in utero and early developmental exposures to environmental chemicals may play a role in the development of obesity later in life (Newbold *et al.*, 2007).

*Autism* is a disorder with multiple dimensions, hence the term autism spectrum disorder (ASD). In a study performed by Larsson *et al.* (2009) of 4,779 Swedish children between 6–8 years old, 72 children (1.5%) had parentally reported ASD (slightly higher than in several other studies). Phthalate exposure (from PVC flooring) early in development of those children appeared to be linked to the reported ASD.

Epidemiological data, supported by experimental studies, point to a possible correlation between phthalate exposure and *asthma and airway diseases* in children, indicating a role for phthalates in the early mechanisms of the pathology of allergic asthma (Bornehag and Nanberg, 2010).

Effects of different EDCs may be additive or even synergistic. Exposure of an adult may have different consequences than exposure to a developing foetus or infant. And there can be a lag between the time of exposure and the manifestation of a disorder. EDCs may affect not only the exposed individual but also the children and subsequent generations. Effects may be transmitted not due to mutation of DNA sequence, but through modifications of factors that regulate gene expression such as DNA methylation and histone acetylation (epigenetic modifications – see Section 3.6.2) (Miller and Ho, 2008).

## 3.5 Oxidative stress

*Oxidative stress occurs when there is an excess of free radicals (stealing of electrons) over antioxidant defences. The removal of electrons from cells through oxidation can create highly reactive oxygen species (ROS). Antioxidants prevent oxidant formation, interact with formed oxidants and repair oxidative damage. Sources of ROS can be both endogenous (inside the body) and exogenous (outside the body). Oxidative stress can damage cellular proteins, membranes and genes, and can lead to systemic inflammation.*

### 3.5.1 Mechanism

Oxidative stress was first introduced by Sies, in 1991. He defined oxidative stress as 'a disturbance in the pro-oxidant-antioxidant balance in favour of former, leading to potential damage' (quoted in Kelly, 2003). Kelly (2003) defined it as 'a potentially harmful process that occurs when there is an excess of free radicals, a decrease in antioxidant defences, or a combination of these events' (Figure 3.7).

Frampton (2011) defines it as 'a mediator of inflammation and injury.' Nadler (2007) states that oxidative stress, as inflammation and free radical damage, are not diseases but are often the by-product of normal cellular processes. The attack and oxidation of other cell components such as lipids (particularly unsaturated lipids), proteins and nucleic acids, by the free radicals,

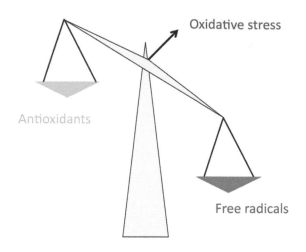

**Figure 3.7** *Oxidative stress: an excess of free radicals over antioxidant defences*
Source: inspired by Kelly, 2003.

can lead to tissue injury and in some cases, the influx of inflammatory cells to the sites of the injury.

A radical ion or a free radical is a molecule with an unpaired electron (see Box 3.8). Free radicals miss an electron and can 'steal' an electron from other molecules, which in their turn can steal electrons from other molecules. Free radicals are therefore unstable and highly reactive. If this stealing process (free radical chain reaction) is not stopped, oxidative stress can occur. Oxidative stress can damage cellular proteins, membranes and genes and can lead to systemic inflammation.

Not all forms of oxidative stress are damaging to the cell: low levels of oxidative stress are involved in signalling pathways within the cell, especially those involved in regulating immune responses (Li *et al.*, 2008). And some ions are essential to life. Sodium, potassium, calcium and other ions play an important role in the cells of living organisms, particularly in cell membranes (see Section 2.2.2). Inert gases such as helium (He), neon (Ne), argon (Ar), krypton (Kr), xenon (Xe) and radon (Rn) cannot oxidize. They are non-reactive gases.

Oxygen plays an important role in this oxidation process. The removal of electrons from cells through oxidation can create highly ROS, chemically reactive molecules containing oxygen.

Some 85 per cent of cellular oxygen is consumed by mitochondria for cellular respiration (see Section 2.2.2), which makes mitochondria the most important cellular source of free radical oxidants (and oxidative stress) (Phull, 2010). Biological antioxidant defence mechanisms keep levels of ROS relatively low. Antioxidants are substances that supply missing electrons.

### 3.5.2 Free radicals and their sources

Reactive oxygen species or ROS are species such as superoxide, hydroxyl radical, hydrogen peroxide and ozone, which are associated with cell damage. It is estimated that 1 to 5 per cent of the oxygen consumed by the mitochondria is converted to ROS (Phull, 2010). ROS include both oxygen radicals and certain non-radicals that are oxidizing agents (see Box 3.9). 'Hence all oxygen radicals are ROS, but not all ROS are oxygen radicals' (Halliwell and Gutteridge, 2007; p. 22).

Sources of ROS can be both endogenous (inside the body) and exogenous (outside the body). Endogenous oxidants can result from (Phull, 2010):

- *Cellular respiration*: in mitochondria (see Section 2.2.2).
- *Immune responses*: immune cells, such as neutrophils and macrophages, release ROS as a necessary part of phagocytosis (see Section 2.4.2). So ROS are also beneficial, as they are used by the immune system as a way to attack and kill pathogens.

## Box 3.8 Molecules, elements and ions

*Molecules* are made up of atoms, and atoms are made up of protons, neutrons and electrons. The core of an atom has a positive electrical charge, determined by the number of protons present. The number of protons is equal to the atom number. For example the atom H has one proton and atom number 1. Oxygen (O) has 8 protons and atom number 8. To keep the atom electrically neutral, a cloud of electrons with equal but opposite (negative) charge is required. Normally, electrons circle around atoms in pairs, having opposite spines. To stabilize the core of the atom, neutrons are present, except in the H atom.

*An element* is a compound comprising atoms of which the cores have the same positive electrical charge. A molecule comprises atoms from one, two or more elements. An element can comprise different types, the so-called isotopes, which only differ in their number of neutrons. The electrons in the electron cloud around the core of an atom find themselves in a fixed configuration, named energy levels. These levels are distributed in so-called peelings: K, L, M, N, O, P and Q (also named 1, 2, 3, 4, 5, 6 and 7). Each level can contain a fixed number of electrons: K has maximum 2 electrons, L maximum 8, M maximum 18 and N maximum 32. The heaviest natural element is uranium and therefore the O-level can in practice contain no more than 22 electrons. The electrons in the peelings are placed in sub-levels: s, p, d and f. Sub-level s can contain maximum 2 electrons, sub-level p 6 electrons, sub-level d 10 electrons and sub-level f maximum of 14 electrons. From this electron configuration in general the ion-configuration can be determined (Figure 3.8). The ion-configuration tells us something of how many electrons an atom can lose or gain. For example the electron configuration of O is $1s^22s^22p^4$ meaning 2 electrons put away at level 1 (or K) in sub-level s, 2 electrons put away at level 2 in sub-level s and 4 electrons put away at level 2 in sub-level p. A total of 8 electrons. O can gain 2 electrons at level 2 in the sub-level p, resulting in oxidation state -2 or an $O^{2-}$ion. H has an electron configuration of 1s and can gain or lose an electron, resulting in an oxidation state of -1 or +1 respectively. The gain or loss of a third electron will not happen so easily because it takes energy to split an electron. The electron affinity is the amount of energy that comes free when an electron is taken up by a neutral atom.

An *ion* is an atom or molecule, which has lost or gained one or more electrons, making it positively or negatively charged.

- *Anions* are negatively charged ions (there are more electrons associated with them then there are protons in their nuclei).
- *Cations* are positively charged ions (have fewer electrons than protons).
- *Dianion*: is a species, which has two negative charges; for example, the aromatic dianion pentalene.
- *Radical ions*: are ions that contain an odd number of electrons and are mostly very reactive and unstable.

Source: Serway, 1996.

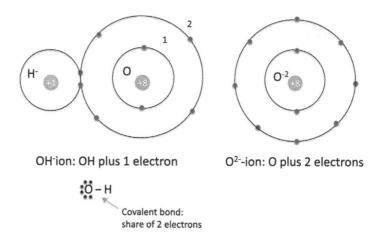

OH⁻ion: OH plus 1 electron          O²⁻-ion: O plus 2 electrons

Covalent bond:
share of 2 electrons

**Figure 3.8** *The electron configuration of OH ion and O²⁻ ion*
Source: Bluyssen; based on Serway, 1996, p. 1289.

- *Fatty acid oxidation*: which also takes place in peroxisomes (an organelle in the cell that breaks down fatty acids in periods of starvation) (see Figure 2.2).
- *Insufficient oxygen supply*: during conditions such as ischemia (local anaemia), anaemia or hypoxia (see Section 2.3.3).
- *Increased energy demands*: for example during high intensity exercise, endogenous production of ROS in skeletal muscle and myocardium due to increased demands on mitochondrial metabolism activates the oxidative processes through tissue injury and inflammation and increased fatty acid metabolism.

Exogenous oxidants arise from (Phull, 2010):

- *Inhaling chemicals*: causing a chain reaction of oxidative stress. ROS can be generated as a direct effect of these substances or by inciting physiological reactions, such as an inflammatory response. Because primary generation of ROS can continue even when the pollutant is removed, previous exposure is an important aspect.
- *UV radiation* (mainly sunlight): during exposure to UV light, ROS are generated. After exposure this generation stops. Prolonged exposure can cause mitochondrial DNA damage leading to poor energy metabolism and oxidative stress.
- *Diet*: can have both a protective and a damaging role.

Oxidative stress can arise from consumption of alcohol, medications, stresses of all kind (including mental and emotional), cold, air pollutants, toxins and radiation (light), and internal metabolic processes. Mitochondrial DNA is highly vulnerable, much more than nuclear DNA, due to both its structure and its function. Mutations caused by oxidative stress may lead to impaired oxidative phosphorylation, inciting further production of ROS. Mitochondria DNA also plays a role in signalling appropriate cell apoptosis (cell suicide), a protective

# Box 3.9  ROS

## Oxygen radicals

### Superoxide $O_2 \bullet -$

Superoxide or hyperoxide is used by the immune system to kill invading microorganisms. In phagocytes, superoxide is produced in large quantities, and superoxide is also produced as a byproduct of mitochondrial respiration as well as other enzymes. Superoxide may contribute to the pathogenesis of many diseases and perhaps also to ageing via the oxidative damage that it inflicts on cells.

### Hydroxyl radical $OH \bullet$

The hydroxyl radical is highly reactive and short-lived. Most notably hydroxyl radicals are produced from the decomposition of hydroperoxides (ROOH) or by the reaction of excited oxygen with water. Hydroxyl radicals are also produced during UV-light dissociation of $H_2O_2$ and likely in Fenton chemistry, where trace amounts of reduced transition metals catalyze peroxide-mediated oxidations of organic compounds.

The first reaction with many volatile organic compounds (VOCs) is the removal of an hydrogen atom, forming water and an alkyl radical ($R\bullet$): $OH\bullet + RH \rightarrow H_2O + R\bullet$

The alkyl radical will typically react rapidly with oxygen forming a peroxyl radical: $R\bullet + O_2 \rightarrow RO_2$

The hydroxyl radical can damage virtually all types of macromolecules: carbohydrates, nucleic acids (mutations), lipids (lipid peroxidation) and amino acids. It has a lifetime of less than one second (Isaksen and Dalsøren, 2011).

Unlike superoxide, which can be detoxified by superoxide dismutase, the hydroxyl radical cannot be eliminated by an enzymatic reaction. Mechanisms for scavenging peroxyl radicals for the protection of cellular structures include endogenous antioxidants such as melatonin and glutathione, and dietary antioxidants such as mannitol and vitamin E (see Section 3.5.3).

## Non-radicals

### Hydrogen peroxide $H_2O_2$

Hydrogen peroxide is a by-product of oxidative metabolism or cellular respiration. The reactions involved in respiration are catabolic reactions that involve the redox reaction (oxidation of one molecule and the reduction of another). Hydrogen oxide is a weak oxidizing or reducing agent and is in general a poor reactive. Despite its poor reactivity, it can be cytotoxic and at high concentrations is often used as a disinfectant.

### Ozone $O_3$

Ozone is an irritating, non-free radical and is formed by the photo-dissociation of $O_2$, which then combines with $O_2$. And it can arise by photochemical reaction between oxides of nitrogen and hydrocarbons or from photocopying machines. Ozone induces inflammation, activating pulmonary macrophages and recruiting neutrophils to the lung. Ozone irritates the eyes and can oxidize proteins and lipids in tear fluid and respiratory tract. Reactions between ozone (in ventilation air) and terpenes (such as limonene, $\alpha$-pinene, styrene present in indoor environments) frequently dominate indoor chemistry (Weschler et al., 2006). Hydroxyl radicals ($OH\bullet$) are formed in these reactions, which in their turn react with other products and form oxidized products. Ozone ($O_3$) is a powerful oxidizing agent, far stronger than $O_2$.

Source: Halliwell and Gutteridge, 2007.

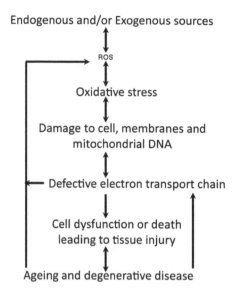

Endogenous and/or Exogenous sources

ROS

Oxidative stress

Damage to cell, membranes and mitochondrial DNA

Defective electron transport chain

Cell dysfunction or death leading to tissue injury

Ageing and degenerative disease

**Figure 3.9**
*Cycle of oxidative stress: ROS are both an indication and a causal factor of cumulative mitochondrial damage and a range of degenerative diseases*
Source: Bluyssen, inspired by Phull, 2010, p. 305.

mechanism to stop replication of damaged cells. Alteration of that function by oxidative stress may cause unhealthy cells to escape apoptosis and healthy cells to be killed (see Section 3.6.1). Figure 3.9 shows the complex interrelationship between oxidative stress, mitochondrial metabolism and energy production.

### 3.5.3 Antioxidant defences

Because free radicals are necessary for life, the body has a number of mechanisms to minimize free radical induced damage and to repair damage that occurs. Antioxidants play a key role in these defense mechanisms. Antioxidants prevent oxidant formation, interact with formed oxidants and repair oxidative damage. Halliwell and Gutteridge (2007) define an antioxidant as 'any substance that delays, prevents or removes oxidative damage to a target molecule' (p. 81). Two categories of antioxidants can be distinguished (see also Box 3.10):

- *Endogenous antioxidants*, including protective enzymes, which help to decrease the cellular oxidants, and antioxidant compounds, synthesized in the body from substances in the diet.
- *Micronutrients antioxidants*, which act directly with cytoplasm, cell membrane and extra-cellular space.

Furthermore, *bilirubin* and *uric acid* can act as antioxidants to help neutralize certain free radicals. Bilirubin comes from the breakdown of red blood cell contents, while uric acid is a breakdown product of purines. Too much bilirubin, though, can lead to jaundice (see Section 2.3.3), which could eventually damage the central nervous system, while too much uric acid causes gout (see also Section 2.3.3).

*Melatonin* reduces oxidative stress by neutralizing the hydroxyl radical. Melatonin stimulates glutathione peroxidase (GSH-PX) activity in neural tissue, which metabolizes reduced glutathione to its oxidized form, thus converting $H_2O_2$ to $H_2O$, thereby reducing the generation of the hydroxyl radical by eliminating its ancestor (Reiter *et al.*, 1995).

Low levels of oxidative stress activate antioxidant defences, whereas higher levels of oxidative stress lead to pro-inflammatory and cytotoxic effects. An example of an antioxidant response is the cellular heme oxygenase-1 (HO-1) expression, which is generated together with a gaseous substance, CO (carbon monoxide), which exerts anti-inflammatory effects in the lung and is exhaled in the expired air (Li *et al.*, 2003).

### 3.5.4 Stressors and effects

Free radical damage, oxidative stress and systemic inflammation are all implicated in serious diseases such as cancer and Alzheimer's (see Figure 3.10). Inflammation helps fight disease and protect parts of the body and it postpones the body's normal immune response and certain metabolic processes. When an

## Box 3.10 Antioxidants

Major antioxidant enzymes are:

- *Superoxide dismutases*: play a role in neutralization of the super oxide radical.
- *Glutathione peroxidases*: remove hydrogen peroxide, lipid/non-lipid peroxides and acids in regeneration of vitamin C.
- *Catalase*: plays a role in elimination of hydrogen peroxide.

*Antioxidant compounds* comprise thiol compounds such as glutathione, cysteine, methionine. They can react directly with hydroxyl radicals in both aqueous and lipid environments. *Reduced glutathione (GSH)* is an antioxidant that protects from free radical damage and carries out many other essential functions, including detoxification, repair of DNA, regulation of enzymes and strengthening of your immune system by increasing production of killer T-cells. It is a vital component of every cell in the body and is made by the body, in the cells of the body (Nadler, 2007).

*Micronutrient antioxidants* comprise three vitamins, vitamin A, vitamin C and vitamin E, and a group of polyphenolic compounds named flavonoids, found in fruits, vegetables and certain beverages. Vitamin E and beta carotene are important in membranes. Vitamin C protects membrane damage directly and it enhances the antioxidant activity of Vitamin E by regeneration of its reduced form. Flavonoids may directly neutralize cellular radicals, prevent cascade of lipid per oxidation in cell membranes and stabilize free oxygen species.

Source: Phull, 2010.

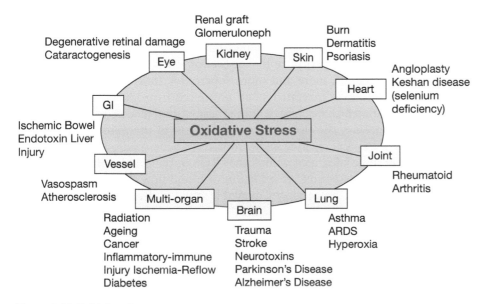

**Figure 3.10** *Oxidative stress*

Source: National Institute of Standards and Technology, www.oxidativestressresource.org (accessed 5 July 2011).

inflammatory stimulus is encountered, cells start to produce inflammatory mediators (e.g. IL-6, IL-8, TNF-α, IL-1β). These inflammatory mediators attract inflammatory cells from the blood circulation towards the site of inflammation in order to clear the inflammatory stimulus from the body (Oberdörster *et al.*, 2005). In the short term this is not a problem, but in the long term it can cause progressive damage. Chronic *systemic* inflammation is not restricted to a particular tissue, but involves the lining of blood vessels and many internal organs and systems. This inflammatory process is often associated with free radical damage and oxidative stress and may not cause pain, as some internal organs do not relay pain (Nadler, 2007). Oxidative stress seems to affect all organs in our body and seems to contribute to multiple health problems.

The pathological consequences depend on the location of the oxidative injury in the body or even within the cell, and the type of physical stressor(s) involved. Damage to circulating triglycerides, for example, has been implicated in cardiovascular disease (see also Box 2.2). Oxidation of LDL cholesterol has been shown to incite inflammatory mechanisms that are detrimental to vascular cells. And ROS can also damage cellular DNA, initiating changes that have been implicated in cancer (Halliwell and Gutteridge, 2007).

## Noise

Oxidative stress also seems to be involved in noise-induced hearing loss (Henderson, 2006; Darrat *et al.*, 2007; Seidman and Standring, 2010). Exposure to noise has been demonstrated to cause a decrease of cochlear circulation (for structure of ear see Section 2.5 in Bluyssen, 2009a) shown by an aggregation of red blood cells, capillary vasoconstriction and stasis (blood flow stops) (Quirk and Seidman, 1999). This vascular compromise may contribute to metabolic changes during acoustic overstimulation. In response to this compromise, free oxygen radicals may be produced (Evans and Halliwell, 1999).

The mechanism of the production of the extra ROS observed in the cochlea is not completely clear. But it is speculated that during the aerobic respiration in the mitochondria (phosphorylation – see Section 2.2.2), to meet the increased cellular energy demands, more and more superoxide (normally formed as an intermediate molecule) is generated as an unwanted byproduct. This is due to the inefficiency with which the mitochondria must work (Henderson *et al.*, 2006). During noise exposure, the electron transport chain of the mitochondria uses large amounts of oxygen, creating large amounts of superoxide, which can then react with other molecules, generating other ROS in the cochlea (Halliwell and Gutteridge, 2007). Noise-induced reduction of blood flow (named ischemia or reperfusion), and thus reduction of available oxygen, leads to an inefficient phosphorylation process (see Section 2.2.2), which in turn leads to the generation of superoxide (Henderson *et al.*, 2006).

Noise exposure can lead to cell death in the cochlea, both necrotic and apoptotic (see Section 3.6.1): noise can lead to overdriving of the mitochondria, excitotoxicity (large amounts of neurotransmitter glutamate lead to cell death) and ischemia effects, which can each lead to increases in ROS. The ROS can damage DNA and the cell membrane (Henderson *et al.*, 2006).

## Air pollution

The oxidative effects of air pollution have been studied mostly for lungs. Eyes are being studied as well as effects in the brain of particles and the effects of skin oil oxidation products on the indoor environment.

### Air pollution and lungs

For over 30 years, research has demonstrated associations between ambient particulate matter (PM) and increased human health outcomes, mortality and morbidity (e.g. Brunekreef and Holgate, 2002; WHO, 2006). In the last decade studies on indoor air pollution also suggest a possible link between exposure to indoor particulates and several health effects (Lewtas, 2007; Weichenthal *et al.*, 2007). PM exposure has been associated with cardiovascular diseases (CVDs), artherosclerosis and problems in the regions of the respiratory tract, such as lung cancer, asthma, and bronchitis (Pope and Dockery, 2006). Particles may be also transported to extra-pulmonary organs, such as the liver, kidney and brain causing potential neurological effects (Oberdörster *et al.*, 2005; Geiser *et al.*, 2005; Peters *et al.*, 2006). Particles generated during combustion processes such as (gas) cooking have been specifically associated with respiratory problems (Hölscher *et al.*, 2000; Varghese *et al.*, 2005) and lung cancer (Yu *et al.*, 2006; Subramanian and Govindan, 2007; Sun *et al.*, 2007).

Air comprises of free radicals (e.g. nitrogen dioxide ($NO_2$) and pollutants that have the ability to drive free radical reactions (e.g. ozone and particulates). Both nitrogen dioxide and ozone can react with substrates present in the lung lining fluid compartment (via reactive absorption), forming oxidized products responsible for the cascade of secondary, free radical derived products which brings the inflammatory cells into the lung. They are unlikely to interact with the pulmonary epithelium (Kelly, 2003). For particulates related health effects the mechanisms are still not fully understood, but it seems that many of the adverse health effects may be derived from oxidative stress, initiated by the formation of ROS at the surface of and within the target cells (Ayres *et al.*, 2008). Components of particles have the potential to generate free radicals in the lung environment and thereby cause oxidative stress, which is an important mechanism leading to inflammation (Kelly, 2003; see also Section 3.2.3). Inflammation plays a key role in COPD (Bucchioni *et al.*, 2003), asthma and CHD (coronary heart disease; Delfino *et al.*, 2005). Lung cancer can also have oxidative stress as an important factor in its causation.

Particles of different sizes, from different sources and with different composition, should be considered (WHO, 2007a; see also Box 3.11). Toxicological evidence provides an indication that aspects of particulates other than mass alone determine toxicity (WHO, 2006, 2007a). Depending on their composition, particles can cause inflammation directly or indirectly. Microbial components on particles can directly cause inflammation (see Section 2.4.2), while organic chemicals and transition metals cause inflammation through oxidation reactions. Organic chemicals and transition metals on particles can generate ROS in the epithelial lining fluid (Dahl *et al.*, 2007), but may also enter the lung epithelial cells and generate reactive oxygen species intracellularly (see Figure 3.11; Li *et al.*, 2008). But also particles themselves (ultra fine particles (UFP) without transition metals) can induce inflammation, indicating that particle size itself may

## Box 3.11  Particulates, size and composition

Many compounds generated indoors are semi volatiles such as phthalates, flame retardants, polyaromatic hydrocarbons (PAHs), chlorophenols, pesticides and organotins, metals (transition metals: such as $Fe^{2+}$, $V^{3+}$, $Cr^{2+}$ and $Cu^+$ – iron, vanadium, chromium and copper), and quinones (oxidized organic compounds derived from aromatic compounds such as benzene or naphthalene). They may also be absorbed to particulate matter present in the indoor air and to house dust. These particles may be inhaled or ingested, depending on their size. Particulate air pollutants thus have very diverse chemical compositions that are highly dependent on their source. SVOCs (i.e. surfactants) and inorganic compounds, which can be absorbed onto PM (particulate matter) surfaces, have been associated with negative impacts on human health (Henvinet, 2009). Physical chemical properties of particles can also be potentially relevant such as hygroscopicity, hydrophilicity, lipophilicity, bioavailability, acidity, redox potential and surface as opposed to bulk chemistry (NRC, 2004).

With respect to the size of the particles, there is a need to identify the role of the ultrafine particle (UFP) fraction (the concentration of particles with an aerodynamic diameter smaller than 0.1 μm or $PM_{0.1}$ (Oliveira Fernandes et al., 2008) because of their greater number concentration, rendering them potentially more biologically active with greater potential for transcytosis across epithelial barriers, thus reaching target tissues beyond the lungs (Reiss et al., 2007). The majority of UFP indoors are produced indoors (50–80%) (Weichenthal et al., 2007). Indoor sources of UFP are home cooking and heating appliances, tobacco smoke, burning candles, vacuuming, natural gas clothes dryers and other household activities (Afshari et al., 2005; Wallace, 2006). In office buildings laser printers are a major source of UFP (Wensing et al., 2009; Morawska et al., 2009). Additionally, products from indoor chemical reactions (e.g. ozone-terpene reactions producing secondary organic aerosols (SOA), a type of UFP) have been found to contribute to adverse effects on human health (Weschler et al., 2006; Carslaw et al., 2009).

**Figure 3.11**
*The lung, oxidative stress and circulation: (1) Organic chemicals and transition metals generate ROS; the presence of oxidants leads to oxidation of biomolecules in the epithelial lining fluid such as lipids and proteins, perceived by TLRs (Toll-like receptors) which signal to the nucleus to produce inflammatory mediators. (2) UFP (ultra fine particles) themselves generate ROS in the lung epithelial cell which causes inflammatory mediators. (3) UFP translocate directly from the lungs into the circulation*

Source: Bluyssen; based on Li et al., 2008; Dahl et al., 2007; Postlethwait, 2006; Kreyling et al., 2006.

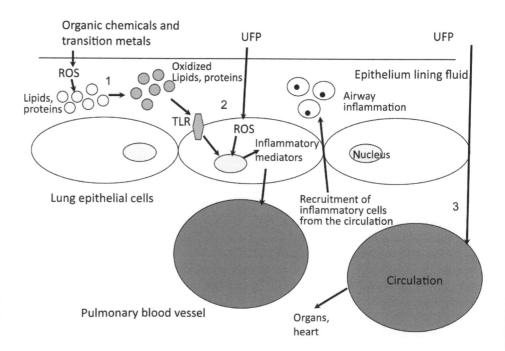

be a critical factor (Kelly, 2003). UFP may even translocate directly from the lung to the circulation by passing the epithelial and capillary cell layers (Kreyling *et al.*, 2006).

Cellular responses to particle induced oxidative stress include activation of antioxidant defence, inflammation and toxicity, also named the three-tier model (Li *et al.*, 2008). Low levels of oxidative stress induce protective effects (tier 1: antioxidant enzymes are induced to restore cellular redox homeostasis). At an intermediate level of oxidative stress pro-inflammatory responses (tier 2: e.g. cytokines and chemokines) are induced. At a higher level of oxidative stress, cytotoxic effects (tier 3) may induce apoptosis of lung cells or necrosis (see Section 3.6.1). Lung lining fluid obtained from the lower respiratory tract contains similar blood plasma to several antioxidants such as reduced glutathione, ascorbic acid (vitamin C), uric acid and alpha topherol (vitamin E), but also antioxidant enzymes such as superoxide dismutase and catalase (Kelly, 2003). Oxidizing species such as transition metals and certain organic compounds located on the surfaces of particles, will interact with and deplete lung lining fluid antioxidants in the same manner as ozone.

## Air pollution and eyes

Besides the normal secretion of tears (to prevent the eye from drying out), tears can be secreted by a reflex in response to a variety of stimuli, for example, irritative stimuli to the cornea, conjunctiva, and nasal mucosa, hot or peppery stimuli to the mouth and tongue or bright lights. Most foreign bodies that enter the eye remain near the surface, causing pain and a flow of tears (washing out the foreign body or not). Unfortunately, many small foreign bodies nestle in the under surface of the upper lid, so that each time the eye blinks, the foreign body rubs on the cornea, causing pain and irritation. High-speed small foreign bodies may penetrate the interior of the eye with very few symptoms: their presence may not even be recognized when inflammatory changes occur. Allergic reactions can comprise of redness and itching of the conjunctiva.

The stability of the pre-corneal tear film (PTF) seems to be an important factor for ocular comfort (Wolkoff, 2010). Alteration of this stability can lead to ocular discomfort (e.g. burning, dry and itching eyes). The total risk of ocular discomfort is intensified by physical alteration of the PTF due to visual tasking and climate conditions (low humidity, high temperature and draft), and by personal factors such as age, gender and use of certain medication also influence the overall stability of the PTF. At least three mechanisms can contribute to ocular discomfort (e.g. burning, dry and itching eyes) (Wolkoff, 2010):

1  *Physical mechanism*: The PTF structure is altered physically by thermal factors and inefficient blink frequency that increases the emission rate of aqua loss resulting in hyper-osmolarity, gland dysfunctions and associated discomfort.
2  *Oxidative stress*: The structural composition of the outermost lipid layer of the PTF is chemically altered by aggressive and oxidative pollutants (facilitated by, e.g., environmental tobacco smoke and traffic pollutants) that facilitate loss of aqua, and possibly chemosthesis.
3  *Trigeminal stimulation*: Strong sensory irritating pollutants (e.g. formalde-hyde and acrolein) cause chemosthesis by trigeminal stimulation (see also

Section 3.2.4). According to Wolkoff (2010), organic and inorganic indoor air pollutant concentrations are in general too low to cause chemosthesis.

### Particles in the brain

For air quality, particulate matter has been extensively studied. Particulate matter has been hypothesized to influence the systems through different routes, both directly and indirectly. The most described route is through oxidative stress and inflammation responses. More recently, it has been hypothesized that particles in the brain may directly disrupt the balance between sympathetic and parasympathetic activity (Pope and Dockery, 2006). Inhaled particles may reach the brain either via translocation into the circulation or via direct translocation to the brain via the olfactory nerve (Oberdörster *et al.*, 2005; Elder *et al.*, 2006). Once in the brain the translocated particles may cause changes to the autonomic nervous system via oxidative stress. Epidemiological studies and controlled human exposure studies have shown particle-related changes in autonomic nervous system function as measured by heart rate and heart rate variability (review by Pope and Dockery, 2006).

### Skin oil oxidation products

A series of experiments in a simulated office environment identified skin oil oxidation products among the most abundant volatile organic species in indoor air when ozone was present at levels of 15–30 ppb (Wisthaler and Weschler, 2010). These studies illustrate that humans, via the reaction of their exposed skin, hair and clothing, directly impact the oxidative capacity of the rooms that they occupy. The same ozone/skin oil chemistry is anticipated to impact levels of hydroxyl and nitrate radicals via the production of various primary and secondary products, including 6-methyl-5-hepten-2-one, geranyl acetone and 4-oxopentanal.

## Radiation (light)

Except for light operating through the visual system (Section 3.2.4) and the circadian system (Section 3.3.2), light can be treated as radiation (see Box 3.12). Light can cause damage to both the eye and skin, through both thermal and photochemical mechanisms (Boyce, 2003).

### UV radiation

In the short term, UV radiation can cause photokeratitis of the eye and erythema of the skin. Prolonged exposure to UV radiation can lead to cataract in the lens as well as skin ageing and skin cancer. But UV radiation is also important for the production of vitamin D in the skin.

Photokeratitis, inflammation of the front part of the eye (with symptoms such as reddening of the eye, tearing, twitching of eyelids and feeling a grit in the eye), occurs because of photochemical reactions (photo-oxidative damage and lipid oxidation causing oxidative stress) to UV radiation at the cornea (Halliwell and Gutteridge, 2007). Long exposure leads to cataract formation, a clouding or opacity that develops in the lens of the eye, which absorbs and scatters light (obstructing its passage) (Boyce, 2003). More than 17 million people worldwide are blind from cataracts (Halliwell and Gutteridge, 2007).

---

### Box 3.12  Light as radiation

Light, electromagnetic radiation can be divided into ultraviolet (UV), visible and infrared (IR) radiation.

- UV components can be divided into three regions: UVA (400–315 nm), UVB (315–280 nm) and UVC (280–100 nm). Part of the UVA region (380–400 nm) stimulates the visual system. (Sunlight UV-C, the most damaging part, is largely filtered out by the ozone layer in the stratosphere.)
- The visible spectrum ranges from 380 to 780 nm.
- IR components can be also be divided into three regions: IRA (780–1,400 nm), IRB (1,400–3,000 nm) and IRC (3,000–1,000,000 nm).

Source: Boyce, 2003.

---

Erythema, redness of the skin, is caused by hyperaemia (increase of blood flow) of the capillaries in the lower layers of the skin. The UV radiation produces a protective response in the skin, migrating pigment to the surface of the skin to decrease the sensitivity of the skin to UV radiation (Boyce, 2003).

UV can also damage dermal mitochondrial DNA and can lead to oxidation of PUFA side chains and cholesterol in skin lipids. DNA damage contributes to skin ageing and cancers induced by UV (basal cell carcinoma and malignant melanoma – see Section 3.6) (Halliwell and Gutteridge, 2007). Damage by UV causes an inflammatory response, including recruitment and activation of lymphocytes and neutrophils, to make ROS. *Skin ageing* or a photo-aged skin is characterized by the presence of massive quantities of thickened, degraded elastic fibres in the skin, which degenerate into amorphous masses (Boyce, 2003).

#### Visible radiation
Visible radiation and near IR-radiation (400 to 1,400 nm) can produce photoretinitis, inflammation of the retina in the eye (a photochemical damage), and it can cause chorioretinal damage (thermal damage of the tissue) (Boyce, 2003). In the latter the temperature of the pigment epithelium at the retina is elevated by the incoming radiation. If the injury is located in the fovea (the part you see with), it will interfere with vision.

#### IR radiation
The eye is reasonably well protected against acute thermal effects of IRB and IRC radiation, which cannot reach the retina. The skin on the other hand can be burned easily. However, prolonged exposure to IR radiation can lead to cataract formation and serious skin burns (Boyce, 2003).

## 3.6  Cell death and alterations

*External stressors can cause unwanted or unintended cell death (cytoxicity) and cell changes. Necrosis (passive form of cell death) can occur when exposed to an intensive physical or chemical insult. And apoptosis (active form of cell death) can be initiated at the wrong time causing crucial cells to die. Other potential consequences of exposure to physical, chemical, biological agents and even psychological stress at the cell level can be mutation (genotoxicity) or epigenetic toxicity.*

### 3.6.1 Cell death

Cell death (cytoxicity) can occur through two different processes (Henderson *et al.*, 2006): necrosis and apoptosis.

#### Necrosis

Necrosis is a passive form of cell death and is associated with cell swelling, which eventually results in rupture of the cell, damage of the surrounding tissue and initiation of an inflammatory response. Necrosis can occur after intensive physical or chemical insult, for example, after traumatic noise exposure or inhalation of toluene. Solvents and chemicals can directly decrease immune cells. For example exposure to toluene (found in many paints and solvents) decreases natural killer cells and T-lymphocytes.

#### Apoptosis

Apoptosis is an active, regulated process that consumes energy. It is a way for the body to eliminate unwanted or damaged/dying cells that could potentially damage neighbouring healthy cells. But apoptosis can be initiated at the wrong time and crucial cells die, for example, during a stroke, heart failure and autoimmune disease (see also Section 3.5.2). Oxidative stress can play an important role in this process.

### 3.6.2 Alterations at cell level

Besides cell death, other potential consequences of exposure to physical, chemical, biological agents and even psychological stress at the cell level can be (Kang and Trosko, 2011):

- Mutation, caused by an error in DNA repair or by an error of DNA replication (genotoxicity).
- Altered gene expression at the transcriptional, translational or posttranslational levels (epigenetic toxicity).

These cell changes can occur through damage, inhibition or mediation of certain functions, epigenetic changes, cumulative effects of certain exposures (e.g. dioxins), and exposure to radioactive substances and certain volatile organic compounds (e.g. formaldehyde and benzene) (Ash, 2010).

#### DNA adducts and mutation

Around 50,000 endogenous DNA adducts (DNA lesions) are normally present in human and animal cells. These adducts can result in mutations if DNA replication takes place before they are repaired (Swenberg *et al.*, 2011). DNA adducts can be caused by exposure to chemical carcinogens. Endogenous DNA adducts can also be the cause of oxidative stress. Fortunately, nearly all DNA adducts and damages are repaired, with only the rare exception leading to a change in sequence and mutation. For the development of cancer (see Box 3.13 for definitions), DNA mutation must occur because of these alterations in the DNA sequence, caused by single nucleotide changes or larger alterations in the chromosomes.

Polymorphins in the enzymes involved in repair processes can lead to differences in susceptibility, just as differences in susceptibility can exist because

---

## Box 3.13 Tumours, cancers, toxicity, carcinogenicity and epigenetics

- A *tumour* is an abnormal lump or mass of tissue, the growth of which exceeds, and is uncoordinated with that of the normal tissue. *Benign tumours* remain at the site of origin. When growing in a solid tissue, they usually become enclosed in a layer of fibrous material, the capsule. The cells of *malignant or cancerous tumours* invade locally, and often pass through the bloodstream and/or lymphatic system to form secondary tumours (metastases) at other sites. Cancers are classified according to the tissues from which they arose; for example, carcinomas arise from epithelial cells and sarcomas from connective tissue or muscle (Halliwell and Gutteridge, 2007).
- *Toxicity* is the degree to which a substance can damage living or non-living organisms.
- *Cytoxicity* is the quality of being toxic to cells. A cytotoxic compound may result in necrosis, a decrease in cell viability (stops growing and dividing) or apoptosis.
- *Genotoxicity* describes an action on a cell's genetic material affecting its integrity. For example benzene is a genotoxic carcinogen.
- A *carcinogen* is an agent directly involved in causing cancer due to the ability to damage DNA directly in the cells or to the disruption of cellular metabolic processes. Usually DNA damage, if too severe to repair, leads to programmed cell death. Cancer is a disease in which damaged cells do not undergo programmed cell death.
- A *mutagen* is an agent that changes the genetic material, usually DNA. A carcinogen is a mutagen.
- Several radioactive substances are considered carcinogens. Their carcinogenetic activity is attributed to the radiation, for example gamma rays and alpha particles, which they emit. Inhaled asbestos, certain dioxins, formaldehyde and benzene are all examples of carcinogens.
- *Epigenetics* is the study of heritable changes in phenotype (appearance) or gene expression caused by mechanisms other than changes in the underlying DNA sequence. Non-genetic factors cause the organism's genes to behave (or express themselves) differently. Mechanisms are (Miller and Ho, 2008):

  - *DNA methylation*: covalent addition of a methyl group to a cytosine residue in a site where a cytosine lies next to guanine in the DNA sequence (see Figure 2.3).
  - *Covalent histone modifications*: including acetylation, methylation, phosphorylation, and so on. Histones are the major protein compounds of chromatin, acting as spools around which DNA winds (see Figure 3.12).
  - *MicroRNA changes*: a microRNA (miRNA) is a short RNA molecule and is a post-transcriptional regulator that bind to complementary sequences on target messenger RNA transcripts (mRNAs).
  - *Chromatin alterations*: chromatin is the assembled histones and DNA (see Figure 3.12).

- Altered gene expression = epigenetic toxicity.

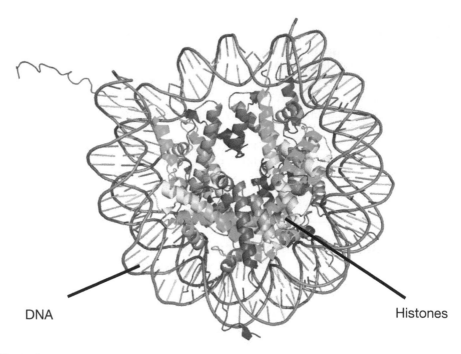

DNA                                                    Histones

**Figure 3.12** *The nucleosome: basic repeat element of chromatin, comprising histones and DNA. A histone is a protein in a cell nuclei that packages and orders the DNA into structural units called nucleosomes. Histones undergo modifications that alter their interaction with DNA and nuclear proteins. Combinations of modifications are said to have a code, the histone code. Histone modifications act in gene regulation, DNA repair etc.*

Source: adapted by Bluyssen from Emw, 'The crystal structure of the nucleosome core particle', https://en.wikipedia.org/wiki/File:Nucleosome_1KX5_colour_coded.png (accessed 14 December 2011).

of variations in metabolic activation and inactivation processes (Cohen and Arnold, 2011). Competing metabolic systems for activation and inactivation of a given chemical are present, but there exists also variability in induction and inhibition of these enzymes because of environmental influences, as well as interaction with numerous other chemicals (Cohen and Arnold, 2011). Formaldehyde, vinyl chloride and ethylene oxide all induce exogenous DNA adducts that are chemically identical to endogenously formed DNA adducts. It should be mentioned that formaldehyde is also an essential metabolic intermediate generated endogenously from serine, glycine, methione and choline, and also produced from the metabolism of xenobiotic chemicals and proteins by demethylation (see Box 3.13) (Swenberg *et al.*, 2011).

Chemicals can be classified based on their ability to generate DNA reactivity (Cohen and Arnold, 2011):

- *Non-DNA reactive* (epigenetic): increase the risk of cancer by increasing the number of DNA replications in the target cell (increased cell proliferation).
- *DNA reactive* (genotoxic): initiation (and progression) is essentially the process of producing irreversible DNA damage, and initiators are chemicals that are DNA reactive, either directly or following metabolic activation.

Diamine dyes, phthalates from plastic and nitrosamines from cured or burnt foods can adduct to DNA (a particular nucleotide on a gene) and thus inhibit DNA replication, RNA synthesis and DNA repair mechanisms (Müller and Yeoh, 2010).

## Epigenetics

Many toxins and toxicants work by 'epigenetic' mechanisms, which play important roles in regulating gene expressions and in the pathogenesis of many human diseases (Kang and Trosko, 2011). Epigenetic changes have been observed to occur in response to environmental exposure (see Box 3.13 for definition). A variety of compounds are considered as epigenetic carcinogens; they result in an increased incidence of tumours, but they do not show mutagenic activity. Examples include hexachlorobenzene and nickel compounds. Epigenetics has the potential to explain mechanisms of ageing human development and the origins of cancer, heart disease, mental illness, as well as other conditions (Beil, 2008). Studies suggest that epigenetic regulation after early (prenatal, post-prenatal) or later environmental exposures (endocrine disruptors, xenobotic chemicals, maternal stress and nurturing, and low-dose radiation) may in part mediate the complex gene–environment interactions that can lead to asthma (Miller and Ho, 2008). It has even been suggested that the risk of ETS exposure for asthma may be transmitted across two generations (Miller and Ho, 2008).

## Inhibition or mediation

Toxic compounds can also lead to indirect changes by inhibition or mediation of certain functional processes. *Mineral toxicity* from mercury, cadmium, lead and aluminium has been implicated in its ability to interfere with neurotransmitter functioning (Puri and Lynam, 2010). Lead competes with calcium, inhibiting the release of neurotransmitters, and interferes with the regulation of cell metabolism by binding to second-messenger calcium receptors, blocking calcium transport by calcium channels and calcium-sodium ATP pumps, and by competing for calcium-binding protein sides and uptake by mitochondria. Dietary deficiencies of calcium, iron and zinc enhance the effects of lead on cognitive and behavioural development (Goyer, 1995). Fluoride acts as a cellular poison because it interferes with calcium metabolism causing raised intracellular calcium and thus disrupting membrane action. Selenium most likely protects from mercury and methyl mercury toxicity by preventing damage from free radicals or by forming inactive selenium mercury complexes (Goyer, 1995).

## 3.7 Learning and memory effects

*Learning and memory effects seem to have an influence on how sensations of environmental stimuli are recognized, organized and interpreted. Physiological effects may be conditioned following the same rules as for classical conditioning, and, the explanation for certain unexplained symptoms seems to originate in the way we process information.*

### 3.7.1 Mechanisms

Perception consists of a complex process by which we recognize, organize and make sense of the sensations we receive from environmental stimuli. Our memory has a crucial role in this process. The types of memory we have can be categorized into different systems and subsystems (see Box 3.14), of which the major division is non-declarative or implicit memory and declarative or explicit memory.

---

## Box 3.14 Memory

'Memory is the way an experience at one time influences us at a future time' (Siegel, 2011, p. 148). The brain comprises of differentiated clusters of neurons that form various groupings called nuclei, parts, areas, zones, regions, circuits or hemispheres. By the firing of neurons in groups, an experience becomes encoded. 'As neurons fire together, they wire together' (Siegel, 2011, p. 40). In other words, we learn by neural firing (those neurons will most likely fire together again in the future). Repetition of this firing together, strengthens synaptic connections and stimulates the production of myelin (the fatty sheath) around axons, which increases the speed of conduction down the neurons length. Human memory can be categorized into the following systems and subsystems (in order of development):

- *Non-declarative memory*: contains implicit and non-declarative information, and memories formed following development of new sensory and motor skills, divided into six domains (perception, emotion, bodily sensations, behaviour, mental models and priming). It is characterized by unintentional learning, or learning without awareness and inability to access conscious recall. No participation of the hippocampus is required. This memory relies on brain areas such as occipital structures (visual planning) and basal ganglia (procedural memory) and can be divided into:
    - *Procedural memory*: memory underlying the acquisition of skills (such as riding a bike) and other aspects of knowledge that are not directly accessible to consciousness (simple associative learning or conditioning; non-associative forms of learning such as habituation and sensitization). Its presence can only be demonstrated indirectly by action (e.g. walking, conditional response).
    - *Perceptual representation system* (PRS): repetition priming, in which the brain readies itself to respond in a certain way, is the unconscious facilitation of performance following prior exposure to a target item or a related stimulus (e.g. visual planning of the way you walk to the toilet when you wake up in the night).

- *Declarative memory*: contains explicit and conscious information, and learned cognitive tasks (e.g. names, forms, symbols, events and words) and begins to emerge and become observed by the second birthday. It concerns learning with awareness, acquisition and retention of information about events and facts. This memory relies on temporal/diencephalic structures and can be divided into:
    - *Semantic memory*: or generic memory, refers to our general knowledge of the world (meanings of the words, concepts and symbols and their associations, as well as rules for manipulating them). Information is stored without reference to temporal and spatial context that was present at the time of its storage.
    - *Working memory*: short-term memory can be divided into primary memory (passively maintained) and working memory. The latter can be divided into a *central executive system* and at least two sub systems, one for verbal and one for nonverbal (visual) information.
    - *Episodic memory*: acquisition and retrieval of information that was acquired in a particular phase at a particular time.

Source: Larsson, 2002, and Siegel, 2011.

Memory is the process of encoding, storing (maintaining) and retrieving (locating and using) information. Learning is a change in a response to a stimulus as a result of experience, which involves three basic processes: acquisition of material, its consolidation and its retrieval from memory (Martin *et al.*, 2010): information enters short-term memory, where it is stored temporarily. If the material is rehearsed long enough, it is transferred into long-term memory (consolidation).

## Storing

It seems that three forms of memory can be distinguished (Martin *et al.*, 2010):

1 *Sensory memory*: in which representations of the physical features of a stimulus are stored for a very short time (a second or less), just long enough to transfer it to the short-term memory. In general we are not aware of our sensory memory (named iconic and echoic memory for visual and auditory stimuli, respectively) because it is difficult to distinguish from the act of perception.

2 *Short-term (and working memory)*: is the memory for stimuli that have just been perceived and stored. It has a limited capacity with regards to the number of items that it can store (7–9 items of information sensed once) and of its duration (about 20 seconds). Short-term and working memory are sometimes used interchangeably. In some definitions working memory may also involve manipulation of the information received, while short-term memory is only for storage.

3 *Long-term memory*: refers to information that is represented on a permanent or near permanent basis (with permanent physiochemical changes in the brain). It has no limits as short-term memory, and it is relatively durable. It involves stimulation of the hippocampus. Long-term memory contains *semantic memory* (conceptual information such as general knowledge; a long-term store of data, facts and information) and *episodic* or autobiographical *memory* (a clustered set of episodic memories; a record of our life experiences), which can interact.

## Learning

Several types of learning, which involve cause–effect relations between behaviour and the environment, and serve as the basis for more complex behaviours, such as problem solving and thinking, can be distinguished (Martin *et al.*, 2010; Siegel, 2011):

• *Habituation*: concerns learning not to respond to an 'unimportant' event that occurs repeatedly (the simplest form of learning) and allows organisms to respond to more important stimuli, such as those related to survival and reproduction. It is different from adaptation, which is loss of sensitivity as a result of prolonged stimulation. Sensitization is an increase in intensity of response upon repeated exposure.

• *Classical or Pavlovian conditioning*: involves *associative* learning about the conditions that predict a significant event will occur, such as flinching (automatic behaviour) by just seeing a over-inflated balloon (neutral stimulus) before the balloon bursts (an unconditioned stimulus). Some responses occur reflexively (unconditioned response), for example salivary excretion by a dog when smelling, seeing or tasting meat. When this response is *conditioned*, for example, such that the dog starts to salivate when just a bell is ringing (without even seeing or smelling the meat), the new conditioned response is caused by associative learning (each time a bell rang, the dog received meat).

- *Operant conditioning*: concerns learning through responding, operating, the environment. For example saying 'hello' after picking up the phone when the phone rings. When a particular action has good consequences the action will tend to be repeated, and when a particular action has bad consequences it will tend not to be repeated.

Classical and operant conditioning complement each other. However, classical conditioning involves a contingency between stimuli, whereas operant conditioning involves a contingency between the organism's behaviour and an appetitive or aversive stimulus. The pairings of neutral stimuli with appetitive and aversive stimuli (classical conditioning) determine which stimuli become conditioned reinforcers (an appetitive stimulus that follows an operant response and causes that response to occur more frequently in the future) and punishers (an aversive stimuli that follows an operant response and causes it to occur less frequently in the future) (Martin *et al.*, 2010).

## Retrieval

*Remembering* is an automatic process: when the appropriate stimulus occurs, it automatically evokes the appropriate response. Retrieval of implicit but also some explicit memories are automatic. *Recollection* is the active search for stimuli that will evoke the appropriate memory, assisted (or not) by contextual variables (retrieval cues), such as physical objects or suggestions (Martin *et al.*, 2010). However, much of what we recall from long-term memory may not be the same as what actually happened before. Learning and recalling are active processes: when we recall the memory, it may contain information that was not part of the original experience. False information can be incorporated into the memory, but also the context in which memory and acquisition take place can influence the recall of events. And some memories may interfere with the retrieval of others. There are two types of interference (Martin *et al.*, 2010):

- *Retroactive interference*: during retrieval other information, learned more recently, interferes. For example recalling your old telephone number because a new one has replaced it.
- *Proactive interference*: previously learned information affects the ability to recall new information.

Some studies even indicate that the explanation for *unexplained symptoms* lies in the way we process information and that this processing may be influenced by our personality, expectations and duration of a certain stress response (Matthews and Mackintosh, 1998; Brown, 2004; Pieper, 2008; Section 3.7.3).

On the other hand physiological effects may be conditioned following the same rules as for classical conditioning (Riether et al., 2008; Section 3.7.2).

## 3.7.2 Conditioning of effects

Experimental data of the last 30 years indicate that *conscious expectation* and *unconscious behavioural conditioning processes* appear to be the major neuro-biological mechanisms capable of releasing endogenous neurotransmitters and/or neurohormones that mimic the expected or conditioned pharmacological effects (Riether et al., 2008). Examples are:

- Asthmatic patients suffer from skin sensitivities to house dust and grass pollen when exposed to these allergens by inhalation: after a series of conditioning trials, they also experienced allergic attacks after inhalation of the neutral solvent used to deliver the allergens.
- Chemotherapy patients often develop anticipatory or conditioned nausea, anxiety and fatigue responses as reminders of chemotherapy, which can also be elicited by thoughts and images of chemotherapy.

Cytokines seem to play an important role during the acquisition time of behavioural conditioning. The CNS is able to detect or 'sense' changes in cytokine concentrations in the periphery, in particular changes in pro-inflammatory cytokines such as IL-1, IL-6 or tumour necrosis factor-alpha (TNT-$\alpha$). The theoretical framework of behavioural conditioning behind these effects seems to follow the same rules as classical or Pavlovian conditioning (see Section 3.7.1; see Figure 3.13).

The brain signals the immune system and vice versa via two pathways:

1 *The peripheral neural pathways*: the sympathetic nervous systems (SNS) innervate secondary limphoid organs, such as the thymus and the spleen, predominantly via noradrenergic nerve fibres, affecting circulation and activity of adrenoceptor-expressing lymphocytes. The neural pathway comprises signalling such as cytokine stimulation of the vagus nerve (see Section 2.6.1). These alterations are immediately transmitted to brain areas via the humoral pathway.

2 *The humoral pathways*: comprising activation of the hypothalamus–pituitary–adrenal (HPA) axis, inducing ACTH secretion from the adrenal cortex via CRH, resulting in elevated cortisol plasma levels (see Section 3.2.1). Alteration in plasma cortisol levels can induce tissue-specific changes in receptor expression of immune cells resulting in impaired cytokine production and gene expression. The humoral pathway implements signalling by way of peripheral cytokines crossing the blood–brain barrier (BBB) via active or passive transport mechanisms.

Based on the bi-directional interaction between the CNS and the immune system, paradigms of behavioural conditioning demonstrate that the brain is capable of detecting signals induced by a substance. The CNS associates these signals with gustatory or olfactory stimuli (CS), and upon CS re-exposure is able to suppress or stimulate immune functions in a way similar to that formerly induced by the drug or substance on demand (Riether et al., 2008).

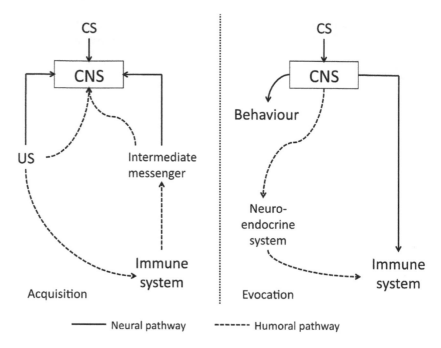

**Figure 3.13** *The theoretical framework of behavioural conditioning: at acquisition time there are two possible unconditioned stimuli (US) associated with a conditioned stimulus (CS). The US that is directly detected by the CNS is named the 'genuine' US, and the indirectly detected US is named the 'sham' US. For any US there are two possible afferent pathways to the CNS: the neural and the humoral afferent pathway. At evocation time there are two possible pathways by which the CNS can modulate immune functions: the humoral efferent pathway (changes in neurohormones that directly or indirectly modify the immune response) and the neural efferent pathway (direct innervations of primary and secondary lymphoid organs)*
Source: Adapted from Riether *et al.*, 2008, Figure 1.

Pavlovian conditioning has also been shown for fear conditioning with an olfactory, auditory or visual CS (Otto *et al.*, 2000). Odours especially adept at eliciting negative emotions in humans, such as disgust and fear, can also evoke positive emotional responses (Stevenson, 2010). It is hypothesized that individuals who suffer from IEI (environmental intolerance) or MCS (multiple chemical sensitivity) have stronger implicit associations between odours and sickness than healthy individuals. Individuals with strong-odour–illness associations interpret ambiguous/unknown odours as more threatening compared with individuals with less strong associations, or that those who have stronger odour–illness associations produce or report more adverse health effects than controls (Bulsing *et al.*, 2009).

### 3.7.3 Inappropriate selection

Cognitive processing has been proposed to comprise of a dual-route system (Matthews and Mackintosh, 1998; Brown 2004; Bulsing *et al.* 2009):

- *An automatic, involuntary route*: enabling rapid approach or avoidant responses.

- *A higher level, conscious and deliberate route*: to make conscious and deliberate evaluations.

Brown (2004) proposed that the automatic, involuntary route is controlled by the *primary attentional system* (PAS), and the deliberate, conscious route by the *secondary attentional system* (SAS) (see Figure 3.14). The *attention control* is said to be the highest control system in the brain. The modules of the system are located in the parietal and prefrontal cortices. This attentional system selects relevant information for further processing and controls action. Before attentional selection, complex perceptual and semantic analysis are performed, integrating the most active information, producing multimodal representations that allow for control of routine behaviour as well as further processing by attentional mechanisms at a higher level of the system (Brown, 2004).

There is evidence that stimuli are processed up to a high level of meaning without being attended to or becoming conscious (Taylor, J., 2006). Bulsing *et al.* (2008) showed that the dual-processing theory could work for olfactory information. Matthews and Mackintosh (1998) showed it for the appraisal of threat and Bar-Haim *et al.* (2007) for the selective processing of anxiety. With the dual-route model, Brown (2004) tried to explain unexplained symptoms: physical illness for which no adequate reason can be found, for example, symptoms of somatic illness such as pain, fatigue, general malaise. According to him unexplained symptoms arise when chronic activation of stored representations in memory causes the PAS to select inappropriate information during attentional selection and the automatic control of processing. Certain

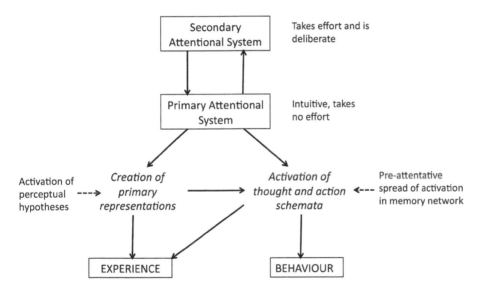

**Figure 3.14** *The dual-route system: generation of experience and control of action by the cognitive system. The primary attentional system (PAS) selects the most active perceptual hypothesis and uses it to organize relevant sensory information into integrated multimodal perceptual units: the primary representations, which provide a working account of the environment for the control of action*

Source: adapted from Brown, 2004, Figure 2.

symptoms, such as those characterized by alterations in experience (e.g. pain, feelings of discomfort, nausea, pseudo-hallucinations), arise when inappropriate (i.e. inconsistent with information provided by the senses) perceptual hypotheses are selected during the creation of primary representations by the PAS. All people possess material about symptoms that could provide the basis for these so-called 'rogue' representations (Brown, 2004). For example, physical components of emotional states (e.g. sweating and headache) leave representations in memory that could provide the basis for the development of unexplained symptoms later in time. Matthews and Mackintosh (1998) even argue that the primary attentional route gives automatic attentional priority to cues that have been associated with a threat. Symptoms characterized by an inability to control perception, cognition or action arise when inappropriate cognitive schemata are, automatically, triggered by the PAS selection.

*Self-focused attention* (as a feature of depression and anxiety) and boring environments seem important factors in this PAS selection process. Negative affect (see Section 3.2.4) influences possibly the actual encoding and storage of rogue presentations directly, but it is also possible that symptoms are acquired as the result of a spontaneous narrowing of attention occurring as a biological response to extreme anxiety of fear (Brown, 2004). Both negative affect and the process of symptom misattribution trigger *illness worry and rumination*, which may play a particularly important role in the development of symptom chronicity (Brown, 2004). Pieper (2008, p. 45) named this process *preserverative cognition*: 'The repeated or chronic activation of the cognitive representation of stress-related content.'

## 3.8 Personal factors

*Personal factors are factors that are related to the person and that can influence the way that environmental stressors are handled. Several types shown to affect that handling can be distinguished:*

- Demography: *demographic variables such as age, gender, ethnicity, social status, income and education.*
- States and traits: *personality in the moment (motivation and emotion) and over a longer time.*
- Life style and health status: *life style related variables such as food pattern and physical activity, and health state variables such as allergies and obesity.*
- Genetics, events and exposures: *such as the risk for certain diseases and disorders, premature birth and low birth weight, substance abuse and attitudes.*

### 3.8.1 Demography

For the demographic variables age and sex, the influence on the effects of exposure to physical stressors have been most investigated. Self-reported annoyance/discomfort has been found to be influenced by age, gender, ethnicity and social status (education and income) (Janssen and van Dongen, 2007). The demographic factors age and socioeconomic status (SES), but also health status and genetic polymorphisms, have been related to an increased risk of PM-related health effects (Sacks *et al.*, 2011).

## Age

Degradation of the eyes, ears and olfactory bulb usually occurs with age. For instance for vision and the related biological clock effects of light, the blue part (wavelength of 450 nm) is exactly the part that is less perceived with age. This is due to the decrement of the pupil size and to the decreasing transparency of the lens and the vitreous body. Degradation of functions of the immune system (Kapit *et al.*, 2000) and hormone secretion of the endocrine system (e.g. oestrogen, testosterone, growth hormone (GH) and dehydroepiandrosterone (DHEA)) also increases with age (Hinson *et al.*, 2010). Heart rate also decreases with age (Antelmi *et al.*, 2004). It has been observed that, compared to young people, the elderly have impaired (Giannattasio *et al.*, 1994):

- Baro-receptor control of the heart rate and cardiac function.
- Baro-receptor modulation of the sympathetic drive to the peripheral circulation.
- Cardiopulmonary stretch receptors, which inhibit sympathetic tone, the renal release of rennin and vasopressin secretion.

These factors may explain at least in part the raised blood pressure and sympathetic activity in the elderly and the reduced ability of elderly people to maintain blood pressure and blood volume and their increased blood pressure variability over 24 hours (Giannattasio *et al.*, 1994). A decrease in glucose metabolism with age is associated with changes in transmitter release and an increase in free radical production, which add to cellular ageing and therefore susceptibility to toxic stress is increased (Hummel *et al.*, 2002).

Several studies have reported lower prevalence of SBS at higher age, some higher and some no relation between age and SBS (Runeson *et al.*, 2006). Burge *et al.* (1987) found that those between 21 and 40 years of age reported more symptoms than either younger or older individuals. However, the evidence in the literature is not consistent with respect to SBS and age (Runeson *et al.*, 2006).

Although Fanger's work (1982) has indicated that age has no effect on thermal dissatisfaction, Choi *et al.* (2010) found that occupants of office buildings over 40 years old are more satisfied than those under 40 years old in the cooling season. Even though metabolism slows down with age, due to loss in muscle tissue and neurological changes, a decrease in thermal sensitivity with age may be more important (Tochihara *et al.*, 1993). Elderly people decrease their ability of cutaneous vasodilatation and vasoconstriction in respectively hot and cold environments. Ageing is associated with (Frank *et al.*, 2000):

- A decreased core temperature threshold (temperature at which the thermoregulatory response is initiated), a decreased maximum response intensity for vasoconstriction (magnitude of response during a given thermal challenge), a decreased total body oxygen consumption and decreased norepinephrine release.
- A decreased vasomotor responsiveness to norepinephrine.
- A decreased subjective sensory thermal perception.

Ages below 30 and ages above 50 seem to relate to less annoyance of noise and odour (Janssen and van Dongen, 2007). Ageing is accompanied by a decrease

---

**Box 3.15 Alteration of neurons in the ageing brain**

- Deficits in the regulation of intracellular calcium levels.
- Increased leakage of synaptic transmitters.
- Changes in neuronal arborisation.
- Decreased amplitude and prolonged excitation of calcium potentials in neurons.
- Loss of plasticity.

Source: Hummel *et al.*, 2002.

---

in intranasal chemosensory sensitivity, decreased trigeminal sensitivity, decreased ability to identify odorants, as well as greater tendencies for olfactory adaptation and slower recovery of threshold sensitivity (Hummel *et al.*, 2002). Olfactory receptor neurons may be subject to alterations similar to those seen in neurons in the ageing brain (see Box 3.15). In some people, functional loss of audition, balance and olfaction may be due to nerve compression caused by narrowing of the holes of the cribriform plate and the bony passage of the acoustic and vestibular nerves with age (Hummel *et al.*, 2002). During adolescence, the period of physical and psychological development between childhood and adulthood starting with the onset of puberty (see Box 3.16), the structure and function of the brain is changing as well, resulting in altered sensory associations, for example, to smell or sight. This process usually begins between ages 8 and 14 years in females (mean age 11), and between ages 9 and 15 in males (mean age 12) (Blakemore *et al.*, 2010).

## Gender

Although there are biological differences due to sex, according to deFur *et al.* (2007) many differences that affect health are socially rather than biologically mediated. Adult women tend to show less physiological reactivity to stress than men. Among children, boys prior to puberty are generally more vulnerable to depression and psychomatic symptoms in response to stressors, while girls are generally more vulnerable after puberty. However, in a study with adolescent school children it seemed that higher endogenous estrogen levels of sexually matured adolescent females (menstruating girls) protected them from the effects of moulds in dust, despite their overall higher symptom prevalence (Meyer *et al.*, 2005).

In questionnaire surveys of indoor environmental quality, the reported symptoms prevalence rate is generally larger in women than in men (e.g. Skov *et al.*, 1987; Bluyssen *et al.*, 1996a; Runeson *et al.*, 2006; Marmot *et al.*, 2006; Bluyssen *et al.*, 2011a). In several studies on the annoyance from outdoor environmental sources, women were more annoyed with odour and men were more annoyed with light, while with noise no difference was found (Janssen and van Dongen, 2007). In an investigation of 402 workers in 20 office buildings (10–15% of population) on satisfaction with the thermal environment, it was

## Box 3.16 Puberty, reproduction and andro/menopause

Puberty

- *Male*: Increasing levels of testosterone result in physical changes (enhanced skeletal growth, wide shoulders and narrow pelvis, enlarged larynx and vocal cords (low-pitched voice), facial and body hair, pubic and auxiliary hair, and receding scalp hairlines and baldness (if genetically susceptible)) and may also result in psychological changes such as active and aggressive attitudes and independence.
- *Female*: Increasing levels of oestrogen result in physical changes (narrow shoulders, wide pelvis, high-pitched voice, non-receding hairlines and soft skin, auxiliary and pubic hair, soft or no facial and body hair, subcutaneous body and deep body fat deposits, which underlie the shape of breasts, buttocks and thighs).

Reproduction

- *Male*: The steady secretion of testosterone maintains spermatogenesis and the secretory functions of the accessory sex organs and glands (regulated by pituitary LH and FSH), maintains male secondary sex characteristics, including muscle and bone mass, and promotes sex drive (libido) and other brain and mental effects. Testosterone is derived from cholesterol.
- *Female*: Oestrogen and progesterone, both derived from cholesterol and regulated by pituitary LH and FSH, are important for the menstrual cycle as well as pregnancy processes. Premenstrual syndrome (PMS) is a collection of physical and emotional symptoms related to the menstrual cycle, such as irritability, tension, unhappiness, stress, anxiety, headache, fatigue, mood swings, increased emotional sensitivity, change in libido, bloating, abdominal cramps, constipation and joint or muscle pain.

Andro/menopause

- *Male*: In men there is an age-related decline in testosterone secretion, named andropause, and some men experience hypogonadal symptoms associated with this, such as poor libido, fatigue, muscle loss, increasing abdominal fat, glucose intolerance, high cholesterol, poor sleep, difficulty concentrating, depression, anxiety and loss of bone mass.
- *Female*: Women go through the menopause at around the age of 50 years and may experience adverse effects of the associated decrease in oestrogen. After menopause, plasma oestrogen is deficient, which has an effect on ageing disorders such as heart attacks due to coronary occlusion and abnormal cholesterol metabolism, osteoporosis and bone fractures. Other effects of low oestrogen may include hot flashes, poor libido, loss of body hair and sleep disturbances.

Source: Kapit *et al.*, 2000; Hinson *et al.*, 2010.

concluded that females are more dissatisfied with their thermal environments than males, especially in the summer season (Choi *et al.*, 2010).

Besides from the physical and physiological differences related to certain periods in life (see Box 3.16) there are also other differences between men and women. For example, females usually have higher heart rates than males (Bjerregaard, 1983; Antelmi *et al.*, 2004).

## Ethnicity

While ethnicity can be seen as a genetic component of heritable physical traits, the social constructs of race can be the source of many of the stress-related factors (deFur *et al.*, 2007). In fact, racial groups represent societally defined categories during a particular point in history and place, while genetically identified groups tend to correlate poorly with socially identified groups because there is more genetic variation within than between groups (Gee and Payne-Sturges, 2004). The racial and socioeconomic composition of communities, predict a broad range of characteristics including housing, transportation, school, occupational structure and more (deFur *et al.*, 2007).

### Social status

Education, occupation and income are mostly used to assess SES. Education relates to social status in early life; occupational status to present state and income describes availability of material resources (Kristenson *et al.*, 2004). Employment grade is used as a measure of individual socioeconomic position, because lower grades command lower salaries, which may lead to more reporting of symptoms (Marmot *et al.*, 2006). Workers are often categorized into three groups: administrative/clerical/secretarial, professional and managerial/executive. The managerial/executive profession in general influences comfort positively (Bluyssen *et al.*, 2011a).

There is a strong relation between the proximity to emission sources, such as hazardous waste and large industrial facilities, and SES (deFur *et al.*, 2007). Low SES could be related to higher psychosocial stress (e.g. Kristenson *et al.*, 2004), and has been linked to higher incidence rates of specific conditions (myocardial infarction) and higher rates of mortality (e.g. van Lenthe *et al.*, 2004). Individuals with a low SES have been reported to be more susceptible to the effects of PM exposure (Sacks *et al.*, 2011). High SES might even be protective against the effects of increasing age on cardiovascular condition (Steptoe *et al.*, 2005). Individuals with low social status report more environmental challenges and less psychobiological resources, which may lead to vicious circles of learning to expect negative outcomes, loss of coping ability, strain, hopelessness and chronic stress (Kristenson *et al.*, 2004; see also Section 3.8.2). Individuals in lower classes tend to have more external locus of control, while people in higher social classes have more internal locus of control.

### 3.8.2 States and traits

Our emotional responses and evaluations of our environment are influenced by our personal traits and state of mind (our motivation and emotions) (Burke *et al.*, 1993; Kuhbander *et al.*, 2009). Personality traits are personal characteristics of the psychological kind, which are stable for a longer time. One's mood of the moment is a state. With *traits* the interest lies in *how respondents 'generally feel'*, while with *states* we are concerned with *how respondents feel 'right now, at this moment'*.

The trait 'anxiety' can be defined as a reflection of frequent past experiences of state 'anxiety', which increases an individual's proneness or sensitivity to experience future state anxiety by interacting with the cognitive appraisal of threatening internal or external stimuli. For example, a state of anxiety is caused when internal (e.g. muscular or visceral activity) or external stimuli (e.g. threat of shock) are cognitively appraised as threatening. Cognitive and behavioural defence processes are activated to combat the anxiety (Grös *et al.*, 2008). Negative mood predominantly induces stimulus-driven processing and positive mood predominantly knowledge-driven processing. This implies that the way the respondent fills in the questionnaire may be affected by their trait and state (Kuhbander *et al.*, 2009).

## Motivation and emotion

Motivation can be defined as a driving force that moves us to a particular action. *Biological needs* (e.g. food, water, heat) can be potent motivators, but also experiences we want to repeat are those that increase, rather than decrease, our level of arousal (all reinforcing stimuli appear to trigger the release of dopamine in the brain). When an organism's behaviour is no longer reinforced, the behaviour eventually ceases or extinguishes. Intermittent reinforcement leads to *perseverance*, even when the behaviour is no longer reinforced, and *conditioned reinforcement* occurs when a person has been regularly exposed to particular stimuli in association with reinforcers (that person's behaviour can be reinforced by those stimuli; Martin *et al.*, 2010).

Psychologists have defined emotion in various ways (see Box 3.5). Martin *et al.* (2010) describe emotions as an expression of behaviours, physiological responses and feelings evoked by appetitive or aversive stimuli, in which both automatic conscious processes play a role. A number of theorists have suggested that there is a group of basic emotions, although the exact number is controversial, as is the notion that there are basic emotions. Mehrabian (1996) even suggested that emotional states can be described by the pleasure–arousal–dominance (PAD) model. For both state and traits the following groups of emotions have been pointed out (see also Figure 3.15):

- Worry, nervousness, fear and anxiety.
- Anger, hostility and aggressiveness.
- Sadness, depression.
- Happiness, satisfaction, joy, ecstasy.

Additional emotions such as disgust/loathing, acceptance/trust, expectancy/anticipation and surprise/astonishment have been introduced (Martin *et al.*, 2010). Additional traits for personality terms have also been used, for example, negative and positive affect; introversion/extraversion; neuroticism, psychoticism; coping skills; self-efficacy and locus of control; intelligence and interest (see Section 6.5.3).

## Personality

According to some psychologists, personality is a set of personal characteristics, so-called *personality traits*, that

**Figure 3.15** *Expression of several emotions*
Source: drawing by Anthony Meertins (12 years old, November, 2011).

determine the different ways we act and react in a variety of situations (Martin *et al.*, 2010). According to Diener (2000) personality is undoubtedly an important contributor to long-term levels of wellbeing, but it is an exaggeration to conclude that circumstances have no influence. People's set-points appear to move up or down, depending on the favourability of long-term circumstances in their lives. Zuckerman (2005) compared several systems of personality structure and identified four major trait dimensions in all of them, of which the first three represent the trait dimensions in the so-called three-factor model of Eysenck (Eysenck and Eysenck, 1985):

- *Extraversion/sociability* versus introversion: outgoing nature and a high level of activity versus a nature that shuns crowds and prefers solitary activities.
- *Neuroticism/anxiety* (or negative effect) versus emotional stability: a nature full of anxiety, worries and guilt versus a nature that is relaxed and at peace with itself.
- *Psychoticism/constraint* versus impulsive sensation seeking/self-control: an aggressive, egocentric and antisocial nature (not a mental illness) versus a kind of considerate nature, obedient of rules and laws.
- *Aggression/hostility* versus agreeableness/cooperativeness.

Another model is the big five framework, which is a hierarchical model of personality traits with broad bipolar factors each summarizing several more specific facets (extraversion; agreeableness; conscientiousness; neuroticism and openness; Goldberg, 1992). Most of the basic personality traits are closely related with emotions: sociability and sensation seeking with positive affects, neuroticism with fear and depression, and aggression with anger (Zuckerman, 2005). Results from meta-analyses suggest that personality traits are not fixed and unchallengeable but are subject to change and evolution, even though this change is in general not pronounced (a 35-year old introvert will most likely not become an uncontrollable 55-year old extrovert, or vice versa; Martin *et al.*, 2010).

Believers of the *social learning approach*, however, claim that personality is determined by both the consequences of behaviour and our perception of them (beliefs or expectancy that an individual has about those consequences). For example, Bandura (1982) introduced the term *self-efficacy* (the expectation of one's own competencies; one's ability of performing a certain action) as an important determinant of whether people will attempt to make changes to one's environment, how much effort they will spend and how they will persist when confronted with obstacles or negative experiences. People with low self-efficacy tend not to try to make changes to their environments, while the opposite is true for people with high self-efficacy.

Additionally, Rotter's research showed that *locus of control* (the extent to which people believe that they can control events that affect them) is also an important determinant of personality. A person who expects to control their own fate, or who perceives that rewards depend upon their own behaviour, has an internal locus of control (Rotter, 1966). People with an internal locus of control will work harder to obtain a goal if they believe that they can control the outcome in a specific situation. They are more likely to be aware of and to use good health practices. But they are more likely to blame themselves when they fail, even when failure is not their fault.

Acquired (learned) expectancies determine how an individual responds to challenges (see also Section 3.7.2). In the case when the relation between exposure and outcome is (Kristenson *et al.*, 2004):

- *Negative*: the individual stores this experience as negative outcome expectancy, and feels 'hopeless'.
- *Non-existing*: the individual develops 'helplessness'.
- *Positive*: the individual learns success (coping); coping has been represented by other terms, such as mastery, sense of coherence, perceived control and self-esteem.

While in the *psychodynamic approach* (mainly represented by Sigmund Freud) it is said that what we do is often irrational and that the reasons for our behaviour are seldom conscious. The *humanistic approach* attempts to understand personality and its development by focusing on the positive side of human nature and people's attempt to reach their full potential: *self-actualisation* (Martin *et al.*, 2010). According to Maslow, one of the humanists, in order to achieve this goal, individuals must first satisfy several basic needs (see Section 6.2.1 of Bluyssen, 2009a).

## Traits and effects

Berglund and Gunnarsson (2000) studied the role of personality (traits) in reporting Sick Building Syndrome (SBS) symptoms and found a significant influence of person-related factors in SBS. In a recent analysis of data from the US BASE-study, multiple personal factors correlated strongly with health and comfort symptoms (Mendell and Mirer, 2009) indicating that statistical control for those factors is required to evaluate possible relationships with other variables.

### Worry, nervousness, fear and anxiety

Pieper (2008) showed that the trait *worry* can have significant immediate and prolonged cardiac effects, even during periods of worry in the absence of a stressor. These cardiac effects were most pronounced for work-related worry and worry about anticipated future stress. Cardiac effects occur during concurrent stressful events, but are also effected by stressors before and by stressors that are anticipated. Worry about the future concerns fear, while worry about the past mostly concerns regret and sadness. Fear or concern has been shown to be related to risk perception (Slovic, 2010). For example, in outdoor environmental studies *fear*, or concern about health, was found to be an important factor with noise, odour and light annoyance (Janssen and van Dongen, 2007).

### Anger, hostility and aggressiveness

*Hostility* (the trait) has been proposed as a marker for psychosocial vulnerability (Kubzansky *et al.*, 2000). Hostility was linked to health outcomes such as the onset of hypertension and coronary heart disease. Hostility has been linked to lower education, occupational rank and income, and it may be that continual confrontation with difficult life circumstances associated with low SES leads to the development of a hostile response pattern. *General hostility*, the combined measure of cognitive, behavioural and emotional aspects of hostility, was strongly associated with allostatic load (when adjusted for age, smoking, alcohol consumption and physical activity). In general hostility increases as a result of frustration when coping is no longer possible (Kristenson *et al.*, 2004).

### Happiness, satisfaction, joy, ecstasy

Research has shown that those who are outwardly *happy* are most likely to have high *self-esteem* and are happy in other aspects of their lives (Diener and Seligman, 2002; Fowler and Christakis, 2009). People who are surrounded by many happy people and those who are central in the network are more likely to become happy in the future (Fowler and Christakis, 2009). Diener (2000) names the total package 'happiness, or subjective wellbeing' (SWB) defined as a person's evaluative reaction to his or her life, either in terms of life satisfaction (cognitive evaluations), or affect (on-going emotional reactions), or cognitive (evaluation of one's life according to subjectively determined standards) and hedonic balance (the balance between pleasant affect and unpleasant affect) (Schimmack, 2006).

### Positive and negative affect

Increased *positive affect* is associated with reduced risk of mortality and reduced risk of physical disease (Ostir *et al.*, 2000; Pressman and Cohen, 2005). High positive affect may act to promote health life style such as increased physical activity and motivation for self-care, and may act to promote one's social support, which beneficially protects health (Ostir *et al.*, 2000). A higher negative affectivity score has been related to an increase in perceived symptoms (Lang *et al.*, 2008). Watson and Pennebaker (1989) have shown that persons with high negative affectivity, experience discomfort under all circumstances, even in the absence of a clear agent that may cause discomfort.

Pressman and Cohen (2005) discuss the dependency of positive and negative affect: if they were bipolar ends of the same construct, benefits of positive affect may merely reflect the absence of negative affect rather than the presence of positive feelings. While if they were independent, positive affect provide would benefit independent of negative affect levels. The dynamic equilibrium model of Headey and Wearing (1992) proposes that people maintain both levels of pleasant and unpleasant affect, which are determined by their personalities: advantageous and disadvantageous events move individuals temporarily away from their personal baselines, but over time they return to them (adaptation). Separate baselines for positive and negative affect are determined by personality tendencies to extraversion and neuroticism, respectively.

### Interest

According to Sommers and Vodanovich (2000) *boredom proneness* may be an important element to consider when assessing symptom reporting. Significant positive correlations have been found with negative effect, such as anxiety, depression, anger, hostility, hopelessness and life dissatisfaction. However, this is also related to impulsivity, procrastination, poor psychosocial development, narcissism and alienation, and with distractibility, low attention control and concentration difficulties (Farmer and Sundberg, 1986).

## Intelligence

Intelligence describes a person's ability to learn and remember information, to recognize concepts and their relations, and to apply the information to their own behaviour in an adaptive way. Intelligence has a hereditary as well as environmental component. Intelligence is often represented by a single score (the intelligent quotient or IQ) (see Figure 3.16), but modern investigators do acknowledge the existence of specific abilities (Martin *et al.*, 2010). Biological and environmental factors can affect intellectual abilities prenatal and postnatal. Factors that impair brain development will necessarily also impair the child's potential intelligence. Studies have shown that people with low IQ are more likely to die earlier than those with higher IQ, whatever social economic class they belong to (Batty *et al.*, 2007; Deary *et al.*, 2008). There may be a relationship between genetics, intelligence and longevity. This relationship may be mediated by socioeconomic status and engaging in healthy behaviour.

### 3.8.3 Life style and health status

Both *life-style-related variables* such as food pattern (diet/nutritional status), drugs (ab)use (smoking, coffee, alcohol, medications) and physical activity (exercise regime), and *health state variables* such as allergies, asthma and obesity (BMI index), are important influencing factors on how we cope with stress (Pieper, 2008) (see also Box 3.17). Pharmacological treatments, such as beta-blockers (a cardio-protective drug), may modify the observed effects (de Hartog *et al.*, 2009; Barclay *et al.*, 2009). Pre-existing diseases may influence susceptibility.

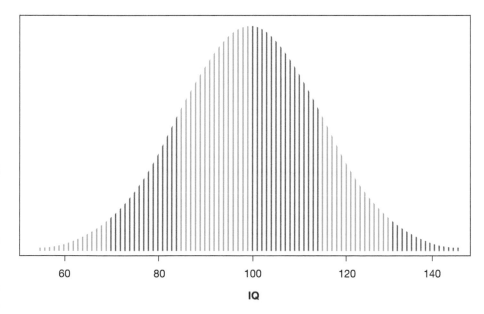

**Figure 3.16** *A normal distribution of IQ scores: the average or mean score is 100, very bright people score around 130 and scores can be higher than 145 (according to the Weschler Intelligence scale, most widely used intelligence test) (Martin et al., 2010)*

Source: Alessio Damato, Mikhail Ryazanov, Wikimedia Commons, http://commons.wikimedia.org/wiki/File:IQ_curve.png (accessed 2 June 2013).

For example, pre-existing cardiopulmonary diseases, as well as diabetes, may increase susceptibility to effects of PM exposure (Sacks *et al.*, 2011).

People with low SES seem to have more unhealthy life styles, for example, they eat more unhealthy food and have more sedentary life styles. People who are depressed or have a low sense of control, tend to drink more alcohol, eat unhealthily and perform less physical activity. Additionally, repeated negative experiences of stressful situations and the use of an unhealthy life style interact and may lead to cumulative effects (Kristenson *et al.*, 2004).

Bio-behavioural variables that activate physiology, such as physical activity and caffeine consumption, are important to consider (Pieper, 2008). Caffeine, a minor stimulant, is present in coffee, tea, soda, several medications and sleep prevention pills. Caffeine can increase blood pressure and enhance cardiovascular activity. The long-term effects appear to be minimal (Green *et al.*, 1996; see also Box 3.17).

Low levels of alcohol consumption (acute intoxication) have been associated with temporary relief of anxiety. On the other hand, chronic alcohol use may promote anxiety disorder and the other way round. Either anxiety disorder or alcohol dependence can serve as a causal stimulus for the other. Anxiety disorder (comorbidity) can contribute to the persistence of pathological alcohol use, and

---

## Box 3.17 Life style: caffeine, chocolate and wine, healthy or not?

### Caffeine

There is growing evidence that a lifetime of habitual caffeine consumption (at least one cup daily) has positive effects on physical and cognitive health. Caffeine acts as a diuretic and is a vasoconstrictor. At cellular level caffeine acts as a neuromodulator: its pre-synaptic action modifies/inhibits the release of neurotransmitters (e.g. it blocks receptors for adenosine which results in CNS stimulation, which is accompanied by the release of several neurotransmitters including noradrenaline, acetylcholine and dopamine). A low dose of caffeine is associated with increased alertness, vigilance and decreased fatigue. High doses can cause anxiety, insomnia and increased sleep latency (the time it takes to fall asleep). Moderate caffeine consumption does not act on the areas of the brain related to addiction, which indicates that typical daily levels of consumption will have a low addictive potential (Thompson and Keene, 2004).

### Chocolate

Besides the pleasure effect of chocolate, chocolate comprises cocoa flavanols (sub-family of flavonoids), which have been demonstrated to be efficient at improving endothelial function and decreasing blood pressure. Flavanols seem to define superoxide anion production and then establish optimal nitric oxide levels and blood pressure (Fraga *et al.*, 2011). However, for the prevention and/or treatment of elevated blood pressure, well-established diet-related life style modifications should be recommended and implemented, such as weight loss in overweight obese patients, reduced salt intake and/or consumption of a diet rich in fruit and vegetables, rich in low-fat diary products and reduced in saturated fatty acids and cholesterol (Rimbach *et al.*, 2011).

### Wine

Numerous epidemiological data suggests that moderate alcohol use (one to two drinks daily for men, one drink daily for women and the elderly) is associated with a beneficial effect on cardiovascular morbidity and mortality. Red wine, specifically, seems to confer a greater benefit than other alcoholic beverages, which may be related to its high concentration of polyphenols (type of flavonoids) (Lindberg and Ezra, 2008).

relapse to drinking among those with both disorders (Kushner *et al.*, 2000; see also Box 3.17).

Besides the addicting effect of cigarette smoking (which stems from nicotine), it has been shown that cigarette smoking exerts an appetite-suppressant effect, an effective means of weight control, particularly applied by women (Freathy *et al.*, 2011). Smokers in general have a lower body-mass index (BMI) than non-smokers.

## 3.8.4 Genetics, events and exposures

They way we experience and process information can lead to unconscious behavioural responses and sometimes unwanted learning effects (psychological and physiological) that influence the production of cytokines and hormones (see Section 3.7.2 and 3.7.3). Past exposures and experiences as well as future anticipations but also genetics seem important factors with these processes.

### Heredity and genetics

The genotype is the genetic makeup of a cell, an organism or an individual. Genes comprise the instructions for the synthesis of protein molecules, the components that are important for the development of the body and all its processes. Genes are found on chromosomes, which consist of DNA and are found in every cell (see Section 2.2.2). Humans inherit from each parent 23 chromosomes, each of which contains thousands of genes, resulting in a recombination of genetic instructions and an enormous genetic diversity, which is required to survive. Alternative forms of genes are named *alleles*. If parents each contribute the same allele for eye colour, the gene combination is called *homozygous*. If they contribute different alleles, the gene combination is *heterozygous*. Brown eye colour is a dominant trait, blue eye colour is a recessive trait. This means that when one allele for eye colour is brown and the other is blue, the child will have brown eyes. The *heritability*, a statistical measure between 0 and 1, that expresses the proportion of the absolute variability in a trait that is a direct result of genetic variability, of many physical traits in most cultures is very high (almost 1), for example, eye colour (Martin *et al.*, 2010).

Variability in all physical traits but also behaviour is determined by a certain amount of genetic variability, environmental variability and an interaction between genetic and environmental factors. For example, the ability to run (running speed) is the joint product of genetic factors that produce proteins for muscle, bone, blood, oxygen metabolism and motor coordination and environmental factors such as exercise patterns, age, nutrition and injuries (Martin *et al.*, 2010).

Genetic factors can also modulate the way we respond to certain exposures, such as ambient PM exposure. It has been found that the presence of null alleles or specific polymorphisms (more than one form) in genes that mediate the antioxidant response, regulate folate metabolism, or regulate levels of fibrinogen may increase susceptibility to PM-related health effects. In some cases, genetic polymorphism may confer protective effects (Sacks *et al.*, 2011).

The sequence of our genome influences not only how we respond to our environment, but also our tendency to seek or avoid the environment. Thorgeirsson *et al.* (2008) found a sequence variant associated with nicotine dependence. This variant did not influence smoking initiation, but carriers of the variant, among smokers, smoked more than non-carriers and did have higher rates of nicotine dependence.

## Past exposures and events

The processing of learning already starts just after we are born and even begins in the womb. Breast-feeding may affect the risk of developing allergies during childhood, in a preventive way, but also in a negative way. Some studies have suggested that breast-feeding may cause an increased risk because of the presence of immuno-toxicants in the milk (van Odijk *et al.*, 2003). Grandjean *et al.* (2010) go a step further back and suggest that immuno-toxicants present in the cord blood causing prenatal exposure may affect sensitization and development of allergic diseases. Developmental immuno-toxicity may predispose children to common diseases of increasing prevalence, such as childhood asthma and allergic diseases. These findings suggest that even the exposure of mothers during pregnancy (exposure in the womb) might be important to consider, but also stressful events, such as being born, can take its toll. The capacity of the brain to change is called *neuroplasticity*. Early experiences can change the long-term genetic regulation within the nuclei of neurons through a process called epigenesis (see Section 3.6.2). Experience creates the repeated neural firing that can lead to gene expression, protein production and change in both the genetic regulation of neurons and the structural connections in the brain. If early experiences are negative, it has been shown that alterations in the control of genes influencing the stress response may diminish resilence in children and compromise the ability to adjust to stressful events in the future (Siegel, 2011).

## Anticipated exposures and events

Many animal species have nervous systems that enable anticipation of events, for example, learning that a flashing light is associated with a reward in a conditioned learning experiment (see Section 3.7.2). But planning for the future seems to be a human prefrontal activity, also named the seventh sense, which allows us to perceive the mind and create representations of time, but also anticipation, worry, rumination and adaptive behaviour, from avoidance of feelings momentarily to long-term shut-offs, or shutdowns (Siegel, 2011).

## Addiction

In the same way that worry about anticipated future stress may influence one's health, seeking too much satisfaction can do that as well. All addictive behaviour involves activation of the dopamine system (Siegel, 2011). Binding of dopamine by receptors in the nucleus accumbens (in the forebrain) and prefrontal cortex causes signals that are experienced as 'satisfaction' (see Box 3.18). Dopamine is released when a certain behaviour needs rewarding. Drugs that enhance the sense of wellbeing, such as morphine and heroin, cocaine, amphetamine, nicotine and ethanol, increase the availability of dopamine in the synapses. Unfortunately, satisfaction cannot be stored: when a signal is prolonged, desensitization occurs. The more often satisfaction is craved for, because of desensitization, the less

---

### Box 3.18 Satisfaction

Our regulatory systems were selected to seek satisfactions that are small and brief. The best satisfaction occurs when our experience exceeds prior expectations, for example, when it contrasts to prior discomfort. Because of this, when an experience is sustained, we soon adapt, and it ceases to satisfy. . . . Adaptation refers simply to the resetting of response sensitivity to a signal.

Source: Peter Sterling, 2004, p. 50.

---

satisfaction will be delivered because the input–output curve is adjusted. More is needed to get stimulated. This 'reward-circuit' can mediate addiction to essentially any source of satisfaction (Sterling, 2004).

## Attitudes

Attitudes are relatively enduring sets of beliefs, feelings and intentions towards an object, person, event or symbol. Attitudes may be learned through mere exposure to the object of attitude, classical conditioning processes and imitation. Attitudes seem to have three components (Martin *et al.*, 2010):

- *Affect*: feelings that an attitude arouses, such as disgust with people from a certain group or exhilaration from dancing on a particular song. These feelings are strongly influenced by direct classical conditioning.
- *Behavioural intention*: an intention to act in a particular way with respect to a particular object, for example, a negative attitude towards smoking can cause the intention not to smoke.
- *Cognition*: a set of beliefs about a person, an activity or physical object, which we acquire through hearing or reading about it directly or which are formed throughout our lives.

The effect of several attitudes towards a certain stressor or source of exposure (e.g. odour, noise, light) on annoyance with the source have been studied (Fields, 1993; Miedema and Vos, 2003; Janssen and van Dongen, 2007; Bluyssen *et al.*, 2011a):

- *Fear*: a fear or concern about health effects of exposure.
- *Preventability and controllability*: a belief that the source could be prevented or should be controlled.
- *Other-stressor annoyance*: an awareness of other stressors related to the same source (e.g. odour from a noise source).
- *Sensitivity*: a general sensitivity with the stressor.
- *Importance or relevance*: a belief that the noise source is not important.
- *Expectation*: a belief that the situation will worsen in future.

# PART II

# Assessment

# 4

# Indicators

*In this chapter a synthesis is presented of methods and techniques available to gather the information required to assess indoor environment quality (IEQ) and explain responses by the human systems. Parameters or indicators used to explain these responses are discussed. For different fields (epidemiology, psychology, clinical medicine, toxicology and product design) research methods are presented in Chapter 5. Chapter 6 shows an overview of collection/sampling techniques that can be applied in those fields.*

## 4.1 Introduction

The human model description in Part I makes clear that all relevant stressors and factors of influence that may affect wellbeing can be potentially important to consider when an attempt is made to pinpoint the effects caused by different stressors (or combination of stressors) in an IEQ field investigation. In Figure 4.1 a compilation of these stressors and factors is presented.

Assessment of occupants' wellbeing is and can be performed by different methods, different collection techniques and different disciplines (e.g. chemistry, physics, biology, physiology, mathematics) in different fields. In *architecture and building engineering* the focus lies on the design, construction and maintenance of healthy and comfortable *buildings and systems* for people, and involves measurement of environmental parameters and characteristics of the built environment using technical instruments as well as occupants of buildings (both subjective and objective assessments).

In IEQ field investigations in general the main focus lies on finding associations between *physical stressors* and *perceived wellbeing*, both in the so-called 'indoor air and thermal comfort community' (Burge *et al.*, 1987; Skov *et al.*, 1987; Preller *et al.*, 1990; Bluyssen *et al.*, 1992; Hedge *et al.*, 1992; Bluyssen *et al.*, 1996a; Caccavelli *et al.*, 2000; Cox, 2005) and in the field of post-occupancy evaluation (POE) (Vischer, 1989; Leaman, 1996; Veitch *et al.*, 2002; Vischer, 2008). In these studies, questionnaires recording levels of dissatisfaction (perceived comfort), Sick Building Syndrome symptoms (perceived health) and sometimes also self-reported productivity are gathered mainly in relation to indoor environmental aspects such as air quality, thermal comfort, light and noise aspects that are monitored either directly via chemical/physical measurements or indirectly via checklists.

Unfortunately, associations found are not clear among other reasons due to the complexity of the relationships between indoor building conditions and wellbeing (health and comfort) of occupants (e.g. Bluyssen *et al.*, 1996a; Jantunen

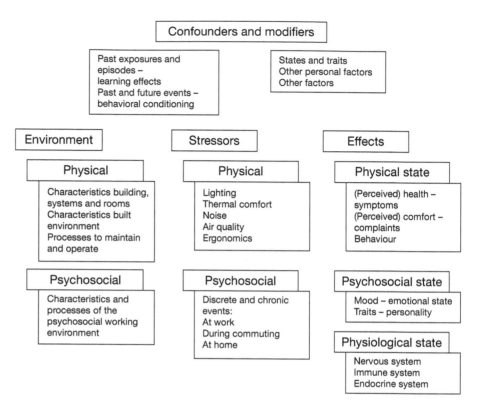

**Figure 4.1** *Stressors, factors, causes and effects*
Source: Bluyssen.

*et al.*, 1998; Apte *et al.*, 2000; WHO, 2003; Bonnefoy *et al.*, 2004). Next to better understanding of the mechanisms behind how and why people respond to external stressors (treated in Part I of this publication), there seems to be a need to improve assessment procedures.

In Part II, 'Assessment', the following question is therefore tackled:

> How can health and comfort (wellbeing) of people in the indoor environment be evaluated?

From different disciplines, assessment procedures are inventoried and discussed (Chapter 5) next to the different techniques that are applied (Chapter 6) to study the human mind and body as well the interactions between them (see Box 4.1).

## 4.2 Assessment protocols

At the start of a study, it is important to have a clear idea about which measures (variables and/or constructs) need to be assessed, which study design is chosen, which methods of sampling will be applied and how measurement of exposure and outcome measures will be performed. The type and size of study samples, the sampling scheme and instruments to be used, and methods for the laboratory

---

## Box 4.1  Dualism: René Descartes

René Descartes (1596–1650), the father of modern philosophy, argued that the mind and body were two separate entities, which interact (dualism). To him the human body was a machine; the behaviour of animals was controlled by environmental stimuli; and the human mind was a uniquely human attribute, not subject to the laws of universe. He was the first to suggest that a link existed between the human mind and its purely physical housing. He proposed a method of studying 'the mind' based on reasoning and not metaphysical analysis.

Source: Martin *et al.*, 2010.

---

analysis of specimens, as well as procedures to monitor the quality of the study activities and data, need to be identified (quality assurance plan). Moreover, the analytical strategy and statistical methods (data analysis plan), including the proposed statistical precision (power) based on the proposed size of study population, needs to be defined (Baker and Nieuwenhuijsen, 2008).

Depending on the discipline(s) involved and the problem or hypothesis to be tackled, different study designs and different collection techniques can be selected (see Box 4.2). Studies differ depending on the questions being investigated and depending on the researchers involved. Studies can be staged or can comprise the simple collection of data in a smart way (e.g. modelling). In not-created analytical studies, relations between measurements of environmental events or exposures, of individuals' physical and social characteristics and of their health status and behaviour are examined to discover causes of the observed effects. In staged (experimental) studies things are created and the experimenter observes the results to identify cause-and-effect relations. An experiment can be performed in a laboratory (most of the variables that can affect the outcome can be controlled) or in the field (participants are observed under 'fairly' naturalistic conditions). In the laboratory, both *in vivo* and *in vitro* studies can be performed. An intervention can be performed both in the field and in the laboratory.

In general, each study performed or undertaken follows more or less similar steps, starting with problem definition (goals and objectives), followed by design and planning (methodology), implementation (the execution of the study) and data management and analysis. Correlating analytical observables in bio-fluids to human health dates back to Hippocrates, who used uroscopy in examination of illness (Kouba *et al.*, 2007). Changes in the human systems can be registered by assessing the contents of the bodily products, such as blood, urine, sweat, stool, saliva, breath and hair (metabolomics). These bodily products contain a whole range of cytokines, hormones and metabolic products as well as physical, chemical or radioactive agents, including antioxidants and reactive oxygen species. Locally, these can be measured in eyes, nose, respiratory tract, skin and other tissues. At cell level, stem cells and gene profiling can be used to inventory changes caused by different exposures. Physical techniques to measure changes in the human systems comprise techniques assessing responses of the autonomic nervous system such as heart rate, blood pressure, lung function and sweating,

> ## Box 4.2 Different fields, different approaches
>
> In *epidemiology* the focus in general lies on an *environment* and involves measurement of exposure in order to explain causes of disease (finding associations) (White *et al.*, 2008) by observations of groups of people (populations) in their natural environment (the real world). Many of these studies do take co-variates – confounders, exposure and effect modifiers – (e.g. *personal factors, psychosocial stressors* and *other factors*) into account.
>
> In *psychology* the focus lies on explanation of *responses and mechanisms of the human mind* (and interaction with the body) and involves measurement of personality, neurological and behavioural aspects by observational and experimental techniques (Martin *et al.*, 2010).
>
> In *clinical medicine* the focus lies on *responses of the human body* and involves measurement of physical and physiological responses that are part of the causal chain by using diagnostic and experimental techniques (Moher *et al.*, 2010). In several fields (e.g. toxicology, epidemiology), medical tests are applied to study the relation between the *physiological state* and exposures, chronic or acute, *psychosocial and/or physical*. Similar tests are applied in *in vivo* and *in vitro* studies.
>
> In *toxicology* the focus lies on the explanation of *mechanisms in the human body* and involves measurement of adverse effects of chemical substances by *in vivo* (within the whole body) and/or *in vitro* (outside the body) experiments (explaining mechanisms), using animals, tissues and/or humans (Hepple *et al.*, 2005; AltTox.org, 2009).
>
> In *(product) design* the focus lies on understanding user's emotional needs together with functional and social needs: the emotional experience of users with a product (Demir *et al.*, 2006). Both quantitative and qualitative research methods are applied in product marketing and design (Sheth *et al.*, 1999). Quantitative methods comprise surveys or experiments, while qualitative methods include focus groups, customer visits, motivation and interpretative research.

and techniques that assess the activities (e.g. electrical activity) and processes in the brain with so-called neuro-imaging techniques. Peripheral nervous system responses comprise several local effects such as endothelial dysfunction and facial contractions.

The measurements of these biomarkers are objective and have the potential to be more specific for relevant exposures than measurements in the environment. Measurements in the environment estimate the available dose of an agent or, at best, the administered or external dose (intake), while measurements of biomarkers estimate the absorbed or internal dose (uptake). Biomarkers may also permit estimation of the active or biological effective dose, namely the dose at the level of the structures (organs, cells and sub-cellular and molecular constituents). On the other hand, measurements at a single point, or even several points, in time may not be able to capture the long-term pattern of exposure. Other types of measurements (e.g. questionnaires or measurements in environment) may be necessary. Additionally, the willingness of the subjects to participate and costs are also important issues. Finally, 'new' techniques focused on epigenetic changes at cell level over time seem to have potential.

The environment of a subject can also be assessed with checklists. Checklists, in combination with several chemical, biological and physical measurements, have been used in several studies to characterize the indoor environment. Additionally, (risk) modelling and simulations are being applied to predict certain health and/or comfort outcomes, in particular for thermal comfort and air quality.

Behavioural monitoring can be used to collect information on bio-behavioural variables that may activate physiology, such as physical activity and caffeine/coffee consumption. It can also be used to 'extract' information from building users on how they interact with their environment, and it can be used to monitor responses, for example, physical reactions of eyes, face and body to events around the study subjects or activities they are focused on. Available techniques to monitor behaviour are a behavioural diary, systematic observation, automatic sensing and logging, behaviour genetic studies and performance tests. Automatic sensing and logging seems very attractive for use in IEQ investigations. Unfortunately, technologies for that particular category are being developed and will be available in the short to medium term, indicating that experience with them still needs to be gained (Brink *et al.*, 2010). These technologies have the potential to replace the more subjective behavioural diary technique, not forgetting systematic observation, which is a time-consuming and tedious technique. Also for performance testing, it can be said that this technique is still in the beginning of its development, but seems interesting to be further explored. Behaviour genetic study techniques are not very easy to use in IEQ studies.

With questionnaires it is in principle possible to gather information on personal factors such as states and traits, descriptive (subjective) information on physical and psychosocial stressors and factors, on past events and exposures, behaviour, perceived wellbeing (or stress) and other responses (physical and psychological) over time.

From these techniques, questionnaires and interviews are usually considered subjective information (White *et al.*, 2008). Nevertheless, those techniques provide us with information from the past that is not always easy to capture with the more so-called objective techniques. Subjective recall of exposure, collected by way of face-to-face or telephone interview of self-administered questionnaire is the predominant method of collection of exposure data in epidemiology. Face-to-face interviews are the best for collection of large amounts of data. Where subjects are widely dispersed and the questionnaire can be kept comparatively brief, the telephone interview may be favoured. Self-administered questionnaires may perform better than the other approaches in eliciting information on sensitive or socially undesirable behaviour, but perform poorly on complex questions, open-ended questions, and branching questions. Self-administered questionnaires on the other hand are less costly, are faster and require less staff in comparison to interviews.

## 4.3  Stress mechanisms and indicators

In Part I of this publication a number of stress mechanisms that can occur in response to external stressors were presented. For each of these stress mechanisms it is possible to identify a number of indicators. These indicators can be categorized into those indicators that tell us something about the physical and physiological state our body is in (physiological, physical and some behavioural indicators), our mental state (psychological indicators and some behavioural indicators) and the state the environment is in (environmental indicators).

## 4.3.1 Physiological and physical indicators

The physical and physiological state of our body can be assessed by a whole lot of different indicators (see Table 4.1). The assessment of the contents of the bodily products, such as blood, urine, sweat, saliva, stool and hair, can focus on different compounds: cytokines, hormones, metabolic products and/or physical, chemical or radioactive agents. The applicability of a certain biomarker or physiological indicator depends on its sensitivity, specificity and its biological intra- and interpersonal variability. In addition there might be financial and ethical limitations. One should be aware that some indicators (e.g. cortisol in blood or saliva) are subject to change over time (i.e. diurnal rhythm) (e.g. Clougherty and Kubzansky, 2009). Most of the physiological indicators presented in Table 4.1 have been used to measure acute responses, i.e. change from the personal baseline, to changes in environmental conditions. It may well be that not only the absolute change but also the rate of recovery (going back to the personal baseline), the peak during the day (e.g. the waking value for cortisol) or an integrated measure (the total amount over time) are important parameters to consider. Additionally, the applicability of indicators to measure response to chronic stress (e.g. cortisol in hair) or effects of acute stress over time (e.g. cholesterol blood level over time) need to be explored to see whether indoor environmental stressors are able to provoke a measurable change from the personal baseline.

Unfortunately, using cytokine levels as markers to identify effects of specific stressors is also difficult because cytokines are involved in various physiological processes (see Table 4.2; Tarant, 2010). Furthermore, as the biological variability of systemic levels of cytokines is substantial, it will be difficult to identify the change caused by a particular cause (stressor), specifically for mild stressors in the indoor environment (Tarant, 2010). Nevertheless, a particular cytokine, IL-6, has shown encouraging associations with stress, depressed mood, and cognitive impairment (Kiecolt-Glaser *et al.*, 2003; Steptoe *et al.*, 2007).

Effects of air pollution have been studied in the eyes, nose, skin and other tissues, using indicators such as tear film breakup time and foam formation for eyes, peak airflows through the nose, and irritation of tissues and skin. For metabolism effects, cholesterol level over time seems to be related to a cholesterol stress response in the past (Steptoe and Brydon, 2005). At cell level, gene expression profiling, stem cells, patterns of damage to DNA and DNA adducts are used. An interesting development here is the pattern analysis of epigenetic changes using micro-arrays, principal component analysis and three-dimensional screening techniques (Afshari *et al.*, 2011; Kang and Trosko, 2011).

Heart rate variability (beat-to-beat variation in heart rate) and blood pressure are important indicators for the responses of the autonomic system. Both have been associated with cardiovascular effects in the long term. Lung function has been studied with expiration indicators (e.g. FEV1, FVC, PEF, FEF and MEF). The pulse wave amplitude (PWA) can be used as an indicator for endothelial dysfunction (Kuvin *et al.*, 2003).

**Table 4.1** *Stress mechanisms and a number of physiological and physical indicators measured in bodily products or by physical techniques (e.g. McEwen, 2004; Halliwell and Gutteridge, 2007; Baker and Nieuwenhuijsen, 2008; White et al., 2008)*

| Indicator(s) | Anti-stress | Circadian rhythm | Endocrine disruption | Oxidative stress | Inflammation, irritation | Cell changes/death |
|---|---|---|---|---|---|---|
| **Endocrine system:** | | | | | | |
| • (Nor)-adrenaline (catecholamines) | X | | | | | |
| • Cortisol | X | | | | | |
| • Dehydroepiandrosterone–cortisol ratio | X | | | | | |
| • Melatonin | X | X | | | | |
| **Immune system:** | | | | | | |
| • Inflammatory cytokines and chemokines (e.g. Il-6, C-reactive protein, and TNF) | X | | | X | X | |
| • Immune response (lymphocyte counts and activity, natural killer cells, immunoglobulins) | | | | X | | |
| • Antioxidants (bilirubin, uric acid, glutathione) | | | | X | | |
| • Reactive oxygen species (ROS): heme oxygenase-1 expression (HO-1) antioxidant response generated together with CO exhaled in expired air; nitrogen oxide in expired air, thiobarbituric acid reactive substances, 8-isoprostane and myeloperoxidase | | | | X | | |
| • Intracellular adhesion molecule, increased sputum neutrophil and esinophilic cationic protein levels | | | | | X | |
| • Protein damage and/or lipid damage | | | | X | X | X |
| **Autonomic nervous system responses:** | | | | | | |
| • Heart rate (variability) | X | | | X | | |
| • Blood pressure | X | | | | | |
| • Sweating | X | | | | | |
| • Skin temperature | X | | | | | |
| • Expiration (e.g. FEV1 (forced expiratory volume in one second), FVC (forced vital capacity), PEF (peak expiratory flow), FEF (forced expiratory flow) and MEF (maximal expiratory flow)) | | | | X | | |
| **Central nervous system:** | | | | | | |
| • EEG | X | X | | | | |
| • Neuro-imaging techniques (e.g. PET (positron emission tomography), MRI (magnetic resonance imaging), fMRI (functional magnetic resonance imaging)) | x | x | | | | |
| **Metabolism:** | | | | | | |
| • Insulin or glucose levels | X | | | | | |
| • Cholesterol; High–Low density lipoprotein ratio | X | | | | | |
| • Waist–hip ratio (abdominal fat) | X | | | | | |
| Physical, chemical or radioactive agents in body and organs | | | X | X | X | X |
| **Local responses/effects:** | | | | | | |
| • Eyes: blinking rate and tear film stability | | | | | X | |
| • Endothelial dysfunction: pulse wave amplitude | X | X | | | | X |
| • Muscle tension | X | | | | X | |
| **Cell level:** | | | | | | |
| • Mutation: DNA adducts | | | X | X | X | X |
| • DNA damage: pattern of damage to DNA bases | | | | | X | X |
| • Signature patterns of altered gene expression | | | | | X | X |

**Table 4.2** *Common modulated blood cytokines associated with physiological responses observed in toxicity studies, including inflammation, immunity and repair*

| Physiological response | Cytokines |
| --- | --- |
| Acute-phase response | IL1β, IL6, TNF-α |
| Cytokine storm/release | IL2, IL6, IL8, IL10, IFNγ, TNF-α |
| Fibrosis | TGFβ |
| Hemophagocytic syndrome | IFNγ, IL1β, IL6, TNF-α |
| Neutrophilic inflammation | IL8, MIP-1, TNF-α |
| Systemic inflammatory response syndrome | IL6, MCP-1, TNF-α |
| Th1 immune response | IFNγ, IL2, IL12 |
| Th2 immune response | IL4, IL5, IL6, IL10, IL13 |

*Note*: Nomenclature: IFN = interferon; IL = interleukin; MIP = Macrophage inflammatory protein; MCP = moncyte chemo-attractant protein; Th = T helper; TGF = transforming growth factor; TNF = tumor necrose factor.

Source: Tarant, 2010.

Short-term indicators for physiological changes evoked by stress are the galvanic skin response (GSR), skin temperature, but also several brain activity indicators (e.g. brain maps of EEG activity) and neuro-images of several neuro-transmitters (e.g. serotonin, dopamine), blood flow, free choline (indicator for deficiency of omega-3 and omega-6 log chain PUFA's) and nucletide triphosphate levels (energy metabolism) by neuro-imaging techniques. The latest technique applied here is the fMRI (functional magnetic resonance imaging). fMRI can follow the fast brain waves that can be related to the conscious and unconscious needs of people (Lindstrom, 2008). Unfortunately, not much experience has been gained yet in the IEQ field with this technique.

### 4.3.2 Behavioural indicators and psychological indicators

In behavioural monitoring several types of information on bio-behavioural variables, on interactions with the environment, and on responses and activities over time is collected. Some of this information can be used directly as an indicator for a disease or disorder (physical state), other information serves as a confounder or modifier. The same can be said for the psychological indicators determined through questionnaires. Table 4.3 shows an overview of the type of psychological and behavioural indicators that have been associated with stress mechanisms as described in Part I.

### 4.3.3 Environmental indicators

Information on the environment we are exposed to can be gathered with environmental indicators. These environmental indicators are generally collected via measurements of the indoor environment conditions (chemical, physical and biological) and are described extensively in Chapter 3 of *The Indoor Environment Handbook* (Bluyssen, 2009a).

**Table 4.3** *Stress mechanisms and a number of psychological and behavioural indicators determined with questionnaires and/or observational techniques*

| Indicator(s) | Anti-stress | Circadian rhythm | Endocrine disruption | Oxidative stress | Inflammation, irritation | Cell changes/ death |
|---|---|---|---|---|---|---|
| **Personal factors:** | | | | | | |
| • States and traits | X | X | | | | |
| • Demographic variables | | | | | | X |
| • Life style and health status | | | X | X | X | X |
| • Events and exposures | | | X | X | X | X |
| **Evaluation of indoor environment quality:** | | | | | | |
| • General acceptability/dissatisfaction | X | X | | | | |
| • Perceived comfort – complaints (odour, noise, light, thermal) | X | X | | | | |
| **Symptoms related to indoor environment quality:** | | | | | | |
| • Perceived health symptoms | | | | | X | |
| • Self-reported breakup time | | | | | X | |
| **Psycho-social stress:** | | | | | | |
| • Work-related stress | X | X | | | | |
| **Performance:** | | | | | | |
| • Reaction time | X | X | | | | |
| • Memory | X | X | | | | |

## 4.4 Link with Part III

Methods applied in IEQ investigations vary from an epidemiological approach, in which questionnaires and health and comfort data may be used either in combination or not with biomarker sample collection (e.g. blood, urine), to field studies in which in general a smaller sample of persons is studied in combination with environmental inventories, and laboratory studies in which persons or animals are exposed to controlled environmental conditions. Health and comfort data are then combined with information on characteristics of the indoor environment in order to find relationships. However, other risk factors that may cause psychological or physiological stress (e.g. major life events), individual differences caused by personal factors (e.g. states and traits), or history and context can all affect the outcome that is being studied. These factors are taken into account only to a limited extent in current methods commonly applied to identify relationships between the health and comfort of people and the physical environment.

In the last hundred years there has been a shift from a focus on physical needs to psychosocial needs. A study performed for the European parliament's committee on employment and social affairs on health risks at work (Houtman *et al.*, 2008) showed that a combination of psychological and physical stressors is the top emerging health risk at work. However, the methods and models applied to identify relationships between needs and the physical and social environment have not been developed accordingly.

From the synthesis in Part II, on the assessment of indoor environment quality, it can be concluded that in order to answer the question 'How can quality of life (wellbeing) in the indoor environment be evaluated and which parameters or indicators can be used to explain the effects or responses?', a different approach is required. The assessment protocols applied in the different fields show large overlaps, but they all struggle with the ultimate goal to find clear cause–effect relations. Pattern analysis both at human cell, human exposure and behaviour level as well as at the environmental level have been proposed as a way forward. From the studies so far on single-dose response relationships, whether psychological, physiological or physical, it seems not possible yet to make predictions hold.

Based on current available performance indicators for healthy and comfortable indoor environments as well as the available information on cause–effect relations for different scenarios, the last part of this book (Part III) discusses what is needed to assess and/or predict the effects or responses of people in their indoor environment, applying different levels of assessment (occupants, dose or environmental parameter, and building and its components).

# 5

# Research methods and analysis

*Health and comfort effects caused by environmental stressors can be assessed in different ways. Depending on the field of expertise, goals and methods applied differ, but in general the steps or stages applied are the same: problem definition, design and planning, implementation, data management and analysis, and reporting. In this chapter these steps are discussed, and an attempt is made to present common study designs for a number of fields (epidemiology, psychology, clinical medicine, toxicology and product design).*

## 5.1 Introduction

To assess health and comfort effects of environmental stressors several research methods exist. Two fundamental approaches can be distinguished: qualitative research and quantitative research.

Qualitative research is an interpretive approach in which one tries to describe and explain how and why people act and make decisions. Qualitative or descriptive research is concerned with explaining the quality of data ('meanings, context and a holistic approach to material') with as little numerical analysis as possible (Martin *et al.*, 2010). For example, in an epidemiological descriptive study the population group of interest is defined, the incidence or prevalence of the disease is estimated, and possible environmental hazards that might have caused the disease are identified. The outcome is useful for creating hypotheses for further study (Baker and Nieuwenhuijsen, 2008).

Quantitative or analytical research on the other hand is concerned with numbers and counting, in which one tries to identify factors or relationships in a sample. The meaning of the collected numbers is the central issue to understand (Langridge and Hagger-Johnson, 2009). In an analytical study a hypothesis about cause and effect is tested or a quantitative relationship between exposure and effect is evaluated.

The border between the two approaches, qualitative or quantitative, is however not always clear. Both can be used in one study. In architecture and building engineering the focus lies on the design, construction and maintenance of healthy and comfortable buildings and systems for people (see Figure 5.1), and involves both measurement of environmental parameters and characteristics of the built

**Figure 5.1**
*A healthy indoor environment!*
Source: European HealthyAir project, Bluyssen
*et al.*, 2010d.

environment using technical instruments and occupants of buildings (both subjective and objective assessments) as well as lab and simulation studies. In fact a whole range of study designs has been applied, ranging from study designs applied in epidemiology, psychology and toxicology to clinical medicine and product design.

Depending on which problem or problems are to be solved, an investigation can have a different character or hypothesis. A hypothesis is a tentative statement about a cause–effect relation between two or more events. A theory, a set of statements designed to explain a set of phenomena, is an elaborate form of a hypothesis. A theory can be a way of organising a system of related hypotheses to explain some larger aspect of nature. Most hypotheses originate from theories (systems of ideas or statements that explain some phenomena) derived from previous inductive research or through intuition and reasoning. '*Induction* is the process by which scientists decide on the basis of multiple observations or experiments if some theory is true or not' (Langridge and Hagger-Johnson, 2009; p. 9). *Deduction* on the other hand starts with a hypothesis (or a prediction), which is subject to some empirical test and deductions made from the results of the test.

Even though the problem definition itself can be different, resulting in a different study design, in general, for each type of study several stages can be defined (MRC, 2000; Baker and Nieuwenhuijsen, 2008; White *et al.*, 2008; Martin *et al.*, 2010):

1  *Problem definition*: identification of the problem to be investigated and/or formulation of hypothetical cause-and-effect relations among variables, approached from different points of view. Depending on the type of problem to be investigated, the disciplines involved, different study designs are and can be selected. Each study design has advantages and disadvantages. A study design in itself is not better than another type. It depends on the question being asked and these questions can vary for epidemiology, psychology, clinical medicine, toxicology and product design (see Sections 5.3.1, 5.4.1, 5.5.1, 5.6.1 and 5.7.1).

2  *Design and planning*: an iterative process in which several steps are taken to develop the study according to the selected study design and problem to be solved (defined in step 1) such as (Section 5.2 and Sections 5.3.2, 5.4.2, 5.5.2. 5.6.2 and 5.7.2):

   a  *Sampling strategy*: selecting the study population(s), the sampling scheme and sample size.
   b  *Data collection and data analysis plan*: selection of measures (and scale of measurement) and techniques needed to collect and analyse the information.
   c  *Quality assurance plan*: quality control to assure validity and reliability and reduce sources of errors (systematic and/or random errors).
   d  Planning of the study with resources and timetable.

3  *Implementation*: in which the study is conducted; data items are collected with the instruments and associated quality control procedures and methods selected in step 2 (examples of studies are presented in Sections 5.3.3, 5.4.3, 5.5.3, 5.6.3 and 5.7.3).

---

## Box 5.1  Contents of a scientific paper

- Abstract (150–200 words): summary of the article.
- Introduction: review of literature and presenting of hypothesis, which the study aims to test.
- Method: who has taken part, what is used (techniques/materials) and procedure applied.
- Results: reports the results and analysis of data.
- Discussion: discusses results found, interprets data in light of earlier findings, defines conclusions and makes suggestions for the future.
- References: lists, usually alphabetically, all studies cited in the text.
- Acknowledgements (optional): in which other contributors than the authors are thanked.
- Appendix (optional): additional information such as questionnaires or checklists.

---

4   *Data management and analysis*: processing and (statistical) analysis of the data items collected (Section 5.8).
5   *Reporting*: communication of the results in for example a scientific paper (see Box 5.1) or in a report.

Box 5.2 presents a number of definitions used in different studies, whether originating from epidemiology, psychology, clinical medicine, toxicology or product design.

---

## Box 5.2  General terminology

Measures

- A *variable* is a measure that varies in a certain way, for example, height, weight, length. Variables can be continuous, such as age, or discrete (or categorical) such as sex. Dichotomous (or binary) variables can take only two values.
- A *construct* is a measure that is not easy to measure, it has to be measured indirectly. For example, anxiety, self esteem.

Levels or scales of measurement

- *Nominal*: Numbers, words or signs are used as labels, for example, label categories of data such as sex or ethnicity.
- *Ordinal*: Phenomena (people, objects or events) are ordered along some continuum to indicate a rank ordering of classes of the variable, but differences between them is not known. For example, 'daily', '4–6 times a week', '1–3 times a week', '1–3 times a month' and 'less than once per month'.
- *Interval*: The difference between items on a scale means something and reflects true differences in the values of the underlying measurements. For example, year of birth or a temperature scale.
- *Ratio*: Also intervals with equal amounts but with a true zero point. It permits valid comparison of measurements by calculation of both true differences and ratios, i.e. one measurement can be some multiple of another. For example, a time scale: at 0 no time can be measured.

*continued . . .*

**Box 5.2 ... *continued***

Errors

- *Constant or systematic errors* occur when we have a systematic effect of some extraneous variable on one (or more) of the experimental conditions. Minimising constant errors is important in experimental design.
- *Random errors* occur due to chance and are a random effect of some extraneous variable on the measurement of the relationship between the independent and dependent variable. Can be reduced by doing a larger study.

Validity and reliability

- *Validity* concerns whether a test (or measure of any kind) is really measuring the thing intended to measure; how accurate does the experiment represent the variable whose value has been manipulated or measured?
- *Reliability* concerns the stability of the measurements: How consistent are the results? Can they be repeated easily?

Sources: Baker and Nieuwenhuijsen, 2008; White *et al.*, 2008; Langridge and Hagger-Johnson, 2009; Martin *et al.*, 2010; Moher *et al.*, 2010.

## 5.2  Design and planning

*In each study it is important to know:*

- *The sampling strategy: who or what will be studied?*
- *Data and collection techniques: what will be measured and how?*
- *Quality assurance and control: how can validity and reliability be assured? (e.g. minimization of errors).*
- *Planning: when, where and with whom shall the study be performed?*

### 5.2.1  Sampling strategy

At the start of a study, it is important to have a clear idea about which measures (variables and/or constructs) need to be assessed, which study design is chosen, which methods of sampling will be applied and how measurement of exposure and outcome measures will be performed. The type and size of study samples, the sampling scheme and instruments to be used, and methods for the laboratory analysis of specimens as well as procedures to monitor the quality of the study activities and data need to be identified (quality assurance plan). Moreover, the analytical strategy and statistical methods (data analysis plan), including the proposed statistical precision (power) based on the proposed size of study population, need to be defined (Baker and Nieuwenhuijsen, 2008).

### Methods of sampling

Several methods of sampling are available (Baker and Nieuwenhuijsen, 2008; Langridge and Hagger-Johnson, 2009):

- *Random or simple sampling*: Each member of the population studied has an equal chance of being included (which is often impossible because access to the total population of interest is difficult). The selection of each participant is independent of the next.
- *Systematic sampling*: The sample is drawn from the population at fixed intervals from the list for example every third birth in a hospital. It is easier than random sampling but more prone to accidentally introducing a bias.
- *Stratified sampling*: Groups within the population studied stratified along sex, age, social class and so on are each sampled randomly or systematically.
- *Cluster sampling*: Sampling of natural groups (such as classes of students) rather than individuals.
- *Stage/multi-stage sampling*: Breaks down the sampling process into a number of stages. For example, in stage 1 a number of classes of children from schools are selected, then in stage 2 a number of children of those classes are selected randomly.
- *Opportunity of convenience sampling*: The most common form of sampling is recruiting participants in any way possible.
- *Snowball sampling*: Through contact with a number of potential participants, participants are recruited by recruiting their friends and their friends of friends. This type of sampling is very unlikely to be representative.

**Figure 5.2** *Several ways of sampling?!*
Source: Bluyssen.

A sample of the larger population is often assumed to be a representative group of participants. The usual way to assign them is randomly from the available list (the differences in personality traits, abilities and other characteristics that may affect the outcome should be equally distributed across the groups). However, the available participants are not always the participants we need or want. They often comprise opportunity or convenience samples: people willing to give up their time. And then a *response bias* can occur because people are responding in different ways to the same experiment. For example, a response bias among different cultures was found by Herk *et al.* (2004): compared to north-western European countries, Mediterranean countries showed a higher tendency to agree than disagree with answers to a questionnaire.

## Sample size

In order to have enough statistical power, the sample size required can be calculated. The uncertainty of measurements or responses can be described by the $(1-\alpha)$ confidence interval for the mean ($\mu$). The $(1-\alpha)$ confidence interval is a stochastic interval that in $(1-\alpha)*100$ percent of the cases includes the true value. If random samples of size $n$ are taken from a normally distributed population then the statistics has a students' distribution with $n-1$ degrees of freedom. For $n \geqslant 30$ the sampling distribution is nearly normal. The 95 per cent confidence limit ($z = 1.96$) for estimation of the population means $\mu$ is given by (Snedecor and Cochran, 1980):

For $n \geq 30$: $X \pm 1.96\sigma/\sqrt{n}$ [5.1]

For $n < 30$: $X \pm t_{.05}S/\sqrt{n}$ [5.2]

Where:

$z = z$ value = 1.96 for a 95 per cent confidence interval

$X$ = estimated mean

$\sigma$ = standard deviation = the square root of the variance

variance = the sum of the quadratic differences between the mean and each single deviation

$n$ = size of sample

$S$ = estimated standard deviation

$t_{.05}$ = t-distribution of 95 per cent confidence interval

## Example 1

An example of how the sample size can be calculated is the determination of the number of subjects required for the evaluation of air quality using the acceptability scale (Bluyssen et al., 1996b). Votes by subjects may be given a value assuming clearly acceptable is 1 and clearly not acceptable is –1, with 0 being the midpoint. Standard deviations of votes by subjects in previous studies are shown in Figure 5.3.

Knowing that the mean standard deviation in the area –0.3 to 0.3 of the acceptability scale is approximately 0.5, a relation between the 95 per cent confidence limit width and the size of $n$ (number of untrained panel members) is presented in Figure 5.4 based on the following equation:

$n = 1.96^2 S^2/L^2$ [5.3]

Where:

width = 2$L$

$L$ = allowable error for a 95 per cent confidence interval

$S$ = standard deviation in previous studies

For example, the numbers of panel members required for 95 per cent of 0.1 and 0.25 width on the acceptability scale are approximately 380 and 70, respectively.

## Example 2

Another example is the determination of the number of respondents in a questionnaire survey in order to get enough statistical power. Assuming that the most important questions in a survey are dichotomous (yes or no), a 95 per cent confidence level, 50 per cent 'yes' answers, and an allowable error of ±4 per cent, the sample size can be calculated with (Snedecor and Cochran, 1980):

$n = 1.96^2 pq/L^2 = 1.96^2 \times 0.5 \times 0.5/0.04^2 = 600$         [5.4]

Where:

$p$ = percentage choosing 'yes'

$q = 1 - p$ = percentage choosing 'no'

$L$ = estimated confidence interval = 0.04.

Response rates from questionnaire surveys in office buildings are generally around 70–80 per cent. Assuming a 70 per cent response rate, at least 860 workers need to be investigated to get enough statistical power. If, on the other hand, the response rate would be disappointing (say around 40 per cent), the number of workers to be investigated need to be at least 1,500 in order to get enough statistical power.

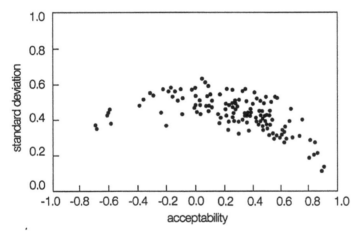

**Figure 5.3** *Standard deviation on acceptability votes with an untrained panel*
Source: Bluyssen *et al.*, 1996b.

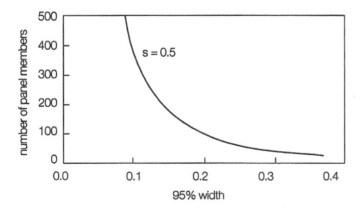

**Figure 5.4** *95 per cent width related to number of untrained panels for an estimated standard deviation of 0.5*
Source: Bluyssen.

## Sample scheme

Next to methods of sampling, various sample schemes can be chosen depending on the type of study design. This can be an *independent samples design*, a repeated measures design or a single-case study (Langridge and Hagger-Johnson, 2008; Martin *et al.*, 2010). *In single-case studies* the behaviour of individuals is observed, not of groups. These can be experiments or correlation studies, and have been used in neuropsychology, for example (the study of the relationship between the activity of the brain and its function) (Lindstrom, 2008).

In an independent samples (or between-groups) design two independent groups take part: a control group, which is not affected by the experiment, and an experimental group on which the experiment is taking place. The control group comprises other people than the experimental group. This study design has the advantage that the same individuals are not exposed to the same condition and are often used in medical treatment tests. In independent samples design the within-group variance will be large, which can be reduced by using matched pairs. Forms of independent samples design are:

- *Matched pairs*: Selecting matched 'controls' for each participant, who are as similar to their matched participant in the experimental condition with respect to those criteria that are likely to have an impact on the dependent variable (DV). The most perfect matched pair designs are those involving monozygotic (or identical) twins.
- *Parallel groups*: A variation on independent samples design; participants are assigned to independent samples.
- *Factorial*: Participants are randomized into four conditions – a quarter undergoes condition A, a quarter undergoes condition B, a quarter undergoes conditions A and B, and a quarter does nothing (the control group). Two interventions or treatments are investigated as well as their interaction.

In a *repeated measures* (or within-groups) design each individual is exposed to the same condition of the experiment but they can be involved at different levels. For example, studying whether men respond differently to a certain environmental condition than women. For statistical reasons this study design might be easier to obtain significant results because the amount of variability introduced by different individuals is eliminated (the within-group variance). Each participant acts as his or her control (completes each condition in the experiment) and there is less variability in the data (such as sex, age and personality differences, etc.). This design is very helpful when it is not possible to randomly assign participants to groups or where the variables needed for matching are unclear. Fewer participants are required than in an independent samples design. However, an *order effect* may occur because participants get exposed to more than one condition. Some forms of repeated measures design are:

- *Cross-over design*: Participants take part in condition A (exposed) and condition B (not exposed), but the order in which they take part in each condition is randomized, which is called counterbalancing. Counterbalancing can control for order effects and practice effects, but a washout period (a gap of time) between conditions is needed to ensure that the condition (or exposure) does not persist in the second condition. The health outcome status of each participant is compared during the exposed and unexposed periods. This approach was, for example, taken by Jaakola *et al.* (1994) in a blind crossover trial to test the hypothesis that recirculated air in mechanically ventilated buildings causes sick building symptoms and perceptions of poor air quality (for the explanation of 'blind', see Box 5.3). Each of the subjects (in two identical buildings) experienced two periods of exposures to recirculated air and two periods of no exposure.

- *Latin square*: is used when there are several different treatments (exposures). It does not work when there is an order effect. An example of Latin square is shown in Figure 5.5 in which three participants are exposed to three conditions.

Other possible study designs are (Langdridge and Hagger-Johnson, 2009):

- *One-group, post-test only*: A single group is given an intervention and then a post-test. Post-test without pre-test is generally useless.
- *Post-test only, non-equivalent groups*: Control group is added to one group, post-test only. Differences measured can be due to intervention or to existing differences between the two groups.

---

## Box 5.3 Single-, double- or triple-blind?

Participants might behave or respond differently during an experiment, while being observed, than normally. Experimenters must remember that their participants do not merely react to the independent variable in a simple-minded way, especially, when the participants might be affected by their knowledge of the independent variable. Two methods to deal with this are (Martin *et al.*, 2010):

1  *Single-blind experiments*: The participants are kept ignorant of their assignment to a particular experimental group. For example, the experimental group gets the drug to be tested while the control group receives a placebo (an inert pill), but none of the participants knows which group they are in.
2  *Double-blind experiments*: Both the experimenter and the participant are kept ignorant. The expectations of experiments can influence results of participants (humans and animals).

Traditionally, blinded RCTs (randomized controlled trials) have been classified as single-blind, double-blind or triple-blind. (In triple-blind tests, even the drug or treatment to be tested is unknown to both the experimenter and the participant.) The 2010 CONSORT statement specifies that those terms should not be used, but instead reports should discuss what was done, who was blinded and how (Schulz *et al.*, 2010; Moher *et al.*, 2010).

---

| Sample | Condition A | Condition B | Condition C | Dependent variable |
|--------|-------------|-------------|-------------|--------------------|
|        | Condition B | Condition C | Condition A |                    |
|        | Condition C | Condition B | Condition A |                    |

**Figure 5.5** *A Latin square design: three participants are exposed to three conditions – one participant does A, B and C; one does B, C and A; and one participant does C, B and A*

- *Pre-test, post-test, single groups*: A single group is given a pre-test, an intervention and a post-test. Limitations of this type of study design include history effects (events occurring apart from intervention), maturation effects (changes in individuals of group over time) and statistical regression effects.
- *Pre-test, post-test, non-equivalent groups*: Close to a true experiment except for random allocation of participant to conditions.
- *Interrupted time-series*: Multiple observations (or tests) of participants (possible with an experimental and a control group) over time with the intervention occurring at some point in this series of observations.
- *Regression discontinuity*: A true experiment but participants are allocated to two or more groups based on some principle, such as scores on a test.

## 5.2.2  Data and collection techniques

### Data collection plan

Depending on the type of information required, different instruments (or techniques) can be selected to collect that information, such as questionnaires, interviews, diaries, medical records, biochemical analysis of blood or other specimens, and physical or chemical analysis of the environment. In Chapter 6 each of these techniques is described.

Besides the selection of the instrument(s), all aspects of the measurement process should be included in the data collection plan, such as instructions for application of measurement method, the method itself and specification of procedures that follow the application of the main method up to presenting 'clean' data for analysis (White *et al.*, 2008).

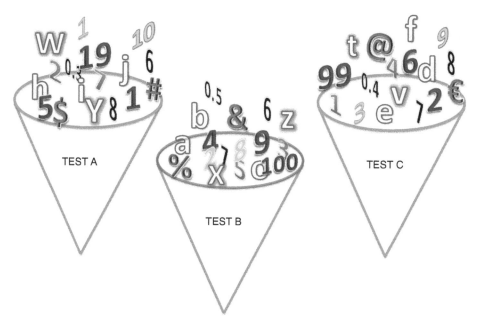

**Figure 5.6** *Collection of data*
Source: Bluyssen.

The type of information required depends on the goals and objectives set at the beginning of the study. Some types of data that can be collected are (White *et al.*, 2008):

- Variables needed to compute exposure dose and effect of timing of exposure: such as ages, frequency of exposure, intensity/dose and so on.
- Potential confounders of behavioural exposure–disease associations: such as risk factors for disease and factors associated with initiation or duration of exposure.
- Potential confounders of biomarker–disease associations: such as body mass and current health.
- Factors to describe the study population: such as age, race, sex, education and/or socioeconomic status and other factors used in subject selection.

And then of course a whole range of exposure variables can be collected. An example is presented in Table 5.1 for the project SINPHONIE (study A), in which, in around 120 schools in 23 countries, exposure and health effect data were collected (see Section 5.3.3 for the description of this project).

When selecting a method or methods the following aspects should be considered (White et. al., 2008):

- *Amount and detail of data required*: Personal interview is often the best method when large amounts of detailed data are required, particularly if past exposure must be documented. Physical and chemical measurements can only provide information on current exposure.
- *Impact and frequency of the exposure*: Low impact and infrequent exposures are almost impossible to measure, because the subject will not remember and chemical/physical measurements (environment or body) will give info with a low variability. Diary methods and observation are ideal for low impact frequent exposures, but in general give information only on present situation, not on the past, and for a short period of time. Frequent exposures of high impact exposures (e.g. recreational physical activity) may be measured by recall methods or records. All other methods might be used as well, but again only give information about present exposure over a short time period.
- *Variability of the exposure over time*: Exposures that do not vary over time can be measured by most methods. Methods that measure only current exposure can be used for exposures that vary in frequency or intensity over time, but their accuracy may be poor unless the measures are repeated for each subject, over multiple periods of time.
- *Accuracy and availability of the method*: Some exposures can only be ascertained by questionnaires, some only by chemical/physical measurements. The accuracy of the methods need to be considered for a particular exposure.

## Data analysis plan

The analytical strategy and statistical methods (data analysis plan) are important aspects of the study protocol. The data analysis plan may identify variables that

**Table 5.1** First inventory of data to be collected in SINPHONIE

| Collection technique → | Checklist | Physical/chemical measurements | Biological measurements | Questionnaires | Health tests |
|---|---|---|---|---|---|
| General building and activities | Year of construction; retrofit/renovation; number of floors; floor area; number of children/ teachers; pesticide treatment; pests; open doors between spaces. | | | For school administrator: on school.<br><br>Registers: absenteeism children. | |
| Construction and building materials | Construction materials; water damage; fire damage; type of glazing; furnishing and fittings. | | | | |
| Building services and special spaces | HVAC system; lighting system; maintenance, copy machines, kitchen. | | | | |
| Classroom 1<br>Classroom 2<br>Classroom 3 | Floor level; number of students Glass%; dimensions; occupation schedule; shading devices; type of glazing; electronic equipment; furniture materials and furnishings; curtains; plants; water damage; mould growth and odour; condensation; air leaks; fire damage; natural ventilation; consumer products (air fresheners, cleaning schedule and products). | *18 parameters average values of 5 (school)days: All schools/all classrooms:*<br>– Passive: formaldehyde, benzene trichloroethylene, tetrachloroethylen, pinene, limonene, $NO_2$, $O_3$, radon.<br>– Active: BaP, naphthalene, PM2.5, PM10 (relative to PM2.5).<br>– Continuous: CO [ppm], $CO_2$ relative humidity, temperature.<br>– Ventilation rate (not for outdoor).<br>*Note: one average value + standard deviation for continuous measured parameter.* | *All schools/all classrooms:*<br>– From settled dust: endotoxin (bacterial marker); ergosterol (fungal marker); 6 specific fungal and bacterial groups; allergens: dog, cat, horse.<br>*All schools/1 room per school:*<br>– Analyte: allergens: house dust mite.<br>*12 schools/3 classrooms:*<br>– Analytes: 6 specific fungal and bacterial groups; from active air sample.<br>*Note: sampling is conducted twice (heating, non-heating season).* | For parents on:<br>– Health of children.<br>– Home and school environment.<br>For children on:<br>– Classroom.<br>– Own health.<br>Circa 30 children per classroom (90 per school).<br>For teachers on:<br>– Classroom.<br>– Own health. | *Level 1 for each child:*<br>– Attention test.<br>– Spirometric testing.<br>*Level 2 and 3: 5 children per class:*<br>– Forced Expiratory Nitrogen oxide.<br>– Tear film break up time of eye.<br>– Skin prick tests.<br>– Acoustic rhinometry.<br>– Lavage of the nasal mucosa.<br>– Exhaled carbon monoxide. |
| Outdoor environment | Location; outdoor sources of air and noise pollution; light obstruction and glare. | *12 schools/1 outdoor location parameter.* | *12 schools/1 outdoor location per school:*<br>analytes: 6 specific fungal and bacterial groups; active air sample; *Note: heating and non-heating season.* | | |

*Note:* A description of the tests methods presented in this table can be found in Chapter 6.

Source: Bluyssen.

are not addressed adequately in the data collection plan or the other way round. It may identify variables in the data collection plan that do not contribute to meeting the objectives of the study. The data analysis plan is a description of the scientific objectives of the study, the data to be analysed and the data analysis approach. A template of an analysis plan is presented in Table 5.2. For details on the statistical analysis see Section 5.8.

## 5.2.3 Quality assurance and control

Quality assurance of any measurements series has two main elements (Berglund et al., 1999b):

- *Quality control*: comprising external quality control, a system for objective checking of (laboratory) performance by an external group, and internal quality control, a set of procedures used by the staff for continuously assessing results as they are produced.
- *Quality assurance*: comprising checks that should guarantee regular quality control and good quality of results. Quality assurance refers to all steps: the collection, transport and storage of samples, the (laboratory) analysis as well as the recording, reporting and interpretation of results.

Quality control is required to assure validity and reliability. Basically, quality control is performed to reduce causes of error and to make sure the study is providing the outcome originally planned for.

### Validity and reliability

*Validity* is about whether a test or measure is really measuring the thing intended to be measured. Validity concerns how accurate the experiment represents the variable that has been manipulated or measured. Validity or *accuracy* is the extent to which the systematic errors are controlled (Langridge and Hagger-Johnson, 2009).

External or ecological validity (generalisability or applicability) is the extent to which the results of a study can be generalized to other circumstances.

**Table 5.2** *Analysis plan template*

| Objectives | Broad objectives and specific objectives (specific aims) |
|---|---|
| *Testable hypothesis* | |
| Variables | A table with columns for variable name, scientific meaning and use in the analysis (e.g. response, exposure, treatment, confounder, effect modifier); measurement units or coding; and type of data (e.g. binary, nominal, ordinal, count, continuous). |
| Other important features | Such as missing data count or pattern, restricted range, limit of detection and recoding scheme. |
| *Other data and study details* | |
| Statistical analysis | Descriptive analyses |
| | Inferential analyses |
| | Sensitivity analyses |

Source: based on Baker and Nieuwenhuijsen, 2008.

Internal validity, which is a prerequisite of external validity, is the extent to which the design and conduct of the trial eliminate the possibility of bias (Langdridge and Hagger-Johnson, 2009).

*Reliability or reproducibility* concerns the stability of the measurements: How consistent are the results? Can they be repeated easily? For example, measuring the thermal feelings of a person sitting behind a computer working when exposed to different thermal conditions is most likely not valid when this desk is placed in a closed container with no windows, no ventilation and bad lighting. The other environmental conditions will influence the outcome of the experiment (i.e. *confounding* of variables is the introduction of one or more unwanted independent variables). To check reliability two or more people can perform a certain task and their ratings or measurements can be compared. Another way is to perform a test twice. The measures should provide similar results on different but comparable occasions. In order to establish validity and reliability of research findings a study needs to be repeated exactly and produce the same results as the original researcher reported (*replication*).

## Sources of error

In principle when dealing with data acquired from subjects (or respondents) or sampling data, two categories of errors play a role: a *systematic or constant* error, which occurs for all data acquired, and an *additional or random* error for a subject or sample, which can differ per subject or sample (Podsakoff *et al.*, 2003). Random errors confuse the effect we are interested in, whereas constant error biases (or distorts) the results. The additional error or natural variation (individual differences in performance), shown by the *within-group variance*, is an unsystematic error or 'noise'. An experimental effect between exposure of two groups to two conditions is shown by the *between-group variance* (if this systematic variance is zero, there is no experimental effect). *Systematic errors* (bias) can be (Baker and Nieuwenhuijsen, 2008; White *et al.*, 2008; Langridge and Hagger-Johnson, 2009):

- *Sampling error or selection bias*: Bias arising from the procedures used to select subjects. Sampling error occurs when your sample is not representative of the population and when the study variables are influenced by characteristics of the population (e.g. socioeconomic status, ethnicity).
- *Assignment error or confounding bias*: Differences in background between exposed and non-exposed groups. Occurs for example when subjects in each experimental condition (or treatment) are different in some way. Subjects may have different personality traits and cognitive skills. Unknown, unmeasured and confounding variables are introduced into the assignment.
- *Measurement error, or information and observation bias*: Measurement error occurs when the measured variable is not measured precisely (unreliable and/or invalid). Bias arising, for example, from misclassification of the study participants with respect to disease or exposure status while measuring. When instruments provide an accurate measure (e.g. skin temperature, blood pressure) the measure is reliable. However, many variables fluctuate and are therefore sensitive. Some variables (e.g. constructs) cannot be measured directly, but only their effects can be observed or questioned (Bradburn *et al.*, 1987).

Two types of measurement bias can be distinguished: non-differential and differential. *Non-differential measurement bias* occurs when the likelihood of misclassification of disease is the same for exposed and non-exposed groups (e.g., taking the distance from a person's home to a roadway as an indicator for exposure). *Differential measurement bias* occurs when the likelihood differs (e.g., recall of an exposure such as passive smoking in people with asthma might differ from that of healthy people). *Differential exposure measurement error* occurs when exposure measurement error differs according to the outcome being studied. For example, self-reported sickness absence rates can introduce an error as compared to absences recorded in employer's registers (Ferrie *et al.*, 2005).

Common sources of measurement bias are (White *et al.*, 2008; Langridge and Hagger-Johnson, 2009):

- *Errors in design of collection instruments*: such as wrong exposures, measure does not reflect exposure or effect, time period is not valid.
- *Errors or omissions in protocol collection instruments*: such as not enough detail, handling of unanticipated situations not included.
- *Poor execution of protocol*: such as data collection not in the same manner, improper handling and/or analysis of samples, bias in the collection or recording of information by study staff.
- *Limitations due to subject's characteristics*: such as differential reporting of information by study participants (cases report differently from controls because of the knowledge or feelings about the disease).
- *Collection/sampling errors*: errors made in measuring study variable, such as an incorrectly calibrated instrument or intra-individual variability effects caused by diurnal or seasonal variation, resulting in an uncertainty.
- *Errors during data capture and analysis*: such as data entry errors, programming errors, conversion errors.

*Randomization* of participants (exposure or treatment is randomly assigned to participants) is used to minimize the effect of constant or systematic error (Baker and Nieuwenhuijsen, 2008; Langridge and Hagger-Johnson, 2009). For reducing confounding and selection bias, other forms are *restriction* (narrowing the ranges of values of the potential confounders, e.g. restricting to students between the age of 20 and 35) and *matching* (on potential confounders, e.g. matching for age, gender and ethnicity) (see Section 5.2.1). Control for confounding is also performed during *data analysis* (Baker and Nieuwenhuijsen, 2008).

## Quality assurance plan

A quality assurance plan for monitoring staff, instruments and procedures is required. Such a quality assurance plan could comprise of the following items (White *et al.*, 2008):

- *Design of collection techniques*: including instructions on how to design forms in order to provide clear and unambiguous forms for collection of data without loss of information, and including a study procedures manual.

- *Preparing for data collection*: including pre-testing of data collection techniques and a training session of the data collectors.
- *Quality control during data collection*: focused on supervision of the data collectors and editing and coding issues of the data.
- *Quality control during data processing*: including data capture and editing by the computer, and data documentation and creation of new variables.

In the European project SINPHONIE (www.sinphonie.eu) a field study has been performed in around 120 schools in Europe (for a description of SINPHONIE, referred to as study A, see Section 5.3.3 and Box 5.7). An example of how quality assurance was tackled for the measurement of chemical and physical parameters in this project is presented in Box 5.4.

ISO 17025 (ISO, 2005b) 'General requirements of the competence of testing and calibration laboratories', is the main ISO standard used by testing and calibration laboratories. If testing and calibration laboratories comply with the requirements of ISO 17025, they will operate a quality management system for their testing and calibration activities that also meets the principles of ISO 9001 (ISO, 2008a). The ISO 9000 family of standards are related to quality management systems and designed to help organizations ensure that they meet the needs of customers and other stakeholders.

## 5.2.4 Planning and resources

Quality of the study findings depends largely on how carefully the investigators plan and prepare for the data collection. 'When, where and with whom shall the study be performed and how much does it cost?', is an important question to document ahead. For each project proposal it is important to establish a list of anticipated activities over time (timetable) and resources required.

### Resources

A detailed estimation of personnel, equipment and financial resources required is an important aspect of almost every proposal. Usually for the acquiring of project budget at the European but also at the national level, this costs overview comprises two main categories: direct costs and indirect costs.

Direct costs are all those eligible costs that can be attributed directly to the project and are identified by the beneficiary as such, in accordance with its accounting principles and its usual internal rules. An overview of the direct costing categories required for example for EU project proposals is presented in Table 5.3.

Indirect costs or *overheads* are all those eligible costs that cannot be identified by the participant as being directly attributed to the project but which can be identified and justified by its accounting system as being incurred in direct relationship with the eligible direct costs attributed to the project. They may not include any eligible direct costs. For example: costs connected with infrastructures and the general operation of the organization such as hiring or depreciation of buildings and plant, water/gas/electricity use, maintenance, insurance, supplies and petty office equipment, communication and connection costs, postage, and so on, and costs connected with horizontal services such as administrative and financial management, human resources, training, legal advice, documentation, and so on.

## Box 5.4  Four levels of quality assurance for the measurement of chemical and physical parameters in study A

*Level 1*: is concerned with the independent duplicate sampling to test the accuracy of an alternative (sampling + analytical) method. Any collected sample that isn't collected or analysed using the preferred method protocol for that compound is collected in parallel and analysed by a reference laboratory. The analysis is performed by a laboratory that is accredited according to ISO 17025:2005 (ISO, 2005b) or that is working according to ISO 17025 quality criteria. It is based on the principle of comparing a method to a reference method or a preferred method. The result, obtained by using the alternative method, is then considered as a representative, valid indoor air quality value, in case this complies with the requirements for expanded uncertainty, as described in prEN 482:2010 (CEN, 2010). This implies that, depending on the reference period and the magnitude of the measurement range compared to the limit values, the relative expanded uncertainty can be 30–50 per cent.

*Level 2*: is an internal quality control of the participating laboratory in order to evaluate the accuracy of the analysis procedure, applied by each laboratory. It involves the analysis of (certified) reference material and is applicable to all laboratories that analyse samples, using preferred methods as well as alternative methods. These referent materials, with known concentrations, are purchased and analysed by each laboratory into the analytical sequence and are reported. For the calculation of the total uncertainty of the analysis procedure, the following formula is proposed:

Total uncertainty [%] = $(|x - X_r| + 2s) \times 100/X_r$          [5.5]

Where:

$X_r$ = the reference value (i.e. the concentration of the reference material)

$x$ = the average of the analysis performed in the laboratory

$s$ = the standard deviation of the individual results.

An uncertainty less or equal to 30 per cent is acceptable; this complies with bias+2s compared to the reference method. It should be noted that the calculation of the uncertainty has to be based on at least six results.

*Level 3*: is an internal quality control of the participating laboratory, which involves duplicate analysis in:

• One classroom per set of three schools (in case three schools are studied).
• Two classrooms per set of five schools (in case five or six schools are studied).

This implies that approximately 10 per cent of the samples are collected in duplicate. A bias less or equal to 20 per cent between the duplicate samples is acceptable.

*Level 4*: relates to recent (less than one-year-old) calibration documents of the used equipment. This is especially important for continuous monitoring, such as $CO_2$.

In the frame of a quality control for sampling in the field a training session was organized. This extended two-day course preceded the sampling campaign and was aimed at training field workers and laboratory technicians in applying the preferred methods for the fieldwork and analysis protocols in their laboratory.

**Table 5.3** *Type of costing categories required for EU proposals*

| Sub categories | Explanation |
| --- | --- |
| Personnel costs | Actual salaries of staff involved in the project directly hired by the beneficiary including social security charges (holiday pay, pension, contribution, health insurance, etc.) and other statutory costs. |
| Subcontracting | A subcontractor is a third party which has entered into an agreement on business conditions with one or more participants, in order to carry out part of the work of the project without the direct supervision of the participant and without a relationship of subordination. |
| Travel and subsistence allowance | Actual travel and related subsistence costs relating to the project may be considered as direct eligible costs, providing they comply with the beneficiary's usual practices and are adequately recorded, like any other cost. |
| Purchase cost of durable equipment | Only equipment purchased for the purposes of carrying out the action can be charged as direct costs. To be considered as eligible, a cost must be determined according to the beneficiary's usual accounting practice and each beneficiary must apply its usual depreciation system for durable equipment. |
| Cost of consumables and supplies | Any consumables necessary for the implementation of the project may be considered as direct eligible costs unless it is the usual practice of the beneficiary to consider consumable costs (or some of them) as indirect costs. |
| Other direct costs | Means direct costs not covered by the above mentioned categories of costs. |

Source: EC, 2012.

## Timetable

Prior to collecting the data, a preparatory phase is often undertaken, in which the details of the study plans are finalized, the study instruments pretested, the study team members trained, and the logistics of any field or community-based data collection planned (Baker and Nieuwenhuijsen, 2008). After organization of material and people, the study (or experiment) is performed and the observations and measurements are recorded. An example of a timetable for the Project SINPHONIE (for description see Box 5.7) is presented in Figure 5.7.

An important part of the planning is the process for getting *ethical approval*, especially when procedures for studying human subjects are involved. Commonly, an application for epidemiological research must be accompanied by proof that approval of the proposed study has been received from an appropriately constituted ethics review committee (Baker and Nieuwenhuijsen, 2008; White *et al.*, 2008). Examples of ethical codes for scientific research are the ethical guidelines for biomedical research involving human subjects by the Council for international organizations of medical sciences (CIOMS) (see Box 5.5) or the ethical guidelines of the International society for environmental epidemiology (ISEE, 2012). These ethics guidelines, covering the general principles of the Declaration of Helsinki (World Medical Association, 2008), are in general focused on the protection for the rights of the person being studied and on the ethical conduct of the researcher.

The following policies are appropriate to follow in this context (White *et al.*, 2008):

- Working only from written research protocols that specifically address ethical issues.

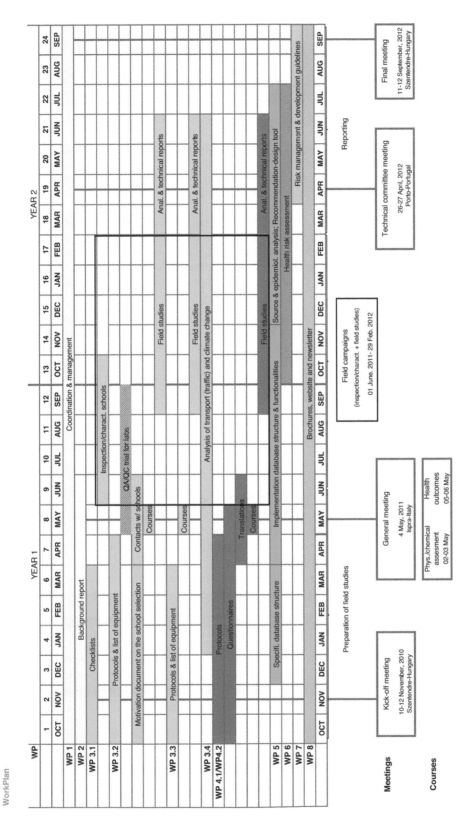

**Figure 5.7** *Example of a timetable for the European project SINPHONIE*

Source: Eva Csobod, personal communication.

---

## Box 5.5  General ethical principles

All research involving human subjects should be conducted in accordance with three basic ethical principles, namely respect for persons, beneficence and justice. . . .

*Respect for persons* incorporates at least two fundamental ethical considerations, namely:

a)  Respect for autonomy, which 'requires' that those who are capable of deliberation about their personal choices should be treated with respect for their capacity for self-determination.
b)  Protection of persons with impaired or diminished autonomy, which requires that those who are dependent or vulnerable be afforded security against harm or abuse. . . .

*Beneficence* refers to the ethical obligation to maximize benefit and to minimize harm. This principle gives rise to norms requiring that the risks of research be reasonable in the light of the expected benefits, that the research design be sound, and that the investigators be competent both to conduct the research and to safeguard the welfare of the research subjects. Beneficence further proscribes the deliberate infliction of harm on persons; this aspect of beneficence is sometimes expressed as a separate principle, non-maleficence (do no harm). . . .

*Justice* refers to the ethical obligation to treat each person in accordance with what is morally right and proper, to give each person what is due to him or her. In the ethics of research involving human subjects the principle refers primarily to distributive justice, which requires the equitable distribution of both the burdens and the benefits of participation in research. Differences in distribution of burdens and benefits are justifiable only if they are based on morally relevant distinctions between persons; one such distinction is vulnerability.

Source: CIOMS, 2002, pp. 17–18.

---

- Ensuring that all research protocols have been approved by an ethical review committee.
- Obtaining fully informed consent from subjects to their participation in research, except where the research is judged to present no more than minimal risk to the subject.
- Making provision for maintenance of the confidentiality of personally identifiable data.
- Ensuring that results obtained during the course of research that may have a bearing on the health and welfare of subjects are communicated to them.

## 5.3  Epidemiology

*Epidemiology is the study of exposures in an environment. Environmental epidemiology involves measurement of exposure in order to explain causes of disease (finding associations) in general by observations of groups of people (populations) in their natural environment (the real world).*

### 5.3.1  Goal and objectives

Although often using the same techniques, specific environment and exposures have lead to several categories of epidemiology (Baker and Nieuwenhuijsen, 2008): environmental epidemiology (dealing with exposures in the environment,

**Figure 5.8** *Doctor investigating a person in its environment*
Source: Anthony Meertins, August 2012.

see Figure 5.8), occupational epidemiology (dealing with exposures of workers in the work environment to potentially harmful agents) and nutritional and infectious disease epidemiology (dealing with respectively diet as a source of health and disease and spread of infections and related disease).

Environmental epidemiology concerns conceptualization of the true exposure and the time period over which it might cause disease (White *et al.*, 2008):

- What is the component of exposure (e.g. active agent, chemical compound or biological activity)?
- What causes (or prevents) the disease of interest?
- What exposures have and do not have the active agent?
- Where on the biological exposure–diseases pathway (from available dose in the external environment to active dose at the target site in the body) should the exposure be measured to best meet the study objectives?

Exposure–response relationships, by comparing disease occurrence in an *exposed* group with disease occurrence in a *non-exposed* group, are the primary focus of epidemiological investigations. Monitoring data are used to estimate *exposure*, exposure is used to estimate *dose*, and dose is used to estimate the biological effective dose at the *target organ*. It is generally assumed that dose is highly correlated with exposure across a *population*. The occurrence of disease is expressed as *prevalence* (based on number of persons that have the disease at a specific point in time) or the *incidence* (based on the number of persons that get the disease during a given time period) (Baker and Nieuwenhuijsen, 2008). Some definitions used in epidemiology are presented in Box 5.6.

The *time period* of the exposure most likely causes the disease (aetiological time window) and is therefore important to know as well as other time periods in the exposure–disease time sequence (induction period and latent period; White *et al.*, 2008; Figure 5.9). Unfortunately, the lengths of the time periods are usually not known with any certainty.

---

## Box 5.6 Some definitions in epidemiology

Exposure–dose response effect

- *Exposure pathway* = physical route a hazard (pollutant, substance) takes from a source to a person.
- *Exposure route* = the way a hazard enters the body (inhalation, ingestion, absorption).
- *Dose* = the amount of exposure; amount of hazard that enters the body.
- *Target organ dose* = uptake of an agent by a specific organ.
- *Cumulative dose* = $\Sigma$ duration$_i$ × frequency$_i$ × intensity$_i$ where: intensity is the dose of the active agent per episode of exposure; $i$ is each type of exposure$_i$ which has the active agent; frequency is frequency of exposure episodes; duration is number of years/hours the person was exposed.
- *Peak exposure* = the highest exposure level (exposure rate) experienced by a subject for some minimal amount of time (e.g. one year).
- *Response* = reaction of a person on a question or another evaluation technique.
- *Effect* = any change in health status or bodily function that can be shown due to exposure.

Population/individual/subjects/participants/respondents

- *General population* = all individuals in a specified area (e.g. country, city).
- *Population at risk* = individuals in the general population who could develop the disease of interest (i.e. not the people who have the disease).
- *Target population* = population at risk with which the study results can be most likely generalized.
- *Source population* = population at risk from which study participants will be sampled (sometimes the same as target population).
- *Study population* = individuals sampled from the source population.
- *Study persons or individuals* = named subjects when it concerns existing data (e.g. from registration); named participants when data collection on people is performed and named respondents when it concerns collection of questionnaire information.

Time/period

- *Risk period* = total time during which the individuals in the source population are at risk of developing an outcome of interest.

*continued . . .*

## Box 5.6 ... *continued*

- *Aetiological exposure time window* = the time period during which a particular exposure is most relevant to causation of the disease of interest, for example, a time period during which a certain physiological state exists (e.g. time in relation to pregnancy or time in relation to menopause) or a time period with a certain age (e.g. in utero, childhood and adolescence).
- *Induction period* = occurs after the aetiological time window; further exposure does not contribute to disease risk.
- *Latent period* = the time interval from the onset of a disease until it is diagnosed. For example when one cell or clones of cells have irreversibly started the path to cancer.

Occurrence of disease

- *Incidence rate* = disease occurrence per unit of time = number of persons with disease/person-time experience of the population.
- *Cumulative incidence or risk* = proportion of study participants who develop the disease (health outcome).
- *Rate ratio* = ratio of incidence rate in the exposed group to that in the non-exposed group.
- *Risk ratio* = the ratio of the cumulative incidence in the exposed group to that in the non-exposed group.
- *Odds ratio* = ratio of the exposure odds in cases to the exposure odds in controls (used in case-control studies; see Section 5.3).
- *Prevalence proportion* = number of existing cases at a point in time divided by the size of the population.
- *Prevalence ratio* = ratio of the prevalence proportion for the exposed to the prevalence proportion for the non-exposed.

Sources: Baker and Nieuwenhuijsen, 2008; White *et al.*, 2008.

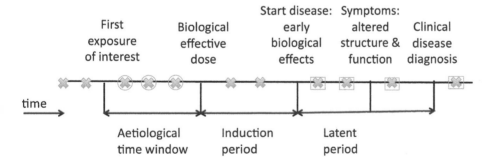

= times of the exposure

= exposures that cause the disease

= exposures that could be influenced by the disease

**Figure 5.9** *Time periods in the exposure–disease time sequence*
Source: Bluyssen; based on Baker and Nieuwenhuijsen, 2008; White *et al.*, 2008.

### 5.3.2 Methodology

The measurement of exposure can involve several techniques (see Chapter 6). Several types of study designs are applied to gather information about exposure and causes of diseases, such as (Baker and Nieuwenhuijsen, 2008):

- *Ecological studies*: studies in which a hypothesized association between environmental exposures and health outcomes are analysed using groups of people. It is determined whether population groups with a higher frequency of exposure tend to be the groups with a high frequency of health outcome occurrence.
- *Cross-sectional studies*: involve the measurement of the *prevalence* of disease in an at-risk population next to the assessment of exposures and other factors that might modify or confound the relationship between exposure and disease. For comparison a non-exposed population, similar to the exposed population, is also surveyed. Or, alternatively, the effects of different degrees of exposure within the whole exposed population are compared.
- *Case-control studies*: associations between exposures and a health outcome are studied by comparing cases within the study population with control subjects (free of the disease) selected from the same study population.
- *Cohort studies*: long-term follow-up of populations who have been exposed to an agent of interest and who are at risk of developing a particular disease or health outcome. Based on exposure status the individuals are divided into groups. The subsequent incidence of the health outcome within each group is studied. When the data is collected as the events occur, the cohort is named prospective. When the data comprises existing records it is named retrospective.

In general a cohort study uses a questionnaire to get the information for making an analysis of risk factors, often together with measurement of physiological endpoints and medical or psychological investigation. The subjects are followed over time. The contribution of the investigated risk factor to the effect is investigated taking relevant co-variates into account. For example, to determine the risks of certain effects caused by long-term exposure to outdoor environmental pollution problems such as noise from traffic during night (Miedema and Vos, 2007; Babisch, 2008; Kluizenaar *et al.*, 2009). Miedema and Vos (2007) based on data from 24 community surveys presented exposure-effect relationships for the association between long-term night transportation noise exposure and self-reported noise-related sleep disturbance.

Epidemiological studies are performed in the field, with people in their natural surroundings. Some advantages and disadvantages of the different field study designs are presented in Table 5.4. Laboratory or clinical experimental studies (see Section 5.4.2 and 5.5.2) are rarely performed in epidemiology.

### 5.3.3 Study A

An example of a *cohort* is the study performed by Dockery *et al.* (1993), in which the effects of air pollution on mortality for a 14- to 16-year follow-up period in 8,111 adults in six US cities demonstrated associations between fine particulate air pollution and mortality.

**Table 5.4** *Advantages and disadvantages of several epidemiological observational field study designs*

| Types of studies | Advantages | Disadvantages |
|---|---|---|
| Ecological | Usually based on existing data and therefore relatively inexpensive to conduct. | Difficult to interpret data because information about factors that could bias is often limited (aggregate measures of exposure may not reflect individual-level associations). |
| | Useful for studying rare diseases caused by relatively rare diseases, because the source population can be large. | Separation of effects of pollutants might be difficult because of high correlation of concentrations in different regions. |
| Cross-sectional studies | Usually costs less than a cohort or a case-control study. | Disease onset may influence subject's recall of earlier exposures and may affect biological markers of exposure. |
| | Suitable for studying factors that do not change as a result of disease, such as ethnicity, genetic markers of susceptibility or environmental exposures measured by objective methods. | Interpretation of associations between exposure and health outcome may be subject to selection bias due to 'selective survival' (e.g. persons with lung disease move to areas with low air pollution). |
| Case-control studies | Efficient for studying rare diseases with a long induction period. Often less expensive than a prospective cohort design. | Identification of appropriate control population is difficult; matching is sometimes necessary. Analysis does not give estimates of health outcome incidence rates, rate differences or attributable risk. |
| Cohort studies | Regarded as the most definitive of the observational designs. Temporal relationship between exposure and health outcome is clearly determined. Allows for multiple health outcomes in relation to exposure. | Difficult to conduct and can be expensive, especially if the latency period between exposure and the health outcome occurrence is long. Can be inefficient if health outcome is rare (a large study population is required). |

Source: based on Baker and Nieuwenhuijsen, 2008.

An example of a *cross-sectional study* is the European project SINPHONIE (see Box 5.7). SINPHONIE is a complex research project covering the areas of health, environment, transport and climate change and aimed at improving air quality in schools and kindergartens. The project was implemented under a European Commission service contract (DG Sanco, Health and Consumer Protection Directorate). Thirty-eight environment and health institutions from 25 countries participated in the SINPHONIE research project in order to implement Regional Priority Goal III of the Children's Environment and Health Action Plan for Europe, which is 'to prevent and reduce respiratory disease due to outdoor and indoor air pollution' (www.sinphonie.eu).

## 5.4 Psychology

*In psychology the focus lies on explanation of responses and mechanisms of the human mind (and interaction with the body) and involves measurement of personality, neurological and behavioural aspects by observational and experimental techniques.*

# Box 5.7  Study A: SINPHONIE – an example of a cross-sectional study

*Objectives*: Respiratory allergies are very common and increasing throughout Europe affecting between 3 and 8 per cent of the adult population, and the prevalence being even higher in infants. More than one in every three children in Europe has asthma or an allergy, and the rate of the respiratory illnesses is increasing year by year. These diseases are the major causes of days lost from school and their socioeconomic costs are very high. Children spend one-third of their normal day in school and they cannot choose the school environment. Moreover, children breathe a greater volume of air relative to their body weights than adults do and their immune system is not ready to respond to environment attacks, presenting higher susceptibility to many air pollutants than adults. In this scenario, it is crucial to assess the problem of indoor air pollution in schools. Therefore, the overall objective of the underlying proposal was 'to produce guidelines on remedial measures in school environment to cover a wider array of situations in Europe and to disseminate these guidelines to relevant stakeholders able to take actions'. The final objective of this project was to recognize the right to breathe clean air in schools as a fundamental health right at all levels.

*Methodology*: The project SINPHONIE comprised eight work packages (WPs).

WP1: *Management/coordination*

WP2*: Background*: To review critically and collate European (and non-European) research on the most indoor air relevant health effects and respective exposure contaminants in indoor air of schools, assessment of the policy relevance of their objectives and conclusions, and identification of epidemiological and toxicological research needs lying on the critical path for knowledge based policy development.

WP3 and WP4*: Assessment of outdoor/indoor school environment* and *Assessment of health outcomes*: representing the field studies in around 120 schools in 23 countries, on physical/chemical and biological aspects and on the health effects.

* To assess the building characteristics and the patterns of the everyday use of the selected classrooms influencing their indoor air quality (IAQ).
* To measure the physical parameters and chemical and biological pollutants of indoor (and related outdoor) air in schools and childcares in all Europe in order to produce new exposure data. Concretely, formaldehyde, benzene, pinene and limonene, naphtalene, nitrogen dioxide, carbon monoxide, carbon dioxide, radon, trichloroethylene, tetrachloroethylene, PAH and BaP, particulate matter and ozone; allergens in dust and mould and bacteria in dust and the air; temperature, relative humidity and ventilation rate in all classrooms.
* To obtain data on health status of children by questionnaire surveys, and also by clinical tests, focusing on asthma, respiratory infections, upper respiratory tract symptoms, cough, wheeze, dyspnea, allergic rhinitis, bronchitis and school performance.
* To evaluate the impact of externalities to the schools, as the traffic and the effects of climate change.

WP5*: Data management, cross analysis and database*: To create a database and to cross analyse the results of the field studies while integrating the WP2 contributions in order to assess:

* The influence of building characteristics, cleaning products and ventilation systems in exposure data obtained.
* The impacts of outdoor air pollution abatement measures, including measures taken in the short term, on IAQ in schools and on the children's exposure in school environments.
* A systematic source apportionment of indoor air pollutants in school environments.
* The influence of mixtures of pollutants in indoor air to the emergence of new pollutants caused by chemical and biochemical interaction.

WP6: *Health risk assessment*: To undertake a comprehensive risk assessment of what is at the core of the problematic of IAQ in schools today.

WP7: *Risk management and development of guidelines/recommendations*: To come up with a set of proposals regarding policies and guidelines that shall reflect wise and well balanced measures that will become the best expression of the current knowledge to be transferred into practice.

WP8: *Communication and dissemination*: To assure the adequate dissemination through timely actions during the project.

Source: www.sinphonie.eu.

### 5.4.1 Goals and objectives

Psychology is the scientific study of behaviour, literally it means the 'science of the mind' (Martin *et al.*, 2010; Figure 5.10). Even though it is a relatively young science, many types of psychology exist (e.g. psychobiology, psychophysiology, neuropsychology, social psychology, cognitive psychology, health psychology, organizational and occupational psychology, behavioural psychology, personality psychology, etc.). All forms of psychology are intended in understanding more about people and human nature, but approach this in different ways and with different observational and experimental techniques.

In psychology both quantitative and qualitative research is applied. Qualitative research is widely used on marketing and product design, and will therefore be discussed in Section 5.7.2. Quantitative research, in both health science and psychology, is mainly concerned with whether the *independent variable* (IV) (the treatment) influences the *dependent variable* (DV) (the outcome) (see definitions in Box 5.8). It is about making claims about the whole population on the basis of what is found in a sample (e.g. women drive better than men; the drinking behaviour of teenagers). For example, if the hypothesis is that higher temperature reduces productivity, the independent variable (temperature) must be controlled (manipulated by the experimenter) so that only the independent variable that is measured is responsible for the changes in the dependent variable (productivity).

**Figure 5.10** *Science of the mind*
Source: Bluyssen.

---

## Box 5.8  Some definitions in psychology

- *Response rate* helps to say something on the effect of representativeness.
- *Ecological validity* revolves around whether results gathered in a laboratory about some phenomenon can be generalized to other settings or places. In other words: Will participants behave in the same way outside the laboratory as they do when subjected to a study within a laboratory setting?
- *Independent variable (IV)* = the variable that the experimenter manipulates in order to measure an effect in the dependent variable.
- *Dependent variable (DV)* = the variable that is not under control, the outcome of the experiment.
- *Conditions* = different manipulations the participants are exposed to.
- *Extraneous variables* = the variables producing constant or random error according to the effect they have on the outcome. Random error obscures the effect investigated while constant error biases (or distorts) the outcome.
- *Confounding* = occurs when a real effect is between IV (or IVs) and the DV is biased or distorted by another variable.
- *Experimental group* = the group which is given some intervention or programme.
- *Control group* = the group who is not given the intervention.
- *Placebo effect* = a person (or group of people) responds to some stimulus as if it were having an expected effect when in fact it should have no effect at all.

Source: Langridge and Hagger-Johnson, 2009.

---

### 5.4.2  Methodology

Two categories of study designs are in general applied in psychology (Martin *et al.*, 2010):

- *Correlation studies*: Generally designs that rely on cross-sectional data from surveys or questionnaires. Variables are not manipulated. Correlations and regression are used to study the associations between them. Relations between measurements of environmental events, individuals' physical and social characteristics and their behaviour are examined to find causes of the observed behaviours. Correlation studies are often used for investigating the effects of personality variables on behaviour. To reduce the uncertainty inherent in correlation studies, *matching* is applied. Instead of chosen randomly, participants in each of the groups are matched on all of the relevant variables except the one being studied (see Section 5.2.1).
- *Created (experimental) studies*: In an experiment things are created and the experimenter observes the results to identify cause-and-effect relations. An experiment can be performed in a *laboratory* (most of the variables that can affect the outcome can be controlled) or in the *field* (participants are observed under 'fairly' naturalistic conditions. The observer does not need to interfere with the people (or animals) being observed.

In an experimental design the central process is to ascertain the relationships between the IV(s) and the DV(s). Several types of experiments can be distinguished (Langdridge and Hagger-Johnson, 2009):

- *True experiment*: Laboratory based investigation with experimental and control groups of participants that are randomly assigned to two (or more) conditions in a double-blind manner:
  - *Independent samples*: Different participants are assigned different conditions. Applied when order effects are likely, IVs are not amenable to repeated measurements, when people are not likely to experience two conditions in real life, and when the DV may be affected by repeated exposure of participants to conditions.
  - *Repeated measures*: Participants take part in both conditions. Used when order effects are not likely, the IVs are amenable to repeated measurements, when the design replicates are experienced in real life, and when individual differences between subjects are likely to have an impact on the DV.
- *Quasi experiments*: Experiments following the true experiment design partly.
- *Field experiments*: Experiment carried out in 'the field', the natural environment of those being studied. The experimenter can manipulate the independent variable(s) (IVs) and measure change in the dependent variable (DV) as in true experiments. Random assignment of participants to conditions is not possible.
- *Natural experiments*: Field experiments where the setting cannot directly be controlled and the IV(s) cannot be manipulated, but measurement of the effect of IV change on a DV is still possible.

Advantages and disadvantages of several psychological study designs are presented in Table 5.5.

**Table 5.5** *Advantages and disadvantages of several psychological study designs*

|  | Advantages | Disadvantages |
|---|---|---|
| True experiments | Accurate measurement and control of variables (minimize errors). Easy comparison, easy analysis of responses and easy replication. | Variables may lack construct validity and ecological validity. Artificial nature may produce artificial findings. |
| Field experiments | Greater face validity and ecological validity. Often cheaper. | Less controlled and greater opportunity for experimental error than true experiments. May raise more ethical concerns than true experiments. |
| Natural experiments | More naturalistic than a true experiment (greater ecological validity). Less potential negative impact on participants. | Less controlled than a true experiment; less (or no) control over the IV or IVs. Greater chance of experimental error and confounding variables. |
| Correlation studies | Some variables cannot be investigated in an experiment because they cannot be manipulated, such as a person's sex, genetic history, income, etc. For each of the group members these variables are measured and are related. | The variables to be kept constant might not all be known and the results can therefore be misleading. |

Source: based on Langridge and Hagger-Johnson, 2009.

### 5.4.3 Study B

An example of a correlation study performed in the architecture and building engineering field is the survey performed in the OFFICAIR study (Bluyssen *et al.*, 2012a). OFFICAIR is a European collaborative project, co-funded by the European Union, in the 7th Framework Program. The project, which comprised 15 partners from 11 countries, started on 1 November 2010 and ran for three years (see www.officair-project.eu). In the first phase of the project, a procedure was prepared for the inventory and identification of associations between possible characteristics of European modern office buildings (building, sources and events) and health and comfort of office workers, via a questionnaire and a checklist including environmental, physiological, psychological and social aspects (see Box 5.9). The survey was executed in 167 office buildings in eight European countries (Greece, Italy, Spain, Portugal, France, Hungary, The Netherlands and Finland) during the winter of 2011–12. The interesting part in this study is the fact that the study included physical, psychosocial, psychological as well as personal aspects.

#### Box 5.9 Study B: OFFICAIR (General survey) – an example of a correlation study

**Objective**: To inventory and identify stressors that have been appointed as possibly related to IAQ problems in European modern offices, via a field investigation (through questionnaires, checklists and IAQ monitoring). This includes, on the emitting side, the building/workspace characterization, the assessment of the physical, chemical and biological parameters and, on the occupant's side, the subsequent exposures and health effects of the time spent indoors.

**Methodology**: The procedure of the survey of the in total 167 investigated buildings comprised: the selection and gathering of the buildings, the survey in the buildings, and the data management.

*Selection of buildings*: In each of eight countries (Greece, Italy, Spain, Portugal, France, Hungary, The Netherlands and Finland) around 20 buildings were selected following pre-determined criteria:

- New or modernized buildings (use of modern equipment and access to Internet).
- Assuming a 70 per cent response rate, at least 860 workers should be investigated in each country (based on a singular calculated sample size of 600), resulting in 10 buildings with at least 70 workers and 10 buildings with at least 30 workers. (More workers were preferred though.)
- Access to basic information on building fabric, services, HVAC-systems (if applicable), cleaning practices, smoking polices, and so on.
- Buildings should have been operating in their current form for a minimum of one year prior to the start of the study (preferably two years).
- No major renovation planned before the autumn of 2012.
- Access to Internet for digital questionnaires and clear point of contact.

*Surveys*: A procedure with a questionnaire and checklist was prepared for the survey of 167 office buildings in the winter of 2011–12. The procedure was based on a three-week pattern in each building:

1 *Information* (Monday first week): A week before the survey took place, an email was sent out preferably by the management, or contact person, in the company, explaining the purposes and the contents of the survey (i.e. a questionnaire would be executed via Internet and the investigating team would visit the building to perform an inspection).

*continued . . .*

---

**Box 5.9 ... *continued***

2  *Invitation letter* (Monday second week): Each worker received an official invitation e-mail in his/her own language containing again a brief explanation of the purposes of the survey, the identification code of the building and the deadline for questionnaire filling in and a link to the digital questionnaire in the country's main language.

3  *Questionnaire*: In principle all workers of a building received a questionnaire. In buildings with more than 100 workers, a part of the building (one or two floors or, if the office buildings comprised several separated buildings, one or more of those) could have been selected. The workers were asked to give an informed consent and the need for a positive answer to further fill in the questionnaire was clearly explained. The worker had the opportunity to select a different language for the questionnaire. In order to maintain confidentiality of data after 90 minutes the answers were erased from the cache and were to be restarted from the beginning. Participants could save the survey at any time and resume it later.

4  *Walk-through with checklist* (preferably Tuesday or Thursday in second or third week): Most information was obtained by a walk-through the building, including plant rooms and circulation areas, as well as offices. This walk-through was in many cases accompanied by a facility manager or equivalent, who could supply some of the information orally. Furthermore, further documentation was obtained (before or after the walkthrough) to complete relevant parts of the checklist (such as maintenance records, cleaning schedules, settings of installations and layout of building).

5  *Reminder* (Wednesday third week): A reminder to fill in the questionnaires was sent on Wednesday morning of the third week to all workers.

6  *End of the survey* (Saturday third week): the deadline of filling in the questionnaire.

*Data management*: All data, from the questionnaire as well as the checklist, were digitally completed and stored in the secure database via a web-link.

Source: Bluyssen *et al.*, 2012a.

---

# 5.5 Clinical medicine

*In clinical medicine the focus lies on responses of the human body and involves measurement of physical and physiological responses that are part of the causal chain by using diagnostic and experimental techniques.*

## 5.5.1 Goals and objectives

In the Declaration of Helsinki, article 7 states:

> The primary purpose of medical research involving human subjects is to understand the causes, development and effects of diseases and improve preventive, diagnostic and therapeutic interventions (methods, procedures and treatments).
>
> (World Medical Association, 2008)

Clinical trials are undertaken to test the safety (information about adverse drug reactions or adverse effects of other treatments) and efficacy or effectiveness of healthcare services (e.g. medicine) or health technologies (e.g. pharmaceuticals, medical devices). Study subjects or patients (after assessment of eligibility and recruitment) receive one or more of the alternative treatments under study (Figure 5.11). (For definitions see Box 5.10.)

**Figure 5.11**
*Doctor examining a
patient*
Source: CSTB; ©
Fotolia/Heinemann.

## Box 5.10  Some definitions in clinical medicine

- *Trials* = an intervention or study in which human subjects receive one or more of the alternative treatments under study.
- *Interventions* = can be treatments (not all interventions are treatments).
- *Participants to a trial* = often named patients.
- *Eligibility criteria* = criteria used to select the trial participants; central to the external validity of the trial.
- *Allocation concealment* = process in which the schedule of allocating participants to the comparison groups is kept a secret for the people involved; persons enrolling participants do not know in advance which treatment the next person will get.
- *Randomization allocation* = process of randomization. Forms of randomization are:

    – *Simple randomization* = pure randomization based in a single allocation ratio analogous to a coin toss.
    – *restricted randomization* = any randomized approach that is not simple randomization, e.g. blocked, restricted and urn randomization.
    – *Blocked randomization* = to ensure that comparison groups are generated according to a predetermined ratio (usually 1: 1 or groups approximately of the same size).
    – *Stratified randomization* = to ensure good balance of participant characteristics in each group (e.g. age or stage of disease).
    – *Minimization* = ensures balance between intervention groups for several patient factors (such as age); first patient is truly randomly allocated; for each subsequent patient the treatment allocation that minimizes the imbalance on the selected factors between groups at that time is identified.

- *Random allocation sequence* = generation by a random-number table or computerized random number generators.
- *Blinding or masking* = withholding information about the assigned interventions from people involved in the trial who may potentially be influenced by this knowledge.

Sources: Baker and Nieuwenhuijsen, 2008; Moher *et al.*, 2010.

Diagnostic and experimental techniques applied comprise mainly physiological and physical techniques (see Sections 6.2 and 6.3).

## 5.5.2 Methodology

In general two types of study designs are applied in medical research: randomized controlled trials (RCT) and non-randomized trails (a quasi-experiment). RCTs are recognized as the standard method for 'rational therapeutics' in medicine. The CONSORT (Consolidated Standards of Reporting Trials) statement comprises a checklist of essential items that should be included in reports of RCTs and a diagram for documenting the flow of participants through a trial. The primary focus is on RCTs with a parallel design and two treatment groups (Moher et al., 2010; Schulz et al., 2010). Random assignment is the preferred method; it has been successfully used regularly in trials for more than 50 years (Moher et al., 2010). Advantages and disadvantages of several clinical medicine study designs are presented in Table 5.6.

### Randomized controlled trials

In a RCT a defined population is divided into groups using a randomized sampling procedure and the investigator assigns or administers an exposure (or treatment) to some of the groups (Baker and Nieuwenhuijsen, 2008). Another group, the comparison group, receives no treatment or receives already a

**Table 5.6** *Advantages and disadvantages of several clinical medicine study designs*

| | Advantages | Disadvantages |
|---|---|---|
| Randomized controlled trial (RCT) or randomized clinical trials | RCTs are considered to be the most reliable form because RCTs reduce spurious causality and bias. Facilitates blinding the identity of treatments, which reduces bias after assignment of treatments. | RCTs can be expensive and external validity (validity of generalized (causal) inferences in scientific studies such as settings, procedures and participants in other populations and conditions) may be limited. |
| | Permits the use of probability theory to express the likelihood that any difference in outcome between intervention groups merely reflects chance. | The conduction of RCTs can take several years. |
| Non-randomized trial (quasi experiment) | Easier set-up than true experimental designs, which require random assignment of subjects. | The lack of random assignment in the quasi-experimental design method may allow studies to be more feasible, but this also poses many challenges for the investigator in terms of internal validity. |
| | Minimizes threats to external validity, as natural environments do not suffer the same problems of artificiality as compared to a well-controlled laboratory setting. | Because randomization is absent, some knowledge about the data can be approximated, but conclusions of causal relationships are difficult to determine due to a variety of extraneous and confounding variables that exist in a social environment. |

Source: Moher et al., 2010.

standard treatment. After the treatment, the groups are followed to determine whether the experimental treatment affects the health outcome over time. Field trials are similar, but they involve disease-free persons who are selected from a community and studied in 'the field' rather than in the hospital.

In a typical randomization process, several steps can be distinguished (Moher *et al.*, 2010):

- *Sequence generation*: generate allocation sequence by some random procedure (see Box 5.10).
- *Allocation concealment*: develop allocation concealment mechanism (such as numbered, identical bottles or sequentially numbered, sealed, opaque envelopes) and prepare the allocation concealment mechanism using the allocation sequence from the sequence generations step.
- *Implementation*: enrol participants (assess eligibility, discuss the trial, obtain informed consent, enrol participant in trial), ascertain intervention assignment (such as opening next envelope) and administer intervention.

RCTs are mainly trials with participants individually randomized to one of two 'parallel' groups. Main alternative designs are multi-arm parallel (> 2 group parallel group trials), crossover (each participant receives two (or more) treatments in a random order), cluster (sampling of groups) and within person (participants receive two treatments simultaneously (often to paired organs)) (Moher *et al.*, 2010; see also Section 5.3.1).

### Non-randomized trial (quasi experiment)

A non-randomized trial or a quasi-experiment is a natural experiment, which can be applied in longitudinal research that involves longer time periods, and can be followed up in different environments.

### 5.5.3 Study C

Interventions can take place both in the field or in a laboratory setting. An intervention study is no more than measuring before and after a change has been made. This change can be administering a medicine or changing environmental (exposure) conditions. An example of an intervention study is the intervention that was performed in one to two building(s) per country during the winter of 2012–13 in the OFFICAIR study (Bluyssen *et al.*, 2012a). The buildings were selected from the 167 buildings investigated in the general survey (see Box 5.9). The intervention included several physiological, psychological as well as chemical and physical measurements of the environment:

- Chemical/physical measurements in the indoor environment.
- A checklist to confirm building characteristics collected in the general survey.
- A time-activity diary (see example in Table 6.2).
- A questionnaire, performance test and a memory test to be delivered online to the 20–30 selected workers.
- Several medical tests of 20–30 workers per building investigated.

## 5.6 Toxicology

*In toxicology the focus lies on the explanation of mechanisms in the human body and involves measurement of adverse effects of chemical substances by* in vivo *(within the whole body) and/or* in vitro *(outside the body) experiments, using animals, tissues and/or humans.*

### 5.6.1 Goals and objectives

Toxicology is defined as 'the study of the adverse effects of chemical, physical or biological agents on living organisms and the ecosystem' and is based on the sixteenth-century principle that any substance can be toxic if consumed in sufficient quantity (AltTox.org, 2009).

Toxicity can arise from reversible or irreversible effects and can affect a range of different organs to different degrees: from minor changes, such as reduced weight gain, small physiological changes or changes in hormone levels, to intermediate effects such as destroying tissue (e.g. skin irritation and corrosion) and adversely affecting tissue function (resulting in pain and suffering), to severe effects such as organ function loss, leading to death (Hepple *et al.*, 2005).

Toxicity is the capacity of a toxic agent to produce injury in an organism and can be acute, adverse health effects occurring within minutes, hours or days (e.g. carbon monoxide poisoning), or can be delayed, from several weeks to many years, even decades (e.g. asbestosis) (Baker and Nieuwenhuijsen, 2008). In Figure 5.12 an example of the toxic effect of exposure of human lung cells (in a medium) to candle smoke as compared to medical air and $NO_2$ is presented by reduction of Alamar blue, which is an indication of cell viability. From the colour difference it can be seen that exposure with candles gave more toxicity than exposure to 30 pmm $NO_2$.

Toxicity studies are conducted (Hepple *et al.*, 2005) to:

• Assess the degree to which substances are toxic (poisonous) for humans, animals or the environment.
• Investigate the mechanism of toxic chemicals.
• Develop new or improved tests for specific types of chemically induced effects.

Important types of studies include (Hepple *et al.*, 2005; AltTox.org, 2009):

• *Acute toxicity studies*: concern examination of adverse effects that may occur on first exposure to a single dose of a substance (usually performed with animals). For example, the effects of contact with the skin and eye (corrosion, irritancy and sensitization; topical or local toxicity – see for definition Box 5.11) and the effects on internal organs of a substance that is swallowed, inhaled, absorbed through the skin or injected (systemic toxicity).
• *Genotoxicity studies*: involve assessment of the potential of substance to interact with genetic material (DNA and chromosomes). Even though the best and most definitive method for demonstrating DNA reactivity is assessment of DNA adducts *in vivo* in the target tissue (Cohen and Arnold, 2011). And even though many have a predictive value for carcinogenesis of less than 50 per cent, nevertheless these studies mostly use *in vitro* assays (see Box 5.11). The reasons can be found in the considerable levels of toxicity

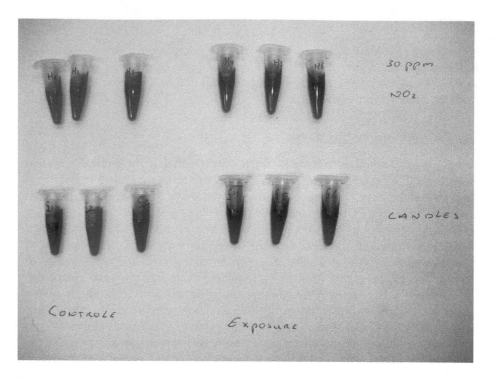

**Figure 5.12** *Toxicity effect of the exposure of human lung cells during 1.5 hours to emissions of scented candles and 30 ppm NO$_2$ shown by the reduction of Alamar blue (part of Study D: see Box 5.12). The control is exposure with medical air. Alamar Blue (10% v/v) was added to the cell layer and medium*

Source: Bluyssen.

and indirect effects on DNA and chromosomes, which cannot be verified *in vivo* (Cohen and Arnold, 2011). Computer analysis relating three-dimensional chemical structure to DNA reactivity are available, and could perhaps take over some of these studies.

- *Repeated-dose toxicity studies*: concern identification of toxicity that occurs after continuous exposure to a substance, identification of the organs most affected and determination of the doses at which each effect occurs. Tests allow an assessment of the NOAEL (see Box 5.11).
- *Carcinogenicity studies*: involve tests that are undertaken to find out whether cancers may develop as a result of exposure to a certain chemical (are usually performed with animals).
- *Safety pharmacology studies*: evaluate the safety of pharmaceutical products and are usually performed with animals.

The vast majority of toxicity testing is carried out in the context of regulatory requirements. This process of regulatory risk assessment can be broken down into three main phases (AltTox.org, 2009):

1   *Hazard identification*: determination of a substance's intrinsic toxicity (e.g. eye irritation, birth defects or cancer) through the use of toxicity tests. Test results are then analysed to determine what, if any, adverse effects occur at

---

## Box 5.11  Some definitions in toxicity studies

- *Acute toxicity* = toxicity produced after administration of a single dose (or multiple doses) in a period not exceeding 24 hours, up to a limit of 2,000 mg/kg.
- *(Chemical) substance* = a form of matter that has constant chemical composition and characteristic properties.
- *NOAEL* = the no-observed-adverse-effect level or the highest dose without significant effect.
- *LOAEL* = the lowest-observed-adverse-effect level (used if no NOAEL is available).
- $LD_{50}$ (the median lethal dose, abbreviation for 'lethal dose, 50 per cent'), $LC_{50}$ (lethal concentration, 50 per cent) or $LCt_{50}$ (lethal concentration and time) of a toxin, radiation or pathogen = the dose required to kill half the members of a tested population after a specified test duration. $LD_{50}$ figures are frequently used as a general indicator of a substance's acute toxicity.
- *In vitro* = using cells and tissues outside the body in an artificial environment.
- *In vivo* = in Latin 'within the living' is an experiment using a whole, living organism as opposed to a partial or dead organism, or an *in vitro* ('within the glass', i.e., in a test tube or petri dish) controlled environment.

Sources: Calabrese and Kenyon, 1991; Hepple *et al.*, 2005; Chapman and Robinson, 2007.

---

different exposure levels (known as a 'dose–response' assessment) and, where possible, to identify the lowest exposure level at which no adverse effects are observed (known as the NOAEL, see Box 5.11).

2   *Exposure assessment*: determination of the extent of human and/or environmental exposure to a substance, including the identification of specific populations exposed, their composition and size, and the types, magnitudes, frequencies and durations of exposure.

3   *Risk characterization*: a composite analysis of the hazard and exposure assessment results to arrive at a 'real world' estimate of health and/or ecological risk.

## 5.6.2  Methodology

The use of living animals to study the potential adverse effects of new drugs, food additives and other substances has been the default method for testing for a long time. The use of animals for research is not described here. For further information one can consult 'The ethics of research involving animals' (Hepple *et al.*, 2005). Instead alternative methods are presented and discussed. The term 'alternative' is used to describe any change from present procedures that will result in the *replacement* of animals, a *reduction* in the numbers of animals used, or a *refinement* of techniques to alleviate or minimize potential pain, distress and/or suffering of animals (Chapman and Robinson, 2007; AltTox.org, 2009).

Methods not involving living animals include (Hepple *et al.*, 2005; AltTox.org, 2009):

- *In silico techniques*: mathematical and computer studies of biological processes including prediction of the biological activity of substances, and the modelling of biochemical, physiological, pharmacological, toxicological and behavioural systems and processes.
- *In vivo testing*: research involving human participants such as non-invasive brain scanning (see Section 6.3.2) and human volunteers for medicine testing.
- *In vitro testing*: research on isolated human or animal cells and tissues in culture, such as (1) sub-cellular (cell-free) fractions (e.g. mitochondria or ribosomes), (2) primary cells and cell lines grown in liquid suspension (cell culture or tissue culture), (3) tissue slices or fragments and even whole organs or (4) purified molecules in the test tube (e.g. proteins, DNA or RNA either individually or in combination).

Advantages and disadvantages of these methods are presented in Table 5.7.

### 5.6.3 Study D

*In vitro* testing, the technique of performing a given procedure in a controlled environment outside of a living organism, provides an additional source of information. Several *in vitro* test methods are available. These methods differ in the cell types used (human or other, primary cells or transformed cell lines, epithelial or other), the exposure method (suspension of collected materials, exposure of conventional cultures to aerosols, or air–liquid interface exposures), and the toxicological endpoints assessed (cell viability (cell death), cytokine production, oxidant stress, cell type-specific function) (Seagrave, 2005).

With the commercially available CULTEX® system (Aufderheide, 2008), it is possible to expose cells cultivated on porous transwell membrane inserts to a dynamic flow of test atmospheres without medium interference. This system was

**Table 5.7** *Advantages and disadvantages of several toxicological studies*

| | Advantages | Disadvantages |
|---|---|---|
| In silico techniques | No use of living animals. | Modelling only. |
| In vivo testing | No use of living animals. No conversion from animal or tissue to real life. | Ethical considerations. Problems caused by variability in human population, controlling variables such as diet and health over long periods and slow rate of human reproduction. |
| In vitro testing | No use of living animals. | Diversity of different tissues and cell types may function and respond in different ways, or to different degrees. |
| | Compared to human studies: faster results. | Cells and tissues interact differently at various locations in the body. |
| | Performed in a lab environment. | Tissue organization (oxygen levels, intercellular communication) affect how cells behave and respond to external stimuli. |
| | Permits simplification. | Extrapolation is a challenge. |

Sources: AltTox.org, 2009; Hepple *et al.*, 2005.

used to test several biological effects associated with human lung cells exposed to several indoor air sources (Bluyssen *et al.*, 2013). In Box 5.12 the goal and methodology is presented. An important aspect of this study was to test whether the CULTEX® system has the potential to study the biological effects of indoor air sources. This was tested by calculating the reproducibility of the toxicological results. How this was done is explained in Section 5.8.2.

---

### Box 5.12  Study D: an example of an *in vitro* study

*Goal*: To test whether the CULTEX® system has the potential to study the biological effects of indoor air sources.

*Methodology*: The emissions of different sources (25 scented candles: ncandles (non-burning) $n = 3$, candles (burning) $n = 5$), two types of sprays (hairspray $n = 5$; water repellent spray $n = 4$)) produced in an over-pressurized climate chamber (4m × 4m × 2.5m) were tested. The internal wall of the climate chamber was covered with aluminium foil to prevent emissions of construction materials. The air supply (via a F7 and a F11 filter) was located in one of the walls, at a height of 2 m next to the chamber door (see Figure 5.13). To prevent pollutants from outside entering the chamber when the door was opened, an extra opening with built-in gloves was created. Via these gloves it was possible to light the candles or to spray without having to enter the chamber (into the hood every 10 minutes for one minute) (see Figure 5.14). Per type of source several repetitions of the test conditions were performed during a time period of 1.5 hours for candles and 1 hour for both sprays. Additionally, one blank situation was tested (empty chamber). Temperature, relative humidity and concentrations of the gases NO, $NO_2$, $SO_2$ and $O_3$ and fine and ultrafine particles (UFPs) were monitored or collected. For some tests, VOC and PAH were assessed. The outdoor airflow was calculated from the pressure difference over the orifice plate (see Figure 5.13). Table 5.8 gives an overview of the measurements and applied equipment. For the health effect assessment of the human lung cells, the Interleukin-8 (IL-8), Lactaat Dehydrogenase (LDH), Alamar blue and RCDC (Reducing agent Compatible Detergent Compatible) Protein assay were measured as toxicological indicators. All results were compared to medical air, also in the blank situation.

Source: Bluyssen *et al.*, 2013.

---

**Table 5.8** *Overview of measurements and equipment*

| | Component | Methods |
|---|---|---|
| (Ultra) fine particles | Fine particles (10–523 nm) | Scanning mobility particle sizer (SMPS) |
| Toxicity | Toxicity human lung cells | CULTEX®: LDH, Alamar blue, RCDC and IL-8 |
| Gasses | Nitrogen oxides ($NO_x$)<br>Sulphur dioxide ($SO_2$)<br>Ozone ($O_3$) | Chemilumescence $NO/NO_x$-analyzer<br>Fluorescence $SO_2$-analyzer<br>Photometric $O_3$-analyzer |
| Chemical pollutants | Volatile organic compounds (VOCS)<br><br>PAH and derivates | Tenax/automated thermal desorption Gas chromatography-mass spectrometry (ATD GC-MS)<br>Berner impactor/ATD GC-MS |
| Other | Temperature (T) [°C]<br>Relative humidity (RH) [%]<br>Ventilation rate [m³/h] | Thermometer<br>Air humidity meter<br>Air pressure measurement |

Source: Bluyssen *et al.*, 2013.

**Figure 5.13**
*Horizontal cross-section of the experimental set-up: measurement points for respectively fine dust, lung cells, $SO_2$ + $NO_2$, (S) VOC + PAH, Temperature (T) + Relative humidity (RH), $O_3$, ventilation + pressure, are noted by numbers 1 to 7. Measurement points 1, 2 and 4 are located next to the wall of chamber. Measurement points 3 and 6 are connected with tubes to continuous measurement equipment outside the chamber. Measurement 5 is performed inside the chamber. F11 and F7 represent two different air filters through which the incoming air is filtered*

Source: Bluyssen.

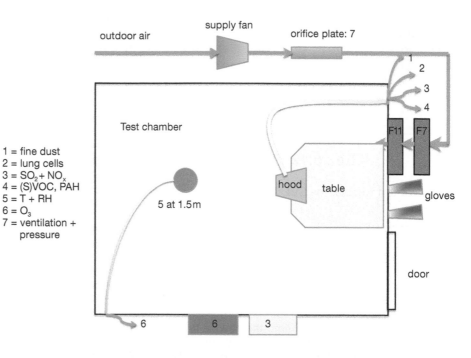

1 = fine dust
2 = lung cells
3 = $SO_2$ + $NO_x$
4 = (S)VOC, PAH
5 = T + RH
6 = $O_3$
7 = ventilation + pressure

a

**Figure 5.14**
*(a) Lighting of the candles with an electrical lighter and (b) spraying*
Source: Bluyssen.

b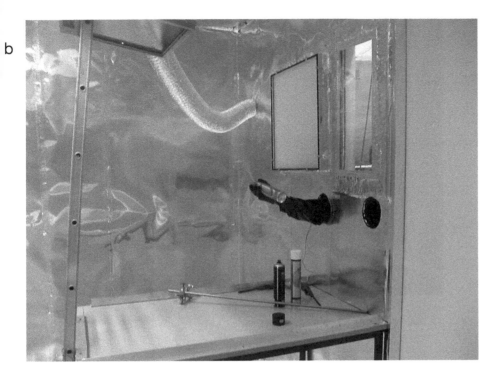

## 5.7  Product design

*In product design the focus lies on understanding users' emotional needs together with functional and social needs: the emotional experience of users with a product. Both quantitative and qualitative research methods are therefore applied in product marketing and design.*

### 5.7.1  Goals and objectives

Design researchers investigate the emotional experience of users with a product within the product experience context (e.g. a car within the context of driving it – Figure 5.15), and how users interact with a product and react emotionally (Karahanglu, 2008). The experience of the user with a product is shaped by both the characteristics of the user (e.g. personality, skills, background) and the product (e.g. colour, shape, texture) (Demir *et al.*, 2006). In addition, the context in which the interaction takes place influences the experience (Desmet and Hekkert, 2007). According to Desmet and Hekkert (2007) three levels of product experience can be distinguished:

1   *Aesthetic experience*: involves a product's capacity to delight one or more of our sensory modalities.
2   *Experience of meaning*: involves the ability to assign personality or other expressive characteristics and to assess the personal or symbolic significance of products.

**Figure 5.15** *A car in vision*
Source: Anthony Meertins, August, 2012.

3   *Emotional experience*: involves those experiences that are typically considered in emotion psychology such as love and anger, which are elicited by the appraisal relational meaning of products.

Product experience is a multi-faceted phenomenon that involves manifestations such as subjective feelings, behavioural reactions, expressive reactions and physiological reactions (e.g. pupil dilation and sweat production). Usability, although not an affective experience but a source of product experience, is relevant for user-centred design approaches. The level of experienced satisfaction (which is a pleasant emotion) is often used as a measure for usability (Desmet and Hekkert, 2007).

Some definitions used in product marketing and design are presented in Box 5.13.

## 5.7.2 Methodology

Both quantitative and qualitative research methods are applied in product marketing and design (Sheth *et al.*, 1999):

- *Quantitative methods* comprise of surveys or experiments.
- *Qualitative methods* comprise of focus groups, customer visits, motivation research and interpretative research.

Along with these specific quantitative and qualitative methods, there are several special research methods: information-processing research, simulation, secondary

---

### Box 5.13  Some definitions used in product marketing and design

- *Customer* = a person or an organizational unit that plays a role in the consummation of a transaction with an organization.
- *Transaction* = an agreement or movement between a buyer and a seller.
- *User* = a person who actually consumes or uses the product or receives the benefits of the service.
- *Buyer* = a person who participates in the procurement of the product.
- *Attitude* = learned predisposition to respond to an object or class of objects in a consistently favourable or unfavourable way.
- *Attributions* = inferences that people draw about the causes of events and behaviours.
- *Beliefs* = expectations that connect an object to an attribute or quality.
- *Non-disguised-structured* = opinions or attitudes on pertinent topics with pre-structured answers such as agree/disagree or true/false.
- *Non-disguised-non-structured* = questioning is open-ended and broad.
- *Disguised-structured* = real intent is disguised, response categories are provided.
- *Disguised-non-structured* = research purpose is not apparent, response categories are not pre-structured. The respondent is asked to interpret a vague stimulus. Techniques used are word association, sentence completion and story completion.

Source: Sheth *et al.*, 1999.

---

research and research using virtual reality and the Internet. These methods range from qualitative (verbal, unordered responses) to quantitative (numerically scalable responses), from exploratory (discovering response possibilities hitherto unknown) to directed (finding out the incidence of previously known responses), from secondary (using published data) to primary (collecting new data), from low-tech (e.g. face-to-face interview) to high-tech (e.g. using virtual reality) and from objective (e.g. measuring eye movement) to subjective (e.g. interpretive research) (Sheth *et al.*, 1999).

### Quantitative methods

The quantitative methods applied are similar in set-up as in psychology (see Section 5.4). Several observation behaviour techniques are applied comprising physical responses (e.g. eye movement, galvanic skin response and brain activity), nonverbal responses (e.g. information boards, visual image profiling) and verbal responses (e.g. concurrent and retrospective protocols).

*Surveys* find out what customers think, know or feel already by asking them to respond to a questionnaire via an interview (on the phone or face-to-face) or self-administered. *Experiments* present some stimuli and seek customer reactions to these stimuli by placing them in a situation that does not 'normally' occur and their behaviour is observed or recorded.

Some major topics researched by quantitative methods are (Sheth *et al.*, 1999):

- *Attitude research*: Attitudes are measured by various scales (e.g. numerical: semantic differential scale, likert scale or pictorial: useful for children) (see Section 6.5.1).
- *Image-self-concept measurements*: Consumers' own self-image, as well as their image of specific products, brands and so on, is measured by using a semantic differential scale (e.g. pleasant to unpleasant; modern to traditional). Such product personality profiling techniques relate to user's self-evaluation of products and emotions (McDonagh *et al.*, 2002). With mood boards or picture cards on the other hand the type of emotions elicited can be studied (Hoem and Bjelland, 2006).
- *Multi-attribute attitude models*: A consumers' attitude towards a product or service is a weighted sum of his or her beliefs about the extent to which the product or service possesses a set of attributes or characteristics. One measures first consumer's perceptions or beliefs about the attribute of the product and then the importance consumers assign to these attributes.
- *Perceptual and preference mapping* (multidimensional scaling): Employed to understand what attributes and criteria consumers use to judge and evaluate alternative brands, products, services and so on. It provides a visual map in a multi-dimensional space that shows how similar or different various brands are considered to be by the consumer.

## Qualitative methods

Qualitative methods do not ask the customer to limit his or her answers to pre-assigned response categories. In *focus groups* a small group of customers (6–15 persons chosen by convenience sampling; see Section 5.2.1) is assembled in a room. A moderator steers the group discussion along certain questions of interest to the marketer. The focus is to understand customers' worldview on a certain topic.

*Customer visits* are applied to hold interviews with customers and observe each customers' experience with the supplier's product (discovery research). Visits are made by cross-functional research teams, comprising people from design, production, logistics and marketing.

To discover the reasons (i.e. motives) of a person's behaviour, *motivation research* can be applied. Motivation research is used to find out the conscious and unconscious reasons that motivate people to buy or not to buy a particular product, service or brand, or to avoid a store, or to accept or reject a marketing communication.

In *interpretative research* (ethnographic studies) a researcher observes a customer or a group of customers in their natural setting, and interprets that behaviour based on an extensive understanding of the social and cultural characteristics of that setting. The researcher becomes a participant observer of the scene (see Section 6.4.2).

Advantages and disadvantages of these qualitative methods are presented in Table 5.9.

**Table 5.9** *Advantages and disadvantages of several qualitative marketing and product studies*

|  | Advantages | Disadvantages |
|---|---|---|
| Focus | A good way to generate ideas and to test new concepts. | Focus groups are not a statistical groups sample of the target population, so findings are not projectable to the entire population of target customers. |
| Customer | A good opportunity to understand customers' requirements (which are generally multifaceted and complex due to the complex nature of products). | Suited for business-to-business visits markets (as opposed to consumer goods companies), especially to producers and vendors of high-tech or otherwise complex products (such as computers, aircrafts, communication networks, etc.). |
| Motivation research | To study reasons that motivate people to buy a product. | Can be difficult to interpret the results. |
| Interpretative | Encounter customer and consumption activities in their natural settings over a longer time; first-hand information. Trust of observer will give more sincere answers. | Requires highly skilled and research well-trained researchers; very time consuming and very expensive. Interpretation of data is too subjective. |

Source: Sheth *et al.*, 1999.

## 5.7.3 Study E

A good example of product research is the way the perfume industry uses panels of professional sniffers to evaluate different smells. Another example is the food industry, in which panels of professional tasters are used to evaluate different attributes of food. Human subjects also have been used to evaluate the air quality in a building (Bluyssen *et al.*, 1996b) and even of a building product or an installation component in laboratory settings (Bluyssen *et al.*, 2003). In many cases, the human senses are superior to chemical analysis for assessing how air is perceived. The striking factor of the human nose is its extreme sensitivity to low concentrations of many chemical substances and its ability to discriminate among them, as compared with the performance of physical/chemical instruments. Furthermore, psychological effects cannot be mimicked by detectors. Therefore, a sensory evaluation of indoor air pollution is often necessary.

The composition of the panel depends on the purpose of the test. Sometimes the test panel is recommended to comprise selected, sensitive or even trained (or calibrated) persons; at other times naïve subjects are preferred (Berglund *et al.*, 1999b). It is a fact though that a truly naïve untrained panel does not exist. A certain level of training is required even for the most simple evaluation and this training is enhanced if the subject participates in several test sessions. One can 'calibrate' test persons for example by using a reference gas and a reference scale (Bluyssen, 1990). The advantage of using trained panel members is that less panel members are required to reach the same accuracy of their mean votes. But comparison of the performance of trained and untrained panels is not simple since the primary voting scales are different. In a study where the performance of both were compared, an average of approximately 280 untrained panel members was required to match a trained panel of around 11 persons (Bluyssen and Elkhuizen, 1996).

In a European project named AIRLESS (Design, Operation and Maintenance Criteria for Air Handling Systems and Components for Better Indoor *Air* Quality and *Lower Energy* Consumption, Pre-Normative Research), both trained and untrained panels were used to evaluate the air quality from several HVAC-system components. AIRLESS ran from 1998 to 2002 with eight institutes from six European countries (Bluyssen *et al.*, 2003). In Box 5.14 the methodology for testing the air polluting effect of air filters exposed to intermittent and continuous flow during 28 weeks is presented.

---

## Box 5.14  Study E: AIRLESS filter study – an example of building product evaluation

*Objective*: Air filters in heating, ventilation and air conditioning systems are used to remove dust, pollen bacteria and other contaminants from the incoming air, in other words clean the incoming air. However, in several studies it was shown that a substantial part of the pollution in ventilation systems mainly originated from the filters (Bluyssen, 1990; Pasanen *et al.*, 1994). Generally, microbial growth in the filters is seen as one of the possible underlying causes. It is however unclear which effects temperature and humidity conditions in an air handling unit have on the growth of microorganisms. From a literature survey the continuity of the airflow through the filter and water content come forward as important factors. Within the framework of the research project AIRLESS, the sensory pollution loads from standard glass fibre filters (filter class F7, 287 x 592 mm, depth 655 mm) in a continuous airflow and an identical filter in an intermittent airflow were compared.

*Methodology*: The set-up of the measurement section is presented in Figure 5.16. Air coming from the outdoor air intake is led to two separated, parallel compartments. Each compartment has its own damper for controlling the flow rate through the compartment. For the sensory evaluations, air from the air-handling unit was extracted through glass tubing and led into a glass cone. The glass construction was chosen to minimize sorption effects. To overcome the under pressure in the air handling unit micro ventilators were installed. An electrical heating wire was added on the outside of the glass tubing. The heating was supplied outside to prevent any disturbance of the sensory and chemical pollution load of the air sample. To prevent pollution from the ventilator to influence the air quality evaluations, it was positioned downstream from the measuring points.

Sensory evaluation, using a trained panel comprising 8 to 12 persons (divided into three groups), and VOC-concentrations upstream and downstream from the filters were measured after 0, 4, 8, 13, 18 and 28 weeks (see Figure 5.17). Microbial growth on the filters was evaluated at the start of the measurements (before the filters were installed) after 13 and after 28 weeks. Temperature and humidity in the air handling unit and pressure difference over the filters were measured continuously. The collected dust on the filters was determined by weighing the filters before they were installed and at the end of the experiment.

During the evaluations the panel members waited in a low polluting room that was placed close to the test unit. The subjects were mostly college students. The percentage of females in the panel was 45 per cent and the percentage of smokers was 35 per cent. Sensory evaluations were performed after 1 week, 4, 8, 13, 18 and 28 weeks. At each measurement day, the panel members assessed all four samples twice, except for the measurement after 28 weeks.

Source: Bluyssen *et al.*, 2003.

**Figure 5.16**
*Experimental set-up of testing of filters (further explained in Box 5.14)*
Source: Bluyssen.

**Figure 5.17**
*Smelling of filter in HVAC-system*
Source: Bluyssen.

## 5.8 Data analysis

*The goal of data analysis is to reveal patterns of interest in the data, ideally without being clouded by bias or random error. In general data analysis comprises multiple steps:*

- *Data management and data cleaning.*
- *Exploration of data and relationships.*
- *Probability testing and/or regression analysis.*

### 5.8.1 Data management and cleaning

After the study has been executed, the data collected has to be 'filed' or stored in such a way that it can be easily called upon for analysis. Before data are 'clean' for analysis (see Figure 5.18), several steps can be taken before, during and after data entry.

### Before data entry

Before data entry it is important to set the rules of the database in which the data are entered, such as pre-defined variable names, codes and where appropriate, multiple entry choices. When data are entered in for example pre-defined Excel sheets, it is very important to provide the persons who have to enter the data with exact instructions, such as:

- Please do not change number of lines or columns or other pre-coding (e.g. do not delete empty rows or columns, or add rows or columns).

**Figure 5.18** *Cleaning and sorting of data*
Source: Bluyssen.

- A number of changes are only possible when you 'unprotect' the cells: unprotecting should be avoided.
- When using copy-paste, you may accidentally change the pre-set options, which will make the file difficult to use centrally, so please, proceed with care.
- Please do not add comments or new lines with comments to the files, as they will not be retained for use in the pooled analysis. Instead, keep the comments in your own files that you keep for your own use and for further checks.

But also in the case when data are entered digitally via a web-based survey, some coding is required, even though the user (in this case the subject) is not aware of it. Important in this case, are the accurate and consistent explanations for the user on the screen.

Agreements on coding before the data are entered is important, especially when you are dealing with data sets in different countries. For example:

- How to deal with missing or incomplete information: do you use '.', nothing (empty) or '0'? Please distinguish between missing on purpose (not available or below detection limit) or forgotten entry.
- About dates: All dates should be coded in the same way, for example as 'dd/mm/yyyy'.

## During data entry

When data are entered manually, it is important to use an independent double entry (with comparison of the results and correction of the inconsistencies). Alternatively, the data can be checked by going through all entries with two persons, controlling the entries against the questionnaires. Other rules can be for example that:

- No personal information (other than the codes and information asked in the questionnaire) should be passed to the central database.
- In a valid record, no cells should be left empty.
- Before completing the submission, please check that the names of the submitted files are the same as those you have received to fill in (please do not change file names).
- Keep backups of the files at every step, and when you upload them.
- Please only submit files that are complete.

## After data entry

After the data entry, a first data cleaning can be performed. This can comprise:

- Checking file names in case of duplicates.
- In each file checking number of columns, codes in column A (and sometimes B), format for date (dd/mm/yyyy).
- Scrolling through the file to see if there are any strange signs ('*', 'xxxx', 'other').

- For files with numbers: decimal divider is ',' or '.'; below detection limit is coded as '0'.
- Non-empty records should have '.' or another sign in every cell.

A second data cleaning round should be performed on the data itself, applying validation rules that are based on acceptable ranges and histograms, and possibly including logical checks. This step is usually part of the descriptive analysis.

### Database

For storing the collected data in a database, many possibilities are available. The most used option is a commercial database management system (DBMS) such as the relational SQL-based Oracle system (www.oracle.com/us/products/database/overview) or its counterpart from Microsoft, SQL Server (www.microsoft.com/nl-nl/Server-Cloud/sql-server/default.aspx). However, this way there might be a risk of a vendor-lock in. Ideally you want to free your data by using a DBMS supporting open standards and even better being open source (free) itself.

And, one could use something even more innovative and choose an open standard semantic web approach with an open source implementation. For this to work for 'normal' data, for example in Excel, the data is transformed to a language this web-server can read. This is called OWL (web ontology language), the World Wide Web Consortium's language (www.w3.org/2004/OWL). Additionally, a tool is required to put that OWL data on a web server and to retrieve the data from the web server with so-called SPARQL queries. An example of an open source system here is Jena (http://jena.apache.org).

If one doesn't mind losing the open standard but wants to retain a free solution, one could use 'Google Fusion Tables', for which positive aspects are web-based functionality (no local software installation needed at all) and that data can be entered online. On the other hand, there is limited validation support and it has a limited number of data types (e.g. text, number, location, date/time). Also the data output needs extra transformation for example to read the data in Excel (www.google.com/fusiontables/Home).

### 5.8.2  Descriptive analysis

*In descriptive or exploratory analysis* the data is summarized using descriptive tables, histograms and calculation of average values, standard deviations or variances in order to (Langdridge and Hagger-Johnson, 2009; see Box 5.15):

- Evaluate for reasonable ranges or frequencies and gain appreciation for broad patterns in the data.
- Look for errors (may inspire additional data cleaning such as out of range data), outliers (should only be omitted if there is a well-founded scientific justification) or other data abnormalities.
- But also to identify important confounding variables and predictors or surrogates of the response in order to characterize the form of functional relationships to be expected and analysed further in the inferential analysis.

## Box 5.15 Calculations of averages, standard deviations and variances

Measures of central tendency

- *Arithmetic mean* (average):

    $$X = \Sigma x/n \qquad [5.6]$$

    Where: $X$ = mean; $\Sigma x$ = all the values of the samples; $n$ = number of samples.

    Best used for interval/ratio measurements.
- *Median* (med): the central value of a set of numbers. If the set of numbers is even, then the median lies in between two of the central values. When the number of values in a data set are few, the median can be unrepresentative. The median is best used for ordinal measurements.
- *Mode* (mo): the most frequently occurring value in a set of numbers; works well for data on a nominal scale, but can also be used for ordinal measurements. The data set: 1,1,2,2,2,3,3,4,4,5,5,5,6,6 is *bi-modal* (2 and 5 are most frequently occurring).

Dispersion (spread) can be measured with:

- *Range*: distance between the highest and lowest values in a set of numbers; can be distorted by extreme values and says nothing about distribution of data.
- *Semi-interquartile range*: distance between the two values which cut off the top and bottom 25 per cent of the scores divided by two.

    $$(Q_3 - Q_1)/2 \qquad [5.7]$$

- *Mean deviation*: the mean absolute difference between a score and the mean for a set of scores.

    $$MD = \Sigma \, |x{-}X| \, /n \qquad [5.8]$$

- *Variance*: the sum of squared deviations scores divided by the number of values in the set.

    $$\text{Variance} = \Sigma(x{-}X)^2/n \qquad [5.9]$$

- *Standard deviation:* the square root of the variance.

    $$SD = \sqrt{\Sigma(x{-}X)^2} \, /n{-}1 \qquad [5.10]$$

    In which: $x$ = individual score; $X$ = mean score; $(x{-}X)^2$ = the squared deviation; $\sigma$ = the sum up sign; $x{-}X$ = the deviation; $n$ = total number of scores in the data set.

Unbiased estimate of a population (generalizing from a sample to a population):

$$\text{Variance} = \Sigma(x{-}X)^2/n{-}1 \qquad [5.11]$$

In which: $n{-}1$ = degrees of freedom (the number of deviations from the mean that are free to vary).

Source: Langridge and Hagger-Johnson, 2009.

## Use of histograms and descriptive tables in study B

In the correlation study presented in Box 5.9, a questionnaire and a checklist were used to investigate IAQ in 167 office buildings (in eight European countries).

From the 167 investigated buildings, most buildings (74 per cent) had a mechanical ventilation system (see histogram in Figure 5.19). Only 14 and 12 per cent respectively have a hybrid/mixed mode and natural ventilation system (openable windows or stack ventilation). Twenty-nine per cent of the buildings were located in a mixed commercial residential area, 18 per cent in the city centre and 13 per cent in the suburbs, with larger gardens (see histogram in Figure 5.20).

**Figure 5.19**
*Type of ventilation system in the investigated office buildings (n = 167)*
Bluyssen *et al.*, 2012b.

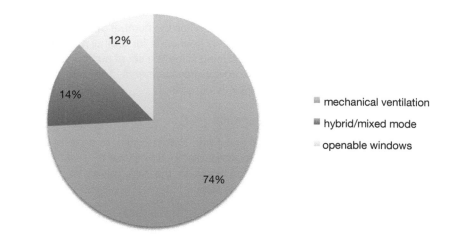

**Figure 5.20**
*Location of the investigated office buildings (n = 167)*
Bluyssen *et al.*, 2012b.

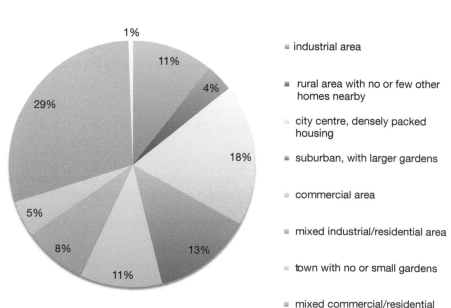

Table 5.10 *Components and existing questionnaires applied in the OFFICAIR questionnaire*

| Components | Sub-components | Existing questionnaires |
|---|---|---|
| Personal data | Age, sex, family status, smoking, education, commuting | Several (see Appendices A.1 and A.2) |
| Work data | Office, job position, work seniority, full/part-time, working hours, type of work, VDU use, daily activities | Several (see Appendix A.3) |
| Psycho-social environment | Work-related stress | ERI-over commitment (Siegrist *et al.* 2004) (see Appendix A.4) |
| Psychological characteristics | Model of affect as one indicator | PANAS short version (10 items) (Thompson 2007) (see Appendix A.5) Emocards (see Figure 6.18) |
| Events | Recently special positive event(s) Recently special negative event(s) | – |
| Physical state | Psycho-physical health Indoor effects and physical environment perception | Adapted from HOPE (Roulet *et al.* 2006) (see Appendix A.6) |

Source: Bluyssen *et al.*, 2012a.

Table 5.11 *Average workers and surfaces of the investigated office buildings per country (n = 167)*

| Country | Number of buildings investigated | Number of workers | Response rate mean [S] [min–max]% | Respondents investigated |
|---|---|---|---|---|
| NL | 20 | 3,570 | 30 (9) (17–51) | 1,014 |
| IT | 21 | 1,222 | 71 (15) (34–97) | 809 |
| ES | 20 | 3,742 | 33 (26) (3–93) | 697 |
| GR | 23 | 3,434 | 40 (24) (6–98) | 1,020 |
| HU | 24 | 4,215 | 45 (21) (14–88) | 1,409 |
| PT | 19 | 2,282 | 35 (21) (3–70) | 508 |
| FI | 19 | 4,746 | 28 (19) (1–70) | 793 |
| FR | 21 | 3,460 | 36 (18) (16–84) | 1,190 |
| total | 167 | 26,671 | 40 (23) (1–98) | 7,440 |

Source: Bluyssen *et al.*, 2012b.

The questionnaire comprised a compilation of existing questionnaires (see Table 5.10). An overview of countries, buildings, number of respondents and number of investigated respondents is shown in Table 5.11.

## Example of descriptive analysis in Study C

In the *in vitro* study presented in Box 5.12, Il-8 as toxicological indicator for inflammation and LDH, Alamar blue and RCDC as toxicological indicators for cell viability, were measured for health effect assessment of the human lung cells when exposed to the emissions of different sources: 25 scented candles of one brand type (non-burning (ncandles) $n = 3$, burning (candles) $n = 5$) and two types

**Table 5.12** *Summary of tests*

| No. | Name | No. | Name |
|-----|------|-----|------|
| 1 | Blank | 10 | Hairspray-1 |
| 2 | Ncandles-1 | 11 | Hairspray-2 |
| 3 | Ncandles-2 | 12 | Hairspray-3 |
| 4 | Ncandles-3 | 13 | Hairspray-4 |
| 5 | Candles-1 | 14 | Spray-1 |
| 6 | Candles-2 | 15 | Spray-2 |
| 7 | Candles-3 | 16 | Spray-3 |
| 8 | Candles-4 | 17 | Spray-4 |
| 9 | Candles-5 | | |

Source: Bluyssen *et al.*, 2013.

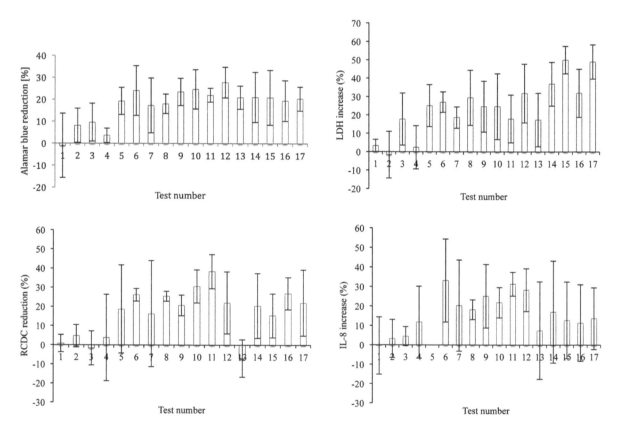

**Figure 5.21** *Study C results: (a) Alamar blue reduction: average and standard deviations per test (n = 12); (b) LDH increase: average and standard deviations per test (n = 9); (c) RCDC reduction: average and standard deviations per test (n = 3); (d) IL-8 increase: average and standard deviations per test (n = 3)*
Source: (Bluyssen *et al.*, 2013).

**Table 5.13** *Average (X) outcome of toxicological indicators between the test exposure and the medical air exposure of the lung cells, the standard deviation (S) and the reproducibility (R) per test situation. R is defined as the standard deviation divided by the average*

| % | Ncandles (n=3) | Candles (n=5) | Hairspray (n=4) | Spray (n=4) |
|---|---|---|---|---|
| *IL-8* | | | | |
| X | 6.5 | 24.1 | 26.0[1] | 13.6 |
| S | 4.5 | 6.7 | 4.6 | 2.4 |
| R | 0.69 | 0.28 | 0.18 | 0.18 |
| *Alamar blue* | | | | |
| X | −7.1 | −20.4 | −26.2 | −20.3 |
| S | 3.1 | 3.1 | 5.9 | 1.0 |
| R | 0.44 | 0.15 | 0.22 | 0.05 |
| *LDH* | | | | |
| X | 6.2 | 24.9 | 25.1 | 41.9 |
| S | 10.3 | 4.0 | 7.6 | 9.0 |
| R | 1.66 | 0.16 | 0.30 | 0.21 |
| *RCDC* | | | | |
| X | −2.3 | −21.5 | −30.4[1] | −21.3 |
| S | 3.4 | 4.3 | 8.2 | 4.8 |
| R | 1.48 | 0.20 | 0.27 | 0.22 |

1 Test 13 is not included because the outcome was considered an outlier (see Figure 5.21), therefore *n* = 3.

Source: Bluyssen *et al.*, 2013.

of nano-particle-producing sprays (hairspray *n* = 5; water repellent spray (spray) *n* = 4; see Table 5.12). In Figure 5.21 the average and standard deviations per test is presented for Alamar blue reduction, LDH increase, RCDC reduction and IL-8 increase, respectively. For IL-8, test 5 is missing, and for both IL-8 and RCDC test 13 can be considered an outlier and was therefore not taken into account for further analysis. Standard deviations for the single tests performed varied on average from 7.7 per cent for tests with Alamar blue, 12.1 per cent for tests with LDH, 12.3 per cent for test with RCDC and 15.4 per cent for tests with IL-8. The lowest standard deviation was 2.6 per cent for an RCDC test on candles and the highest standard deviation was found for an RCDC test of candles (27.7 per cent). Average outcome of the toxicological indicators (in per cent difference with medical air), the standard deviation and the reproducibility is presented in Table 5.13 per test situation (ncandles, candles, hairspray and spray). The blank situation was statistically identical to medical air. The exposure results with candles, hairspray and spray showed a reproducibility (R) between 0.05 and 0.30 (where R = 0.01 is excellent and R > 1 is very bad). The non-burning candle situation was not as reproducible as the other situations (range of 0.44 for Alamar blue to 1.66 for LDH).

## 5.8.3 Inferential analysis

*Inferential analysis* allows for generalization from the sample (study population) to the broader population by identifying and explaining associations, or discriminating between individual samples and predicting new observations based on these features. Table 5.14 presents a summary of the inferential analysis that can

**Table 5.14** *Which test now?*

| Number of conditions/ variables? | Differences or relationships? | Parametric Normally distributed | Non-parametric (ranks) Non-normal data | Frequency data |
|---|---|---|---|---|
| Two variables | Differences | Unrelated: independent t-test Related: related t-test | Unrelated: Mann-Whitney test Repeated measures: Wilcoxon test | Unrelated: Chi-square Related: McNemar |
| | Relationships | Pearson correlation coefficient | Spearman correlation coefficient | Chi-square |
| Two or more variables | Differences | Unrelated: independent ANOVA Related: Related ANOVA | Unrelated: Kruskal-Wallis test Repeated measures: Friedman test | Unrelated: Chi-square Related: Cochran Q |
| | Relationships | Multiple regression | Multiple regression | Chi-square |

*Note*: ANOVA = analysis of variance

Source: based on Langridge and Hagger-Johnson, 2009.

be performed, depending on the type of data (normally distributed, rank or non-normal distributed, or frequency data), the number of conditions (independent variables), depending on whether differences or relationships are studied, and depending on whether the samples taken are related or unrelated to each other.

Tools used for inferential analysis comprise of probability testing (looking for differences) and regression analysis (looking for relationships).

## Probability testing

In probability testing inferences are made about the likelihood of the findings occurring. What are the probabilities (likelihood) that the relationships/differences occurring between variables within our data set are significant? Assuming a normal distribution of a continuous variable, a t-test is often applied (see Figure 5.22 and Box 5.16 for explanations of terms).

## Example probability testing for Study C

For the toxicological indicators (Alamar blue, LDH, RCDC and IL-8), it was analysed how different the average outcomes for the different exposure situations (non-burning candles, burning candles, hairspray and water repellent spray) are, using the 95 per cent confidence interval for a t-distribution (see Figure 5.23). The confidence intervals show a relevant difference between the reduction in Alamar blue and RCDC for non-burning and burning candles, but not for LDH and just not for IL-8.

## Correlations between two variables

A t-test can be used to show a difference (or not) between two samples. While t-tests are based on the mean, the Mann-Whitney and Wilcoxon tests are based on ranks and are concerned whether populations are the same or not. They are

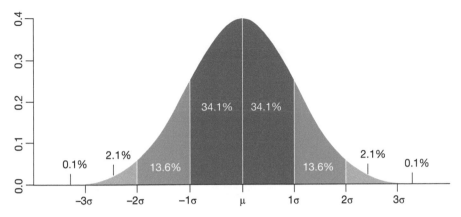

**Figure 5.22** *Normal distribution of a continuous variable with standardized scores (z-scores): dark blue is less than one standard deviations and accounts for about 68 per cent; two standard deviations (to be exact 1.96 times the standard deviation (σ) on both sides; medium and dark blue) represented the 95 per cent confidence interval and three standard deviations account for 99.7 per cent*

Source: Ainali, Wikimedia Commons, http://commons.wikimedia.org/wiki/File:Standard_deviation_diagram_micro.svg (accessed 2 June 2013).

---

## Box 5.16 Probability testing

A *significance level* of 5 per cent (*p* = 0.05) tells us that 5 out 100 (1 in 20) times the effect observed is due to random error and not to the independent variable tested. Thus, a 1 in 20 chance hat a type I error (α) occurs. A type II error (β) occurs when the independent variable has no effect on the dependent variable and in fact it is really responsible for the change in the dependent variable. This can occur when *p* is taken too high (e.g. 0.0001). A type I error can occur when the level of *p* is taken too low.

In general a distribution of scores on a continuous variable demonstrate a particular pattern (not random distributed). The most common distribution is the *normal* or *Gaussian distribution curve*. Standard scores, or *z-scores*, are produced by using the deviation from the mean in terms of standard deviation units:

$$z = x/S \text{ where } x = |X - \bar{X}| \qquad\qquad [5.12]$$

Z-scores for 5 per cent (*p* = 0.05) significant level is the same as a distribution that falls between 1.96 (times the standard deviation) and the mean on both sides (see Figure 5.22). Z-scores for 1 per cent (*p* = 0.01) fall between 2.58 and the mean.

A *t-test (Student's t-test)* is a test for difference between two mean scores. The t-test takes two factors into account: *t* = difference in mean/standard error of difference in means. Standard error = standard deviation / $\sqrt{n}$ (standard deviation of a set of means). No response would give 0 ($\mu_0$).

A *null hypothesis value* is the value the investigator wants to disprove (e.g. $H_0$: $\mu$=0). The probability that a statement about the parameter is true, given the null hypothesis value of the parameter is correct. For example: Does the mean fall within the 95 per cent confidence interval around 0. If yes, then reject. If no, then accept. Is *x* outside the limits $\mu_0 \pm 1.96 \ \sigma / \sqrt{n}$ ? (see equation 5.1 for $n \geqslant 30$ and equation 5.2 for $n < 30$).

Source: Langridge and Hagger-Johnson, 2009; Baker and Nieuwenhuijsen, 2008.

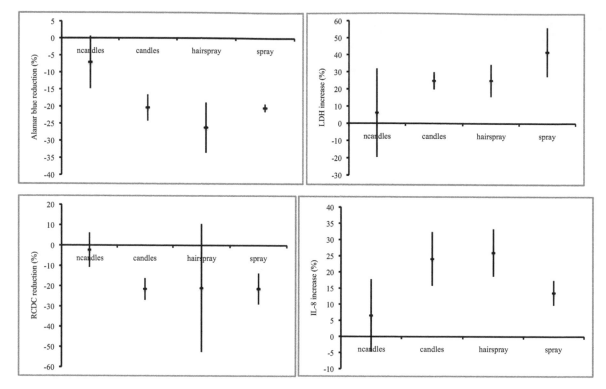

**Figure 5.23** *95 per cent confidence interval for Alamar blue, LDH, RCDC and IL-8 during tests with candles, hairspray and water repellent spray (for n < 30: $X \pm t_{.05}S \div \sqrt{n}$, with X = average (per cent difference with medical air); n = number of tests; S = standard deviation; $t_{.05}$ = t-distribution for 95 per cent confidence = 4.303 for n = 3; 3.182 for n = 4 and 2.776 for n = 5)*
Source: Langridge and Hagger-Johnson, 2009, Appendix A.6.

non-parametric equivalents of the independent group and related t-tests. Mann-Whitney is used when you have independent groups and Wilcoxon is used when you have related samples. Use t-tests unless data are rank data or if data are obviously non-normal or there are large differences in variances between the two groups/conditions (Langridge and Hagger-Johnson, 2009).

The Chi-square ($X^2$) test is a test of association suitable for use with frequency data (data in the form of counts, categorical variables), specifically for unrelated samples. For example, the number of subjects male versus female is used to look at whether this has a relation with having a certain disease or not. The McNemar and Cochran Q are respectively applied for related samples with two variables and more than two variables (Langridge and Hagger-Johnson, 2009).

Relationships between two variables (bivariate relationship) can be shown for:

- Pairs of continuous variables (parametric data) by the Pearson correlation coefficient ($r$).
- Spearman ($\rho$) correlation coefficient for non-parametric data, defined as the Pearson correlation coefficient between ranked variables.

For relations between two or more variables, analysis of variance (ANOVA) can be used for parametric variables and for non-parametric variables the Kruskal-Wallis or Friedman test can be applied (Langridge and Hagger-Johnson, 2009). The ANOVA is very much like the t-test, also comparing the means of the variables being tested to see if they are sufficiently different and also includes measures of variation of scores within each set of data. It uses the variance ratio ($F$):

$$F = \text{between conditions variance/error variance} \qquad [5.13]$$

The error variance is the within-groups variation. $F$ is designed to identify differences between means (between-groups variation) and difference in scores within the groups (within-groups variance).

The Kruskal-Wallis test is the non-parametric equivalent of the independent ANOVA test and the Friedman test the equivalent of the repeated measures one-way ANOVA. In a two-way ANOVA test for two factors the significance of difference between conditions is calculated. In a one-way analysis of variance only one factor is considered (Langridge and Hagger-Johnson, 2009).

## Multivariate analysis

Multivariate analysis of variance (MONAVA) can be used when there is *more than one dependent variable* in the study as for example occurs in cross-sectional studies. Alternatives to MANOVA are (Langridge and Hagger-Johnson, 2009):

- *Factor analysis*: a data reduction technique in which a large set of variables can be reduced to a more manageable set of variables (using terms such as Eigen value and Cronbach alpha).
- *Correlation and regression analysis*: using relationship between two or more variables for prediction rather than for description. With one independent and one dependent variable it is named simple regression, and with two or more independent and one dependent variable it is named multiple regression.
- *Treating each dependent variable separately*: conducting univariate analysis (e.g. t-test, ANOVA, regression) on each.

In a linear regression model (Baker and Nieuwenhuijsen, 2008):

$$y = \text{dependent variable} = a + bx + e \qquad [5.14]$$

Where:

$a,b$ = intercept and slope of the line of best fit

$x$ = independent variable

$e$ = error = variance in $y$ that cannot be explained by the other terms in the regression equation.

Multiple regression refers to multiple predictors (Langridge and Hagger-Johnson, 2009):

$$Y = a + b_1 x_1 + b_2 x_2 + \ldots + b_n x_n + e \qquad [5.15]$$

Regression models normally used are (Baker and Nieuwenhuijsen, 2008):

- *Linear*: studies with continuous outcome, such as cross-sectional or panel studies.
- *Logistic*: studies with binary outcomes, including case-control and cross-over studies.
- *Poison*: studies with count outcomes, including air pollution time-series studies.
- *Proportional hazards*: survival studies, studies with time to event outcomes (survival outcomes).

### Example correlation analysis in Study C

It was investigated whether a correlation between Alamar blue/LDH/RCDC/IL-8 on the one hand and particle concentration on the other existed. Additionally, it was investigated whether a correlation could be found between Alamar blue/LDH/RCDC/IL-8 and the total amount of particles to which the lung cells were exposed. In Table 5.15 those correlations are presented for respectively the single tests performed and the average outcome of the different exposure situations (non-burning candles, burning candles, hairspray and water repellent spray). Alamar blue and IL-8 gave the best correlation for the average outcome per exposure situation, for both the average concentration of particles as the total number of particles, the cells were exposed to. RCDC is the only endpoint that gives a reasonable correlation with the single test comparison in which each test is taken independently.

**Table 5.15** *Correlation coefficients (r) of particle concentrations and total number of particles for the single tests and for the average outcome of the exposure situation with Alamar blue, LDH, RCDC and IL-8 results*

|  | Alamar blue | LDH | RCDC | IL-8 |
|---|---|---|---|---|
| *Average* |  |  |  |  |
| Number of Particles/m³ | 0.46 | 0.005 | 0.14 | 0.45 |
| Total number of particles | 0.47 | 0.005 | 0.15 | 0.48 |
| *Single tests* |  |  |  |  |
| Number of particles/m³ | 0.11 | 0.00003 | 0.40 | 0.17 |
| Total number of particles | 0.13 | 0.0002 | 0.43 | 0.19 |

# 6

# Data collection techniques

*Chapter 6 shows an overview of collection techniques that can be applied to investigate indoor environment quality in the field or in laboratory studies. Physiological techniques to measure the contents of the bodily products, physical techniques to measure changes in the human system, behavioural monitoring to collect information on what people do and how they interact with their environment, questionnaires to inventory personal data and monitoring of the environment people are exposed to.*

## 6.1 Introduction

There are three main categories from which information may be considered to collect in an indoor environment quality (IEQ) field investigation:

- *The human being and systems*: from which several types of information can tell us something on how the human body copes with or responds to stress.
- *The indoor and built environment*: characteristics and processes creating the physical stressors and possible factors of influence.
- *The psychosocial environment*: characteristics and processes creating the psychosocial stressors and possible factors of influence.

A variety of subjective and objective methods of exposure measurements directed towards present or past personal attributes or environmental contacts can be used to collect this information: personal interviews (face-to-face or by telephone), self-administered questionnaires, diaries of behaviour, reference to records, physical or chemical measurements on the subject, physical or chemical measurements in the environment, direct observation of the subject's behaviour by the investigator, and, when the subject is too young, too ill or deceased, a proxy respondent. In addition to costs, the choice of method is influenced by the type of study to be undertaken, the amount of detail required by the study's objectives, the frequency and level of exposure over time, and the accuracy of the methods that are available to measure the topic of interest (e.g. specific exposure or effect).

Several assessment techniques are available or are currently being investigated. These assessment techniques can be categorized into techniques that:

- Measure *physiological and physical changes* in the body or the bodily products: In this category information on physiological and physical states (objective responses) of the body and systems is gathered with techniques focused on:

- changes in the bodily products (see Section 6.2);
- responses of the nervous system, both autonomic and in the brain (see Section 6.3);
- changes at cell level (e.g. epigenetics; see Section 3.6).

• Observe or ask the subject or subjects about their *psychological and behavioural responses and psychosocial and physical stress*: This category concerns gathering information on the emotional state and trait and other personal factors, descriptive (subjective) information on physical and psychosocial stressors and factors, on past events and exposures, and to some extent on behaviour and responses over time (e.g. diseases and disorders). The following techniques are applied:

- behavioural monitoring techniques (see Section 6.4);
- questionnaires and interviews (see Section 6.5);
- records: such as occupational, medical, birth and death records (not discussed here).

• Measure in the environment of the subject with *environmental monitoring*: In this category information on potential stressors and factors caused by indoor and built environment (physical) and psychosocial environment is gathered by:

- checklists (see Section 6.6.1);
- chemical, biological and physical measurements (see Section 6.6.2).

The use of subtle physiological responses to capture physiological changes as a response to acute or chronic stress is a relatively new approach in the field of IEQ. So far, the focus has merely been on clearly visible physical responses or self-reported symptoms and complaints. In other fields of expertise there is a long history in using such an approach (e.g. medical science and stress research). For example, one research group defined 10 measures reflecting 'wear and tear' on the bodily systems as a first attempt to operationalize allostatic load (for definition see Section 1.1) presented in Table 6.1 (Seeman *et al.*, 1997). In subsequent work, information on parameters of inflammation (immune system)

**Table 6.1** *Ten measures reflecting 'wear and tear' on the bodily systems*

| Measure | Explanation |
| --- | --- |
| Waist–hip ratio | Marker of metabolic dysregulation |
| Serum high-density lipoprotein (HDL) Total cholesterol levels | Indexes of long-term atherosclerotic risk |
| Blood plasma levels of total glycosylated haemoglobin | Integrated measure of glucose metabolism during a period of several days |
| Systolic and diastolic blood pressures | Markers of cardiovascular activity |
| Serum dehydroepiandrosterone sulfate (DHEA-S) | A functional HPA-axis antagonist |
| 12-hour urinary cortisol excretion | An integrated measure of 12-hour HPA-axis activity |
| 12-hour urinary norepinephrine excretion levels 12-hour urinary epinephrine excretion levels | Markers of sympathetic nervous system activity |

Source: Seeman *et al.*, 1997.

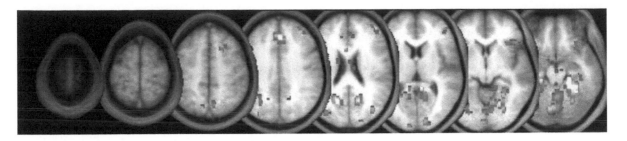

**Figure 6.1** *fMRI (functional magnetic resonance imaging) scans: yellow areas show increased blood flow when a person is finding their way around in a virtual environment*
Source: Tom Hartley, Department of Psychology, The University of York.

was added (Seeman *et al.*, 2010). The use of such markers could provide new opportunities for IEQ and are now slowly being introduced in IEQ studies.

Examples of local responses that have been used in IEQ field studies include tear film stability (Rolando, 1984) and self reported break up time (SBUT) (Wieslander *et al.*, 1999; Bakke *et al.*, 2008) as markers of sensory (eye) irritation, but also allergic reactions or other diagnosed symptoms of diseases and disorders. Various experimental techniques are used in both humans and laboratory animals to study chemical-induced irritation consisting of examinations of functional changes, for example, alterations in breathing frequency and pattern, nasal, bronchial and pulmonary function parameters, nasal mucosal swelling, acoustic rhinomanometry, eye blinking frequency and chemosensory evoked potentials (Kjaergaard and Hodgson, 2001; Arts *et al.*, 2006).

One could also think more out of the box and use other techniques such as facial expression, human behaviour (observation) (Akimoto *et al.*, 2009), a panel of independent visitors that is trained and calibrated to evaluate comfort of an indoor environment (trained panels have been used to evaluate the perceived air quality in a building (Bluyssen, 1990)), or even neuroscience (Lindstrom, 2008) (see Figure 6.1 and Section 6.3.2).

## 6.2 Physiological techniques

*Changes in the human systems can be registered by assessing the contents of the bodily products, such as blood, urine, sweat, stool, saliva, breath and hair. These bodily products contain a whole range of cytokines, hormones and metabolic products as well as physical, chemical or radioactive agents, including antioxidants and reactive oxygen species.*

### 6.2.1 Metabolomics

Metabolomics is 'the comprehensive and quantitative analysis of all metabolites' (Fiehn, 2001). Metabonomics and metabolic profiling are names used also for this type of analysis, which can be used interchangeably (Robertson *et al.*, 2011). Metabolomics is the systematic study of the unique chemical finger-prints that specific cellular processes leave behind, specifically the study of their small-molecule metabolite profiles. The metabolome presents the collection of all metabolites in a biological cell, tissue, organ or organism, which are the end products of cellular processes. Metabolites are the intermediates and products of metabolism. The analytical goal of metabolomics is to achieve a comprehensive measurement of the metabolome and how it changes in response to stressors (Robertson *et al.*, 2011).

Metabolomics approaches that can be used are (Fiehn, 2001; Robertson *et al.*, 2011):

- *Metabolic fingerprinting*: Comprises analytical fingerprints determined by spectroscopy or chromatography that contain thousands of single data points related to the composition of the sample. Multivariate statistical tools such as principal component analysis (PCA) can be applied to analyse these data points.
- *Non-targeted metabolomics*: Goes beyond fingerprinting and looks for as many individual chromatographic or spectroscopic peaks as possible. Changes in components can be mapped to specific pathways and provide biomarkers and/or mechanistic information (see Table 6.2).
- *Targeted metabolomics*: Analytes, which have been pre-selected, are measured usually to address certain specific biological questions within a study.

Analytical techniques most used are mass spectrometry (MS) and nuclear magnetic resonance (NMR).

Metabolomics has been propagated as an approach for mechanistic under-standing and biomarker discovery for neuroscience applications, oncology, cardiovascular disease, tuberculosis, cystic fibrosis and metabolic syndrome (Robertson *et al.*, 2011). Metabolomics has been applied among others to study cardiovascular disease (Goonewardena *et al.*, 2010), hypertension (Holmes *et al.*, 2008), respiratory diseases (Basanta *et al.*, 2010), autoimmune diseases (Seeger, 2009), and even eye diseases (Young and Wallace, 2009). For example, Hsu *et al.* (2012) found that levels of indoor dust-borne BBzp (benzylbutyl phathalate) and DPB (dibutyl phthalate), as well as the metabolites MBP (mono n-butyl phthalate) and MEHP (mono-2-ethyl phthalate) found in urine samples, were associated with increased risk of having specific health outcomes for asthmatic and allergic children, while indoor fungal exposure remained a significant risk factor as well.

Metabolomics can be performed on bodily fluids readily accessible to the physician, including blood, saliva, breath condensate, sweat, ascites, amniotic and cerebrospinal fluid. For example, by metabolomic analysis of urine or serum changes in metabolites can be revealed, both related to immune system changes as well as endocrine system changes (see respectively Sections 6.2.2 and 6.2.3).

**Table 6.2** *Some selected metabolomic-derived biomarkers in humans*

| Biomarker | Disease/toxicity/physiological change |
| --- | --- |
| Fatty acid profiling | Type II diabetes |
| Propionyl carnitine | Methylmalonic and propionic acidemias |
| BCAA (branched chain amino acid) | Insulin resistance |
| $\alpha$-hydroxybutyrate | Insulin resistance |
| Medium chain acylcarnitines | Moderate exercise |

Source: adapted from Robertson *et al.*, 2011.

## 6.2.2 Immune system

Toxicity tests to assess toxic metal exposure or load comprise of toxic metal tests on stool, urine, hair and sweat, fat biopsy, blood levels of metals, pesticides, solvents and cholinesterase activity, and last but not least serum levels of DNA adducts, indicating the presence of pro-mutagenic markers (see Section 6.2.4; Müller and Yeoh, 2010).

Several cytokines in blood, associated with inflammation, immunity and repair regulated by *the immune system*, have been monitored to investigate the response of the immune system (e.g. Tarant, 2010). Besides cytokines, C-reactive protein (CRP), fibrinogen (glycoprotein), lymphocyte counts and activity and immunoglobulins (antibodies) have been measured as markers of inflammation or immune responses (Steptoe and Poole, 2010).

Oxidative stress can cause damage to major bio-molecules. To assess oxidative stress, affected tissues can be screened for the presence of *biomarkers of attack of ROS* upon bio-molecules or the *antioxidant status* of tissues in terms of their glutathione, ascorbate, uric acid and α-tocopherol content as well as enzymic activities, catalase and glutathione peroxidase (Evans and Halliwell, 1999). But perhaps also preceding responses, such as *calcium response* before the oxidative stress responses (e.g. formation of ROS, protein carbonyls and the induction of heme oxygenase-1) (Haase *et al.*, 2012). It is also important to realize that there are different types of damage, which can be caused by different 'biomarkers of attack' (Evans and Halliwell, 1999) (see Box 6.1).

Several bodily fluids have been studied in relation to changes in immune system functioning, such as exhaled breath, nasal lavage and urine. For example,

---

### Box 6.1  Types of damage

- *DNA damage*: the reactive oxygen species (ROS) hydroxyl radical OH• (see Section 3.5.2) can damage all four of the DNA bases (see Section 2.2.1). The presence and concentration within DNA of the modified bases, the pattern of damage, can give an indication of the ROS causing the damage.
- *Protein damage*: comprising different ways of damage, such as:
  - Conversion of certain amino acid side chains to carbonyl groups: which can be assayed by derivatizing these groups with dinitrophenylhydrazine to form the yellow chromogen dinitrophenylhydrazone (which can be determined colorimetrically and by other methods).
  - Reactive nitrogen species induce nitration of protein tyrosine groups to give nitrotyrosine residues: that can be quantified by several methods including enzyme-linked immunosorbent assay (ELISA).
  - Products of oxidation of specific amino acids such as methionine sulphoxide: which can be detected by high-performance liquid chromatography (HPLC) and the amino acid phenylalanine has been used to detect the presence of OH•.
- *Lipid damage*: to estimate the degree of lipid peroxidation in tissue and body fluids several methods are used, such as the measurement of lipid hydroperoxides, aldehydes (malondialdehyde may arise as a product of free-radical attack on certain carbohydrates such as deoxyribose), hydrocarbon gases, conjugated dienes and isoprostanes (as end products of the nonenzymic peroxidation of lipids).

Source: Evans and Halliwell, 1999.

Lu *et al.* (2007) investigated whether symptoms within the so-called SBS and indoor air pollution for office workers (389 in 87 offices) were associated with oxidative stress indicated by urinary 8-hydroxydeoxyguanosine (8-OHdG). 8-OHdG is an important product of oxidative stress in DNA (Loft *et al.*, 1992). The results indicated that the 8-OHdG level was significantly associated with SBS complaints after controlling for air pollution and smoking.

But also several tissues and organs (e.g. skin, respiratory tract, nose and eyes) can give information in relation to a specific immune related disease or disorder.

## Skin

In a skin allergy test a microscopic amount of an allergen is introduced to a subject's skin by a prick or scratch test or a patch test. In a prick test, the skin is pricked with a needle or pin containing a small amount of allergen, while in a patch test, a patch with allergen is applied to the skin. If an immuno-response is seen in the form of a rash or worse, it can be concluded that the patient has a hypersensitivity (or allergy) to that allergen. If no reaction occurs, it cannot be concluded that the patient is not allergic, it only indicates that the body gave no response at the concentration of allergen applied. As a control, histamine or glycerine are applied as well: most people show a reaction with histamine, while no reaction with glycerine. An example of a patient that received a number of pricks on his forearm is presented in Figure 6.2.

Next to the prick and patch tests, an intradermal test can be performed. In this test intradermal injections of allergens at increasing concentrations are used to measure the allergic response.

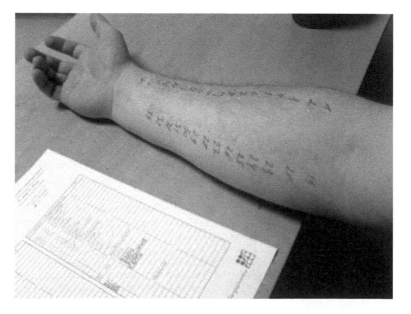

**Figure 6.2** *Skin allergy test*

Source: Wolfgang Ihloff, Wikimedia Commons, http://commons.wikimedia.org/wiki/File:Allergy_skin_testing.JPG (accessed 2 June, 2013).

## Respiratory tract

Airborne chemicals capable of stimulating nerve endings in the respiratory tract can be classified as follows (Arts *et al.*, 2006):

- *Sensory irritants*: stimulate trigeminal nerve endings when inhaled via the nose and evoke a burning sensation in the nose and inhibit respiration in that way. These are in general highly water soluble and/or reactive substances (e.g. formaldehyde).
- *Pulmonary irritants*: stimulate nerve endings within the lung and increase respiratory rate with a decreasing tidal volume. These are in general less water-soluble (like ozone).
- *Broncho-constrictor agents*: induce an increase in resistance to airflow within the airways of the lung, caused by a direct effect on the smooth muscles of the conducting airways, by a neural reflex or by liberation of mediators such as histamine (can be caused by inhalation of allergens, but can also be exercise-induced).
- *Respiratory irritants*: can act as a sensory irritant, broncho-constrictor and pulmonary irritant.

Assessment of the irritation caused by these airborne chemicals can be based on examination of tissues (histopathology: the Greek word 'histo' means tissue) of the lungs or complete respiratory tract of animals repeatedly exposed by inhalation to the compound under investigation. Other approaches are the use of subjects to determine sensory effects (see Section 6.5.2), questionnaires and interviews (see Section 6.5), spirometry (see Section 6.3.1) and the analysis of *exhaled breath condensate* (EBC).

EBC is a biological fluid that contains trace amounts of secreted pulmonary proteins and is a simple, non-invasive source for studying the biochemical and inflammatory molecules in the airway lining fluid (Bloemen *et al.*, 2009). Subjects breathe through a mouthpiece and a two-way non-rebreathing valve, at normal frequency and tidal volume, while wearing a nose clip (Laurentiis *et al.*, 2008). The exhaled breath is condensed, typically via cooling using a collection device (commonly to 4°C or sub-zero temperatures using a refrigerating device). EBC reflects changes in the respiratory fluid that lines the airways. EBC has been studied for the analysis of inflammation and oxidative stress markers (Corradi *et al.*, 2002; Montuschi *et al.*, 2002) by applying ELISA or NMR-spectroscopy (Laurentiis *et al.*, 2008). NMR studies molecules by recording the interaction of radiofrequency electromagnetic radiation with the nuclei placed in a strong magnetic field. ELISA is an analytic biochemistry assay that uses one sub-type of heterogeneous, solid-phase enzyme immunoassay to detect the presence of a substance in a liquid or wet sample.

Nitric oxide (NO) in the exhaled air or fractional concentration of exhaled nitric oxide (FeNO), after exposure to indoor air pollutants, has been studied as a marker of airways inflammation or adverse respiratory health effects (e.g. Kolarik *et al.*, 2009; Smith *et al.*, 2005). The NO concentration in exhaled air (measured by a chemiluminescence NO analyser) was used to objectively assess human response to indoor air pollutants in a climate chamber (Kolarik *et al.*, 2009). NO is produced inside the epithelium cells of the airways, the smooth muscle is relaxed by stimulation of NO and the cross section of the airways increases. A small increase of NO in the exhaled air was shown after exposure to indoor air pollutants. Similar effects were shown by Graveland *et al.* (2011) in a study on short-term changes in air pollution exposure in children. They found that short-term changes in ambient $PM_{10}$ largely attributable to biomass burning were associated with increased levels of exhaled NO in 812 children from nine Dutch schools within 400 m of motorways.

## Eyes

Ocular discomfort (e.g. burning, dry and itching eyes; see Section 3.5.4) is an often occurring disorder in the indoor environment. Measurements for ocular discomfort include (Kjaergaard and Hodgson, 2001):

- *Tear film stability*: determined by measuring tear film breakup time, assessed by non-invasive methods based on a grid or equidistant circles of light that are blurred by tear film breakup, or by using vital staining with sodium fluorescein. Tear film stability is the time the film remains intact on the eye when no blinking occurs.
- *Self-reported break up time* (SBUT) (see Section 6.5.2).
- *Epithelial damage*: measured by rose-bengal or lissamine-green-B staining of the epithelial cells (see Figure 6.3). Dry eyes are indicated when the stain is taken up by the exposed cellular surface indicating lubrication from the tear film was not sufficient.
- *Foam formation*: measured by slit-lamp microscopy. The absence of foam in the eye canthus is believed to be an indicator of bad tear film quality.
- *Inflammation*: measured by biomarkers such as eicosanoids and interleukins in the tear fluid or epithelium.

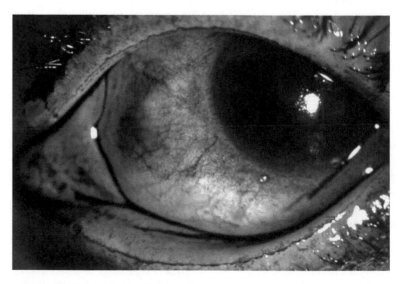

**Figure 6.3** *Ocular surface damage shown with lissamine-green-B staining of the epithelial cells in a patient with dry eye disease*
Source: Barabino *et al.*, 2012.

Other indicators are blinking frequency (Noejgaard *et al.*, 2005) and hyperemia of the conjunctiva (red eyes). Both can be determined using observation techniques (e.g. direct inspection, photographs and video techniques).

## Nose

Besides ocular discomfort, nasal inflammation is another symptom of concern encountered indoors. Nasal mucosal irritation leads to a swelling of the nasal mucosa (a lower nasal patency) and an increase in cytokines and several emissions from activated granulocytes, such as eosinophil cationic protein and lysozyme (Wieslander *et al.*, 1999; Norbäck *et al.*, 2000; Kjaergaard and Hodgson, 2001). Nasal inflammation can thus be studied by examining the thickness of the nasal mucosa or by detecting several biomarkers in the nasal lavage fluid. Other ways to evaluate nasal inflammation or irritation include the measurement of peak airflows through the nose (inspiratory and expiratory), and asking patients or participants to evaluate the sensory irritation they experience with the use of scales (see Section 6.5.2).

The thickness of the nasal mucosa can be determined by *rhino-stereometry*, which measures the thickness on the anterior nasal turbinate with a specialized microscopic method, and by *acoustic rhinometry* (see Box 6.2), a non-invasive diagnostic method for accurately measuring the 'patency' or openness of the airway (Kjaergaard and Hodgson, 2001). *Acoustic rhinometry* has been used to measure the degree of swelling of the nasal mucosa of students exposed to different levels of indoor air pollutants. Increased levels were associated with increased swelling (Wållinder *et al.*, 1997).

The method NAL (nasal lavage) has been used to measure different biomarkers. NAL is a useful method for the detection of nasal inflammation in occupational settings where comparison can be made using test subjects as their own controls (Roponen *et al.*, 2003). On a group level, NAL may be useful for monitoring effects of exposure and interventions (Quirce *et al.*, 2010). Nasal lavage fluid has been assayed for an array of cytokines (IL-1, TNF, IL-6, IL-8)

## Box 6.2 Acoustic rhinometry

Acoustic reflection technology is a non-invasive diagnostic method for accurately measuring the 'patency' or openness of the airway. Different forms of device are used for measuring the upper airway (the pharyngometer) and the nasal airway (the rhinometer) (see Figure 6.4). The patient sits upright and places the end of the device into their mouth or nasal cavity. Sound waves are then directed into the airway and a number of simple breathing exercises are performed to locate the most unstable and narrow portions of the airway. Different jaw positions are then tested to determine the best possible position. Measurements and impressions are taken to record this position and the oral appliance is constructed to these specifications to ensure maximum effectiveness.

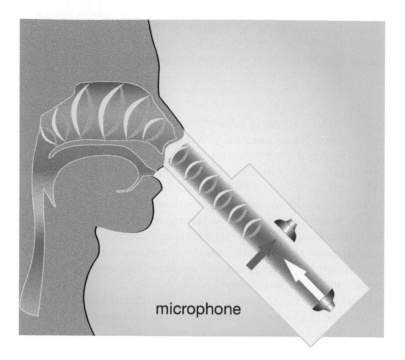

**Figure 6.4** *Rhinometer*
Source: www.sleeptherapyclinic.com.au/acoustic-reflectometry.

by means of a multiplex cytofluorimetric assay (Wieslander *et al.*, 1999) and for detecting eosinophil cationic protein and lysozyme (Norbäck *et al.*, 2000).

Both NAL and acoustic rhinometry have been applied in the European SINPHONIE project (see Section 5.3.3).

### 6.2.3 Endocrine system

Considering *the endocrine system*, different hormones in different bodily fluids have been studied in relation to environmental conditions. Urinary adrenaline, nor-adrenaline and cortisol, ACTH (adrenocortico tropin hormone) and prolactin in blood have been investigated in relation to various symptoms such as skin complaints (Berg *et al.*, 1992), electromagnetic field of VDU (video display unit; Arnetz and Berg, 1996), burnout (Melamed *et al.*, 1999), self-reported alertness (with cortisol), tenseness and irritability (with adrenaline and nor-adrenaline; Fibiger *et al.*, 1984). The ACTH levels in serum (blood) increased significantly during VDU work (Arnetz and Berg, 1996). Salivary cortisol has been used as a measure for HPA-activity (Melamed *et al.*, 1999). Melatonin has been studied in relation to entrainment of the human circadian system by light (Duffy and Wright, 2005; see Section 3.3.2). Cholesterol, but also body mass index (BMI), waist–hip ratio, glycose and insulin are used as markers for the metabolic processes in our body, while blood pressure is a marker for cardio-vascular functioning. It has been demonstrated that the cholesterol level over time may be related to a cholesterol stress response years earlier (Steptoe and

Brydon, 2005). In a recent study of work stress and cardiovascular disease risk (Vrijkotte, 2010) it was found that from the atherosclerotic risk blood parameters that were investigated, fibrinogen showed the largest effect. The results indicated that people with too large a commitment (problems) at work had a bad fibrinolysis, which makes the occurrence of blood props easier.

## Cortisol

Cortisol as an indicator for stress (see Figure 6.5) has been studied in most of the body fluids we have, both related to psychological and physical stress. In studies of humans with chronic psychosocial stress, and low SES, cortisol baseline levels were raised, and the cortisol response to acute stress attenuated. Low job control was associated with insufficient recovery of catecholamines and cortisol, and a range of negative health effects (Kristenson *et al.*, 2004). Nabkasom *et al.* (2005) studied excretions of cortisol and epinephrine (adrenaline) in urine in relation to the exercise regimen. Excretions of cortisol and epinephrine in urine were reduced due to the exercise regimen.

However, there are some points to make. First of all, corticosteroids (e.g. cortisol) in blood or saliva are difficult to interpret because stress-related HPA-function changes lead to cortisol deregulation, not simply increased cortisol production. It seems therefore that cortisol is a better indicator for acute than chronic stress (Clougherty and Kubzansky, 2009). Second, cortisol is secreted in a diurnal rhythm, with higher levels in the morning and lower levels in the afternoon and evening. Additionally, levels can differ per person and average cortisol levels over 24 hours tend to increase with age as well as changes in cortisol levels in response to stress (Steptoe *et al.*, 2005). Therefore, mean changes among a population studied doesn't always lead to simple conclusions. And third, most serum cortisol is bound to proteins. Free cortisol passes easily through cellular membranes, where they bind intracellular cortisol receptors. On the other hand, it has been indicated that free cortisol response (in saliva) to awakening can serve as a useful index of HPA activity (Wüst *et al.*, 2000). It also shows good intra-individual stability across time and appears to be able to detect subtle changes in stress responses. In a study with 105 schoolteachers, Steptoe *et al.* (2000) found that job strain is associated with elevated free cortisol concentration in saliva early in the working day, but not with reduced cortisol variability.

For chronic exposures *cortisol in hair* as a biomarker has been explored. Several studies have demonstrated an association between high levels of hair cortisol and chronic stress in humans (Sauvé *et al.*, 2007; Yamada *et al.*, 2007; Pereg *et al.*, 2010). Hair grows at an average of around 1 cm/month (0.35 mm per day); therefore, hair analysis can accurately reflect long-term endogenous production of cortisol (Pereg *et al.*, 2010). Unfortunately, some notes with this method have to be made as well: a lot of hair is required for an analysis, and coloured or bleached hair cannot be used.

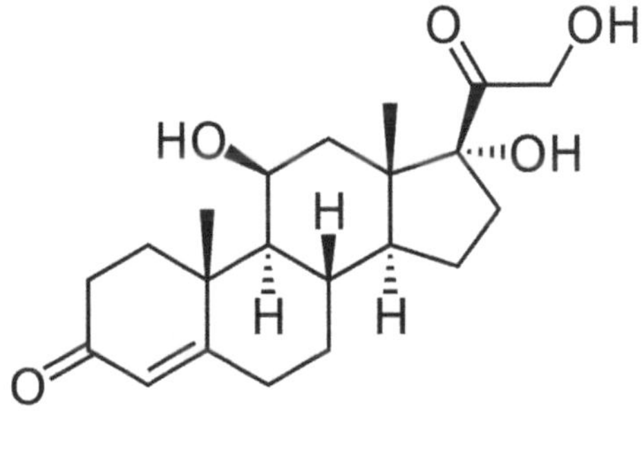

**Figure 6.5**
*Chemical structure of cortisol or 11β)-11,17,21-trihydroxypregn-4-ene-3,20-dione or hydrocortisone*

Source: Foobar, Wikimedia Commons, http://commons. wikimedia.org/wiki/File:Cortisol.png (accessed 2 June 2013).

## Saliva sampling

Because of the non-invasive way of sampling, saliva sampling is becoming more and more applied in both lab and field studies. The sampling of saliva can be performed in several ways. One can use the oral swab, in which a small cotton swab is placed under the tongue of the participant or patient for two minutes, and then placed in a small plastic tube. One can use a liquid that stays in the oral cavity for two minutes before it is spit out into a beaker, and from there put into a tub or tubes. Or one can just passively drool saliva into a test tube. Depending on the sampling method, the samples need to be stored frozen at −70°C (passive drool) or at room temperature (oral swab) awaiting the analysis.

Salivary α-amylase activity (SAA) is used as a non-invasive biomarker for the sympathetic nervous system (Nater and Rohleder, 2009). SAA has been used as an index of sympathetic nervous activity (plasma norepinephrine concentration) to evaluate the comfort level of a living environment. According to Tahara *et al.* (2009), lower values of SAA should indicate that the sympathetic nervous activity has been inactivated and thus a comfortable condition is experienced. They found a difference in SAA with 12 healthy young female subjects exposed to two different air conditioners. Wargocki *et al.*, 2009 found a concentration decrease of salivary α-amylase and cortisol with increased thermal comfort.

Next to SAA, the level of chromogranin A (CgA) in saliva has been used as a non-invasive biomarker. CgA was related to noise exposure. CgA, a member of the granin family of neuroendocrine secretory proteins, is considered a substitute measure for catecholamines (e.g. norepinephrine and epinephrine). Individual differences of CgA correlated well with the score on somatic symptoms (Miyakawa *et al.*, 2006).

### 6.2.4 Cell level

At cell level, several techniques are available to evaluate or test the effect of certain stressors:

- *In vitro* testing (see Section 5.6.2).
- The use of *stem cells* (embryonic or adult) to trigger epigenetic changes in the targeted tissue, *in vivo*.
- *Gene expression profiling* to visualize the effects of certain molecular events.

## Stem cells

Next to *in vitro* assays, stem cells (see Box 3.6) are being used more and more for detection of toxicities of physical, chemical and biological toxins or toxicants. According to Trosko and Upham (2005), using stem cells it seems is a good option because *in vitro* assays are in general not consistent, do not measure what they are supposed to measure, generate data from abnormal cells, are extracted from cell types not relevant to the human system, are derived from the wrong conditions for measuring toxicities that occur *in vivo* at different conditions, and in many cases are misinterpreted.

The potency of a stem cell determines the potential to differentiate into different cell types:

- Totipotent stem cells can construct a whole organism.
- Pluripotent stem cells can differentiate into nearly all cells.
- Multi-potent stem cells can differentiate into a number of closely related cells.
- Oligopotent stem cells can differentiate into only a few cells.
- Unipotent can produce only one type of cell, but can renew themselves unlike non-stem cells.

Mutating genes in stem cells, altering the numbers (increasing or decreasing) and abnormally modifying the normal expression of genes in stem cells, can affect homeostatic control of tissue development, causing various stem cell-dependent diseases. Epigenetic changes can occur both in proliferating and non-proliferating cells (cell proliferation is the increase in the number of cells as a result of cell growth and cell division (National Cancer Institute, 2012)). So, it is important to assess the epigenetic potential of a chemical's contribution to human diseases such as cancer (Trosko and Upham, 2005). Determination of how exogenuous agents might induce specific intracellular signalling to trigger epigenetic changes in the targeted tissue, *in vivo*, is the challenge. Kang and Trosko (2011) therefore propose instead of *in vitro* assays to use *in vitro* three-dimensional screening assays that can assist in their use to assess toxicities of chemicals.

---

### Box 6.3 Stem and progenitor cells

*Stem cells* are biological cells that can divide and differentiate into diverse specialized cell types and can self-renew to produce more stem cells. A stem cell is a cell that can divide either symmetrically or asymmetrically after being given an appropriate external signal. It can differentiate into diverse specialized cells.

*Progenitor cells* are biological cells that have the tendency to differentiate into a specific type of cell, and are already more specific than a stem cell. They are in between stem cells and fully differentiated cells, but will be pushed to differentiate into its 'target' cell.

Stem cells can replicate indefinitely, while progenitor cells can divide only a limited number of times. In an adult organism, stem cells and progenitor cells act as a repair system for the body, replenishing adult tissues.

---

### Gene expression

Individual gene expression profiles are used to study the effects of certain molecular events (see an example in Figure 6.6). The no observed transcriptional effect level (NOTEL) is the dose (or concentration) of a compound or stressor that does not elicit a meaningful change in gene expression (i.e. the threshold of the dose/concentration that elicits minimal mechanistic activity) (Afshari *et al.*, 2011).

Tracking of molecular events at a whole genomic level across multiple doses and time points is nowadays done by microarray chip technology (Afshari *et al.*, 2011). The result is a miniature array with the expression pattern of

**Figure 6.6** *Gene expression profile reflecting an anti-viral response: hierarchal cluster representation of gene expression changes induced in human fibroblasts treated with the anti-viral agent Interferon-alpha for 4, 8 or 24 hours as compared to untreated cells. Differential gene expression distinguishes up- (red) or down-regulation (green) and gene clusters according to their expression profile over time*

Source: Hans A.R. Bluyssen: www.lhmg.amu.edu.pl.

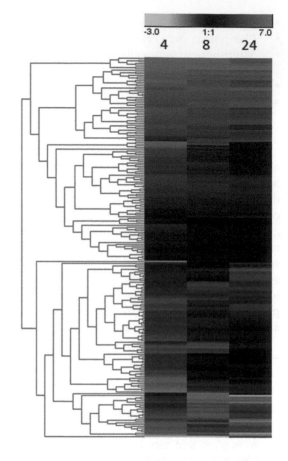

thousands of genes (Tamayo *et al.*, 1999). Self-organizing maps (Tamayo *et al.*, 1999), clustering of gene expression data, extracting patterns and identifying co-expressed genes (Chou *et al.*, 2007) and principal component analysis (PCA) can be used to interpret the massive data sets (Afshari *et al.*, 2011).

It is hypothesized that it would be possible to define signature patterns of altered gene expression that indicate specific adverse effects of chemicals, drugs or environmental exposures (Paules, 2003). Studies have demonstrated the capability of identifying signature patterns of altered gene expression that can be used to predict the classes of chemicals an organism was exposed to based on an initial training set of chemicals. An idealistic expectation is that gene expression data suffice as a 'digital pathology' representation of the phenotype of toxicity for a given toxicant (Paules, 2003).

## 6.3 Physical techniques

*Physical techniques to measure changes in the human systems comprise techniques assessing responses of the autonomic nervous system such as heart rate, blood pressure, lung function and sweating, and techniques that assess the activities (e.g. electrical activity) and processes in the brain with so-called neuro-imaging techniques. Peripheral nervous system responses comprise several local effects such as endothelial dysfunction and facial contractions.*

## 6.3.1 Autonomic nervous system

Responses of the autonomic nervous system that have been evaluated or monitored upon exposure to an environmental stressor comprise heart rate, blood pressure, lung function and sweating.

### Heart rate

The parasympathetic nervous system influences the heart rate via the release of acetylcholine by the vagus nerve (during resting and relaxed conditions). The sympathetic nervous system influences the heart rate via release of epinephrine and norepinephrine (during exciting and tensional conditions; see Section 3.2.1). Therefore, the heart rate could be an important physiological parameter. Heart rate, heart rate variability (HRV), but also PEP (pre-prejection period), a measure for the contraction strength of the heart muscle, have been used as indicators of (para)sympathetic activity.

The heart rate variability (HRV), heart-to-heart beat alterations, an index of the parasympathetic influence on the heart, has been studied intensively as an important indicator of the overall balance between parasympathetic and sympathetic activity. HRV has been related to particulate matter, indoor air temperature and noise, suggesting effects on the autonomic nervous system (Pope and Dockery, 2006; Graham *et al.*, 2009; Yao *et al.*, 2009). According to several studies decreases in heart rate variability can be predictors of mortality (Nolan *et al.*, 1998; La Rovere *et al.*, 2003).

In a recent study of work stress and cardio-vascular disease risk (Vrijkotte, 2010), an increase in heart rate was strongly related to the work situation, even after correction for physical activity (activity diaries were coupled with physiological measurements and repeated measurements were performed during a whole working week). Vrijkotte (2010) also found a strong relation between over-commitment and reduced variation in PEP.

Heart rate is measured by finding the pulse of the body, for example, at the wrist on the side of the thumb or the neck. A more precise method of determining the pulse is *electrocardiography* (ECG). ECG is used for measuring electrical potentials generated by the heart: it detects and amplifies the electrical changes on the skin that are caused when the heart muscle depolarizes during each heartbeat (the influx of $Na^+$ and $Ca^{++}$ ions; see Section 2.6.2). For an ECG normally a patient or participant receives 10 electrodes on the skin: two on the pulses, two on the ankles and six on the chest (see Figure 6.7).

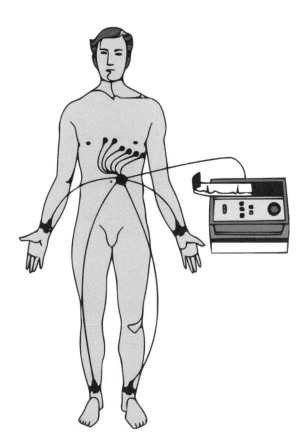

**Figure 6.7**
*A patient connected to 10 electrodes for measurement of an ECG*
Source: Wikimedia Commons, public domain.

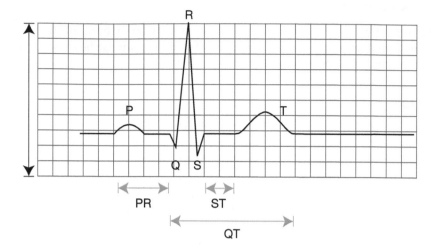

**Figure 6.8** *Heart beat wave with different deflections P, Q, R, S and T*
Source: www.hartstichting.nl/hart_en_vaten/medisch_onderzoek/onderzoek_hart/hartfilmpje (accessed 18 January 2013).

With an ECG the activity of the heart is seen in a typical electrical wave, made up of various characteristic deflections (the directions of the wave characteristics; see Figure 6.8; Martin *et al.*, 2010):

- *P wave*: small deflections before the contraction of the atria.
- *QRS complex of waves*: produced during depolarization prior to contraction of the ventricles.
- *R wave*: the largest wave.
- *T wave*: small blip-like deflection as a result of activity in ventricles.

It takes around 160 milliseconds from P to QRS, 300 milliseconds from QRS to T and 370 milliseconds until the next contraction.

## Blood pressure

Raised blood pressure has been pointed out as an important risk factor for cardiovascular diseases (see Section 2.3.3). An impaired or an excessive blood pressure fall at night and a steeper blood pressure rise at the time of morning awakening might be associated with a higher rate of cardiovascular effects. It seems that circadian blood pressure and heart rate changes, including a fall in these variables at the time of night sleep and their rise when wakening up in the morning, represent the quantitatively most important components of 24-hour blood pressure and heart rate variability (Parati, 2004).

Blood pressure is usually measured by a device around a patient's upper arm, measuring the systemic arterial pressure expressed in the systolic (maximum) pressure over the diastolic (minimum) pressure in millimetres of mercury (mmHg). Normal values for blood pressure of adults lie between 90–120 mmHg for systolic pressure and 60–80 mmHg for diastolic pressure.

## Lung function

The amount (volume) and/or speed (flow) of inhaled and exhaled air are indicators for pulmonary lung function. With spirometry, an important tool for the screening of general respiratory health, these indicators can be tested (see Figure 6.9). The patient or participant is asked to take a deep breath and then exhale into the sensor as hard as possible, for as long as possible. During the test, nose clips may be used to prevent air from escaping through the nose.

The principal parameters measured in spirometry are (Miller *et al.*, 2005):

- *Forced vital capacity (FVC)*, which is the total volume of air expired forcefully and completely after a full inspiration.
- *Forced expiratory volume in one second (FEV1)*, which is the volume of air expired in the first second FVC manoeuvre.
- *Peak expiratory flow (PEF)*, which is the maximal expiratory flow achieved during the maximally forced expiration.

**Figure 6.9** *Desktop spirometer*
Source: Wikimedia Commons, author Cosmed www.cosmed.com/ponyfx.

- *Forced expiratory flow (FEF)*, which is the flow of air coming out of the lung during specific intervals of a forced expiration, is usually 25–75 per cent of FVC manoeuvre (FEF25–75%).
- *Maximal expiratory flow (MEF)*, which is the instantaneous forced expiratory flow when 75 per cent or 50 per cent or 25 per cent of the FVC remains to be expired (MEF75%, MEF50%, MEF25%).
- *FEV1/FVC*, which is the ratio of FEV1 to FVC.

These parameters can for example indicate whether a patient suffers from obstructive airway disease. FEV1, FVC and FEV1/FVC are most often used. FEF25–75% seems however a good alternative. Figure 6.10 shows average values for FVC, FEV1 and FEF25–75%, which, according to a study in the United States in 2007 of 3,600 subjects aged 4 to 80 years (Stanojevic *et al.*, 2008).

Spirometry was also applied in the European SINPHONIE project (see Section 5.3.3) on children.

## Sweating and skin temperature

*Infrared thermography* (IRT) is a highly accurate non-contact method of measuring changes in skin temperature arising from vasodilation and constriction, and can be used to visualize and measure infrared radiant energy (Jenkins *et al.*, 2009). IRT uses a camera that is sensitive to infrared radiation. Thermography has been applied in medical and ergonomic research to analyse various physical conditions, and it has been explored to monitor levels of frustration and physiological changes evoked by stress or fear inducing stimuli. Jenkins *et al.* (2009) used IRT to test whether there is a relationship between IRT and EEG to support the hypothesis that forehead temperature dynamics are a reflection of internal 'states of mind'.

Finger temperature has been used as a measure for thermal sensation (Stroem-Tejsen *et al.*, 2009; Lan *et al.*, 2010). The typical skin temperature for the human body is about 32–3°C. However, in a lab study skin moisture was found to be a more suitable parameter to characterize physiological effort to elevated temperature conditions than skin temperature (Nöske *et al.*, 2011). The so-called galvanic skin response (GSR) or electrodermal response (EDR) measures the skin resistance to electric current. Sweat reduces skin resistance and allows conductance to occur on the skin.

Skin conductance changes can be influenced by experience of positive and negative emotions and in perceptual awareness (Martin *et al.*, 2010). GSR activity may therefore give information on a patient's response to a stimulus (the polygraph or lie-detector uses e.g. GSR). Skin conductance increases with increased stress, arousal and cognitive activity and reduces when activity level is low. During states of anxiety, sweating

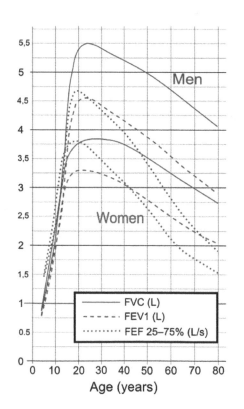

**Figure 6.10**
*Normal values for FVC, FEV1 and FEF25–75: y-axis is expressed in litres for FVC and FEV1, and in litres/second for FEF25–75 per cent*

Source: Mikael Häggström, Wikimedia Commons, ttp:// commons.wikimedia.org/wiki/File:Normal_values_for_ FVC,_FEV1_and_FEF_25-75.png (accessed 2 June 2013).

occurs especially in the palms, while, with physical stimulation such as heat, forehead, neck and back of the hands usually respond more intensely.

GSR can be measured with a simple inexpensive ohmmeter: a current generator sends a constant current through the electrodes and measures the changes in conductance.

## End-partial CO$_2$ and arterial blood oxygen saturation

Shallow breathing (hypoventilation) in poor air causes the CO$_2$ level in the blood to increase (respiratory acidosis), which is known to cause headache, easy fatigue and sleepiness (Lan *et al.*, 2011). The concentration of end-tidal partial CO$_2$ (ETCO$_2$), the partial pressure of CO$_2$ at the end of expiration, can be monitored with a non-invasive capnographic monitor to approximate arterial CO$_2$ non-invasively. Normal values of ETCO$_2$ are 35–45 mmHg (Lan *et al.*, 2011). Arterial blood oxygen saturation (SPO$_2$) can be measured with a monitor of pulse oximetry. Low blood oxygen saturation has been associated with fatigue (Sung *et al.*, 2005) and decreased cognitive functions (Andersson *et al.*, 2002).

## 6.3.2 Activities in the brain

Activities in the brain can be investigated with the use of electrical impulses, so called *electroencephalography*, or one can use several *neuro-imaging techniques*, which allow to visualize and obtain images of brain functions and structure. They do not measure neural functioning, they measure the processes associated with neural function such as blood flow or oxygen and glucose consumption.

## Electroencephalography

EEG is a technique with which the brain's electrical impulses are recorded. EEG is defined as the difference in voltage between two recording locations plotted over time (Jenkins *et al.*, 2009). Electrodes can be placed on the scalp (non-invasive) or in the brain. EEG activity is seen in the form of a line-tracing or brainwave (Martin *et al.*, 2010). EEG data can be converted to brain-maps (two-dimensional representation of EEG activity), coloured or in grey-scale (representing areas of high and low activity; Figure 6.11). There are different types of EEG waves, called frequencies, which are thought to represent different psychological states (see Section 2.7.2). With the EEG technique a stimulus or a task can be associated with brain activity (such as sleep). EEG has been applied to investigate circadian changes in brain temperature in response to thermal changes of the environment (Yao *et al.*, 2009).

## Neuro-imaging techniques

With *positron emission tomography* (PET), and, to a lesser extent, single-photon emission (computerized) tomography (SPECT or SPET), direct investigations of the functional levels of some CNS

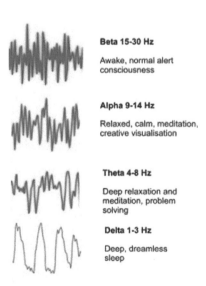

**Beta 15-30 Hz**

Awake, normal alert consciousness

**Alpha 9-14 Hz**

Relaxed, calm, meditation, creative visualisation

**Theta 4-8 Hz**

Deep relaxation and meditation, problem solving

**Delta 1-3 Hz**

Deep, dreamless sleep

**Figure 6.11** *EEG brain waves*
Source: www.meditations-uk.com/images/information/brain_waves.jpg.

neurotransmitters, including dopamine and serotonin can be carried out (Puri and Lynam, 2010). Both types of investigations are expensive and expose the patient to ionizing radiation.

With a PET scan the amount of oxygen consumed by, or blood flow travelling to, neurons is examined. The radioactive part of glucose (that is injected in the arm and accumulates in particular regions of the brain) emits positrons, which are detected by a PET scanner (Martin *et al.*, 2010). Healthy subjects were exposed to aversive olfactory stimuli while measuring regional cerebral blood flow (rCBF) with positron emission tomography. The findings support a critical role of the human amygdala in either the processing of aversive olfactory stimuli or the transduction of neural signals from smells into emotional responses (Zald *et al.*, 1997).

*Computerized tomography* (CT) displays the structure of the brain in slices (*tomos* means cut). The scanner sends a beam of X-rays through the head and the computer calculates the amount of radiation passing through at various points along each angle (Martin *et al.*, 2010; see Figure 6.12).

An alternative functional neuro-imaging method is *neurospectroscopy*. This technique does not use ionizing radiation but uses magnetic fields. Different uses of this technique are (Puri and Lynam, 2010):

- *Proton neurospectroscopy*: allows levels of free choline to be measured. A rise in free choline is consistent with a functional deficiency of omega-3 and omega-6 log chain PUFAs, indicating a chronic viral infection.

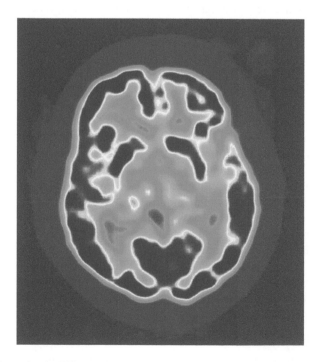

**Figure 6.12** *Example of a PET scan*

Source: Jens Lagner, Wikimedia Commons, http://commons.wikimedia.org/wiki/File:PET-image.jpg (accessed 2 June 2013).

- *Radioactive phosphorus neurospectroscopy*: allows nucletide triphosphate levels to be measured and to study energy metabolism.

Other neuro-imaging methods using magnetic fields are (Martin *et al.*, 2010):

- *Magnetoencephalography* (MEG) measures magnetic fields generated by neurons with a superconducting quantum interference device (SQUID) immersed in liquid helium. MEG can be used to localize sources of activity fairly well and can be plotted on a three-dimensional image of the participants head. Various functions, from language to olfaction, can be studied.
- *Magnetic resonance imaging* (MRI) uses magnetic fields and radio waves rather than X-rays in the CT scan and gives much more detailed images of the structure of the brain (see Figure 6.13). When a magnetic field passes over the head, reverberations are produced by hydrogen molecules, which are picked up by the scanner and converted to a structural image.
- *Transcranial magnetic stimulation* (TMS) is a relatively new, non-invasive technique for studying localization. It involves modulating cortical activity by passing alternating magnetic fields across the scalp. Electrical currents are induced in the cortex and the excitability of the cortex is subsequently increased or decreased. This can take a few seconds or minutes to a few weeks. It can produce transient impairment or improvements in cognition non-invasively.

Both PET and MRI have good spatial resolution, but they have poor temporal resolution: it is difficult to match the psychological neural event in time precisely because a number of scans are recorded and then averaged.

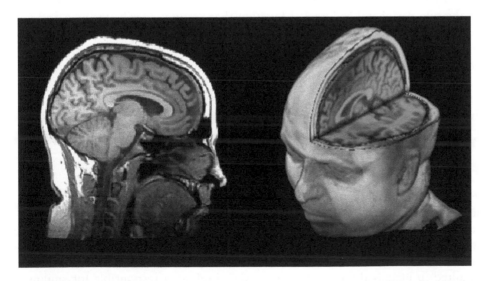

**Figure 6.13** *Structural MRI scan: detailed images of the brain allow anatomical structures to be measured – the size or grey matter density within a given structure may relate to different skills and abilities*
Source: Tom Hartley, Department of Psychology, The University of York.

The latest technique applied is fMRI (functional magnetic resonance imaging), which follows the fast brain waves, and can be used to identify conscious and unconscious needs of people (Lindstrom, 2008). fMRI has been used in the cognitive neurosciences to observe brain activity while cognitive tasks are taking place (Jenkins *et al.*, 2009). Hartley *et al.* (2003) investigated way-finding in humans by applying fMRI to investigate the neural bases of different cognitive processes (see Figure 6.1).

With these techniques it is possible to see what is happening in the amygdala (the region of the brain that generates dread, anxiety and the fight-or-flight impulse) and it is possible to detect when the body craves something – the 'craving spot' nucleus accumbens lights – even when one claims to need nothing (Lindstrom, 2008). Thanks to fMRI, we know now the extent to which the senses are intertwined: that fragrance can make us see, sound can make us smack our lips, and sight can help us imagine sound, taste and touch, that's if it is the right pairing of sensory input. According to Lindstrom (2008) brain scans predict accurately, while questionnaires do not; what we say we like is not what the scans say.

### 6.3.3 Local effects

#### Endothelial dysfunction

Endothelial dysfunction is a process that results in abnormal regulation of blood vessel tone and the loss of the atheroprotective properties of normal endothelium (Celermajer, 1997). Evaluation of endothelial function could be performed with a non-invasive non-operator-dependent and relatively easy to perform technique, the pulse wave amplitude (PWA), that correlates (Kuvin, 2003) with the flow-mediated dilation (FMD) of the brachial artery. FMD is a validated method for evaluating peripheral endothelial function that requires specific training and is highly operator dependent. Changes in PWA during reactive hyperemia can be examined with a specially designed finger plethysmograph (*peripheral arterial tonometry* (PAT); see Figure 6.14). The peripheral endothelial function as a cardiovascular risk factor has been measured with PAT in relation to mental stress (Goor *et al.*, 2004) and in relation to indoor air pollution (Wu *et al.*, 2010). In the latter study the cardio-ankle vascular index (CAVI) was applied to assess changes in arterial stiffness.

#### Facial contractions

*Electromyography* (EMG) is used for measuring skeletal muscles electrical activity. EMG activity is recorded by electrodes (circular disks of around 10 mm in diameter) from the surface of the skin. In a study using EMG with 30 male and 30 female subjects, Schwartz *et al.* (1980) found that women tend to be more facially expressive than males. Additionally, imagining pleasant thoughts resulted in increased muscle activity in the cheek area responsible for smiling (the zygomatic muscle), while the corrugator muscle at the eyebrow was more active during imagining of unpleasant thoughts.

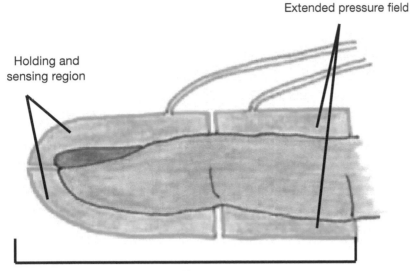

**Figure 6.14** *Schematic diagram of the sensor's structure for peripheral arterial tonometry (PAT): finger arterial pulse wave amplitude measurement (pulsatile volume changes) by a finger probe. The sensor is partitioned into two separate sections that are independently pressurized to keep the venous transmural pressure negative to prevent venous pooling and distention and to ensure that only arterial volume changes are recorded. The sensor cap has a split thimble design that imparts a two-point clamping effect to lock the sensor to the fingertip while measuring*
Source: Bluyssen: adapted from Chouraqui *et al.*, 2002.

## 6.4 Behavioural monitoring

*Behavioural monitoring can be used to collect information on bio-behavioural variables that may activate physiology, such as physical activity and caffeine/coffee consumption. It can also be used to 'extract' information from building users on how they interact with their environment, and it can be used to monitor responses, for example, the physical reactions of eyes, face and body to events around the study subjects or activities they are focused on. Available techniques to monitor behaviour are a behavioural diary, systematic observation, automatic sensing and logging, behaviour genetic studies and performance tests.*

### 6.4.1 Behavioural diary

In a behavioural diary the subject reports which actions/activities he or she conducts as a response to certain events, or vice versa. Diaries are detailed prospective records of exposure kept by the subject (White *et al.*, 2008). Diaries are for example used to measure physical activity (La Porte *et al.*, 1979) and dietary intake (Dolecek *et al.*, 1997), but also to measure symptoms, minor illnesses, medication use and medical care (Verbrugge, 1980). In general, diaries are open-ended and have the form of a booklet (or journal) in which the subject records each occurrence of a particular behaviour, with one entry per line and columns indicating the details needed. An example is shown in Table 6.3.

**Table 6.3** *Example of diary of daily activities in minutes (presented here for Monday only). For each activity, please fill the corresponding time windows*

| | Monday | | | | | | | | | | | | |
|---|---|---|---|---|---|---|---|---|---|---|---|---|---|
| | 7 | | 9 | | 11 | | 13 | | 15 | | 17 | | 19 |
| | | 8 | | 10 | | 12 | | 14 | | 16 | | 18 | |
| **Transport to work** | | | | | | | | | | | | | |
| By bike | | | | | | | | | | | | | |
| By bus/train | | | | | | | | | | | | | |
| By car | | | | | | | | | | | | | |
| Walking | | | | | | | | | | | | | |
| **Work activity** | | | | | | | | | | | | | |
| Computer work | | | | | | | | | | | | | |
| Paper work | | | | | | | | | | | | | |
| Copying/printing | | | | | | | | | | | | | |
| Meeting in other room | | | | | | | | | | | | | |
| Meeting in other building | | | | | | | | | | | | | |
| Meeting in own room | | | | | | | | | | | | | |
| **Personal time** | | | | | | | | | | | | | |
| Coffee with machine nearby | | | | | | | | | | | | | |
| Coffee in bar/restaurant | | | | | | | | | | | | | |
| Coffee outside building | | | | | | | | | | | | | |
| Smoking in separate room within building | | | | | | | | | | | | | |
| Smoking outside building | | | | | | | | | | | | | |
| Smoking in own office | | | | | | | | | | | | | |
| Lunch in bar/restaurant within building | | | | | | | | | | | | | |
| Lunch in own office | | | | | | | | | | | | | |
| Lunch in own office while working | | | | | | | | | | | | | |
| Lunch outside building | | | | | | | | | | | | | |
| Other hot drinks and cold drinks with machine nearby | | | | | | | | | | | | | |
| Other hot drinks and cold drinks in bar/restaurant | | | | | | | | | | | | | |
| Other hot drinks and cold drinks outside building | | | | | | | | | | | | | |

Source: Bluyssen.

Diaries can also be closed-ended or partly closed-ended. Electronic diaries can be used via touch-tone telephones or hand-held electronic devices ('palm' computers), which can be programmed to remind subjects to record information in their diary (Kubey *et al.*, 1996). Hybrid forms are:

- *Diary-interview*: when subject is phoned, one or several times per day, to report a certain activity.
- *Objective measure-diary*: electronic or mechanical devices that automate data capture with little subject involvement such as motion sensors that record physical activity (Bassett *et al.*, 2000).

Automated diary methods are in general more accurate than paper diaries and completeness of record keeping has been found to be higher as well (Jamison *et al.*, 2001).

There are techniques available, such as the experience sampling method (ESM), in which users are asked to reflect upon their actions at the moment or right after they have performed an activity (Intille *et al.*, 2003). An example is the study performed by Brink and Spiekman (2009), in which office workers were asked to indicate why they made a certain intervention (such as opening a window) by a pop-up screen on their PC (personal computer). By connecting the behaviour monitoring to any notification means (e.g. a pager, PDA (personal digital assistant) or PC), users can be asked to reflect upon any action of interest immediately. The big advantage, gaining information at the time the action is performed, is also the largest drawback of the system. The notification can be intrusive, intensive, repetitive and inconvenient (Intille *et al.*, 2003). Because of these reasons, 'lost data' can be a problem. Response ranges around 80 per cent are feasible, but there are also examples of less than 30 per cent (Khan *et al.*, 2009).

A complementary method to ESM is the day reconstruction method (DRM), which asks participants to elaborate on their events, activities and emotions in detail at the end of the day (Khaneman *et al.*, 2004). When data of ESM and DRM are combined (ESRM), researchers can gain insight into the feelings and experiences of users, coupled with factual information about those experiences (Khan *et al.*, 2009). This technique seems worthwhile to investigate in combination with checklists that help to inspect the indoor and built environment (see Section 6.6.1).

However, with diaries it should be kept in mind that the act of keeping the diary may affect the behaviours being recorded, such as changes towards more socially desirable or health conscious behaviours or changes in behaviours in order to reduce record keeping (White *et al.*, 2008).

## 6.4.2 Systematic observation

In systematic observation a researcher observes real time or video recorded behaviour of users and notes the user actions according to a predefined schedule. Data can be recorded through taking notes, video recording and/or coding (symbols or shorthand codes: each time some type of behaviour occurs it is indicated). Three types of systematic observation can be distinguished (Langridge and Hagger-Johnson, 2009):

- *Fully structured observation*: involves collection of data in a systematic structured manner for example with a grid recording system that notes how often a particular behaviour occurs. This type of observation often takes place in a laboratory setting to control extraneous variables, and often with video recording equipment.
- *Naturalistic field (uncontrolled) observation*: involves the observer studying people in their natural environment.
- *Participant observation*: in which the observer participates. It varies from full participant observation (full member of group – observers keep identity secret), participant as observer (identity not secret, but kept quiet), observer as a participant (researcher has to earn trust) and full observer (is equal to naturalistic observation).

Some advantages and disadvantages of the three types of systematic observation are presented in Table 6.4.

With systematic observation not only the activity patterns of subjects are monitored, but in many cases also their nonverbal responses related to the activity they perform (as for example is performed in the restaurant of the future – see Section 8.3.2 in Bluyssen, 2009a). In product design this type of technique is used as well in combination with (Sheth *et al.*, 1999):

**Table 6.4** *Advantages and disadvantages of fully structured observation, naturalistic field observation and participant observation*

| Fully structured observation | |
| --- | --- |
| **Advantages** | **Disadvantages** |
| Systematic and permanent collection of data. | May lack ecological validity. |
| Setting and extraneous variables controlled. | Behaviour may not be spontaneous or realistic. |
| Replication possible; minimized observer bias. | May not be possible (where participants not cooperate) or ethical in some situations. |

| Naturalistic field observation | |
| --- | --- |
| **Advantages** | **Disadvantages** |
| Greater ecological validity. | Difficult to conduct and to record; difficult to be unobtrusive. |
| Realistic, spontaneously occurring behaviour. | Extraneous variables poorly controlled. |
| Useful where it is not possible to observe participants in a laboratory. | Replication difficult; observer bias present. |

| Participant observation | |
| --- | --- |
| **Advantages** | **Disadvantages** |
| High ecological validity. | Researcher needs to rely on memory. |
| | Highly subjective. |
| Detailed but highly contextual data. | May be non replicable and difficult to generalize results. |
| Development of trust may produce information inaccessible by other means. | There may be problems and even dangers in maintaining cover if full participant. |

Source: Langridge and Hagger-Johnson, 2009.

- *Information boards*: Participant is asked to make a brand decision by uncovering the cells of a matrix of brand names in the rows and attributes in the column, in whatever order and as many or as few. The sequence of uncovering the cells is recorded and used to say something on the evaluation criteria customers' use.
- *Visual image profiling*: Nonverbal method using a set of pictures illustrating facial expressions that respond to a range of human emotions (see also Section 6.5.3).

*Biometric behaviour genetic studies* could be considered as a form of systematic observation. These studies on humans use samples of twins and adoption to separate effects of genetic and environmental variation. Other studies use biologically related siblings or parents and children to estimate total familial effects combining genetic and environmental sources of variance. More recent studies have combined the twin and adoption methods using pairs of twins separated at or near birth and adopted into different families. Quantitative genetics is based on the assumption that if genetic factors affect a quantitative trait, resemblance of relatives should increase with increasing degree of genetic relatedness (Zuckerman, 2005).

These biometric methods can answer the question of the relative effects of heredity and environment in a given population of a given age at a given time, but they cannot identify the actual genes associated with personality traits or illnesses. *Identification of such genes* is important for understanding the biological variants produced by the genes, and it is these biological variants that most likely directly affect the behaviours defining the trait (Zuckerman, 2005).

## 6.4.3 Automatic sensing and logging

In automatic sensing and logging the observation procedure is automated by using sensors, which record specific behavioural or context data stored in a database. Technologies that can be used for automatic sensing and recording that have been or are being developed (Brink *et al.*, 2010) are as follows:

- Sensors that record the use of a climate system (e.g. window switch, sensor on thermostat knob). These sensors are readily available or can be designed especially for any of the components of the applied climate systems.
- Motion sensors to record presence and pressure sensors to record shifts in seating positions (Figure 6.15).
- Stereo cameras to record movement patterns. The disadvantage is that (for now) the camera's cover is only a small surface.
- Face reader to recognize six basic emotions. The disadvantages are that usually emotions in a non-social context are not distinct enough to recognize that only the six predefined basic emotions can be recognized, and that there are tight conditions for the positioning of the face. A three-dimensional face reader is being developed that is better suited for use in a field environment.
- Face reader to recognize dynamic patterns (specific actions). This technique will save the researchers the labour of manually scoring the actions. However, this technique is still in its infancy.

**Figure 6.15** *Motion sensor next to lighting system*
Source: Bluyssen.

- Eye-tracking to record attention. Eye tracking has also been used to record pupil diameter, which is an indication of cognitive load (Bradley *et al.*, 2008).
- Percentage of eye closure. The eye's closure over a specific interval of time is the most popular oculometric for detection of fatigue and is used successfully in the military field where, for example, there is a need to accurately determine when pilots become fatigued to dangerous levels (McKinley *et al.*, 2011). The percentage of eye closure can be measured with cameras mounted on a set of eyeglass frames. These cameras not only measure the eye closure, but also the pupil position and size.

By monitoring behaviour it can be assessed what the study subjects do. However, the reason why they behave the way they do is left to the interpretation of the researcher. To gain more information about the reasons for action (for instance if it is comfort-driven) the user can be asked about these reasons or the environment can be monitored.

When using these behavioural measure methods, it is very important that they are unobtrusive, continuous, complete and flexible. Most of the mentioned techniques above are already in use in the field, or in pilots. Technically, deviations from optimal situations and processing the data are the biggest problems. For research purposes, integrating the data into meaningful information and models is the biggest challenge that has to be overcome.

An application of these behavioural automatic sensing and logging techniques are social agents or robots. Through social interaction they can provide reminders and guidance on daily life activities too, for example, the elderly with mild dementia by communicating what needs to be done and how to do it, in an intuitive and at times persuasive manner (Spiekman, 2012).

## 6.4.4 Performance tests

In performance tests users (occupants) are asked to perform a task or respond to a certain question in order to receive more information on how the environment influences them. Productivity has for example been measured this way (Clements-Croome, 2002; see also Section 8.3.2).

Performance can be measured for example by measuring the speed of working and the accuracy of outputs with reaction time tasks or it can be measured by memory tasks. The Wechsler Adult intelligence Scale (WAIS-III) and the Wechsler Memory Scale (WMS-II) are widely used scales for this (Martin *et al.*, 2010; see also Section 3.8.2). Lan *et al.* (2011) applied seven computerized tests in their study on the effects of thermal discomfort in an office on perceived air quality: mental orientation (a spatial orientation test), grammatical reasoning (a logic reasoning test), digit span memory (a traditional test of verbal working memory and attention), visual learning memory (a picture memory task measuring spatial working memory), number calculation (a mental arithmetical test), stroop (a test of attentional vitality and flexibility owing to perceptual/linguistic reference) and choice reaction time (a sustained attention task measuring response speed and accuracy to visual signals). In Box 6.4 an example of some memory tasks and reaction time tasks are presented.

---

### Box 6.4  Performance tests

Memory tasks (Wadsworth *et al.*, 2005):

- *Immediate free recall task* (measuring episodic memory). A list of 20 words is presented on the screen at a rate of one every two seconds. At the end of the list, the participant has two minutes to write down (in any order) as many of the words as possible (20 blank spaces).
- *Delayed free recall task* (measuring episodic memory). Towards the end of the test session participants have unlimited time to write down (in any order) as many of the words from the list of 20 presented for the immediate free recall task as possible.
- *Delayed recognition memory task* (measuring episodic memory). At the end of the test session, participants are shown a list of 40 words, which consists of the 20 words shown at the start of the session plus 20 distracters. Participants have to decide as quickly as possible whether each word was shown in the original list or not.

Deary-Liewald reaction time tasks (Deary *et al.*, 2011):

- *The simple reaction time*: in which a participant has to press a button or key in a response to a single stimulus (appearance of a diagonal cross within a square on the screen).
- *Four choice reaction time*: in which a participant has to press the button that corresponds to the correct response, choosing from four possibilities (four white squares correspond to four keys on the computer keyboard; the stimulus to respond is the appearance of a diagonal cross within one of the squares).

## 6.5 Questionnaires

*Many researchers from different disciplines have designed and applied questionnaires for different purposes. With questionnaires it is in principle possible to gather information on personal factors such as states and traits, descriptive (subjective) information on physical and psychosocial stressors and factors, on past events and exposures, behaviour, perceived wellbeing (or stress) and other responses (physical and psychological) over time.*

### 6.5.1 Design of questionnaire

Questionnaires can be used in all types of study design, including experiments, and can be applied in different ways (White *et al.*, 2008; Langridge and Hagger-Johnson, 2009), such as:

- *Personal interview*: An interview, either face-to-face or by telephone, is the most commonly used method to obtain data about subjects themselves or their environments, both on past and present exposure (see Figure 6.16). It is very expensive to execute, but more data can be collected in the course of an interview than by means of a self-administered questionnaire. Interviews can be *unstructured* (e.g. therapy session), *semi-structured* (questions are usually open-ended) or *structured* (fixed and ordered set of questions, used for example with children that cannot complete a questionnaire).

- *Self-administered questionnaires*: Questionnaires are usually applied to gather information from a group of people, a representative part of a population, or even an individual, in a systematic way. In general in a questionnaire less detailed information can be gathered than with an interview, but the costs are often low(er). Response rates of mailed or digital questionnaires (see Figure 6.17) are generally low. Recall bias can occur as with a personal interview.

- *Use of proxy respondents*: Proxy or surrogate respondents are people who provide information on exposure instead of the subjects themselves, because the subjects are too young, deceased or suffering from dementia.

- *Diaries*: Diaries are used for the collection of present behaviour or experiences. The use of diaries is more accurate than recall methods, but more difficult to analyse. The burden for the respondent is high, which might lead to difficulties in recruiting a representative sample.

**Figure 6.16** *A personal interview is the most commonly used method to obtain data about subjects themselves or their environments*

Source: Sebastian Meertins eight years old, January 2013.

Some general principles for designing a questionnaire are presented in Box 6.5.

0%  [                    ]  100%

In order to maintain confidentiality of data after 90 minutes the answers will be erased from the cache of your PC and the survey should be restarted from the beginning if not previously saved.
You will be able to save the survey at any time and resume it later by clicking on the 'RESUME LATER' button.

---

1 ID_01
* **Please provide the BUILDING CODE you have been communicated by email (you can also find this information on your time-activity diary).**

[                    ]

---

2 nq_01
* **Please provide the OFFICE CODE you can find on your time-activity diary.**

[                    ]

---

3 nq_03
* **When did you finish your last break from work (e.g. lunch, coffee break)?**

[                    ]

---

1 nq_04
* **How long did this break last?**

[                    ]

---

1 nq_05
* **Which floor is your workstation located?**

[ Please choose...          ]

---

[ Next >> ]

[ Exit and clear survey ]

**Figure 6.17** *Example of start page of a digital questionnaire used in the European OFFICAIR project (study B, Chapter 5, Box 5.9; Bluyssen et al., 2012b)*

## Box 6.5 General principles of a questionnaire

- The first rule of questionnaire design: don't reinvent the wheel. Use existing validated questionnaires.
- Keep it as short as possible; quick and easy to complete, and keep it readable. No passive sentences: school grade level age 12–13, max. 13–14, sometimes 8–9 is necessary.
- Include a participant information sheet, which tells participants about the study, who you are (including contact details), what will be done with the data and whether anonymity and/or confidentiality is assured. If necessary include a consent form, in which the participant agrees to take part.
- Always pilot the questionnaire. Test out on a small number of people before widely distributed. Feedback is very important to improve the questionnaire, with respect to layout, contents, relevancy of questions, easiness to complete and so on.
- Avoid using open-ended questions. Only if there are good reasons to include them (to collect richer data or to give an opportunity for more responses than asked for in closed questions). Open-ended questions need to be focused to avoid essay responses. Close-ended questions enable reliable information easily analysed.
- Ask one question at a time ('how do you feel about the temperature and the control of it' are two questions in one) and avoid ambiguous questions (questions that have more than one meaning), double negatives (e.g. 'not be banned') or technical terms and jargon.
- Tell respondents if your questions are going to be sensitive (increasing the sensitivity of questions once people have started the study is generally not ethically acceptable).
- Questions should be neutral rather than value-laden or leading (e.g. 'How often do you smoke?' instead of 'Would you generally refrain from smoking?' The phrase 'How often' suggests to the respondent that it is acceptable to report that they smoke; the word 'refrain' leads the participant towards a desired response.

Source: Langridge and Hagger-Johnson, 2009.

### Response rates

In addition to the principles presented in Box 6.5, in a postal questionnaire response rates can be maximized by including a pre-paid envelope with postal surveys, sending a reminder after one or two weeks have passed, giving advance warning – by a letter or postcard, and offering a small incentive if possible (Edwards et al., 2002). Tailoring the questionnaire to the needs of the respondents (e.g. moving forwards in case certain questions are not relevant, without showing this to the respondent) is a possibility with digital questionnaires that is recommended (Zagreus et al., 2004).

The prevalence of reported symptoms strongly depends on the design of the questionnaire (Brauer et al., 2008) and on the information provided at the beginning of the study to the participants (Brauer and Mikkelsen, 2003). Study design, for example, recall period, symptom frequency and symptoms definition, strongly influences the outcome (Tamblyn *et al.*, 1992; Raw *et al.*, 1994; Bluyssen *et al.*, 1996a; Chao *et al.*, 2003). In a double blind experimental crossover design to study interventions, a decrease of complaints was found within a six-week period (Tamblyn *et al.*, 1992). Bluyssen *et al.* (1996a) observed significant lower complaint rates at the same moment the questionnaire was answered in comparison with a three-month recall period. In a longitudinal study of 21 offices in four buildings, Chao *et al.* (2003) found the highest symptom prevalence at the beginning of the study and observed that people were most enthusiastic at the beginning of the study.

In order to reduce non-response rates in a web-based survey, Vicente and Reis (2010) recommend a number of design features for a 'good' questionnaire that can influence respondents' participation during the filling out of the questionnaire (increase the percentage of respondents who complete the questionnaire of all those who started the questionnaire (*overall completion rate*)) but have limited impact on the initial decision to participate in a web survey (reduce *the item non-response rate*, the unanswered questions as a percentage of the total number of questions in a questionnaire):

- Choose short over long questionnaires and scroll designs over screen designs.
- Include progress information and do not include forced response procedures.
- Apply a 'plain' design (as opposed to a 'fancy' design).

## Scale

Questionnaires can comprise questions with different scales:

- A *Likert scale* is a five (seven or sometimes more) point scale where respondents are able to express how much they agree or disagree with a statement. There are an equal number of positive and negative statements on the scale (e.g. from strongly disagree (1) to strongly agree (7)).
- A *semantic differential scale* is an alternative to the Likert scale that takes more of an indirect approach and relies on the ability of people to think in metaphors (e.g. good (1) to bad (7) or strong (1) to weak (7)).

Because it is questionable whether bad and good ratings are of the same descriptor (comfort not always being the opposite of discomfort) (Rouby and Bensafi, 2002), a safe way of rating any descriptor is to use single ended scales, that is ranging from neutral to either bad or good. Another advantage of single ended scales is the fact that subjects tend to give central (or neutral) ratings on bidirectional scales, giving a range reduction in effect. If only a limited range of ordinal values is then given, no effect may be found, even if there is.

## Response bias

Response bias to questionnaires can comprise of *social desirability*, respondents attempts to 'look good', or *response acquiescence*, the tendency to agree rather than disagree with statements.

A possibility to deal with social desirability is to include a social desirability (or lie) scale. The lie scale consists of a series of questions that if responded to consistently show someone acting in a 'saintly way' (e.g. that 'you never lose your temper' or 'you always say 'thank you'), which is just not realistic. A person who responds like that should potentially be excluded from the analysis.

A way to handle response acquiescence is to include both positively and negatively worded questions. In the SDR-5 (see Box 6.6) three of the five questions refer to 'underside' behaviour, so people with response acquiescence would score neutrally.

### Existing questionnaires

In questionnaires originating from 'the indoor air and thermal comfort community' and in questionnaires used in the field of *post-occupancy evaluation* (POE) in general the main focus lies on finding associations between *physical stressors* and *perceived wellbeing* in field investigations (see Section 6.5.2). In these questionnaires percentage of dissatisfied (perceived comfort), sick building symptoms (perceived health) and sometimes also self-reported productivity are gathered mainly in relation to indoor environmental aspects such as air quality, thermal comfort, light and noise aspects. Sensory evaluation techniques applied in the perfume and food industry comprise a similar approach (see Section 6.5.2).

Cohort studies originating from the 'outdoor environmental community' have used questionnaires to assess the relationship between environmental risk factors (e.g. noise and air pollution) and certain health or wellbeing effects (see Section 6.5.2). Many of these studies do take co-variates – confounders, exposure and effect modifiers – (e.g. *personal factors, psychosocial stressors* and *other factors*) into account (see Sections 6.5.3 and 6.5.4).

In the field of *personality psychology*, questionnaires have been developed and applied that focus on *emotional states* and *traits*. For these aspects many (standardized) questionnaires and scales are available (see Section 6.5.3).

Questionnaires specifically focused on *major life events*, on *daily event* assessments, and/or on *psychosocial stressors and factors)* are being applied in the *social and behavioural science* field (see Section 6.5.4).

---

**Box 6.6  Social desirable response set five-item survey (SDR-5)**

How much is each statement TRUE or FALSE for you? (responses are presented as 'Definitely true' (1), 'Mostly true' (2), 'Don't know' (3), 'Mostly false' (4) and 'Definitely false' (5)).

I am always courteous even to people who are disagreeable (1 = 1 point).
There have been occasions when I took advantage of someone (5 = 1 point).
I sometimes try to get even rather than forgive and forget (5 = 1 point).
No matter who I'm talking to, I'm always a good listener (1 = 1 point).

If the participant's score is 5 it is recommended to remove him/her from the analysis.

Source: Hays *et al.*, 1989.

In the *medical world* they make use of questionnaires that contain both physical and mental health aspects. An example is the SF-36 Health outcome survey (36 items), which measures a general mental component summary score and physical component summary score (Hays *et al.*, 1995). Eight subscales provide scores on different and important aspects of health: physical functioning, role physical, role bodily pain, general health, energy vitality, social functioning, role emotional and role mental health.

And last but not least, perhaps a stranger in this list, are the questionnaires developed for *product design*, including those not comprising questions in words but of pictures. A questionnaire can thus comprise of questions related to physical stressors, but also to psychosocial stressors, personal factors, other factors of influence and events (see Table 6.5).

**Table 6.5** *Possible components and examples of sub-components of a questionnaire*

| Components/stressors | Examples of sub-components |
|---|---|
| Physical environment | Characteristics of building, systems and rooms, such as: windows, view, services (heating, lighting systems), individual control and cleanliness. |
| Psycho-social environment | Individual, such as: marital problems, family composition, access to health care and financial stress; working i.e. job strains, such as high demands and low control, working hours; commuting, such as travel time and queuing. |
| Physical state | Perceived health – symptoms (such as SBS symptoms) and perceived comfort – complaints (such as feeling cold, finding the environment smelly, boring or dirty). |
| Psychological states and traits | Personality to determine one's personal baseline and mood of the moment. For both state and traits, in general the following basic emotions are distinguished: 1. Worry, nervousness, fear and anxiety; 2. Anger, hostility and aggressiveness; 3. Sadness and depression; and 4. Happiness, satisfaction, joy and ecstasy. Additional traits or personality terms that have been used are: negative and positive affect; introversion/extraversion; coping skills, self-efficacy and locus of control; intelligence and interest. (See Section 3.8.2). |
| Other personal factors | Gender, age, (pre-existing) health status (e.g. allergies and asthma), genetics, SES (Socio Economic Status), diet/nutritional status, education, obesity (BMI index), drugs (ab)use (smoking, coffee, alcohol), marital status, intelligence, environmental sensitivity, crowding (home), family structure, life style, work status, physical activity. |
| Other factors of influence | Neighbourhood quality, safety (crime and violence), crowding (neighbourhood), time of day, week or month, social support. |
| Events and exposures | Previous exposure and major life events (how far back depends on the aims and the design of the study: such as smoking history, episodes of depression and anxiety), previous events (causing expectations and worries) and habits (daily events – activity pattern; working hours, sleeping pattern, etc.). |

Source: Bluyssen *et al.*, 2011b.

## 6.5.2 Physical stressors and effects

Studies focused on finding associations between physical stressors and perceived wellbeing (including health and comfort) can be categorized into studies in which questionnaires are used:

- *For diagnostic purposes* in combination with extensive checklists and several biological, chemical and physical measurements to inventory environmental stressors (Burge *et al.*, 1987; Skov *et al.*, 1987; Preller *et al.*, 1990; Hedge *et al.*, 1992; Bluyssen *et al.*, 1992; Bluyssen *et al.*, 1996a; Caccavelli *et al.*, 2000; Cox, 2005). It is investigated why certain people complain or are having symptoms at their workplace or in their homes.
- To evaluate a certain built environment after a building has been occupied with the so-called *post-occupancy evaluation (POE)*, used in architectural and interior design disciplines, and human perception on indoor environment (Vischer, 1989; Leaman, 1996; Veitch *et al.*, 2002; Vischer, 2008).
- To make *a risk analysis of exposures over time* in population studies (e.g. Babisch *et al.*, 2006; Miedema and Vos, 2007; Beelen *et al.*, 2009; Kluizenaar *et al.*, 2009; see Section 5.3).

In Appendix A.6 an example is presented of questions applied for self-reporting of health symptoms and comfort problems in the European HOPE project (Roulet *et al.*, 2006a). For subjective symptoms of fatigue, a questionnaire is available consisting of three groups of questions, each consisting of 10 questions describing 'drowsiness and dullness', 'difficulty in concentration' and 'lack of physical integration' (Tanabe and Nishihara, 2004). Specific questionnaires are available to study certain health aspects as for example eye health with the self-reported break-up time (SBUT) (Bakke *et al.*, 2008) or the Ocular Surface Disease Index (OSDI) (Schiffman *et al.*, 2000):

- *SBUT*: Assessed by recording the time the subject could keep his or her eyes open without blinking, when watching a fixed point on a wall. This method has been used previously (Bakke *et al.*, 2008) and it has been shown to correlate well with the fluorescein method for break-up time (BUT) (Wyon and Wyon, 1987).
- *OSDI*[c]: The questionnaire consists of 12 items about eye health (symptoms, etc.). Each item could be scored from 4 (corresponding to 'All of the time') to 0 ('None of the time'). The OSDI is a valid and reliable instrument to measure dry eyes disease (normal, mild to moderate, and severe) and effects on vision-related function (Schiffman *et al.*, 2000).

There are several sensory evaluation techniques available in which the subjects have to answer a certain question or combination of questions for different exposure conditions in a chamber or field study (e.g. noise, thermal comfort, lighting and/or air quality) (see Box 6.7).

## 6.5.3 Psychological states and traits

Psychometric tests are standardized (tested on many persons) forms of questionnaires designed to measure particular traits or personality types, comprising

---

### Box 6.7 Sensory evaluation techniques

The attributes of perception that can be measured for all sensory modalities are:

- *Limit values* for absolute detection (sensory threshold) or for discrimination (minimum difference between two perceptions that makes them distinguishable).
- *Intensity* of a defined perception (e.g. odour intensity, sensory irritation intensity): obtained by various psychological matching and scaling methods. Magnitude estimation, with one or several references, has been applied in both field and chamber studies (e.g. the use of a trained panel (Bluyssen, 1990)).
- *Quality* of a defined perception: performed by value judgement, based on intensity and/or perceived quality of the sensory stimulation. The result can range from hedonic tone to acceptability.

For the latter classification and descriptor profiling has been widely used in surveys as well as in chamber studies. A typical example of *classification* is counting the persons reporting on a binary yes–no response task, for example, whether they have a symptom or not, resulting in a prevalence of that symptom. In *descriptor profiling* a set of attributes is evaluated, such as freshness, preference, irritation, and so on. Next to that, *category scaling* is often applied, such as a voting scale from clearly not acceptable to just not acceptable, just acceptable and clearly acceptable.

Source: Berglund *et al.*, 1999b.

---

*personality inventories*, *cognitive ability tests* (measure quality of intellectual functioning, such as IQ – see Section 3.8.2) and *measures of mental and physical health status, including mood states*.

## Personality

Examples of objective tests of personality include (Martin *et al.*, 2010):

- *The Eysenck Personality Inventory* (Eysenck and Eysenck, 1985). The Eysenck Personality Questionnaire is a rather short scale (14 items) which is widely used in stress research because it records the way people usually react (i.e. are stressed) in the face of external events.
- *The NEO-PI (neuroticism, extraversion and openness personality inventory)* follows the 5-factor model (neuroticism, extraversion, openness, agreeableness and conscientiousness; see explanation in Section 3.8.2) and is available in different versions: for example, NEO-PI (180 items), NEO-PR-R (240 items) and NEO-FFI (120 items). Free versions are available on http://ipip.ori.org/ipip. Mini-markers (respondents are asked to rate themselves on 40 adjectives) (Saucier, 1994) as well as very brief versions containing only 10 items (Gosling *et al.*, 2003; Rammstedt and John, 2007) work well for research purposes. An example of a brief version is presented in Appendix A.7.

---

### Box 6.8  Examples of questionnaires focused on one or more traits

- *Hostility*: Cook–Medley hostility scale (Cook and Medley, 1954; Barefoot *et al.*, 1989): which has been proposed as a marker for psychosocial vulnerability (it was linked to health outcomes such as the onset of hypertension and coronary heart disease and was strongly associated with allostatic load; Kubzansky *et al.*, 2000). Comprises hostile attribution (individual's tendency to interpret the behaviour of others as intended to harm), hostile affect (negative emotions associated with social relationships) and aggressive responding.
- *Worry*: The Penn State Worry Questionniare (Meyer *et al.*, 1990; measures the tendency for excessive uncontrollable, pathological worry) and the Worry Domain questionnaire (Tallis *et al.*, 1992; quantifies worry over different areas of content). The trait worry was related to significant immediate and prolonged cardiac effects (Pieper, 2008).
- *Boredom proneness*: The boredom proneness scale (BPS) (28 items (7-point highly disagree to highly agree)) may be an important element to consider when assessing symptom reporting. Significant positive correlations have been found with, among others, negative effect (Sommers and Vodanovich, 2000) and concentration difficulties (Farmer and Sundberg, 1986).
- *Locus of control*: The Pearlin Mastery Scale (Pearlin and Schooler, 1978; comprises 7 items scored on a 5-point scale from (1) completely agree to (5) completely disagree. High score indicates a high locus of control) and the Rotter scale (Rotter, 1966; 23-item forced choice items and 6 filler items scale).

---

There are questionnaires focused on one or more traits. Some of these traits have been suggested as important indicators of personality aspects relevant for the response to environmental stressors (see Box 6.8).

### Mood states

There are questionnaires with which specific mood states can be determined or more general states, such as the positive and negative affect based questionnaires *positive and negative activation scale* (PANAS; Egloff *et al.*, 2003). The PANAS comprises a 10-item negative and positive affect scales (5-point scale: 1 = very slightly or not at all; 2 = a little; 3 = moderately; 4 = quite a bit; 5 = extremely). The I-PANAS-SF (international PANAS short form) is presented in Appendix A.5.

The indicator *negative affect* has been suggested as a cumulative indicator of mental distress and psychological stress (Clougherty and Kubzansky, 2009). The PANAS scale seems attractive because it assumes that all negative emotions (states and traits) are covered with one factor, negative affect (NA), and all positive emotions are covered with positive affect (PA). Specifically, measures of trait NA were strongly correlated with neuroticism; conversely measures of trait PA were more strongly correlated with extraversion (Watson *et al.*, 1999).

Diener *et al.* (1985) claim that subjective wellbeing (SWB) comprises three separable components: positive affect, negative affect and life satisfaction. SWB is defined as a person's evaluative reaction to his or her life – either in terms of life satisfaction (cognitive evaluations) or affect (on-going emotional reactions), or cognitive (evaluation of one's life according to subjectively determined standards) and hedonic balance (the balance between pleasant affect and unpleasant affect) (Schimmack, 2006; Diener (2000)). The *satisfaction with life scale* (Diener *et al.*, 1985) can be used for this.

The *sense of coherence scale*, developed by Antonovsky (1993) has frequently been used in studies of the role of personality. Sense of coherence, a psychological measure of a life attitude, can detect personal susceptibility in relation to suspected environmental stress (Runeson *et al.*, 2003). The sense of coherence is defined as a global orientation that expresses the extent to which one has a pervasive, enduring though dynamic feeling of confidence that:

- Stimuli derived from one's internal and external environment in the course of living are structured, predictable and explicable.
- The resources are available to meet the demands posed by the stimuli.
- These demands are challenges worthy of investment and engagement.

The *Profile of Mood States* (POMS; McNair *et al.*, 1971) measures six identifiable mood or affective states scoring on a 5-point Likert-type scale ranging from 0 (not at all) to 4 (extremely): tension–anxiety; depression–dejection; anger–hostility; vigour–activity; fatigue–inertia; and confusion–bewilderment. The total mood disturbance can be computed by adding the scores for tension, depression, anxiety, fatigue and confusion, with vigour scores subtracted. The higher the score, the more negative the mood. POMS standard form contains 65 items and POMS brief form contains 30 items on a 5-point scale.

Questionnaires, adjective checklists and emotional scales are used to measure the level of emotions elicited by the product (Dormann, 2003). For example, the *PAD emotion* scales (Mehrabian, 1996) and several pictorial scales, such as:

- The *self-assessment manikin* (SAM; Bradley and Lang, 1994), in which the aspects pleasure/valence (positive or negative), arousal and dominance (measure for control of situation) are directly monitored with pictures of a manikin with different facial expressions (pleasure), heart beat (arousal) and size (dominance).
- *Emocards* (Desmet *et al.*, 2001) comprising 16 cards (8 for men and 8 for women) with facial expressions for 8 emotions (combinations of pleasant/unpleasant and calm/excited) based on Russels circumplex of affect (Russel, 1980; see Figure 6.18).

For *specific mood states* depression and anxiety a whole range of questionnaires is available, such as the State-Trait Inventory for Cognitive and Somatic Anxiety (STICSA; Grös *et al.*, 2008), the depression anxiety stress scales (DASS; Lovibond and Lovibond, 1995), the State-Trait Anxiety Inventory (STAI; Bieling *et al.*, 1998), the CES-D (Centre for Epidemiologic Studies Depression) rating scale (Radloff, 1977) and the GHQ (general health questionnaire; Goldberg, 1972).

**Figure 6.18** *Emocards (Desmet et al., 2001): please choose the image which best fits your mood now*
Source: P. Desmet.

### 6.5.4 Psychosocial stressors and other factors

As described in Chapter 1, stressors can be divided into physical stressors and psychosocial stressors, stressors caused by the physical and psychosocial environment respectively. Additionally, several other factors influence the effect of those stressors at the moment or over time: demographic variables; states and traits; life style and health status; genetics, events and exposures. Psychosocial factors comprise discrete stressful events (e.g. major life events and minor life events) and chronic stressors originating at the community level (locally, nationally or worldwide) or at the individual level (e.g. at work, at home or during commuting conditions; see Section 3.2.3).

### Personal factors

In general most questionnaires include a number of questions on demography (e.g. age, gender, ethnicity, social status, income and education), health status (e.g. allergies, obesity), life style (e.g. food pattern, physical activity). An example of such questions is presented in Appendices A.1 and A.2, taken from the OFFICAIR project (Bluyssen *et al.*, 2012a).

### Events

Questionnaires that specifically focus on major life events, daily event assessments or on psychosocial stressors and factors originate mainly from the social and behavioural sciences. Some examples of questionnaires (in fact they are checklists) including events are presented in Table 6.6.

**Table 6.6** *Questionnaires including events (major life events, daily life events)*

|  | Content | Reference |
|---|---|---|
| **Major life events** | | |
| Social readjustment rating scale | A checklist approach for rating major life events. | Holmes and Rahe, 1967 |
| Recommendations for a checklist | Tailoring of 30–50 negative life events (not chronic) to fit the risk status of populations, with a minimum of a one-year time frame, including the time course of each event. | Cohen *et al.*, 1995 |
| **Daily life events** | | |
| Daily life experience checklist | Contains 78 events organized in five major domains (i.e. work, leisure, family and friends, financial, other). Respondents are asked to rate the desirability (6-point scale from extremely undesirable to extremely desirable) and the meaningfulness (3-point scale from slightly to extremely) of those events that 'happened since you first awoke this morning'. | Stone and Neale, 1982 |
| Hassles scale | Consists of 117 items covering the areas of work, health, family, friends, the environment, practical considerations and chance occurrences. Yielding a frequency score and a severity score. | Delongis *et al.*, 1982 |
| Daily stress inventory | A 58-item questionnaire to assess the sources and magnitude of minor stressful events. Has been used to measure physiological responses to stress and daily fluctuations in symptoms associated with psychomatic disorders such as asthma. | Brantley and Jones, 1993 |

### Work-related stress

In studies of work stress, it is interesting to know how often workers are exposed to various types of daily problems such as work overloads and interpersonal conflicts with co-workers (Cohen *et al.*, 1995). Questionnaires to assess work demands, work control and work support have been developed. Such as the job content questionnaire developed by Karasek *et al.* (1998) and the effort–reward imbalance (ERI) at work (Siegrist *et al.*, 2004) based on the model of ERI (Siegrist *et al.*, 1997). The job content questionnaire is designed to measure scales assessing psychological demands, decision latitude, social support, physical demands and job insecurity. The ERI work-related stress is identified as non-reciprocity or imbalance between high efforts spent and low rewards received. The short version of the ERI questionnaire is presented in Appendix A.4. The questionnaire consists of 10 ERI and 6 over commitment items.

## 6.6 Environmental monitoring

*The environment of a subject can be assessed with the use of checklists and through several chemical, biological and physical measurements. In several studies these techniques have been applied together to characterize the indoor environment.*

### 6.6.1 Checklists

Extensive checklists have been developed and applied in combination with protocols for biological, chemical and physical measurements in several international IEQ studies. Examples of such studies are the BASE study in the US (Apte *et al.*, 2000; EPA, 2003), the European Audit project (Bluyssen *et al.*, 1996a), EPIQR (Bluyssen *et al.*, 1999), TOBUS (Caccavelli *et al.*, 2000), HOPE (Cox, 2005), and more recently OFFICAIR (Bluyssen *et al.*, 2012a) and SINPHONIE (www.sinphonie.eu).

The checklists applied focus mainly on the characteristics and processes creating the environmental stressors and can comprise of a part for the investigator but also a part for the facility manager and/or owner of the building. Currently lacking in most studies is the (objective) inventory of the psychosocial environment (see Table 6.7), because that type of information is usually gathered via interviews or questionnaires.

Associations between the measurements of the indoor environment conditions and wellbeing (health and comfort) of occupants are not clear. It seems possible however to gather so-called *short-cuts* with the use of checklists. In a short-cut, the building characteristics (such as having a HVAC (heating ventilating and air conditioning) system or measures taken (such as a maintenance or cleaning schedule) are directly related to comfort or health responses of occupants (Bluyssen *et al.*, 2011b). Several associations have been found, but further studies are required (see Section 8.4.2).

## 6.6.2 Chemical, physical and biological measurements

### General (standardized) techniques

Environmental stressors include physical, chemical and biological components or agents of the general environment (soil, air, water), the local environment (home, workplace, recreational sites), and the personal environment (food,

**Table 6.7** *Suggested components and sub-components of a checklist in an IEQ field investigation*

| Components | Sub-components |
|---|---|
| *The indoor and built environment:* | |
| Characteristics of the built environment | Description of outdoor environment i.e. location and surroundings. An example is presented in Appendix B.1. |
| Characteristics of building, systems and rooms | Description of building (e.g. construction information, building materials and furnishing). An example is presented in Appendix B.2. Description of building services (e.g. type of HVAC-system, lighting system and control system). An example is presented in Appendix B.3. Description of room(s) and interior. An example is presented in Appendix B.4. |
| Processes to maintain and operate the building and its activities | Description of building use (e.g. maintenance of HVAC-system, cleaning activities/schedule, renovation and retrofitting activities). An example is presented in Appendix B.5. |
| *The psychosocial environment:* | |
| Working environment (offices) | e.g. organizational structure, working hours and social working conditions. |
| Living environment (homes) | e.g. number of persons per home and social background occupants. |
| Learning environment (schools) | e.g. number of children per class and social background children. |
| Neighbourhood | e.g. neighbourhood quality. |

Source: adapted from Bluyssen *et al.*, 2011b.

drinks, cosmetics, drugs). Measurements of the environmental stressors represent an alternative or complementary approach to questioning or monitoring the subject.

Measurement methods vary from rating by skilled observers (e.g. the dustiness of air in the workplace) to measurements made in the field by simple (see Figure 6.19) or complex instruments (e.g. concentration of dust in air) to measurements in the laboratory of the concentration of substances in samples taken from the environment. In general, the purpose of sampling is to measure the exposure of individuals.

**Figure 6.19** *Measurement of airflow from a supply grille with a flow finder*
Source: Bluyssen.

Two approaches exist in environmental monitoring or sampling (White *et al.*, 2008):

1   *Measurements at fixed points* (static sampling) within the environment. The individual exposure is determined from the concentration of contaminants or quantity of the stressor (e.g. lux for lighting, decibel for noise, temperature for thermal comfort) measured in the parts of the environment covered by each sampler.
2   Measurement of the immediate and continually changing environment of individual subjects by some form of *personal sampling*.

In the case of air pollutants, these can be active samplers, the air is actively drawn through a filter or medium with a pump, or passive samplers, based on the diffusion or permeation of gases and vapours into some chemical trap. Analysis of the air samples is usually performed in a laboratory.

Measurements are also performed to study the effect of a process or change in the environment. For example, Tani and Hewitt (2009) explored the uptake of aldehydes and ketones by plants using a small-scale flow-through apparatus, employing proton-transfer reaction mass spectrometry as the analytical method.

Except for biological measurements, Chapter 3 in the *Indoor Environment Handbook* (Bluyssen, 2009a) provides an overview of general (standardized) methods that are applied to measure thermal, acoustical, lighting and indoor air parameters directly, *in vivo*, in the environment under study. With *in vitro* techniques, presented in Section 5.6.2, the conditions of exposure are simulated.

Box 6.9 presents an overview of the biological sampling performed in study A (description in Box 5.7). Those sampling techniques were focused on sampling microbes and allergens found in dust and air.

More specific biological measurements can be performed in environments where water is an important factor, such as shower cabins or bathrooms and ventilation systems (i.e. legionella in shower heads and humidifiers).

Studies can also be built around modern microbiological sampling and analysis methods, such as those designed to measure bacterial and fungal DNA (e.g. the non-culture based method polymerase chain reaction (PCR; Peccia and Hernandez, 2006)), or other molecular components and by-products of bacteria or fungi (e.g. airborne protein, endotoxin and $(1\rightarrow3)\text{-}\beta\text{-D-glucan}$ (Chen and Hildemann, 2009)). It should be noted that significant discrepancies between sampling methods regarding indoor microbial exposures exist (Frankel *et al.*, 2012). There is no standardized sampling methodology available (Adan and Samson, 2011).

## Calculation of total uncertainties or errors

What should be emphasized is that some measurements of stressors to measure exposure are used as indicators for other parameters. As for example the concentration of carbon dioxide ($CO_2$) is often used to determine the ventilation rate of a space. In this case it is important to realize that the measurement error or uncertainty will be a calculated value of the individual errors per parameter as well.

---

### Box 6.9  Biological sampling in Study A (SINPHONIE)

Environmental sampling for microbial and allergen determinations was performed by:

- *Settled dust sampling*: Determination of microbial contaminants from settled dust samples. Settled dust represents a long-term integrated sample and represents airborne exposure.
- *Floor dust vacuum sampling*: In parallel to settled dust, floor dust was vacuumed from the study classrooms. Regular vacuum cleaners were equipped with nozzles and filters in order to facilitate standardized sampling of dust for analyses of allergens.
- *Indoor and outdoor air sampling*: Active air sampling for the assessment of human exposure.

Analysis of samples comprised:

- Ergosterol as a general marker for fungal exposure in settled dust.
- Endotoxin was assessed from settled dust as a general marker for bacterial exposure.
- For more specific measurements of microbial exposure levels in the school environment as well as outdoors, defined fungal and bacterial groups (*Cladosporium*, *Penicillium/Aspergillus*, *Stachybotrys*, *Aspergillus versicolor*, *Streptomyces*, gram positive/negative bacteria) were measured via quantitative polymerase chain reaction (QPCR; Peccia and Hernandez, 2006) using settled dust and air filter samples.
- Selected indoor relevant allergens such as house dust mites, cat and dog allergens were measured from vacuumed floor dust samples.

---

*Measurement uncertainty* is defined as a 'parameter associated with the result of a measurement, that characterizes the dispersion of the values that could reasonably be attributed to the measurand (particular quantity subject to measurement) (ISO, 2007). The uncertainty on each value is required to determine the total uncertainty on a calculated value that comprises of several parameters (values). If for example the total calculated value ($Z$) comprises measured parameters $A$, $B$ and $C$:

$$Z = (A - B) \times C \tag{6.1}$$

And uncertainty is defined as:

$$\text{Uncertainty}^2 = (\Sigma((\delta Z/\delta X_i) \times dX_i^2) \tag{6.2}$$

Where:
$Z$ = total calculated value
$X_i$ = parameter to calculate the total value
$dX_i$ = uncertainty of parameter

Then the total uncertainty of the total value can be calculated as follows (Bluyssen *et al.*, 1995):

$$\text{Uncertainty}^2 = ((\delta Z/\delta A) \times dA)^2 + ((\delta Z/\delta B) \times dB)^2 + ((\delta Z/\delta C) \times dC)^2$$
$$= C^2 \times dA^2 + (A - B)^2 \times dC^2 + C^2 \times dB^2 \qquad [6.3]$$

### 6.6.3 Simulations and modelling

Predicting how people feel or how people like an indoor environment has mostly been focused on modelling one environmental aspect, in particular thermal comfort and air quality (Bluyssen, 2009a).

From thermal comfort models, computer simulation programmes were developed from around the 1980s. They offer the possibility to predict building performance aspects during the design phase, taking historical data (such as temperature and humidity). Next to that thermal manikins are used to determine heat transfer and thermal properties of clothing, and heat transfer coefficients for the human body systems, to predict human responses to extreme or complex thermal conditions, to determine air movement around the human body in closed spaces, and for the evaluation and assessment of the thermal environment. The increased use of thermal manikins has, for example, led to the development of a thermal manikin that can simulate breathing and has been used to assess the amount of re-inhaled exhaled air (van Hoof *et al.*, 2010).

For health-related aspects, in particular air quality, risk modelling is in general used to determine whether a certain level of exposure or duration of exposure is allowed or not (see Section 9.5.1). Studies performed by Berglund and Cain (1989) demonstrated that perceptions of indoor thermal conditions and air quality are closely linked. Some investigations have been undertaken in a laboratory set-up to determine the effect of different parameters on the acceptance of the indoor environment (e.g. Hygge and Knez, 2001; Witterseh *et al.*, 2004; Chen *et al.*, 2007; Clausen and Wyon, 2008; see Section 9.4.3). To predict the integral building performance as experienced by the end-user or the integral health effects of being in a building is still a bridge too far.

# PART III

# Analysis

# 7

# Needs and opportunities

*Based on Parts I and II, the need for a different view on IEQ is emphasized, including 'new' assessment procedures and models as well as the need to consider the built indoor environment as a system with sub-systems. A view in which IEQ is approached in an integrative multi-disciplinary way, taking account of possible problems, interactions, people and effects, focusing on situations rather than single components. While currently available performance indicators for healthy and comfortable indoor environments are presented and discussed in Chapter 8, Chapter 9 presents available information for defining a conceptual framework of cause–effect relations (patterns) for different scenarios (homes, schools and office buildings) and for major end-points (e.g. asthma, learning ability, productivity, and dry eyes and skin).*

## 7.1 Introduction

For most of the time, science has relied on the optimization of single factors such as thermal comfort or air quality. The realization that the indoor environment is more than the sum of its parts, and that its assessment has to start from human beings rather than benchmarks, has only been gaining ground in recent years. Research and practice merely focused on single components of the indoor

## Box 7.1  Claude Bernard (1865)

I am convinced that, since a complete equation is impossible for the moment, qualitative must necessarily precede quantitative study of phenomena.

(p. 130)

In physiology we must never make average descriptions of experiments, because true relations of phenomena disappear in the average; when dealing with complex and variable experiments, we must study their various circumstances.

(p. 135)

By statistics, we get a conjecture of greater or less probability about a given case, but never any certainty, never any absolute determinism.

(p. 138)

environment: thermal, lighting, air and sound quality. Much time was spent on identifying objective relations between indoor environmental parameters and the human reactions (dose–response). The outcome can be seen in comfort models, for example, for thermal comfort (Fanger, 1982), quantitative recommendations in the form of indices (Bluyssen, 2001) and/or criteria/limit values for temperature, light, noise, ventilation rates and certain substances in the air (e.g. CEN, 2002a, b; ASHRAE, 2004a, b; EU, 2004; ISO, 2005b; WHO, 2003, 2006). Only in the last decade of the twentieth century, a first try was made through epidemiological studies to approach the indoor environment in a holistic way (Burge *et al.*, 1987; Skov *et al.*, 1987; Preller *et al.*, 1990; Bluyssen *et al.*, 1996a; Jantunen *et al.*, 1998; Apte *et al.*, 2000). The scientific approach towards the evaluation and creation of a healthy and comfortable indoor environment developed from a component-related to a bottom-up holistic approach (trying to simply add the different components).

The research following these studies was still component related but the relation with other components was better taken into account. In the 1990s, it was acknowledged that complaints and health effects related to the indoor environment are not caused by one single parameter. Findings of studies performed with a larger population and a wider spread showed a complex link between present-day housing conditions (thermal comfort, lighting, moisture, mould and noise) and human health and wellbeing (European Audit project (Bluyssen *et al.*, 1996a) and EXPOLIS (Jantunen *et al.*, 1998) in Europe, BASE study in US (Apte *et al.*, 2000); WHO study (Bonnefoy *et al.*, 2004)). It was concluded that the Sick Building Syndrome is a multifactor cause–effect problem, of which the causes may only be indicators of others. The mechanisms behind it are not fully understood. It is known that besides the physical factors, confounding factors (age, sex, working position, social status, etc.) and psychological factors are involved.

Nevertheless, the indoor environmental parameters – i.e., thermal, air, lighting, and sound quality – are still described with quantitative dose-related indicators, expressed in number and/or ranges of numbers assumed to be acceptable and healthy for people. Unfortunately, these indicators are only valid when a clear relation has been established between the parameter under study and a certain health of comfort effect, and when interactions with other parameters have been identified. As long as we do not know:

- Which sources/causes/pollutants are responsible for certain health effects?
- Where and how those originate?
- How and when these health effects occur?
- How these effects have to be measured in relation to responsible compounds or stimuli, time, different conditions (scenarios) and other indoor environment parameters?

It will continue being difficult to proceed in this way.

In this Part III, 'Analysis', an attempt is made to give answer(s) to the following question:

What do we need to assess and/or predict the effects or responses?

It seems that both the way indoor environment quality is looked upon and assessed as well as the way relations between the different components are established need to be reconsidered.

## 7.2  A different view on IEQ

Previous studies have shown that the relationships between indoor building conditions and wellbeing (health and comfort) of occupants are complex (e.g. Jantunen *et al.*, 1998; Apte *et al.*, 2000; WHO, 2003; Bonnefoy *et al.*, 2004). There are many indoor stressors (e.g. thermal factors, lighting aspects, moisture, mould, noise and vibration, radiation, chemical compounds, particulates) that can cause their effects additively or through complex interactions (synergistic or antagonistic). It has been shown that exposure to these stressors can cause both short-term and long-term effects. In office buildings, a whole range of effects have been associated with these stressors such as Sick Building Syndrome (SBS), building-related illnesses and productivity loss (Bluyssen, 2009a). People in the Western world in general spend 80–90 per cent of their time indoors (e.g. at home, at school, at the office, etc.). The increased asthma prevalence in most countries in the past decades – it has become the first chronic disease in childhood (Eder *et al.*, 2006) – seems to place the finger of suspicion on the indoor environment of schools and homes. More recent studies have indicated that indoor building conditions may also be associated with mental health effects (Houtman *et al.*, 2008), illnesses that take longer to manifest (e.g. cardiovascular disease (Lewtas, 2007; Babisch, 2008), a variety of asthma-related health outcomes (Fisk *et al.*, 2007) or obesity (Bonnefoy *et al.*, 2004).

Although previous studies have shown associations between indoor stressors and comfort, health and productivity, relevant relations between measurements of chemical and physical indoor environmental parameters and effects have been difficult to establish ('Review' in Bluyssen, 2009b). This may be explained by the following (Bluyssen, 2010a):

- Many exposure–response relationships have not yet been (sufficiently) quantified.
- Little is known about the complex interactions between risk factors (or parameters) in the indoor environment and effects are not all known (ASHRAE, 2010a).
- Factors other than indoor environmental aspects (e.g. social and personal factors) may influence the effects.
- Exposure and response may be time dependent (e.g. daily, weekly and seasonal patterns).

These findings point out that *the built environment and its indoor environment with occupants is a complex system*: characterized by feedbacks, interrelations among agents and discontinuous non-linear relations. So far, the focus has been on isolating factors, primarily dose-related indicators, which may be causes of a particular disease state. Performance indicators can be looked upon from:

- *The dose or environmental parameter*: concentrations of certain pollutants, ventilation rate, temperature level or lighting intensity.

- *The occupants or end-user*: such as sick leave, productivity, number of symptoms or complaints, health adjusted life indicator or specific building-related illnesses.
- *The building and its components*: certain characteristics of a building and its components, such as possibility for mould growth or even labelling of buildings or its components.

Most of these indicators are not at all easy to assess and in many cases far from applicable due to incomplete or wrong information applied (review in Bluyssen, 2010a). Each of these groups of indicators has pros and cons, depending on which situation it is applied to and for which purpose. The first group, the dose or environmental parameter, is used most frequently in guidelines and standards as well as in the commercial building assessment tools. Unfortunately, a discrepancy of current standards with end-users' needs is observed: indoor environmental quality as experienced by the occupants is often not acceptable and even unhealthy, even if standards and guidelines for individual environmental parameters are met. In the second group of indicators, the occupant-related indicators, indicators for the current physiological and biological indicators applied are thinkable, based on 'new' techniques such as genomics and metabolomics. In the last category, the building-related indicators, short-cuts i.e. characteristics of a building or measures taken, have been directly related to comfort or health responses (Bluyssen *et al.*, 2011b). Patterns of those factors, including interactions among them, have shown potential in both directions, negative and positive, stimulating or influencing health and comfort. Moreover, it seems that factors influencing satisfaction can be divided into basic factors, so-called minimum requirements, and other factors that can either have a linear or a non-linear relation with satisfaction (Kim and de Dear, 2012).

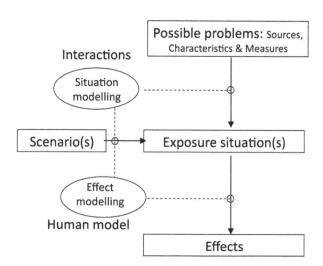

**Figure 7.1** *A different view on indoor environment quality*
Source: Bluyssen.

**Table 7.1** *Some implications of working with a systems approach for indoor environment quality*

| From | To |
| --- | --- |
| Insight in single dose and response relationships | Insight in interacting parameters |
| Attention directed mainly to negative impacts | Attention to positive and negative impacts |
| Distributed knowledge on effects on indoor environment quality | An integrated framework on indoor environment quality effects |
| Ad hoc collection of recommendations to improve IEQ | An integrated approach for IEQ improvement |
| Management of incidents | Integrated risk management |
| Ad hoc communication of possible roles for different stakeholders | An integrated approach which provides insight in the tasks for all stakeholders |

For the assessment of health and comfort risks for people staying indoors, it is clear that a different view on IEQ is required. A view in which the focus is on users (situations) instead of single components and in which the goal is to improve quality of life as opposed to only preventing people from getting ill or feeling bad. A 'different view on indoor environmental quality' could help to better understand the indoor environment and the effects on people. A view in which indoor environment quality is approached in an integrative multi-disciplinary way, taking account of possible problems, interactions, people and effects, focusing on situations rather than single components (see Figure 7.1).

Additionally, it seems that if we want to make a step forward in assessing the health and comfort (potential) of a building, all types of indicators at different levels are potentially important to consider: dose-, occupant- and building(object and non-object)-related indicators; as well as both positive and negative effects.

Some of the implications of working with such a *systems approach* for indoor environment quality are described in Table 7.1.

## 7.3 IEQ assessment

Unfortunately, our assessment models are not capable yet of taking account of the combined effects of stress factors on people, the interactions and dynamic behaviour of people and buildings in different scenarios (e.g. 'homes and energy efficiency', 'children and school environments' and 'workers and office buildings'). This model needs to be developed. To develop such a model the following questions need to be answered:

- *What type of information is required to perform this IEQ assessment?* A selection of indicators at different levels (building, dose, occupant) is required that can be used to define a framework of cause–effect relations (patterns) for different scenarios (e.g. homes, schools and offices) for major end-points (e.g. asthma, learning ability, productivity and dry eyes/skin).
- *How can the 'right' information be gathered?* The outcome should comprise an assessment protocol for the selected indicators, making use of a

combination of measurement techniques (e.g. questionnaires, checklists and physical tests).

- *What type of analytical model or models can be used to determine the relationships between patterns of indicators and end-points?* Models that can test scenarios under different conditions, rather than simply observing associations within finite and specific datasets, need to be tested on available information from previous studies as well as 'new' studies.

## Selection of indicators

Studies on different scenarios (homes, schools and offices) so far have resulted in incomplete and in many cases inconclusive information on associations between occupant-related indicators and dose- and/or building-related indicators (see Table 7.2). The reasons for this are:

- The focus of those studies has been in general on one occupant-related indicator (e.g. asthma, satisfaction and/or performance) and one environmental factor (e.g. thermal, lighting, air and sound quality).
- Except for a few lab studies (for example simulating an office environment) interactions have not been taken into account.
- Studies have been different in set-up, which make comparison difficult.
- There is not enough 'data' to make the ultimate analysis of all factors and indicators per scenario possible.
- The analysis techniques applied have been mostly linear-based.

A selection of indicators at different levels (building, dose, occupant) that can be used to define a framework of cause–effect relations (patterns) for different scenarios (e.g. homes, schools and offices) and major end-points (e.g. asthma, learn ability, productivity and dry eyes/skin) needs to be made. Indicators that can be used to perform a pattern or situation analysis need be selected for the different scenarios based on the outcome of previous studies performed on health

Table 7.2 *A framework of building-, dose- and occupant-related indicators*

| Building (situation) | Dose (exposure) | Occupant (effect) |
|---|---|---|
| Characteristics | Sound quality | Non-communicable diseases |
| • Building materials | | |
| • Control systems (heating etc.) | Thermal quality | Satisfaction/annoyance |
| • Furnishing and decoration | | Symptoms (dry eyes, headache) |
| Processes | Air quality | Productivity/performance |
| • Building materials | | Behaviour/activity |
| • Control systems (heating etc.) | Lighting quality | Physiological markers |
| • Furnishing and decoration | | Physical markers |
| Psychosocial factors | Environmental factors | Personal factors |
| Age | Time | History/future |
| Interactions | Interactions | Interactions |

and comfort including recent/running projects and new studies, both field and/or (semi-)lab investigations.

Additionally, little is known on how indoor building conditions can contribute in a positive manner. What is known is mostly related to single aspects as for indoor stressors. How are interacting IEQ aspects contributing to feelings of well-being, health, productivity and/or recovery? If we are serious about improving the IEQ, indoor stressors are important as a means to prevent possible harm, but opportunities to contribute in a positive manner should not be overlooked.

## Assessment protocol

The goal is to define an assessment protocol for the selected indicators, making use of a combination of measurement techniques (e.g. questionnaires, biomarkers, checklists). How we evaluate and respond to our environment does not only depend on the external stressors involved (physical and psychosocial), but also on personal factors and processes that occur over time (memory and learning) influenced by past events and episodes. They all determine the way external stressors are handled at the moment or over time. *The 'right' package of information* is thus required: not only enough data to be able to perform the analysis on, but also data that have been collected in a uniform way and preferably for more than one scenario. Then the selection of the analytical and/or statistical models to be applied is the next step.

## Analytical models

To be able to perform a situational analysis, *the 'right' model or algorithm* is required; a model that is suitable for determining patterns and interactions, and that can take account of dynamic behaviour. The impact of determining this model is enormous. With that model, a realistic risk assessment can be made of situations that are to be renovated or new. Eventually, with the outcome it will be possible to not only assess the health and comfort of people in different indoor environments with more accuracy, but it will form the basis for improving existing indoor environments and designing truly healthy and comfortable, even stimulating, buildings. Prevention is better than curing, but it is even better to turn this around in a positive way. Attempts with different models, such as meta-analysis, Bayesian networks, structural equation analysis and neural networks and fuzzy logic, are being explored (e.g. Babisch, 2008; Brauer *et al.*, 2008; Bell *et al.*, 2009; Durmisevic and Ciftcioglu, 2010), but still with a modest level of factors.

# 7.4 Next steps

## Pragmatic approach

Besides the *lack of knowledge* shown by the discrepancy between standards and end-users' wishes and needs, there also seems to be a discrepancy between what end-users want and what they get. The latter is often blamed to be related to the complex communication and the fragmented structure of the building sector, leading to lack of coherency, lack of life-cycle orientation and slow take-up of innovation. Additionally, from consultation with different stakeholder groups (architects, producers of construction products and end-users represented by housing corporations) in a number of European countries as well as at European level, it was concluded that the general awareness of what indoor environmental quality (specifically indoor air) is, how you can improve it and who should or could undertake actions is poor (Bluyssen *et al.*, 2010d). The dynamic process of managing the indoor environment, involves many stakeholders, such as the owner, the end-user and the contractor, but also the persons that maintain the indoor environment. If those stakeholders do not understand each other, problems can occur (Bluyssen *et al.*, 2010b). This discrepancy between what end-users want/need and what they get points not only to a lack of knowledge but also to an *inefficient or wrong use of existing knowledge*.

It is clear that we cannot wait until we fully understand all the interactions or mechanisms taking place between the sources that produce/cause the stimuli, among the stimuli, and between the stimuli and the exposed persons. We need to make use of existing data: the use of *short-cuts* is a possibility. A *framework of short-cuts* determined in previous studies and projects, and other bottom-up information conceived during the life-cycle, could serve as a database of knowledge during the whole building life-cycle. This framework or verification matrix (see Figure 7.2) can also contain information on the end-user wishes and needs both present and future, the (social) context, and factors of influence on health, comfort, sustainability and other aspects; and information on the interactions at all interfaces of human being, indoor environment, building (elements) and outdoor environment (over time). That information can be used to make optimal choices, also in relation to other values (e.g. sustainability, affordability; see Figure 7.3).

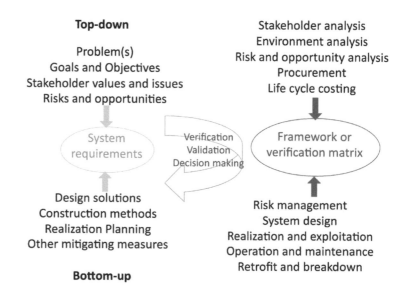

**Figure 7.2** *Values, processes and requirements*
Source: Bluyssen.

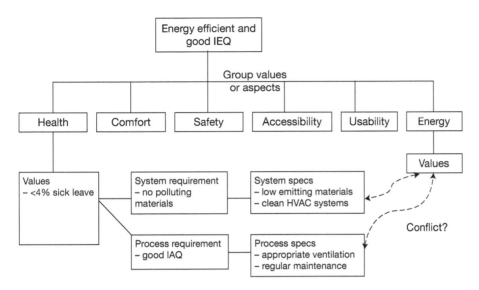

**Figure 7.3** *Pathways for optimal choices*
Source: Bluyssen.

## Life-cycle analysis

Once the framework of information (or verification matrix – see Figure 7.2) is accomplished, the question remains 'How to embed this outcome into the life-cycle of a building?' Life-cycle-based systems engineering as an integrated management process is a possible way to go. Systems engineering can be applied to guarantee that all parties involved in a project, work together in achieving predefined goals with respect for the environment and stakeholders' values (see Table 7.3).

The life-cycle-based systems engineering method described by ISO (2008c) as an integrated project management process procedure is a possibility to apply. In this method, a set of processes is applied throughout the life-cycle of the systems created by humans through the involvement of all interested parties (stakeholders) with the ultimate goal of achieving customer satisfaction. In order to set and meet the system requirements during the whole life-cycle, the framework and the processes can together help to identify:

- Goals and objectives.
- Stakeholders and their values and issues.
- Risks and opportunities.
- Possible solutions, methods and measures to be taken.

The framework and the processes can help with decision-making and communication processes, and optimization of selected options. Verification, validation and decision-making based on the information gathered in the framework is crucial in all phases of the life-cycle, such as information specific for the project, but also information previously gathered in other projects, interactions, and so on.

**Table 7.3** *Possible goals and stakeholder values*

| | Human being | Building | Control/services | Total life-cycle management |
|---|---|---|---|---|
| Goals | Healthy, safe and comfortable indoor environment for all people | An energy-efficient, flexible and accessible building | A controllable and maintainable indoor environment | A sustainable and affordable building |
| | *Wellbeing and society* | *Energy-efficiency and functionality* | *Control and maintainance* | *Sustainability and economics* |
| Values | Health and comfort Safety and security Usability and accessibility Aesthetics and image Cultural and social | Energy efficiency Adaptability and expandability Obsolescence and degradation | Operational reliability and maintainability Inside and outside services Security and emergency | Environment and energy management Water and waste management Investment and life-cycle costs |
| Information type | Basic criteria occupants and stakeholders | Basic criteria building structure and materials | Interactions occupant and building | Interactions building and environment |

For translation of the indoor environmental requirements to technical performance requirements of the built environment, the interactions are of utmost importance, together with the applied communication process in the top-down approach and the realization that performance requirements as well as wishes and demands can change over time and are context-related. Eventually, the wishes and demands of end-users have to be translated into real building products (building and elements) and processes (e.g. maintenance, energy use, security, environmental services) by the stakeholders involved in the whole life-cycle of the indoor environment of concern, both component related and holistically at the same time. In Table 7.4 an example is given for the general design requirement 'How to create a good indoor air quality'; different design requirements can be translated into technical requirements for different phases of the building cycle (Bluyssen *et al.*, 2010b).

## From people to buildings to cities and back

In the underlying publication, the focus lies on people in their indoor environments. A notable but under-researched issue is the increasing urbanization and densification of cities, their impacts on consumption, environmental degradation, health and social wellbeing, and the potential for addressing these concerns (Lorch, 2011). Since outdoor has an important impact on indoor, it is important to take the 'growing' outdoors on board as an important factor of influence on wellbeing of people. Context becomes leading and this requires a change in mindset and worldview as well tools and methods.

**Table 7.4** Examples of design and technical requirements for the general design requirement to create a good-quality indoor environment. The column 'life-cycle phase' shows the phase of the life-cycle of the building the technical requirement is relevant for. The column 'stakeholder' shows for which stakeholder the requirement is important, and column 'type' indicates whether the technical requirement is related to a parameter, an element or a procedure

| Value group well-being | | | | | | Design requirement | Technical requirement | Life-cycle phase | | | | | Stakeholder | | | | | | Type | | |
|---|---|---|---|---|---|---|---|---|---|---|---|---|---|---|---|---|---|---|---|---|---|
| Health | Comfort | Safety | Security | Accessibility | Usability | | | Design | Production | Operation | Maintenance | Break-down | Architect | Contractors | Owner | End-user | HVAC consultant | HVAC service | Parameter | Element | Procedure |
| | | | | | | *Indoor air quality* | | | | | | | | | | | | | | | |
| X | X | X | | | | Only use products that do not emit dangerous pollutants | Do not apply asbestos containing materials | X | | | | | X | | | | | | X | | |
| X | X | X | | | | Use a HVAC system that cleans the incoming air | HVAC system should include a filter for fine dust, etc. | X | | | | | | | | | X | | | X | |
| X | X | X | | | | Products should not get wet during the construction phase | Cover products when laying outdoors | | X | | | | | X | | | | | | | X |
| X | X | X | | | | HVAC system should be clean before installed | Clean ducts (i.e. oil) | | X | | | | | X | | | | | | | X |
| X | X | X | | | | HVAC system should be regularly checked during operation | Maintenance schedule for HVAC system | | | | X | | | | | | | X | | | X |
| | | | | X | | HVAC system should be accessible for maintenance | HVAC system should be accessible for maintenance | X | | | | | | | | | X | | | X | |
| X | | | | | | Products should be recyclable | Products should be recyclable | | | | | X | | | | | | | | | X |

Source: Bluyssen *et al.*, 2010b.

A major challenge for the next 40 years will be the delivery of capabilities for implementing change (Leaman *et al.*, 2010). A fundamental change is needed in how professionals define themselves (Lorch, 2011). There is increasing recognition that social solutions (including defining acceptable/appropriate behaviours) may play an equal, if not greater, role in creating change than technological ones. The scale of environmental events and where intervention to affect positive change occurs – global, national, communal or individual – are, therefore, central to this discussion, particularly in the way it has influenced performance requirements and strategic approaches to buildings (Raymond, 2011).

# 8
# Performance indicators

*In this chapter current available and applied performance indicators for indoor environments are presented and discussed. Performance indicators can be looked upon from the dose or environmental parameters, from the occupants or end-users and from the building and its components. To really assess the health and comfort (potential) of a building, a new approach is required, which can potentially include all types of indicators at different levels: dose-related, occupant-related and building(object and non-object)-related indicators, in different phases of a building; and in which both positive and negative effects are included.*

---

**Box 8.1 Hunter Patch Adams (1998)**

Our job is improving quality of life, not just delaying death.

---

## 8.1 Introduction

Wellbeing (health and comfort) is an important aspect of the quality of life of an occupant. In the late 1980s and during the 1990s, the WHO concept of health became significant for identifying the concept of a 'healthy building' in terms of building performances (i.e., indoor air quality, thermal comfort, lighting quality and acoustics). A healthy building is free of hazardous material (e.g. lead and asbestos) and capable of fostering health and comfort of the occupants during its entire life-cycle, supporting social needs and enhancing productivity. A healthy building recognizes that human health needs, and to some extent comfort needs, are priorities. On top of that a healthy building should be ready for the future, adaptable to 'new drivers' such as climate change, the change towards a multifunctional and diverse society, the increasing individualization and the observed change in the type of end-users' wishes and demands (WHO, 2002; EU, 2007a, b). Most national, European, nationwide and even worldwide organizations agree that indoor environments, including work and living spaces, can be a threat to one's health and that the indoor environmental parameters themselves can contribute to that threat.

The last decades' multiple concepts and tools have been developed and used to evaluate the performance of the built environment, buildings, building parts

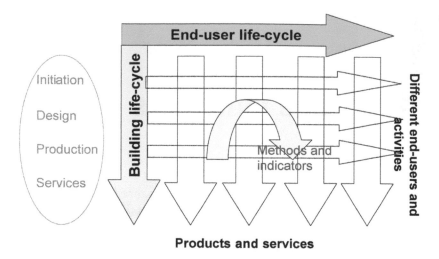

**Figure 8.1** *Methods and indicators for different phases, different end-users and activities and different products and services*
Source: Bluyssen, 2010a.

or specific aspects of buildings. The focus of these concepts and/or tools varies (technical, functional, etc.), as well as the target group for which they are meant. For the different phases of a building's life, different methods or concepts can be used and have been presented (see Figure 8.1).

In most approaches towards sustainability assessment of the building and its components (e.g. the preEN 15643 of CEN TC 350 (CEN, 2008) and the ISO/DIS 21931 of ISO/TC 59/SC17 (ISO, 2008b)) the aspects or issues included are divided into three categories:

1   *Economic*: such as financing and management, whole life value and externalities.
2   *Environmental*: such as climate change, biodiversity, resource use and environmental management.
3   *Social*: including occupant wellbeing (health and comfort), accessibility, security and social and cultural value (i.e. the quality-of-life indicators).

Of these categories, the *quality-of-life indicators* are the most difficult to define, especially the ones that most influence the local adopted architecture solutions (Boschi, 2002). The criteria related to the social aspects and sub-aspects are not always known or completed, for example, with respect to indoor environmental quality (IEQ) (Bluyssen, 2009b).

The health and comfort indicators available today can be looked up from (Bluyssen, 2010a):

•   *The dose or environmental parameter*: concentrations of certain pollutants, indicators such as ventilation rate or $CO_2$ concentration, temperature and lighting intensity (see Section 8.2).

- *The occupant or end-user*: such as sick leave, productivity, number of symptoms or complaints, health-adjusted life indicators or specific building-related illnesses (see Section 8.3).
- *The building and its components*: certain characteristics of a building and its components, such as possibility for mould growth or even labelling of buildings or its components (see Section 8.4).

Of these groups of indicators, the first one is used most frequently in guidelines and standards as well as in the commercial building assessment tools used at national level and in some cases more and more on an international level, such as BREEAM (BRE Environmental Assessment Method) in the UK (www.breeam.org), LEED (Leadership in Energy and Environmental Design) in the USA (www.usgbc.org), CASBEE (Comprehensive Assessment System for Built Environment Efficiency) in Japan (www.ibec.or.jp) and Green Globes in Canada (www.greenglobes.com).

To improve health and comfort conditions of people in the indoor environment, however, it seems that it is not only the indicators that are important, but also the way these indicators are applied during the whole life-cycle of a building (see Section 8.5).

An interactive top-down approach introduced in *The Indoor Environment Handbook* (Bluyssen, 2009a) was recommended. This approach could help all parties involved in a project, including all actors and stakeholders, to work together in achieving predefined goals and values.

## 8.2 Dose-related indicators

*The indoor environmental parameters thermal, lighting, air and sound quality are in general defined and identified with quantitative indicators, mostly expressed in a 'for humans' (assumed) acceptable number or range. But guidelines and regulations on maximum allowable concentrations or ranges of acceptable doses are only valid if a clear relation has been established between the parameter regulated and health or comfort effect, and when the interactions with other parameters are known.*

### 8.2.1 Dose–response relationships

The indoor environment as such can be described by the so-called environmental factors or (external) stressors:

- *Indoor air quality*: comprising odour, indoor air pollution, fresh air supply and so on.
- *Thermal quality*: such as moisture, air velocity and temperature.
- *Acoustic or sound quality*: comprising noise from outside, indoors, vibrations and so on.
- *Visual or lighting quality*: such as view, illuminance, luminance ratios and reflection.

## Standards and guidelines

Current standards and guidelines for indoor environment in general make use of dose-related indicators for certain sub-factors of those factors. In the European project Perfection (performance indicators for health, comfort and safety of the indoor environment), a database of associated standards, regulations, technologies, research activities and policy documents related to the indoor environment was created (Lupisek *et al.*, 2009). The selected standards and guidelines were linked to relevant indicators for the indoor environment (e.g. health, comfort, safety, accessibility, energy and economic indicators) and sorted by the applicable building type (e.g. office, residential, commercial, educational and others). Additionally, performance indicators for acoustic comfort, visual comfort, indoor air quality and thermal comfort in a building (so-called core indicators) were reviewed and target values were presented, based on the following definitions (Lupisek *et al.*, 2009):

- A *performance indicator* is a property of a product, building component or building that closely reflects or characterizes its performance (state or progress towards an objective) in relation to the performance requirement that has been set. The indicator should be a quantitative, qualitative or descriptive parameter that can be readily assessed.
- A *core indicator* defines an essential aspect of a building. The core indicators *acoustic comfort*, *visual comfort*, *indoor air quality* and *thermal comfort* are characterized by respectively 4, 7, 4 and 5 performance indicators (see Appendix C1).
- A *target value* is a quantified value (range) for the performance indicator in order to adhere to the performance requirement set.

The standards and guidelines for these types of parameters have been and are still being developed with the traditional 'bottom-up' approach (Bluyssen, 2008). Focusing on defining threshold values for indoor environmental parameters, different subsequent steps are taken, i.e.:

Step 1   Identification of sources and other influencing factors.
Step 2   Definition of dose–effect relationships.
Step 3   Establishing threshold values for recognized exposures.
Step 4   Assimilating or integrating all factors into end-user satisfaction.

Except for health threatening exposures, sub-factors, complexity and number of the indoor environmental parameters and lack of knowledge make a performance assessment using only threshold levels for the single parameters difficult and even meaningless. Most standards are based on averaged data, thereby overlooking the fact that buildings, individuals and their activities may differ widely and change continuously. Furthermore, considering both the numerous sub-factors (in particular for indoor air quality) and the lack of a solid scientific basis, it appears implausible to make the final and complex integrating step.

On top of that, in practice these regulations are very difficult to comply with (measurement in homes cannot be performed on a regular basis and the concentration as well as the types of, for example, indoor pollutants may vary widely as a function of both time and space). And it is seen that the indoor environmental quality as experienced by the occupants is often not acceptable and even unhealthy, even if standards and guidelines for those individual environmental parameters are met (Cox, 2005; Bluyssen, 2009a). The dose–response mechanisms are not straightforward. Ventilation rate is a good example of this.

## Ventilation rate

For most of the twentieth century, appropriate ventilation was considered to be the only means to create acceptable indoor air quality. Recommendations for good indoor air quality were therefore always related to ventilation rate. Based on either $CO_2$ as an indicator for bioeffluents or on certain emissions of building materials, minimum ventilation rates have been (and are still being) discussed for almost 200 years now (see Figure 8.2).

In a literature review of 27 papers published in peer-reviewed scientific journals, Sundell et al. (2011) concluded that relatively few studies explicitly confirm the relationship between building ventilation rates and health outcomes, although higher ventilation rates in offices, up to 25 l/s per person, are associated with reduced prevalence of Sick Building Syndrome (SBS) symptoms. According to Sundell et al. (2011), uncertainty as to the form of the ventilation–health relationship is an important limitation in establishing rational ventilation standards. Major issues they pointed out are:

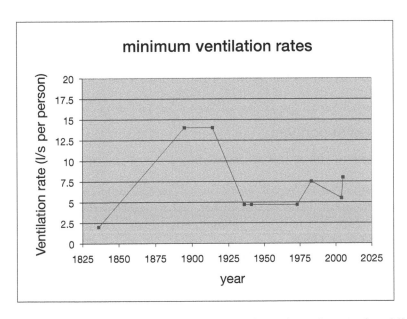

**Figure 8.2** *The recommended minimum ventilation rate changed over the years: from 2 l/s per person by Tredgold in 1836, to 14 l/s per person by Billings in 1895 (Billings et al., 1898), back to 4.7 l/s person in 1936 (Yaglou et al., 1936), to 7.5 l/s person in 1983 (Cain et al., 1983; and Fanger and Berg-Munch, 1983), to 2.5 l/s per person added with 0.3 l/s per $m^2$ for single person cellular offices (10 $m^2$) by ASHRAE (2004b) and 4 l/s per person added with 0.4 l/s per $m^2$ for single person cellular offices (10 $m^2$) by CEN (2005)*
Source: Bluyssen, 2009a, Figure 5.2, p. 111.

- When evaluating the effects of ventilation rates on health, it is important to be mindful of the quality of the outdoor air used to provide ventilation. Any recommendation to increase ventilation must address the issue of poor ambient air quality and not just assume that more outdoor air will necessarily improve indoor environmental quality.
- Factors related to the heating ventilating and air conditioning (HVAC) system that can influence the air quality such as the hygienic state of the system, the type of ventilation system applied, air intake location, fraction of recirculated air, and filtration and humidification.
- Time available for homogeneous reactions among indoor pollutants is another aspect to consider when evaluating the effects of ventilation rates.

Ventilation is the method to control exposures occurring indoors when all other methods have already been implemented (Wargocki *et al.*, 2012). Increasing ventilation can lead to increased indoor $PM_{2.5}$ concentration and can therefore increase population exposures to ambient particulate matter (Hänninen *et al.*, 2012). Badly maintained mechanical ventilation systems can cause incoming air to be polluted (Bluyssen *et al.*, 2003). Some studies have shown associations of increased sick leave with lower levels of outdoor air supply (Milton *et al.*, 2000). Occupants in buildings with low outdoor air supply may have an increased risk of exposure to infectious droplet nuclei emanating from a fellow building occupant (Myatt *et al.*, 2004). But these relations are likely to be related to the number of persons per office (crowding); overcrowding may increase the likelihood of disease transmission via direct contact, and airflow directions (Li *et al.*, 2007).

Additionally, the use of $CO_2$ as a performance indicator should be addressed. $CO_2$ is a good indicator for the presence of people. In spaces where people are the main pollution source, for example children in a school class, the concentration of $CO_2$ is therefore in general a good indicator for the ventilation rate (see Box 8.2).

### 8.2.2 Issues of concern

Issues of concern with indoor environmental parameters as indicators are the following:

- The *relationships* between certain risk factors (and their sub-factors) and a certain indoor-environmental(building)-related health effect (dose–response) are far from complete.
- The risk factors or *parameters* known to have a relationship with a certain building-related health effect are most likely incomplete (most likely all factors that have an effect are known).
- The *interactions* of the risk factors or parameters at different levels in the indoor environment (with occupant, parameter, building) but also with the outdoor environment are not all known.
- The *changes* taking place, static versus dynamic (occupant behaviour, crowding and changing conditions and adaptation).
- The differences on *person* level and the type of knowledge on the occupants themselves that should be taken into account is unclear.

## Box 8.2 Some facts about $CO_2$

Air comprises roughly 78 per cent nitrogen, 21 per cent oxygen. Carbon dioxide is a trace gas in dry air at a concentration of 0.039 per cent by volume. Small amounts of other compounds comprise, for example, noble gases such as Argon and Neon, but also other greenhouse gases such as methane, nitrous dioxide and ozone. Air also contains a variable amount of water vapour, on average around 1 per cent.

The main compounds that people introduce to the indoor air are water and carbon dioxide. With each breath (around 0.5 litre) $CO_2$ is produced, while oxygen is used (see Table 8.1). Breath also contains water, but sweat is the main contributor to the water production of people indoors. The more intense your activity, the more water is produced.

**Table 8.1** *Water ($H_2O$) and carbon dioxide ($CO_2$) production of an average adult person*

| Activity | $CO_2$ production (l/h) | $H_2O$ (l/h) |
|---|---|---|
| Sedentary | 18 | 0.002 |
| High | 170 | 0.8 |

With each breath oxygen decreases, but your body cannot detect the absence of oxygen. It is not the lack of oxygen that makes you feel uncomfortable. It is also not the excess of $CO_2$ that makes you feel uncomfortable (see Table 8.2). It is most likely a result of all other compounds produced in small amounts by people or by other sources that make you feel as you feel. In fact, the $CO_2$ has to increase up to 1 per cent (10,000 ppm) before you feel drowsy because of the excess of $CO_2$. The amount of oxygen in the air would then still be enough to keep functioning.

**Table 8.2** *$CO_2$ and health effects*

| Effect | $CO_2$ concentration | |
|---|---|---|
| | (%) | (ppm) |
| Drowsiness | 1 | 10,000 |
| Suffocation | 7–10 | 7–100,000 |

## Parameter–health relationships

Besides ventilation rate, indoor air quality is now being evaluated on its substances (WHO, 2000, 2006). Even after many years of research, an understanding of the causal associations between indoor environment (building)-related symptoms and indoor air pollutants – whether microbial, organic or particulate – are incomplete (Clausen *et al.*, 2011). For some pollutants, health implications have been well established or documented (e.g. formaldehyde, carbon monoxide, radon, particles in outdoor air). However, knowledge about outdoor particles indoors and of new chemicals that are continuously added is missing. The screening processes for new chemicals are inadequate for detecting their subtle yet important effects. The situation is complicated by the fact that chemical transformations in indoor air, on indoor surfaces and even on human skin alter indoor exposures (Clausen *et al.*, 2011).

But also for other parameters the relationship with health, and even comfort measures, is unclear. For example, in the European HOPE project, no significant relationships could be found between the measured parameters for thermal, lighting, air and sound quality, and perceived health and comfort of the occupants of the investigated buildings (Cox, 2005). Associations found are not clear, among others, due to the complexity of the relationships between indoor building conditions and wellbeing (health and comfort) of occupants (e.g. Bluyssen *et al.*, 1996a; Jantunen *et al.*, 1998; Apte *et al.*, 2000; WHO, 2003; Bonnefoy *et al.*, 2004).

## Other and/or new parameters

Many of the products of indoor chemistry are 'stealth compounds' – compounds that are present but that cannot be seen using traditional analysis techniques: free radicals, other short-lived highly reactive species, thermally labile compounds, and compounds with multifunctional groups that are difficult to chromatograph without derivatization (Weschler, 2011). It is known that (Weschler, 2011):

- Under the right conditions, indoor chemistry can significantly alter the types of chemicals found indoors and their concentrations.
- Indoor chemistry can be the primary determinant of short-lived, highly reactive species indoors.
- Secondary pollutants are often of greater concern than primary pollutants (such as formaldehyde, aldehydes and $NO_2$ produced with ozone reactions indoors; see Section 9.3.1 in Bluyssen, 2009a).

## Interactions

It is a fact that the indoor environment is more than the sum of its parts. To predict the integral building performance as experienced by the end-user is complex. The whole environment is more than the sum of its constituents stimuli and therefore hard to predict (de Dear, 2004).

Interactions not only occur between different air compounds (indoor chemistry), but also between parameters from different core indicators, for example, between *light* and *indoor air* through photolysis. Photolysis rates of key indoor species over the range of wavelengths commonly observed indoors are needed as well as information on flux at different wavelengths coming through windows and emitted by various types of indoor lighting (Weschler, 2011). Another example is the interaction between *light* and *thermal comfort*, when sunlight heats up the interior surfaces. Additionally, the psychological factor of (day)light, including not only colour and illuminance, but also view or contact with the outdoor environment, should not be underestimated (Meerdink *et al.*, 1988; van den Hout, 1989; Vroon, 1990).

Interactions also occur between *thermal conditions* and *indoor air*. Emission rates of most VOCs increase with increasing temperature and when relative humidity increases, hygroscopic constituents from particles can sorb water. Submicron airborne particles typically comprising 30–50 per cent water-soluble salts (Finalayson-Pitts and Pitts, 2000). Sorbed water can influence hydrolysis reactions, acid-base chemistry, and oxidation-reduction reactions (corrosion reactions). Compounds that are subject to hydrolysis include phthalate esters (used as plasticizers), phosphate esters (plasticizers, flame retardants and pesticides) and the texanol isomers (coalescing agents in latex paint; Weschler, 2011). A higher water activity at the surface can result in mould growth and decay of building materials (Adan, 1994). A higher humidity in combination with cold surfaces can lead to condensation and increase the water activity at the surface.

The interactions between *acoustics* and *indoor air* should not be forgotten, via the introduction of ventilated air by HVAC systems, which can produce equipment as well as air flow noise, or via natural ventilation through open windows, bringing in the noise produced outdoors indoors.

Indoor air quality, thermal environment, acoustics and illumination are all interconnected in the indoor environment (ASHRAE, 2010a). Even within one factor, different parameters can interact (e.g. indoor chemistry). Perception is also influenced by the indoor environmental conditions. For example the sense of smell is more sensitive at higher temperatures, but also humidity affects human responses. For example, an odour that may be acceptable when thermal conditions are cool and dry, may be annoying when thermal conditions are warm and humid but still within the thermal comfort zone. Acceptability of air increases with decreasing enthalpy of air (Fang *et al.*, 1998). Human reactions to VOCs at constant concentrations are greater at higher temperatures (Mølhave *et al.*, 1993), but human reactions are also influenced by psychological factors, and indoor environmental parameters interact with the building and its environment and last but not least with the occupants in it (human behaviour; see Chapter 9, Bluyssen, 2009a).

An important issue worth mentioning, although not further explained here, is the interaction of indoor and outdoor conditions on human behaviour and sustainability or green performance of the built environment, caused by *climate change* (Aries and Bluyssen, 2009; Wilde and Coley, 2012).

## Changing conditions

While climate change in itself is a long-term process, changing indoor environmental conditions (as a consequence of climate change or other just local climatic changes) can be considered short-term events (during the day or even in hours or minutes): for example, entering an office building with air conditioning while outside it is a tropical summer day of around 30°C and a humidity of 70 per cent, or when the sunlight increases and decreases during the day, or the sudden change when a cloud is passing by. The indoor environmental conditions vary with time of day and season, as well as geographic location and the nature of the building itself.

Besides climate change, there are also other drivers for health and comfort in the indoor environment, which are different from 100 years ago, leading to an increase in complexity. We see (Figure 8.3):

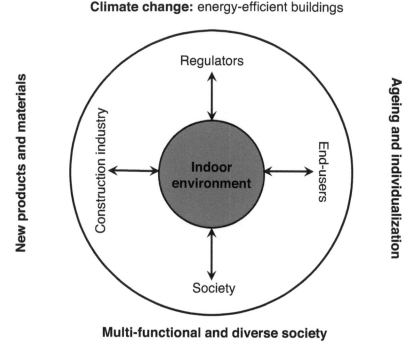

**Figure 8.3** *Current drivers for health and comfort in the indoor environment*
Source: Bluyssen.

- A change from family-oriented to multifunctional and diverse society.
- Individualization/ageing population leading to other/new needs and demands.
- New products and materials (building and other) leading to new emissions and other behaviour.

## 8.2.3 Future standards

It is clear that there is a need for adapted and/or new standards and guidelines that take account of the individual needs and demands, other factors and parameters as well as the sources of concern.

### Individual needs

The European standard EN 15251 sets up a series of indoor conditions to optimize thermal, lighting, air and sound quality to enable professionals and consultants to make energy calculations for buildings (CEN, 2005). In their critique of this standard, Nicol and Wilson (2011) point out a number of problems, one of them being the fact that the standard is focused mainly on office buildings with mechanical cooling:

- The standard is worded in such a way as to encourage the notion that mechanical cooling is the starting point and that natural ventilation is only to be used in circumstances where a 'second-rate' environment is acceptable. Therefore, caution should be used when applying it in different building types where expectations will be different.

• The methodology applied is based on the specification of design environments (and ranges around them) and results in the definition of categories as acceptable ranges: the higher the category, the narrower the limits of the ranges.

According to Nicol and Wilson (2011), the alternative approach, that comfort is a goal that occupants should be enabled to seek, is not acknowledged. Wyon (1996) for example suggested that target values for optimal IEQ could be specified as ranges of parameter values within which individual control should be used to attain each occupant's optimum. ASHRAE (2010b) reviewed its standard 55 (ANSI/ASHRAE 55–2010) on comfort and defined two types of comfort: one with air conditioning and another without air conditioning, also designated 'adaptive'. With this change in approach, many other options are envisaged in the future, introducing flexibility in what are regarded as concepts, levels and types of ventilation and a larger diversity in types of heating and cooling techniques, processes and systems. It also opens up the way comfort is defined. Comfort will rely on the adaptive capacity of each person and thus personal perception and corresponding capacity of personal response, reducing the number of cases where strict indoor environmental conditions are prescribed, opening up a wider spectrum of options regarding systems and also control strategies.

## Other factors and parameters

Considering the fact that the overall comfort in buildings or perceived wellbeing can only partly be explained by the perceived comfort of the four factors normally taken into account (see Section 8.2.2), it might well be that we need to consider other factors. It could even well be that different metrics have to be used for different types of buildings (Clausen *et al.*, 2011).

In the European project LEnSE, fully titled 'Methodology Development towards a Label for Environmental, Social and Economic Buildings', a methodology for the assessment of the sustainability performance of existing, new and renovated buildings was developed (LEnSE partners, 2007). In order to do so a long list of possible environmental, social and economic issues was set-up. The short-list for occupants' wellbeing already comprised more factors than the four we normally use (see Table 8.3).

## Sources as indicators for mixtures of pollutants

For indoor air quality it is not uncommon to take the presence of a certain material as an indicator for a potential harmful concentration of a certain pollutant or mixture of pollutants (see also Section 8.4.2, 'Short-cuts'). There are a number of sources of pollutants in the indoor environment including combustion of gas, coal and wood for heating and cooking, environmental tobacco smoke from smoking of cigarettes, vapour from vehicles in attached garages (benzene), burning of candles and incense, emissions from printers and photocopiers and

**Table 8.3** *Short-list for occupants' well being in the LEnSE project*

| Sub-issue | Potential indicators |
|---|---|
| Lighting comfort (artificial and natural) | Recommended maintained lighting levels (lux)<br>Provision of daylight (average daylight factor) |
| Thermal comfort | Degree and type of thermal comfort analysis carried out<br>Performance standards for avoidance of overheating |
| Ventilation conditions | Litres of fresh air/second/person |
| Acoustic comfort | Internal noise levels<br>Reverberation time<br>Sound insulation levels |
| Occupant satisfaction | Post completion monitoring<br>Occupant satisfaction survey |
| Private space | Proximity (m)<br>Size (m²)<br>Type/facilities |
| Outdoor space | Proximity (m)<br>Size (m²)<br>Type/facilities |
| Materials/substance exclusion | VOC levels<br>Eco-labels<br>Materials exclusion clauses |
| Indoor air quality | $CO_2$ levels (ppm)/ventilation controls<br>Design of humidification systems<br>Location of air intakes and extracts |
| Quality of drinking water<br>Building safety assessment | Best practice design of domestic hot water system<br>— |

Source: LEnSE partners, 2007.

some types of air 'cleaners' (ozone). In addition there are a wide range of other substances including those released from the large range of building, furnishing and consumer products present indoors, by the occupants themselves, by microbiological growth, and present in soil gas that enters the building from the ground. Box 8.3 presents an overview of the main groups of substances known to cause indoor pollution and their sources.

## 8.3 Occupant-related indicators

*The performance indicators, which are applied to describe the diseases and disorders caused by the indoor environment, can be divided into:*

- Directly expressed occupant-related indicators: *Which are the complaints and symptoms and who is susceptible?*

- Indirectly expressed occupant-related indicators: *What is the effect in the short term (i.e. financial indicators) and long-term (i.e. health-adjusted life indicators)?*

---

### Box 8.3 Main groups of substances and their sources known to cause indoor air pollution

*Endocrine-disrupting chemicals*: including phthalates, pesticides, PCBs, etc. (see Box 3.7).

*Radon*: is a naturally occurring radioactive gas that can enter buildings from the ground and the amount of ingress depends upon a number of factors including local geology, the type of foundation, the positioning of service pipe work and internal ventilation levels. Measures such as installation of gas proof membranes in the foundations of new buildings can significantly reduce levels of radon gas.

*Inorganic gases*: carbon dioxide ($CO_2$), carbon monoxide (CO), nitrogen oxides ($NO_x$), sulphur dioxide ($SO_2$), inorganic gases and particles from biological origin, cooking or sources outside.

*Volatile organic compounds (VOCs)*: are emitted over periods of weeks or years from construction and furnishing products and have the potential to cause poor air quality. VOCs are also released from consumer products including electrical goods such as computers and printers as well as cleaning products and air fresheners. Environmental tobacco smoke (ETS) contains a complex mixture of organic compounds and while smoking is banned in the workplace and public buildings in some European countries it remains a significant source in many homes.

*Formaldehyde*: is a very volatile organic compound (VVOC) that has been widely studied because of its release from a range of building and consumer products.

*Semi-volatile organic compounds (SVOCs)*: have a relatively low vapour pressure and therefore tend to occur at lower concentrations in indoor air than the more volatile VOCs. They include plasticizers used in polymeric materials such as vinyl floorings and paints, pesticides such as DDT and penta-chlorophenol, and polyaromatic hydrocarbons (PAHs) produced during fuel combustion and present in coal tar and in tobacco smoke. Many compounds which are generated in the indoor environment are semi volatiles such as phthalates, flame retardants, PAHs, chlorophenols, pesticides, organotins and metals, which may absorb to particulate matter present in the indoor air and to house dust.

*Microbial volatile compounds (MVOCs)*: are compounds formed in the metabolism of fungi and bacteria.

*Ozone*: is primarily a pollutant of ambient air produced by photochemical reaction. It undergoes reaction indoors with surfaces and airborne pollutants to produce new organic compounds and particles.

*Ultrafine and nanoparticles*: both ultrafine particles and nanoparticles are sized between 1 and 100 nanometers, and are suspected to cause considerable health risks. While nanoparticles originate from nanomaterials, which are composed of nanoparticles, fine particles and ultra fine particles are in general generated during combustion, such as burning a candle or smoking a cigarette.

*Asbestos fibres*: are a particular type of particle, and the use of asbestos in buildings has been an important route of worker and population exposure. Asbestos is present in many buildings and presents a risk of cancer if fibres are inhaled. Strictly controlled removal by specialist contractors is required but it is often sufficient to manage the risk adequately by ensuring the material is not disturbed.

More information on substances and sources can be found in Section 5.1.3 of Bluyssen (2009a).

Source: Bluyssen, 2010c.

---

## 8.3.1 Symptoms and complaints

The number and type of symptoms and/or complaints from occupants of a building administered via a questionnaire has been the most applied method to inventory the health and comfort status of an occupant in a building. The Building Symptom Index (mean number of symptoms reported by occupants) and the Building Comfort Index (based on complaints for different questions/aspects related to thermal comfort, IAQ, light and noise) are good examples of performance indicators based on this method (see Bluyssen, 2009a, Section 5.3.4, for a description). For trouble shooting the identification of number and type of symptoms and/or complaints had been found a very good way to locate the problem areas in a building. For identifying relations between symptoms or complaints of large groups of occupants and certain parameters of the indoor environment in the field, this has not been very successful, even though exposures in laboratory settings (e.g. Witterseh *et al.*, 2004; Balazova *et al.*, 2008; Clausen and Wyon, 2008) give promising results (see also Section 8.2.1). This can indicate that in the studies performed in the field perhaps:

- The 'wrong' environmental parameters were focused on.
- Interactions of the parameters were too complex.
- Confounding factors were dominating.
- The occupants were not able to 'correctly' inform us about their symptoms and complaints.

The way we evaluate our environment (perception), the way we respond to our environment (behaviour) and the way we believe we are responding are different processes (Taylor, J., 2006). This might explain why there is often a discrepancy between what people tell us what they need or want and what their behaviour tells us, or what they tell us what the cause is of certain complaints and what the real problem is. The majority of our responses are unconsciously influenced by habits, our surroundings (context) and other people. People make unconsciously false statements (Schultz and Schultz, 2006). With surveys it is possible to measure wishes but in general not why people want them. From years of investigations with questionnaires it follows that the relation between objective aspects and satisfaction is not clear, also the relation between objective aspects and the experience of people is not clear: the relation between experience and satisfaction is much clearer (Perner, 2007).

In the applied self-administered office questionnaire of the Health Optimisation Protocol for Energy-efficient Buildings (HOPE) project (Cox, 2005), employees were asked to rank the perceived quality of several indoor environmental parameters such as thermal, lighting, air and sound quality. In a re-analysis of the office part of the HOPE project data (see description of study F in Box 8.4) the outcome of the PCA (principal component analysis) showed that for most persons the outcome of those different environmental parameters was strongly correlated (Bluyssen *et al.*, 2011a). The reason for this was unclear, but several possibilities were brought forward (Bluyssen *et al.*, 2011a):

- The occupants could not distinguish between those different parameters, i.e. they do not know the meaning of the questions. In a recent European project named HealthyAir in which 104 parties were interviewed, it was concluded that most of the interviewed persons did not even know the meaning of air quality, i.e. what specific aspects one would share under the general term 'air quality' (Bluyssen *et al.*, 2010d).

**Figure 8.4** *Sick person*
Source: Bluyssen.

# Box 8.4  Study F: re-analysis HOPE project

From 2002 to 2005 the European project HOPE, sponsored by the European Commission in the fifth framework, was conducted by 14 organizations in nine European countries. In this study 164 buildings were investigated using checklists addressing the building characteristics and self-administered question-naires asking about perceived health and comfort. From these buildings 69 were office buildings and 95 apartment buildings. The data collection took place in the winter of 2003–04. In 32 of these 164 buildings, detailed measurements of chemical, biological and physical parameters were performed in 2004 (not reported here). In the investigation of the 164 buildings, three assessment methods were used (Roulet *et al.*, 2006b):

1   An inspection of each building according to a checklist, providing data on the building and its environment.
2   Interviews with building management.
3   Questionnaire surveys of occupants, providing information on how they feel and perceive their indoor environment.

The re-analysis focused on office buildings only. The total amount of respondent cases that was considered for analysis amounted to 5,732 cases from 8 countries (Germany, Switzerland, Italy, Finland, Denmark, Portugal, The Netherlands, UK) in 59 office buildings (see Table 8.4).

The office environment survey comprised a self-completion form for office occupants, and was designed to be completed in 10 minutes or less. The subjects covered included acute health symptoms, productivity and environmental comfort, as well as background information about the respondent and their office space.

Table 8.4 *Number of office buildings and questionnaires per country*

|             | Buildings | Questionnaires |
|-------------|-----------|----------------|
| Germany     | 8         | 257            |
| Switzerland | 7         | 1,194          |
| Italy       | 4         | 611            |
| Finland     | 8         | 784            |
| Denmark     | 8         | 580            |
| Portugal    | 10        | 859            |
| Netherlands | 9         | 941            |
| UK          | 5         | 506            |
| **Total**   | **59**    | **5,732**      |

Supplementary data, collected by means of a checklist, were added (one per office building), which contained information about the buildings such as dimensions, building materials, HVAC system, lighting installations, use of building, and so on. Some parts of the checklist were completed by the research team in advance (either from documentation obtained about the building or by site visit), and some were completed with the help of a building manager or equivalent. A protocol for fieldwork (guidance and instructions on carrying out the fieldwork), sample selection (guidance on selecting a sample from within a large population) as well as a model letter to occupants (each country translated the questionnaire and the letter in their own language) can be found in the final report of the European HOPE project (Cox *et al.*, 2005; the public final report can be downloaded at hope.epfl.ch.).

Source: Bluyssen *et al.*, 2011a.

- The questions were asked in the 'wrong' way, i.e. asked in a particular order that influences the answers of the next questions, or, in an attempt to make questionnaires as short as possible, the clarity might have been compromised and the questions might have been too general/not specific enough.
- Occupants' answers were lead by one's mood of the moment (emotional state; Kuhbander et al., 2009; De Dear, 2009) or in general by one's personality traits (such as locus of control (Rotter, 1966) or negative affectivity (Berglund and Gunnarsson, 2000)).

### 8.3.2  Financial indicators

Besides productivity (quantitative and/or qualitative work output of people (product or service they deliver)), health-related financial indicators are sick leave (number of days sick, away from work place, per year) and estimates of life expectation. Productivity has received a lot of attention in recent years (EU, 2007b). With the numbers provided by Fisk (2000), it was shown that the biggest potential financial gains can be found in direct improvements of worker performance. Reduction of building-related diseases and SBS symptoms are well below this but still interesting to follow up on. Many studies have shown that productivity at work bears a close relationship to the work environment, and more specifically to the comfort status of the occupant (Clements-Croome, 2002). List of measures at building level that promote productivity have been established (named creative working environments), but the question remains how to really measure this phenomenon in practice (see Box 8.5). Many theories are available but not much proof has been given (yet).

---

**Box 8.5  Assessing productivity**

Productivity depends on many aspects: wellbeing, mental drive, job satisfaction, technical competence, career achievements, home/work interface, relationship with others, personal circumstances, organizational matters, and so on – and last but not least environmental factors (indoor and outdoor environment; Clements-Croome, 2002).

Productivity can be measured:

- *Objectively*: by, for example, measuring the speed of working and the accuracy of outputs by designing very controlled experiments with well-focused tests (e.g. productivity effects as related to thermal comfort (Wyon, 1993), air quality (Wargocki et al., 2000); see Section 6.4.5).
- *Subjectively*: by using self-estimated scales and questionnaires to assess the individual opinions of people concerning their work and environment (Raw et al., 1990; see Section 6.5).
- *With combined measures*: using, for example, some physiological measures such as brain rhythms to see whether variations in the patterns of the brain responses correlate with the responses assessed by questionnaires (e.g. alertness and light (LHRF, 2002); see Section 6.3.2).

Assessing the effect of a certain factor on productivity in a controlled environment, whether it is performed with one or more of the methods above, is complex. First of all because productivity as a performance indicator is related to so many aspects. Second for each of the methods described in Box 8.5 critical points can be made that makes the relevance of the outcome and the purpose questionable.

The methods that are meant to be objective, such as typing a standard text, in general comprise monotone activities that are only part of the normal activities of a certain person. Only if the tested activity is the dominant activity of that person in normal life (which means they probably don't learn to type faster during the test), can the test be found relevant for measuring the output of that person, such as the number of characters typed, which is some form of productivity.

For the second category of measurements, the subjective methods, including self-estimated productivity, next to the influence of other factors over time, the interpretation of one's own productivity might differ among different persons. For example one might estimate his own productivity low, while in fact it is much higher than for a person that has estimated his productivity high. There is a lack of a reference point for 100 per cent productivity (Wiik, 2011). In this case the absolute numbers will not be relevant at all, and even the difference from one situation to another can be irrelevant due to other uncontrolled factors such as personal circumstances.

Additionally, honesty of subjects is an important issue to take on board. The social desirability effect describes people's tendency to exaggerate their own skills and performance in an effort to present themselves in a favourable light. Only when a factor has a real dominant effect, a sudden extreme change in environmental conditions, might a difference in productivity be relevant, under the conditions that enough test persons are involved. Direct self-assessment appears to be less sensitive to indoor stimuli that are not clearly associated with specific symptoms (Witterseh et al., 2004).

The relationship between productivity, but also comfort, and the psychological and physical status of human wellbeing are still under investigation. In a study of 26 office buildings in 5 European countries with more than 4,500 desk visits, self-assessed productivity was significantly related to satisfaction with the various aspects of the office environment, while the relation with the measured physical conditions was indirect and weak (Humphreys and Nicol, 2007). Also, in the European HOPE project (study F; Box 8.4), the 'self-estimated productivity change' (in summer) was related to 'comfort overall in summer' (see Figure 8.5).

These results imply that to relate productivity directly to environmental variables, without first considering the overall comfort of the worker, is likely to be misleading. According to Sundstrom and Sundstrom (1986) several processes can account for short-term influences of the work place on the performance of individual workers (see Box 8.6).

## 8.3.3 Health-adjusted life indicators

Public health focus has gradually changed from life expectancy to health expectancy (EU, 2007a), which does not include mortality but rather aspects of quality of life (Hollander and Melse, 2006) such as:

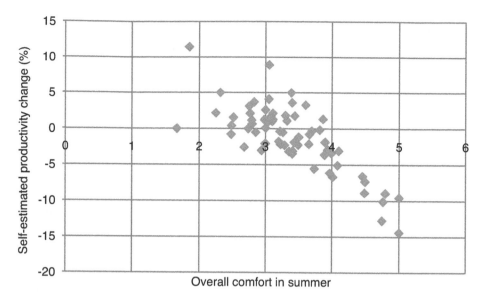

**Figure 8.5** *'Self-estimated productivity change' related to 'comfort overall in summer' in 69 office buildings from the HOPE project. The seven-point scale for 'comfort overall' ranged from satisfactory (1) to unsatisfactory (7)*
Source: Bluyssen.

---

## Box 8.6 Theories of short-term environmental influence

*Arousal* refers to a person's general state of alertness or excitation, both physiological and psychological, and it can range from drowsiness through alertness to extreme agitation. Increased arousal may increase heart rate, blood pressure, blood flow, perspiration, skin temperature and muscular tension. The level of arousal optimal for performance of a task depends upon its complexity. The so-called Yerkes–Dodson Law suggests that optimal arousal is lower for complex tasks than for simple ones. Arousal can be created by heat, noise, light or odours. Arousal can either help or hinder performance.

*Stress* is difficult to distinguish from arousal; stress usually refers to a stronger or more intense reaction to environmental conditions. Mild stress can improve the performance of simple tasks, while severe stress may impair even the simplest task. The environment can create stress by posing a perceived threat to wellbeing, resulting in arousal and narrowing of attention.

*Distractions* can divert someone's attention from a task, which may cause lapses in performance. *Sensory overload* refers to excessive stimulation that contains no specific meaning for the individual, such as the flickering of a light. *Information overload* refers to sources of stimulation that carry meaning and ask for response. Overload occurs when information comes faster than it can be assimilated and dealt with. This can lead to coping responses such as ignoring low-priority inputs.

Source: Sundstrom and Sundstrom, 1986.

- Aggravation of pre-existing disease symptoms (e.g. asthma, chronic bronchitis, cardiovascular or psychological disorders).
- Severe annoyance, sleep disturbance, reduced ability to concentrate, communicate or perform normal daily tasks.
- Feelings of insecurity or alienation, unfavourable health perception and stress in relation to poor quality of the local environment and perceived danger of large fatal accidents.

Examples of health-adjusted life indicators are the DALY (disability adjusted life-year) and the QALY (quality adjusted life-year; Carrothers *et al.*, 1999). Probably the most well-known and most used health adjusted life-year indicators is the DALY, which combines years of life lost (YLL) and years lived with disability (YLD), standardized by means of severity weights. Originally developed by the WHO and used in the Global Burden of Disease project (Murray and Lopez, 1996), it is a measure to assess the global disease burden, which is increasingly common in the field of public health and health impact assessment (see Box 8.7).

The general procedure of a quantitative health impact assessment at which the DALY concept is applied can be described in six steps (Knol and Staatsen, 2005):

1  *Selection of health endpoints*: sufficient proof should be available for a causal relationship with the risk factor.
2  *Assessment of exposure*: DALYs are determined for a general population, since exposure–response relationships are usually not based on individual exposure assessments.
3  *Identification of exposure–response relations*: selected (recent) exposure–response relationships based on well-founded epidemiological cohort or laboratory studies.
4  *Estimation of the extra number of cases with the specific health state*: prevalence data taken from national mortality and morbidity registries.
5  *Selection of severity weights and duration of health effects*: severity (or disability) weights determined by expert panels (doctors and scientists); duration is set to one year.
6  Computation of the total health burden of all risk factors: summarizing all DALYs.

The DALY concept has been applied at different levels (see Figure 8.6), in which not all of the above mentioned steps are applicable:

- *(Global) population level*: to compare relative burdens among different diseases and among different populations (Morrow and Bryant, 1995).
- *Building level*: to estimate the effect of unhealthy buildings on the burden of disease (Smith, 2003).
- *Building component level*: to compare the effects of certain building components or systems expressed in DALYs (such as mechanical ventilation systems (Pernot *et al.*, 2003)).
- *Risk factor level*: to assess the impact on health of one risk factor (such as dampness-asthma (Knol and Staatsen, 2005)) or parameter. In the European

projects EnVIE (coordination action on indoor air quality and health effects) and IAIAQ (promoting actions for healthy indoor air), the DALY was used to estimate the environmental burden of disease caused by indoor air quality (de Oliveira Fernandes *et al.*, 2008, Jantunen *et al.*, 2011; see Box 8.8).

---

## Box 8.7  Disability adjusted life-years (DALYs)

For each disease, a DALY could be assigned related to different risk factors or parameters (e.g. particulate matter concentrations, lighting, noise, radiation, temperature, etc.). A DALY measures the number of healthy years that a population will lose due to a particular disease. Simplified, the equation comes down to an exposure–response function (AB) representing the relationship between an exposure concentration and a disease. Health effects are calculated by multiplying the DALY for the disease (S × D) with its exposure–response function. DALYs can be calculated using the equations below (Knol and Staatsen, 2005):

$$DALY = AB \times D \times S \qquad\qquad [8.1]$$

$$AB = AR \times P \times F \qquad\qquad [8.2]$$

$$AR = (RR' - 1)/RR' \qquad\qquad [8.3]$$

$$RR' = ((RR - 1) \times C) + 1 \qquad\qquad [8.4]$$

Where:

*AB* = attributable burden: number of persons with a certain disease as a result of exposure to a risk factor, not corrected for co-morbidity.

*D* = duration of disease: for morbidity duration is one year in case of prevalence numbers, for mortality the duration of time lost due to premature mortality.

*S* = severity: reduction in capacity due to morbidity using severity weights (between 0 (perfect health) and 1 (death), determined by experts).

*AR* = attributive risk: risk of getting a specific disease as a result of exposure to a certain factor.

*P* = base prevalence for morbidity; number of deaths for mortality.

*F* = fraction of population exposed to the risk factor.

*RR'* = adjusted relative risk; and RR = relative risk.

*C* = concentration of risk factor.

In these equations 'age weight' (relative importance of healthy life at different ages) or 'time preferences' (value of health today compared to the value in the future) are not taken into account.

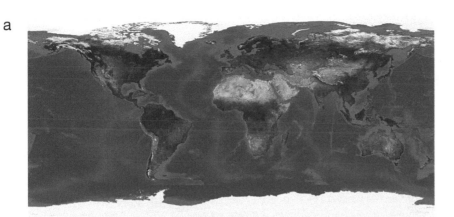

a

b

c

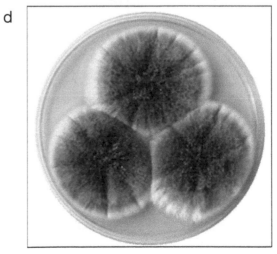

d

**Figure 8.6**
*Health assessment at (a) global/regional, (b) local/building, (c) building component and (d) agent level*
Source: Bluyssen and Aries, 2009.

---

**Box 8.8  Environmental burden of disease
caused by indoor air quality**

In the European projects EnVIE and IAIAQ, the annual burden of disease-related to bad indoor air quality was estimated to be two million disability adjusted life years (DALY)/year in EU-26 (Malta was excluded because of lack of exposure data).

For each exposure and disease (allergy and asthma, lung cancer, cardio-vascular diseases, chronic obstructive lung disease, respiratory infections/symptoms and acute toxication) a predefined attributable burden of disease was applied (WHO, 2012b). Exposures comprised a list of exposure agents (tobacco smoke, combustion particles, carbon monoxide, radon, dampness, mould and bioaerosols, and (S)VOCs including indoor chemistry products). Health risk was also assigned to the main sources of these agents (outdoor air, building/equipment/ventilation, consumer products and occupant behaviour). The main sources of indoor exposure data from across Europe comprised the outcome of several European projects, national surveys and European data surveys.

A whole of two thirds of the indoor air quality (IAQ) associated burden of disease was caused by exposure to fine PM, originating mostly from outdoor air. Other significant exposures were building dampness (11%) and bioareosols from outdoor air (8%). The underestimated role of VOCs (only 2%) is due to the fact that only for very few VOCs, benzene, naphthalene and formaldehyde (Kotzias et al., 2005) data about specific disease dose/response are available, the effect of VOC mixtures is largely unknown, and population representing indoor air VOC exposure data only exist for two countries (Germany and France) and a number of cities studied.

Source: Jantunen et al., 2011.

---

However, using DALYs as a health indicator seems only valid when the exposure–response function for the risk factor and the disease under investigation is known and preferably valid for the population under investigation. To make the DALY calculation meaningful:

- The exposure–response function for the risk factor (valid for the population under investigation) has to be known and preferably be statistically relevant.
- If this risk factor comprises sub-factors as in the case of indoor air pollution, for each of these sub-factors the relationship has to be known and be statistically relevant.

Additionally,

- The other risk factors (and sub-factors) involved for the disease under investigation should be known to estimate comorbidity.
- The interactions of those other risk factors should be known to make an adequate estimation of the investigated risk factor on the investigated disease for the population under investigation.

So when the (sub) risk factors are not all known (e.g. in the case of Sick Building Syndrome) or when the exposure response function (in fact the dose–effect relation) for a certain risk factor or parameter is not known, or the interactions between risk factors causing the same disease are not known, the exact calculation of DALYs for a certain condition or disease becomes very difficult or even meaningless. Many assumptions have to be made to make a calculation at all possible, and the uncertainties of the outcome can increase tremendously.

An additional factor is the determination of the severity for a disease expressed in a number between 0 and 1. This is performed by experts and can therefore be highly volatile as well. Next to that, DALYs are calculated per disease (e.g. sleep deprivation, brain cancer) and not per health outcome (e.g. headache), meaning that for a person with more than one disease, the combination of DALYs can result in a number over 1, which means more severe than death (Knol and Staatsen, 2005).

Because the risk factors or parameters involved in causing health and comfort problems in the indoor environment are numerous and not clearly defined, the application of DALYs to compare the health effects of buildings or building components seems impossible at the moment. On the other hand, for single indoor risk factors that have been shown to have a dominant relationship with a certain disease (other risk factors are minorly important), DALY can be used to show effects over time or even to show differences between different populations (or buildings).

### 8.3.4 'New' indicators

#### One index

From the 'dose-related indicator' point of view, questions have risen to introduce an index for overall satisfaction, based on different factors (thermal, lighting, air and sound quality). But this seems quite impossible, since these different factors would acquire different weightings under different circumstances or for different individuals (ASHRAE, 2010a). A universal index could do no more than average the responses for several diverse activities or social settings.

Data from an environmental survey of 26 offices in Europe showed that dissatisfaction with one or more aspects of the indoor environment does not necessarily produce dissatisfaction with the environment overall (Humphreys, 2005). Conversely, satisfaction with one or more environmental aspect does not necessarily produce satisfaction with the total environment. The correlation of the average satisfaction of the aspects included with the overall comfort assessment was only 0.23 ($r^2$, $p < 0.001$), indicating that 77 per cent of the variation of the overall comfort evaluation remained unexplained. Some of this unexplained variation is probably attributable to individual differences and some to aspects not included.

Occupants seem to balance the good features against the bad to reach their overall assessment (Humphreys, 2005). It could be that there is some kind of 'forgiveness factor' involved that operates if people feel they can control their environment (de Dear, 2009). For example, an excellent lighting scheme might cause one to forgive a poor thermal environment and poor acoustics, or the other way round. Deuble and de Dear (2009) define the forgiveness factor as a variable derived by dividing scores for the variable 'comfort overall' by the average of the summary variables for temperature in summer and winter, ventilation/air in summer and winter, noise and lighting.

The study reported by Humphreys (2005) also showed that even though room temperature and air quality seemed to be the most important contributors to overall comfort, the relative importance of the various aspects differed from country to country (in Europe). In the US, Bayer *et al.* (2012) found that thermal comfort and satisfaction with IAQ were not strong predictors of perceived wellbeing, while lighting was. These findings make it impossible to develop an internationally valid index to rate office environments by means of a single number. Assessment of each of the aspects separately might be better (Humphreys, 2005). It is becoming clear that increases in occupants' overall satisfaction do not correspond uniformly to improvements of individual IEQ factors (Kim and de Dear, 2012).

Minimizing all pollutants (chemical, physical and biological) or other environmental stresses (light pollution (glare, distorted spectral distribution) and thermal conditions that are far from the central range of thermally acceptable conditions) will reduce the likelihood of interactions that result in an unacceptable indoor environment. However, even when a design conforms to the standards and guidelines for the four major environmental factors, an unacceptable indoor environment quality may occur (ASHRAE, 2010a).

## Another indicator

Perhaps another indicator than perceived comfort via self-reported questionnaires should be used. Suggestions such as emotional state, prior experiences and the forgiveness factor (e.g. Vink, 2004; de Dear, 2009; Deuble and de Dear, 2009) have been brought forward. Or should one use the person-related factors as a correction factor on perceived comfort as suggested by Berglund and Gunnarsson (2000)? They recommended developing a new questionnaire method for practical use, in which person-related causes for occupant symptoms are separated from building-related causes. Personal characteristics of the psychological kind that are stable over a longer time, named personality traits, in particular negative affectivity have been found to be associated with self-reports of dissatisfaction (Burke *et al.*, 1993). Also one's mood of the moment has a possible effect (Kuhbander *et al.*, 2009): it is well established that mood modulates the general style of processing information. Negative mood induces predominantly stimulus-driven processing and positive mood predominantly knowledge-driven processing. For filling in questionnaires this could imply that when an occupant is in a negative mood (just had a quarrel with his/her spouse) the questions are answered without really considering the past experiences, but only referring to the now in a negative way. While, when an occupant is in a very positive mood, everything in the past is considered and depending on their personality trait they will respond more positively or negatively. Aries *et al.* (2010) found a direct relation between depression sensitivity and perceived comfort: the more positive an employee is about his/her environment, the less discomfort was reported.

Other indicators for the current physiological or biological indicators applied (see Section 6.2) are also thinkable. The larger objective for those measurements is to use biological indicators that reflect possible malfunction across multiple systems (e.g. HPA-axis, sympathetic and parasympathetic system, immune system). It would be important to have a limited set of biomarkers in place for each system, but those that are sensitive to broadly based dysregulation while not being overly burdensome when employed in large population studies. According to Singer *et al.* (2004), an obvious limitation of the extant allostatic load measurement and scoring schemes is that they are based on a small number of indicators of dysregulation in complex dynamical systems, where breakdown can occur at many sites. In particular, they are not capable of monitoring the diverse possibilities for dysregulation that are present in metabolic, immune, neuro-endocrine and autonomic nervous system networks. Singer *et al.* (2004) propose therefore to relate gene expression to phenotypic outcomes (see Figure 8.7). Techniques to achieve this are:

- Genomics and transcriptomics, which examine genetic complement and gene expression, respectively (see Section 6.2.4).
- Metabolomics and metabonomics (see Section 6.2.1).

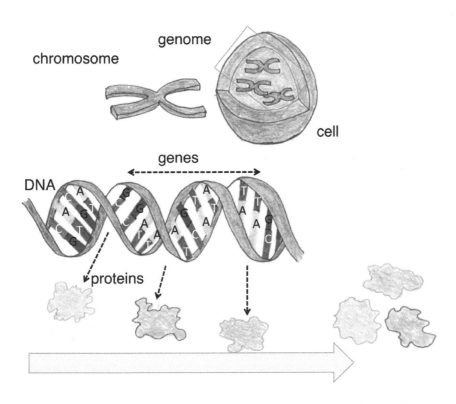

**Figure 8.7** *Cell with genetic information: genes contain instructions for making proteins, proteins act alone or in complexes to perform many cellular functions. DNA comprises of base pairs: adenine (A) and thymine (T); guanine (G) and cytosine (C) (see also Figure 2.3)*
Source: Bluyssen.

## 8.4 Building-related indicators

*With respect to building-related indicators two types of indicators can be distinguished:*

- *Labels of some kind to compare buildings or components.*
- *Measures for prevention or problem solving: for example, in a checklist or examples of good practice. These can be product- or process-related indicators.*

### 8.4.1 Labels

#### Policies and standardization

While *regulations* are understood as broader set of regional, national and international legal documents that, in hand with standards, forms the way in which buildings are being designed, constructed and operated, a *policy* is considered to be a deliberate plan of action to guide decisions and achieve rational outcome(s). Policy differs from regulation but can be considered as a complement to it. While regulations can compel or prohibit behaviours, policy merely guides actions towards those that are most likely to achieve a desired outcome (Lupisek *et al.*, 2009). Labelling is in fact a policy measure that makes it attractive for stakeholders to include indoor environment quality in their daily business. It is a way of raising awareness, and it helps in making choices. At the European level, labelling of buildings linked to sustainability is promoted by TC 350 European standardization organ CEN (2008). Parallel to that CEN TC 351 is developing a harmonized testing scheme for emissions of construction products, which could form the base for a construction product labelling system (EU, 2005).

The CEN and ISO standards (CEN TC 350, ISO TC 59 SC17) are based on a common life-cycle analysis based approach, supplemented by additional environmental and technical information. The standards only fully address environmental performance assessment, while work is continuing to address more fully the social performance of buildings (SuPerBuildings, 2010a).

In CEN TC 350 Sustainability of Construction Works, the developed standards (for all types of buildings) provide a European system for the assessment of environmental, social and economic performance of buildings based on a life-cycle approach (e.g. PrEN 15643 series; PrEn 15978; CEN, 2008). This TC has determined groups of indicators for the assessment of environmental impacts and aspects, with the starting point that all indicators must be capable of being applied at building level (also indicators used for the assessment at product level). For social performance in Table 8.5 an overview is presented for the indicators determined.

In Standard 15643–1 from CEN TC 350 (CEN, 2010) a general framework is defined with several levels:

- *Concept level*: environmental, social, economic, technical and functional performance.

**Figure 8.8** *Are we ready for a health label?*

Source: European HealthyAir project, Bluyssen *et al.*, 2010d.

- *Framework level*: prEN 15643–1 Sustainability Assessment of Buildings – General framework, prEN 15643–2 Framework for Environmental Performance, prEN15643–3 Framework for Social Performance, prEn 15643–4 Framework for Economic Performance.
- *Building level*: prEn 15978 Assessment of Environmental Performance and WI 003 Use of Environmental Product Declarations, WI 015 Assessment of Social Performance, WI 017 Assessment of Economic Performance.
- *Product level*: prEN 15804 Environmental Product Declarations, prEN 15942 Comm. Format, and CEN/TR 15941. Technical information related to social and economic performance are included under the provisions of prEN 15804 to form part of environmental product declarations.

ISO TC 59 SC17 – Building Construction – Sustainability in building construction is responsible for the development of standards relating to the sustainability of construction works, which are the main precursors to the work of CEN TC 350. Standard ISO CD 21929–1 (Building Construction Sustainability in Building Construction – Sustainability indicators – Part 1 – Framework for the development of indicators for buildings and core indicators) describes and gives guidelines for the development of sustainability indicators related to buildings and defines the core indicators of buildings for three levels: location, site-specific and building-specific indicators (ISO, 2006).

**Table 8.5** *Indicators determined by TC350 for social performance*

| Building-related data | User- and control system-related data |
|---|---|
| Health and comfort | Health and comfort |
| • Thermal performance<br>• Humidity<br>• Quality of water for use in buildings<br>• Indoor air quality<br>• Acoustic performance<br>• Visual comfort | • Thermal performance<br>• Humidity<br>• Indoor air quality<br>• Acoustic performance<br>• Visual comfort |
| Safety and security | Safety and security |
| • Resistance to climate change<br>• Fire safety<br>• Security against intruders and vandalism<br>• Security against interruptions of utility supply | • Security against intruders and vandalism |
| Accessibility | Maintenance |
| • Accessibility for people with specific needs (prams, children, etc.) | • Maintenance requirement |
| Maintenance | Loading on neighbourhood |
| • Maintenance requirement | • Noise<br>• Emissions |

Surce: adapted from SuPerBuildings, 2010a.

The Energy Performance of Buildings Directive (EPBD) 2010/31/EU (EU, 2010a) promotes the improvement of the energy performance of buildings, taking into account outdoor climatic and local conditions, as well as indoor climate requirements and cost-effectiveness. Unfortunately, nothing is indicated on how to achieve indoor climate requirements (EU, 2010a).

Additionally, there exist several organizations or initiatives around the world that develop common metrics for measuring, such as the SBA (Sustainable Building Alliance) and UNEP SBCI (Sustainable Buildings and Climate) initiatives. *The SBA* is a not-for-profit membership organization that develops common metrics that can be used within any building performance assessment and rating system, and that will provide a means to monitor and compare ecological and sustainability performance of buildings. The metrics are limited to six (SuPerBuildings, 2010a):

- carbon emissions
- energy consumption
- water consumption
- waste production
- indoor environment quality (represented by thermal comfort and indoor air quality).

The SBCI initiative is hosted by UNEP Division of Technology, Industry and Economics (DTIE). DTIE has the mandate to encourage decision makers in industry and government to develop and implement policies, strategies and practices that are cleaner, safer and make efficient use of natural resources. A forum for the building sector has been provided by UNEP-SBCI to provide globally applicable common metrics for measuring and reporting the energy use in and GHG (greenhouse gas) emissions from existing buildings operations (the common carbon metric for buildings; SuPerBuildings, 2010a).

ASHRAE (the American Society of Heating, Refrigerating and Air-Conditioning Engineers) developed in cooperation with the USGBC (US Green Building Council) and the CIBSE (Chartered Institution of Building Services Engineers) performance measurement protocols for commercial buildings (ASHRAE, 2010c).

Also at the product level several labelling systems, directives and regulations are available. For example:

- *The voluntary European EcoLabel scheme*, which was established in 1992 by the European Commission (EC) to encourage businesses to market products and services that meet high standards of environmental performance and quality (http://ec.europa.eu/environment/ecolabel).
- *Regulation No 305/2011* of the European Parliament and the European Council that replaces the Construction Products Directive (CPD; 89/106/EEC; EU, 2011). It covers very broadly the finished building as well as the components and equipment and the construction materials. It also concerns the impacts on IAQ, thermal comfort and indoor noise and health.
- *Green public procurement* (GPP), a voluntary instrument in which the process to seek to procure goods, services and works with a reduced environmental impact throughout the life-cycle is defined (EU, 2008). The GPP

**Table 8.6** *The most well-known building evaluation tools BREEAM, CASBEE and LEED*

| Name of the tool | Country of origin | Date of first entry into operation |
|---|---|---|
| BREEAM and Code for Sustainable Homes | UK | 1990 |
| CASBEE | Japan | 2001 |
| LEED | USA | 1998 |

Source: adapted from SuPerBuildings, 2010a.

considers construction (the overall environmental profile of an entire building), but also building materials and furniture.

* *General product safety directive* (2001/95/EC), intended to ensure a high level of product safety for consumer products that are not covered by specific sector legislation (e.g. toys, chemicals, cosmetics, machinery; EU, 2001). It only covers risks to human health and safety.

## Building evaluation tools

The most well-known building evaluation tools are BREEAM, LEED and CASBEE (see Table 8.6). In the European project SuPerBuildings an overview was made of the most well-known, with the addition of some of the national tools used (SuPerBuildings, 2010a). As with the CEN and ISO standards, the issues and indicators with the different building evaluation tools are divided into three groups: environmental, economic and social. The social indicators are split into three sub-groups: comfort and health; accessibility of the building and access to transport; and safety and security (SuPerBuildings, 2010a). The social issues and indicators commonly used for health and comfort as inventoried in the European project SuPerBuildings from the different building evaluation tools are presented in Tables C.1 to C.4 in Appendix C for the issues of visual comfort, thermal comfort, acoustic comfort and air quality. What can be seen from this inventory is that indicators presented are not only indicators at environmental parameter level, but also at building level (see also Section 8.4.2).

## Critical note

Labels linked to sustainability as they exist today do not include the detailed information required to identify sources of exposures encountered in the indoor environment that influence the quality of life (health and comfort). Besides the indicators related to life-cycle assessment (LCA) of the building (such as energy (generation and use), environmental impact of infrastructure and materials, recycling and reuse (e.g. water)), indicators of quality of life related to life-cycle of the users for different type of users (i.e. the elderly, children, etc.) should be included as well. At the moment these so-called social aspects or indicators are not covering all aspects of quality of life. The assessment is limited to some aspects, based mainly on current standards for indoor environment quality, and the interrelations with other categories of indicators is not identified.

In the project SuPerBuildings (2010a) additional social indicators related to health and comfort were proposed, namely:

- Consideration of user's needs.
- Quality of the building as a place to live and work:
    - Efficient use of space: expressed in $m^2$ working space/person of $m^2$ room space/bed; size and layout of rooms.
    - Control by users of ventilation, sun protection devices, anti-glare devices, temperature, daylight and artificial light, and user-friendliness of these control systems.
    - General conditions within the building (overlap with comfort and health indicators).
    - Nuisances: limit nuisances by ideally locating services inside buildings and by taking adequate measures to limit nuisances to building users.
    - Provision of facilities.
    - Actual level of occupant dissatisfaction through post-completion monitoring.
- Individual life styles and preferences usability.

Additionally, these different indicators and aspects may have conflicting interests. Cautions must be adopted to avoid the danger that improvement in sustainability may cause a decrease in quality of life (health and comfort of the occupants or end-users).

For *indoor air quality*, a label on product level will at least ascertain that the total emissions of products are controlled or reduced and therefore the total exposure is better controlled. Since a 'complete picture' of the effects of exposures on each other and on people is missing, the best that can be done is to reduce or control exposure and therefore reduce or control the sources of emissions. Research is needed to understand the health effects of product mixtures, to understand dominant exposure pathways and uptake to design and test optimal labelling strategies, and to even understand how additional composition and labelling information affects public behaviour with respect to the purchase and use of products (Clausen *et al.*, 2011).

*Indicators* can be structured according to the quality versus load (see in particular CASBEE), in different life-cycle phases, of the building (e.g. construction or use phase); scale of evaluation (building–site–location); dimensions of sustainability (economical, environment, social); and object (building characteristics) or process (e.g. construction, management, commissioning) related. Environmental indicators are usually structured according to impacts (e.g. climate change, resource depletion) or pressures (e.g. materials, energy, water). Moreover, indicators related to functional quality, health and comfort are often grouped in a distinct (sub)category, so that these results are clearly visible at semi-aggregated stage. Technical quality is rarely included, and only a limited number of systems cover the economic dimension (SuPerBuildings, 2010b).

### 8.4.2 Measures

#### Single parameter control

The control of the indoor environmental factors has merely been focused on the prevention or curing of the different related observed physical effects in a mostly isolated way: thus trying to find solutions for thermal, lighting, air and sound quality separately (Bluyssen, 2009a; see summary in Box 8.9). What has been

# Box 8.9  Measures to control IEQ factors

*Thermal comfort* can be controlled via so-called heating, cooling and air conditioning systems. Heating can be provided through convection, conduction, radiation and air systems. Regulation of the relative humidity can be provided through (de)humidification systems via an air conditioning system or locally. Additionally, one can adjust one's clothing and type of activities.

The best way to control exposure to pollutants (*air quality*) is to perform source control, i.e. to minimize the emission of either primary or secondary pollutants to the air, which we are exposed to. Besides source control there are three other ways to control the exposure, directly or indirectly: ventilation, air cleaning and activity control, such as designating smoking areas in a non-smoking building. What is often underestimated now is the contribution of ventilation systems to the indoor air pollution encountered in the indoor environment. Design and maintenance of such systems are important matters. Furthermore, the detection and interpretation of the thousands of pollutants in the indoor air is difficult and complex. Besides the fact that these pollutants interact with each other (indoor chemistry), they also cause interactions in the sensations and perception of the human body. Therefore, providing guidelines for levels of permissible concentrations will always be questionable, unless a direct relation has been found.

*Comfortable light* doesn't cause blinding (of lighting systems or direct sun light), any flickering or stroboscopic effects, reflections or glare (in computer screen), but does create good colour impression and an equal distribution of light. Positioning and intensity of lighting systems, surface area treatment (mat surface area and colours), solar screens and solar reflecting glazing are means for this. Comfortable light also means controllability and healthy light (day–night rhythm). The latter can be provided by offering the right variation on light intensity and colour temperature at the right time. With automatic or manual dimming or intensifying this light, an appropriate integration of artificial light and daylight can be reached. Transparent parts in the enclosure of a space play a pivotal role in the human need for visual contact with outdoors (visual comfort) and daylight entrance on the one hand, and thermal comfort on the other. For thermal comfort, in particular, (one-sided) direct sun radiation can be perceived as uncomfortable, even though the air temperature may be at a comfortable level. Furthermore, the issue of energy is inherently involved, particularly in relation to the influx of solar energy and the demand for cooling. Therefore, a great effort is made to develop so-called integrated systems.

Control strategies can be performed to prevent *noise* from entering a space or approaching a person, or, to make the space perform acoustically better. With respect to the latter, besides the reverberation time and the speech–background noise ratio, speech audibility is influenced by the speaker, the communication channel and the listener. Speech intelligibility can therefore be improved by the speech–background noise ratio, by shortening the reverberation time, but also by a clearer or softer speech. By introducing absorbing material and/or decreasing the volume, the reverberation time can be shortened. Introduction of absorption material also decreases the sound pressure level and suppresses echo. Prevention or reduction of noise entering a space can be established by preventing/closing of sound leaks, prevention or reduction of contact sound transmission, and/or by applying active (noise) control. Besides the level and duration of the noise disturbance, the positive or negative feeling one has with the type of noise, vibrations seem to affect people as well. Additionally, the health effects reported from exposure to traffic noise during the night are very serious.

Source: Bluyssen, 2009a: Section 1.3.

largely underestimated, however, is the interactions between the four main factors. Measures to control one factor can deteriorate another. For example, to control acoustic high surface area materials (so-called 'fleecy' materials such as carpet, fibrous or highly textured ceiling tiles, and textiles) are introduced. These materials are, in general, not ideal for indoor air quality: they are sinks for pollutants and not easy to clean. And to control the thermal environment better, thermal storage provided by interior surfaces of high thermal capacity ('thermal mass') are often applied. But these high thermal capacity materials can be deficient in sound absorption leading to unacceptable acoustics. Other examples are presented in Table 8.7. It is thus not enough to define a list of measures to control IEQ: interactions should be made visible. But also within one group one has to pay attention: i.e., in the factor indoor air quality, it was shown that portable ion generators that are intended to clean the air of particles can emit ozone and aldehydes as byproducts (Waring and Siegel, 2011).

## Short-cuts

Because we simply do not know all the interactions or mechanisms taking place between the sources that produce the exposure parameters and the exposure parameters themselves, and the exposure parameters and the exposed persons, in IEQ investigations *short-cuts* have been taken. In a short-cut, the characteristics of building (such as having an HVAC-system) or measures taken (such as a maintenance or cleaning schedule) are directly related to comfort or health responses of occupants. For example, in the BASE-study a significant relationship was found between type of filter media and building-related symptoms (Buchanan *et al.*, 2008; Apte, 2009). In a study on prevalence of childhood asthma

**Table 8.7** *Measures and effects of IEQ improvements to parameters that influence IEQ*

| Measures taken on | Leading possibly to failures in |
|---|---|
| INDOOR AIR QUALITY<br>Increased ventilation rate<br>Reduced humidity to reduce micro-organisms growth | THERMAL QUALITY<br>Potential draughts<br>Inadequate humidity levels |
| INDOOR AIR QUALITY<br>Increased air supply rate<br>Elimination of many synthetic materials | ACOUSTIC QUALITY<br>Inadequate noise levels<br>Reduction of available sound-absorbing materials |
| THERMAL QUALITY<br>Reduced infiltration to improve thermal comfort<br>Use humidifier | INDOOR AIR QUALITY<br>Inadequate outdoor air rate<br>Potential micro-organisms growth |
| THERMAL QUALITY<br>Use solar protection to reduce overheating | VISUAL QUALITY<br>Decreased day lighting efficiency |
| VISUAL QUALITY<br>Increased glazed surfaces | THERMAL QUALITY<br>Potential overheating |
| VISUAL QUALITY<br>Use glazed partitions | ACOUSTIC QUALITY<br>Potential sound reflecting |
| ACOUSTIC QUALITY<br>Use of sound attenuation in air inlets | INDOOR AIR QUALITY<br>Insufficient fresh air flow rate |

Source: Bluyssen, 2009a: Table 9.3, p. 218.

and allergy, a low ventilation rate in combination with a mouldy odour showed an increased risk for allergic symptoms (Hågerhed-Engman *et al.*, 2009). Stephens and Siegel (2012) found a significant and negative correlation between particle penetration (from outdoor to indoor) into buildings and year of construction. Additionally, in a study performed by Aries *et al.* (2010), statistically significant relationships were found between combinations of view type, view quality and social density, and perceived discomfort.

The prerequisite for successfully performing such a short-cut or pattern recognition is, however, that the 'right package' of information is gathered. From a re-analysis of part of the European HOPE project data it was clear that this gathering of information was not good enough to perform the pattern recognition (finding the short-cuts) as was originally planned (Bluyssen *et al.*, 2011a). It seemed that the 'wrong' questions had been asked, the questionnaire was most likely 'incomplete' and even the techniques to get the 'correct' answers could have been improved. The outcome of the analysis of the HOPE study (Bluyssen *et al.*, 2011a) could also indicate that perhaps important questions (or tools) to collect the required information were missing. Several investigators recommend focusing much more on the satisfaction with environmental features: work station location, flexibility of workstation (Choi *et al.*, 2009); office layout, amount of office space, satisfaction with other facilities and cleanliness (e.g. Vink, 2004; Vischer, 2008), which is not new in the POE (post-occupancy evaluation) community. Hansen *et al.* (2008) recommended including physiological stress indicators as a supplementary measure to questionnaires when studying the relationship between the work environment and building-related symptoms.

Methods applied in IEQ investigations varied from an epidemiological approach, in which questionnaires and health/comfort data may be used either in combination or not with biomarker sample collection (e.g. blood, urine), field studies in which in general a smaller sample of persons is studied in combination with environmental inventories, to laboratory studies in which persons or animals are exposed to controlled environmental conditions. Health and comfort data are then combined with information on characteristics of the indoor environment in order to find relations. However, other risk factors that may cause psychological or physiological stress (e.g. major life events), individual differences caused by personal factors (e.g. states and traits), or history and context can all affect the outcome that is being studied. These factors are taken into account only to a limited extent in current methods commonly applied to identify relationships between health and comfort of people and the physical and social environment. *To be more successful in determining the health and comfort effects of certain indoor environmental aspects there seems a need to improve procedures applied in IEQ investigations.*

## Building design characteristics

According to Evans and McCoy (1998) building design has the potential to cause stress and eventually affect human health. They introduced five dimensions of the designed environment that potentially could affect human health by altering stress levels:

1 *Stimulation*: The amount of information in a setting or object that affects the human user. Lack of stimulation can lead to boredom, while over-stimulation can lead to stress (see also Box 8.6). Loud noise, bright light, unusual or strong smells, and bright colours can increase stimulation. But also crowding and interpersonal distances increase stimulation. Too much complexity or mystery makes interiors confusing and hard to analyse, too little makes prediction unimportant.

2 *Coherence*: The clarity or comprehensibility of building elements and form. Ambiguity, disorganization and disorientation are major impairments to coherence. Multiple repetitive features, underlying expression of rules and thematic continuity, all contribute positively to coherence. Conflicting information from adjacent design elements or abrupt shifts in size, colour, texture or stimulation levels can increase stress.

3 *Affordances*: Mis-affordances occur when we are unable to effortlessly distinguish the functional properties of a space, building or technological function. When a building user cannot see what or how something in the space functions or when confronted with cues about purpose or use that are vague or in conflict, human reactions are likely to include frustration, annoyance and sometimes even hostility or helplessness.

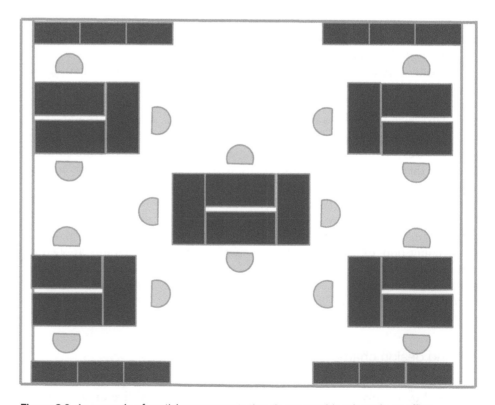

**Figure 8.9** *An example of spatial arrangements that do not provide privacy in an office environment (sense of control)*
Source: Bluyssen; inspired by Evans and McCoy (1988).

4   *Control*: Ability to either alter the physical environment or regulate exposure to one's surroundings. Physical constraints, flexibility, responsiveness, privacy, spatial syntax, defensible space and certain symbolic elements are key design concepts relevant to control (see also Box 9.5). An example is presented in Figure 8.9.
5   *Restorative qualities*: The potential of design elements to function therapeutically, reducing cognitive fatigue and other sources of stress. Design elements include retreat, fascination (e.g. window views, burning fireplace and moving water) and exposure to nature.

## Kano's satisfaction model

Kim and de Dear (2012) modified and applied the Kano's satisfaction model (see Box 8.10) on a total of 43,021 respondent samples from 351 different office buildings (including educational, public administration and research organizations) in various climate zones in different countries (Australia, Canada, Finland and the USA) to estimate individual impacts of 15 IEQ factors on occupants' overall satisfaction, depending on the building's performance in relation to those IEQ factors. These empirical analyses identified nonlinearities between some

---

### Box 8.10 The Kano model

The Kano model is a model of customer satisfaction based on a classification of the type of relationship between specific product qualities and overall satisfaction. Different qualities or factors impact overall customer satisfaction in different ways: some in a positive way, some in a negative way, and some in both directions. Adapting the Kano's satisfaction model to the building context, indoor environment quality factors can be classified into three categories (Kim and de Dear, 2012; see Figure 8.10):

*   *Basic factors*: minimum requirements, which can cause dissatisfaction when they are not fulfilled. They are only noticed when they are deficient or defective in some way. Impact on overall satisfaction from underperformance is greater than from positive performance.
*   *Bonus factors*: go beyond minimum expectation. They can have a strong positive effect on occupant's satisfaction. Impact on overall satisfaction resulting from positive performance is greater than that resulting from underperformance.
*   *Proportional factors*: occupant's satisfaction level changes proportionally according to the performance of these factors. Relationship between occupant's overall satisfaction and the performance of these factors is linear.

These relationships are dynamic: they can differ between groups and change over time. A basic factor for one group could be a bonus factor for another group, depending also on the scenario. For example, an occupant located in a spacious private office would have a higher expectation for IEQ than an occupant located in a dense, open-plan office.

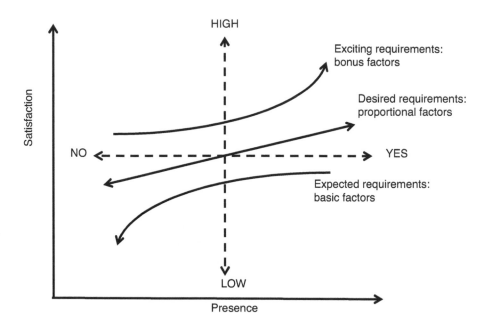

**Figure 8.10** *The Kano satisfaction model*
Source: adapted from Kim and de Dear, 2012.

IEQfactors and occupant satisfaction. They identified basic and proportional factors (bonus factors, see Box 8.10 for explanation, were not identified):

- *Basic factors*: IEQ factors that had a predominantly negative impact on occupants' overall satisfaction when the building underperformed, comprising 'temperature', 'noise level', 'amount of space', 'visual privacy', 'adjustability of furniture', 'colours and textures' and 'workplace cleanliness'.
- *Proportional factors*: IEQ factors that had a predominantly linear relationship with overall satisfaction, including 'air quality', 'amount of light', 'visual comfort', 'sound privacy', 'ease of interaction', 'comfort of furnishing', 'building cleanliness' and building maintenance'.

### 8.4.3  Positive stimulation

#### Positive indicators

What is interesting to notice, besides the fact that in general all of the indicators pointed out before can be divided into product or object-related and process or non-object-related indicators, is that most of them are concerned with negative effects or the risk that something negative will happen (discomfort, illness, disorder, pollutants, etc.).

In the European project Perfection (Desmyter *et al.*, 2010) an attempt was made to inventory indicators for positive stimulation among others in the commercial environment (see Box 8.11). In the ProWork project, a ProWork-

---

**Box 8.11 Some indicators for positive stimulation in the commercial environment**

- *Fragrance stimulation*: for example by perfuming an aisle to push the consumer to buy a product; or by use of fragrance on a product to improve its perception; or by baking food to release mouth-watering smells.
- *Visual stimulation*:

    - By lighting: use of special lamps to give the impression of greener vegetables or more colourful bakeries, meat, and so on, is a powerful technique to increase sales. Under normal supermarket lighting fruits often look dull. Lighting combined with flashy colours is also used to get the consumer's attention in shopping malls' galleries.
    - By packaging: use of bright and large packaging.

- *Hearing sense stimulation*: audio advertisement, background music.
- *Indoor space arrangement*: spatial arrangement of goods is crucial to maximize sales.

Source: Desmyter *et al.*, 2010.

---

toolbox was developed, comprising two levels based on a framework to study productive knowledge work as a function of the key elements 'task factors, contextual factors and process factors' (Nenonen *et al.*, 2009):

- *Level A*: for identification of success factors and measurable indicators;
- *Level B*: comprising the methods which can be used for alternative ways of investigating physical tangible), virtual (electronic collaboration) and social (interactions for building shared mental spaces) workplaces.

To be able to say something on the effect of certain parameters, one should not only automatically strive for a neutral environment, in other words focus on discomfort and diseases only. Both positive and negative effects of stimuli should be included.

While studies on the positive effects have hardly started, preferences (trade-offs) of certain environmental aspects, for example, have been studied on a small scale by Santos and Gunnarsen (1999) and by Clausen and Wyon (2008); studies on negative effects are well ahead. Positive stimuli such as natural odours can increase the balance of people's behaviour (Von Kempski, 2003).

In general we do not ask the occupant what did he like or why did he like it, but always what he didn't like, in order to prevent diseases and disorders in the future. Taking into account the positive effects of certain parameters should be just as important. To study that with the human being as the ultimate sensor at component level seems an important direction for the future.

### Positive places

To support the creation of spaces that provide people with sensory environments that help them reach, or sustain, their fullest possible potential, Barett and Barett (2010) proposed a framework that captures the essence of an individual's holistic experience of spaces, comprising three themes:

1 *Role of naturalness*: significant evidence of positive impacts deriving from aspects of naturalness such as day lighting, sound, temperature and air quality.
2 *Opportunity for individualization*: particularization, accommodating the functional needs of very specific types of users; and personalization, individuals preferences resulting from their personal life experiences of spaces. For schools this could be summarized as choice, flexibility and connection.
3 *Appropriate levels of stimulation*: for given situations with respect to complexity, colour and texture.

This framework could help to optimize designs around user outcomes: for example, schools in which pupils' wellbeing is improved and where they achieve enhanced academic results or offices in which workers are more efficient and productive; and homes in which people can live more happily. Six built environment design parameters, namely colour, choice, connection, complexity, flexibility and light were linked to the learning improvements of pupils from 34 varied classrooms in seven different schools in the UK (Barett *et al.*, 2013). It must be noted that the information on the design parameters was gathered by the researchers themselves and not by asking the pupils or teachers.

## 8.5 A new approach

*The system engineering approach used in other industries is discussed based on the results of three European and one national project. The outcome reveals that the applied top-down approach seems an improvement for the building process, but we still have a long way to go. It was concluded that to be really usable, the top-down approach needs to be facilitated by a clear framework that makes the links between system requirements, design and technical requirements for the different phases of a building clear to all stakeholders, and which separates generic from the to-be-specified requirements for a specific building.*

### 8.5.1 Design and construction process

#### Over-the-bench

In design and construction processes, the players are diverse and many (see Figure 8.11). Since the majority of players are not in direct contact with end-users, it is not strange to notice that briefing, and identification and communication of end-users needs, is a problem in everyday practice. Briefing is the most important activity of the whole life-cycle of a building. If briefing is already

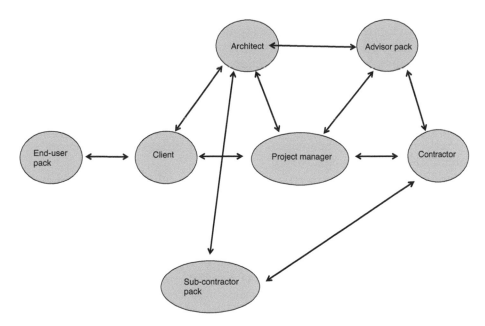

**Figure 8.11** *Communication between players in a traditional design and construction process*
Bluyssen *et al.*, 2010b.

a problem, then the translation of end-users' needs into technical requirements is likely not to result in the necessary requirements. Consequently, those involved in the design and construction process experience a dramatic loss in failure costs (Brokelman and Vermande, 2005), failures at delivery and severe problems (Layard and Glaister, 1994).

In a traditional design and construction process, the parties involved often use an 'over-the-bench methodology', i.e. when one has finished his work it is given to the next without looking back ('no longer my responsibility'). A real team is not formed. Parties do not understand each other's stakes or products. As far as back in the times of Vitrivius (100 BC) and Palladio (1570), effective communication between the different disciplines involved was pointed out as a necessity in their 10 and 4 books of architecture, respectively. In those books they both established a clear link between architecture, occupant's health, comfort and social needs with beauty within a specific (cultural) context. In fact nothing has changed, only the number of disciplines and experts involved in the design and construction process have increased, making the communication process even more complex.

Additionally, the demands and constraints on these processes are increasing. We, as a society, demand higher quality levels (comfort, health and safety), quicker outcomes, life-cycle focus (adaptability to change, design for maintenance and exploitation) and a decrease in the use of resources (material, energy, emissions and waste). On top of that end-users demand more influence (choice, flexibility of use, changeability; Clements-Croome, 2004).

If the same processes are used, more problems and more failure costs will arise. To escape this situation, a fundamental change in design and construction processes is necessary: fulfilment of all the new requirements, and more, will then come within reach. An interactive top-down approach is required, both for the communication process required to facilitate the design, construction, maintenance and occupation of a building, and, for the establishment of end-users' wishes and demands (requirements and needs).

Using a top-down approach will ensure that the assessment is comprehensive and treats all sustainability issues adequately, because existing indicator 'bottom-up driven approaches' often do not cover the full range of issues, may be overlapping (double counting) and may be of different value in terms of significance (SuPerBuildings, 2010b).

Luckily we do not have to start from scratch. In other industries and even in construction there is already a lot of knowledge and experience available such as the theories on lean construction (Jørgensen, 2006), performance-based building (CIB, 1982), the open building approach (Kendall and Teicher, 2000), the value-domain model (Rutten and Trum, 1998) and the system engineering approach (Blanchard, 2004). For each of those theories, it can be said that an underlying structure is used (organizational structure, model or even a contract) to accommodate the communication process, making it more effective and efficient while reducing the risks that overall project goals are not achieved.

## System engineering as the starting point

The system engineering approach facilitates the definition and translation of end-users' requirements in the appropriate way. Its application suggests a different form of cooperation between demand and supply including all the major stakeholders and possibly also the involvement of parties who normally are not consulted. System engineering requires (Blanchard, 2004):

- A top-down approach, viewing the system (the building) as a whole.
- A life-cycle orientation.
- A better and more complete effort to relate the initial identification of system requirements to specific design goals, the development of appropriate design criteria, and the follow-on analysis effort to ensure the effectiveness of early decision making in the design process.
- An interdisciplinary effort (or team approach) throughout the system design and development process to ensure that all design objectives are met in an effective manner.

When considering the building as a system and applying this system engineering approach, there seem to be great possibilities to improve the current approach in the building process. Apart from the division of the building life-cycle into steps, two important characteristics should be emphasized: the team approach and the always looking back (or up) to the basic system requirements set at the beginning of the process. The system engineering approach is all about setting requirements and translating these to processes and products of the different phases of the design and construction life-cycle. In the national project Omnium and the European projects ManuBuild, InPro and SWOP, the system engineering way of thinking was applied and/or tested at different levels (Bluyssen *et al.*, 2010b):

- In *Omnium* an integrated design of social housing was formulated based on pre-determined system requirements by a team of experts.
- In *Inpro* a general framework was developed in which the different requirements could be placed to facilitate the process of defining requirements.
- In *ManuBuild* the need for a different business model was identified resulting in more and different roles and new concepts of design and manufacturing supported by intelligent configuration tools.
- And in the *SWOP project* a new industrial construction concept was specified, and, a semantic modelling approach and supporting software tools were developed.

From the analysis of those projects it was concluded that there is a need for some kind of a platform of common understanding, which makes the facilitation of the communication and the translation process possible and feasible during the system engineering process. To be really usable, the top-down integrative approach needs to be facilitated by a *framework* that makes the links between system requirements, design and technical requirements for the different phases of a building clear to all stakeholders (see Section 8.5.2). Setting-up a new set of system requirements each time is not only taking a lot of time, but also causes a lot of irritation and misunderstandings with the parties involved.

An up-coming challenge is the *interactions* taking place. The interactions between the requirements following from the captured values are very complex. A way to deal with the complex interaction of requirements on product, process and service level is the division of the design at different levels (area – building – infill), in line with current practice, along with differentiating between generic designs (see Section 8.5.3). This division helps to structure information, tools, stakeholders and decision makers.

And last but not least, from the analysis of the projects, it was concluded that it is possible to create the *ICT software* for the new (system) engineering part of the building process (see Section 8.5.4). The input data for such a software system determines at the end the output and its usability. Therefore, the challenge remains to define the requirements and interactions at different levels in the production process but also in and with the operational process. One can easily envision a system where not only the data aspects but also the software functionalities acting on them are integrated in interoperable ways, constantly learning and reusing product knowledge as it evolves: functionalities that not only assess the current product definitions/configurations, but also guide the engineers (product developers and configurators) in their choices.

It is thus important to consider developing more effective feedback systems, which can provide useful evidence for future designs. In addition to setting performance targets, creative integration and a shared view from the start of the process is required in order to allow much more systems thinking and analysis than has been customary.

### 8.5.2 System requirements

#### Framework

In the European project InPro (Sormunen *et al.*, 2009), in which the final objective was to develop an 'open information environment (OIE)' supporting the early design phase for the construction life-cycle of a building, a general framework for system requirements was defined, with which it is possible to:

- Capture the wishes and demands (the goals) of different stakeholders over the life-cycle of a building and express them in system requirements, also named the 'brief' in the construction industry (strategic briefing).
- Translate the identified system requirements into possible design requirements (operational briefing).
- Translate design requirements in the next step into technical requirements (technical briefing).

In the InPro project, an attempt was made to define an exemplary list of stakeholder goals and values, which can be used in future projects as a starting point. Knowing that for each future project the number and type of selected goals can differ, stakeholders can use this list to identify their own goals more

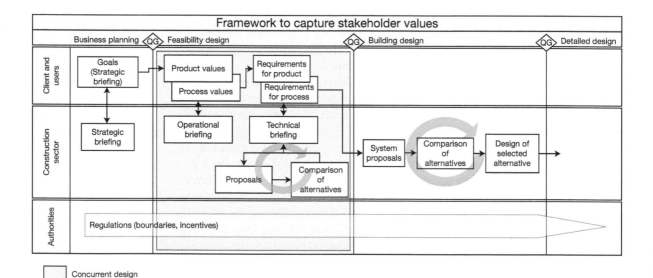

**Figure 8.12** *The InPro framework for capturing stakeholder values and their implementation in the InPro life-cycle design (LCD) process. The framework is an attempt to apply system engineering using the language applied in the construction industry. Strategic briefing defines the common project goals in non-technical terms (the basic system requirements). Values are then derived from these goals and translated into design requirements (operational briefing). Values in a project are both product and process values. The product (or the building) is evaluated according to architectural, economical, material and functional aspects. Related value factors in the product value category are for example usability for the end-user, life-cycle costs, technical performance and environmental impact. Process-value-related factors can be of the following character: time, ways of cooperation, use of innovative techniques, quality management, feedback of experiences, efficient use of resources, transparency of management and decision making, and so on*

Source: Sormunen *et al.*, 2009.

**Table 8.8** *General list of goals, group values and values*

| Possible goals/ Value group | Values | Possible goals/ Value group | Values |
|---|---|---|---|
| Healthy and comfortable indoor environment for all people/ **Human wellbeing** | Health Comfort Safety Security Usability Accessibility | Sustainable building/ **Sustainability** | Materials life-cycle Energy management Water management Waste management Biodiversity Environment management |
| A safe and secure building/ **Safety and security** | Fire safety Emergency procedure Operational reliability Maintainability Structural safety Security | Flexible and adaptable building/**Functionality** | Adaptability Expandability Outside services Inside services |
| Impression: i.e. open, united, friendly, interaction with environment/**Aesthetics** | Architectural quality Image | In line with local constructional tradition/ **Cultural and social** | Constructional traditions Architectural styles Life style |
| Affordable construction and operation of building/ **Economics** | Investment costs Life-cycle costs | | |

Source: Bluyssen *et al.*, 2010b.

easily, avoiding the need to start from scratch with every new project. Table 8.8 presents the generalized goals, value groups and values. Design parameters were identified as well but are not presented here.

## General design requirements

Setting the system requirements (the goals and values of the team members involved) is not an easy task. It is much easier to just assume the traditional requirements and/or the current guidelines and standards and then go from there. The number of requirements, and the fact that they are influencing each other and are sometimes conflicting, makes it rather complicated. Because stakeholders can have different goals, their requirements can differ as well. If only the over-the-bench methodology is applied, certainly the focus will only be on the demands and wishes of the most dominating party.

From these design requirements, the next step is to define technical requirements for the building and its products and materials, which can be:

- Technical specifications of a product (such as the minimum allowable emission rate of an inner wall).
- Specifications for a parameter (such as temperature of the air in a room).
- Procedures for how to use a product or part of building (such as regularly cleaning of an inner wall to prevent it from emitting certain pollutants).

These technical requirements can be different depending on the building phase and also for the stakeholder of relevance. For each value or requirement defined at the initiation phase, such an analysis can be made. In this way for each product used in the building, a list of technical specifications can be defined, interactions can be identified and choices can be made. The same can be done for the parameters and procedures.

A problem can occur when we do not know how to reach certain design requirements, i.e. which parameter could be regulated to reach a certain quality or performance or which procedure will maintain or create that performance. In some cases, performance indicators have been used to overcome that problem. For example, in the case of the concentration of $CO_2$, which can be used as an indicator for air quality when the major pollution sources in a room are people. In the initial design, models are applied to predict or estimate the required technical specifications of products, parameters and procedures (Alwaer et al., 2008).

### 8.5.3 Requirements at different levels

#### New design concept

ManuBuild, which stands for Open Building Manufacturing, is a European project in which effort has been put into developing four key elements essential for radical integration and industrialization of the design, production and delivery process: open industrial building systems, new business models, ICT support systems and manufacturing methods (www.manubuild.net). An inventory of the requirements of the different stakeholders in the process, including also the clients and end-users (representatives) across Europe, formed the basis and were used to test and steer the outcomes of the four key elements.

In Manubuild a clear differentiation between at least three levels of design were made when it comes to buildings: area, base building and infill. This differentiation, which originates from Open Building, is made to facilitate change and the involvement of different parties throughout the process. In addition a distinction is made between generic and specific designs (see Table 8.9).

At the generic level, different concepts can be designed and, to use the building level as an example, translated into building systems and platforms in

**Table 8.9** *Different forms of design*

|  | GENERIC | SPECIFIC |
| --- | --- | --- |
| Area | Concepts<br>Maintenance scenarios and additional services | Specific plan layout, maintenance scenario and service contract |
| Building | Building concepts<br>Building system and additional services | Specific building<br>Building specific configuration and contract |
| Apartment | Templates<br>Infill systems and additional services | Specific apartment<br>End-user specific configuration and contract |

Source: Bluyssen et al., 2010b.

combination with services proposals. These generic designs are used as a starting point for a specific assignment. The requirements for a specific building assignment will help to choose between alternative generic concepts. As a second step the relevant concepts will be used for a project-specific design. On the basis of specific requirements due to location, culture, use, and so on, the necessary choices, changes and additions are made. This two-stage design process means buildings are no longer treated as a new design each time. Not only the quality of the building as a product can be improved this way, as more requirements can be addressed and more knowledge included in the design, it will also become possible to provide more service. ICT tools delivering financial benefits or energy use for a specific situation, for example, can help clients and end-users to choose between alternatives. Overall, the design of buildings according to the wishes and demands of end-users will be easier to accomplish. Nevertheless, integration, understanding and translation of requirements at different levels become a major focal point.

## Interactions

From the InPro framework it can be seen that it is likely that certain values link to more than one value group, such as security, and this is even more the case for design and technical requirements. Interactions occur between all. If you define a design requirement for the value health this most likely has consequences for several design requirements of the value energy management under the value group sustainability. This process of going back and forth and trying to optimize the design for both values, called system engineering, should eventually lead to a set of optimized technical requirements (measures and criteria) for products or elements of the building, parameters of the indoor environment and the processes required to design, build, operate and maintain the building.

Besides the above-mentioned missing translations for certain values (from design to technical requirements), another issue is the most likely occurring interactions between the different values and requirements. A first attempt to define those interactions was explored in the retrofitting tool TOBUS, in which the general state of an office building is diagnosed and the actions for improvement are defined (Caccavelli *et al.*, 2000). The decision-making procedures are applied at the retrofitting scenario level. The result of the TOBUS method is a proposal for a refurbishment strategy, the corresponding global actions along with their typical cost and impact on energy savings and the improvement of IEQ. The interesting part of TOBUS is the basic division of a building in building objects or elements. After the application of the tool and the methods of TOBUS, for each of those elements several actions are presented. Those actions will help the building manager, the architect or the engineer to improve (or replace) this element for a certain aspect. Four aspects are included: physical state of degradation of building elements, functional obsolescence of building services, energy consumption and IEQ.

Based on relations between characteristics of buildings and systems and the use of the building, different possible causes for the problems were identified, possible actions for improvement are then selected:

- Object-related actions, for example:
  - Observation: hindering of noise due to traffic because of use of natural ventilation (e.g. grills).
  - Action: use sound attenuation in grills.
- Non-object-related actions, for example:
  - Observation: stuffy/bad smell inside due to smoking (smoking permitted in all rooms).
  - Recommendation: use separate smoking zones with good exhaust or ban smoking.

### 8.5.4 ICT architecture

#### Different levels

The European SWOP (Semantic Web-based Open Engineering Platform) project was concerned with business innovation when specifying products (buildings) to suit end-user requirements (www.swop-project.org). There are two main business drivers behind this innovation:

- To reduce wasted effort in terms of cost and time in re-designing and re-specifying products when, for the most part, the work has been done before, and
- To configure solutions from pre-defined partial solutions ('modules') rather than design from scratch. When there are choices, product configurations are optimized in SWOP by applying genetic algorithms, so that the resulting product is not just a valid solution but even a near-optimal solution that can be achieved following design constraints, end-user requirements and optimization criteria.

As in Manubuild, in the SWOP project the notion that the design, realization and operation of a building can no longer be approached as a project each time but needs to be life-cycle and process oriented, and lead to the clear definition of a different building process being divided into two levels (see Figure 8.13):

- Product development.
- Product configuration.

Normally, the balance between development and configuration effort is different for each type of business. While for the construction industry the configuration level is currently not or hardly present: the design of a building starts from scratch each time (all effort goes to client-specific development or product configuration). By contrast, for example, the assembly of a desktop computer needs minimal development. An informed end-user could configure and assemble his own, because almost everything has been standardized (at product development level). Configuration doesn't take much effort in this case. Typically the truth for a particular company is somewhere in the middle following some rule like 80 per cent via configuration of reference products and

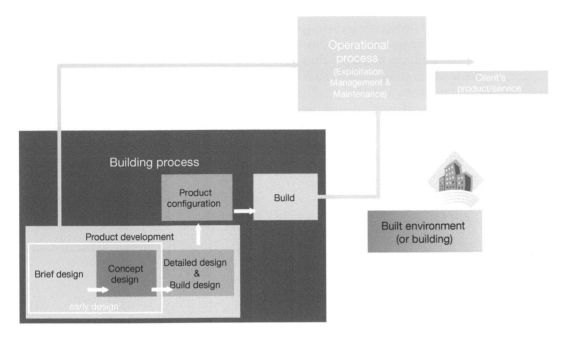

Program … Design ……………........ Build…… Operate

**Figure 8.13** *A different process for the design, realization and operation of a building*
Source: Bluyssen *et al.*, 2010b.

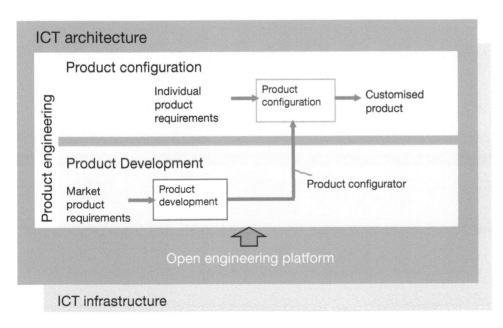

**Figure 8.14** *System architecture involving product development and product configuration*
Source: Bluyssen *et al.*, 2010b.

20 per cent via client-specific development. This means that 20 per cent of the *number* of orders needs client-specific development (not 20 per cent of *each* order needs client-specific development!).

The building process is thus interpreted as a workbench of related software tools supporting the product development layer resulting in actual product configurators and the product configuration levels. In an ideal world the 'development part' of such workbench would be a kind of 'configurator' of configurators: you input the choices and constraints for a product and its (often externally supplied) parts and a specific web application is automatically generated to be used by a particular client over the web to configure his product (Figure 8.14).

# 9
# Scenarios

*There is a need for a risk assessment model that takes account of the combined effects of stress factors in buildings on people, comprising a framework of indicators, interactions and dynamic behaviour over time per scenario. Based on recent and past studies, information for defining a conceptual framework of cause–effect relations (patterns) for different scenarios (homes, schools and office buildings) for major end-points (e.g. asthma, learning ability, productivity, and dry eyes and skin) is gathered. The information presented per scenario shows the need for other risk modelling and risk analysis than the current mostly applied linear modelling.*

---

**Box 9.1  Frank Duffy (2008)**

Architecture and other professions in construction, with few examples, have not developed a tradition of practice-based, user research, preferring to outsource both user research and teaching almost entirely to the universities. This fractured history may be the reason why environmental design has never developed to the equivalent of the science of epidemiology – which, in its macabre way, is an essentially user-based, feed-back rich discipline.

(p. 657)

If there were two fields of intellectual endeavour that certainly never can be separated from either context of values, they are architecture and design. . . . Consequently, environmental research is inevitably complex and must combine scientific, technical and moral (not to mention aesthetic) issues in every investigation.

(p. 656)

---

## 9.1 Introduction

The challenge of today lies in the accomplishment of a sustainable and *low-energy built environment* and at the same time a *healthy, comfortable, accessible and safe built environment*. Health and sustainability are interrelated in many ways. In the built environment a major reduction of the fossil fuel consumption should be achieved in order to meet the Kyoto targets. The existing stock is, however, far from the currently discussed low-energy standards and the path towards future low-energy use, or even energy autonomy or energy-positive buildings, is seriously hampered by the fear of introducing a negative impact on human health. No consensus understanding of this relationship between energy efficiency and IEQ (Indoor Environment Quality) exists. Even though, the understanding and processes needed to achieve a healthy, comfortable, safe and accessible indoor environment has been an issue among architects, engineers and scientists for centuries, guidelines for energy-efficient refurbishment of buildings generally consider energy efficiency in isolation from IEQ (including health, comfort, safety, usability and accessibility). Moreover, optimization of IEQ and energy efficiency is hampered by a lack of information regarding which indicators, criteria and interrelations need to be considered (see Chapter 8). For the assessment of health and comfort risks people have when staying indoors, a different approach or procedure seems inescapable as well as other or better assessment techniques – *a risk assessment approach that is not solely based on single dose–response relationships, and that takes account of different scenarios, interactions, different end-points and sources (starting points).*

---

## Box 9.2 Systems thinking

The essential properties of an organism, or living system, are properties of the whole, which none of the parts have. They arise from the interactions and relationships among the parts. These properties are destroyed when the system is dissected into isolated elements. Systems cannot be understood by analysis. The properties of the parts can be understood only from the organization of the whole. Systems thinking is *contextual*. Analysis means taking something apart to understand; systems thinking means putting it into the context of a larger whole.

(p. 29)

Different *systems levels* represent levels of differing complexity. At each level the observed phenomena exhibit properties that do not exist at lower levels. There are no parts at all. What we call a part is merely a pattern in an inseparable web of relationships.

(p. 37)

The key criteria of a living system are:

- *Pattern* of organization: the configuration of relationships that determines the system's essential characteristics.
- *Structure*: the physical embodiment of the system's pattern of organization.
- *Process*: the activity involved in the continual embodiment of the system's pattern of organization.

(p. 161)

To understand a pattern we must map a *configuration of relationships*. In other words, structure involves quantities, while pattern involves qualities.

(p. 81)

Source: Capra, 1996.

|  human being |  indoor environment |  control |  holistic and integrative top-down approach |
|---|---|---|---|
| All health and comfort aspects are important | A sustainable indoor environment that guarantees a high basic level of health and comfort | Performance on demand, anticipating wishes and needs during different activities and over time | End-user focused, multi-disciplinary, life-cycle oriented, sustainable |

**Figure 9.1** *The building as a system*
Source: Bluyssen.

This 'new' approach seems to be not far from the way Capra (1996) and others have introduced 'systems thinking' (see Box 9.2). Together with the ability to create and innovate while remaining practical, integration requires the interactive consideration of people, processes and products within systems (Clements-Croome, 2011). So, the built indoor environment is considered a system with sub-systems that do matter, but the system will only function if all sub-systems (components) are optimized along with the total system, whether this is related to health, comfort or sustainability issues (see Figure 9.1).

Next to that, 'systems thinking' is contextual (Capra, 1996). Translating this thinking to the built environment, this means that situations or scenarios are important. People differ in their responses, environments differ in their conditions. For different scenarios, different interactions occur between different factors of importance. Patterns of factors or indicators seem therefore important to identify, even though those patterns are likely to change over time, creating new patterns. In the words of Capra (1996: p. 80): 'From the systems point of view, the understanding of life begins with the understanding of *patterns*.' To be able to understand the patterns of concern in the built environment, different scenarios thus need to be studied.

Other important aspects of patterns and their interactions are the feedback concept and the fact that the relationships in a network pattern are in general nonlinear. A network pattern is capable of *self-organization*. Again, in the words of Capra (1996, p. 85): 'Self-organization (by self-organizing systems) is the spontaneous emergence of new structures and new forms of behaviour in open systems far from equilibrium, characterized by internal feedback loops and described mathematically by nonlinear equations.' So, if we choose to look at a building (the built environment) as if it were a system, there is most likely a need to describe the relationships in the patterns present with non-linear algorithms.

In this chapter, for different scenarios, studies performed so far in the field are inventoried:

- *Homes and energy efficiency* (see Section 9.2): because in general people spend most of their time in their homes and the optimization of energy efficiency and IEQ is the challenge for the coming years.
- *Children and school environment* (see Section 9.3): because schools are in second place when it comes to time spent indoors for children.
- *Workers and office buildings* (see Section 9.4): because office buildings have been studied the most when it comes to the health and comfort of office workers in relation to building-related and dose–response-related indicators.

Studied associations between both building-related indicators and dose-related indicators, and several occupant-related indicators, are presented for each scenario. The outcome of the inventory shows that, besides the need for more 'unified' and 'broader' assessment protocols and more data, there is a need for other risk modelling and risk analysis than the current mostly used linear modelling. In Section 9.5 several options are presented.

# 9.2 Homes and energy efficiency

*The IEQ in a large part of the residential buildings is subjected to increased health effects, energy efficiency improvements and possibly climate change effects. People spend more than 60 per cent of their time at home. The challenge is retrofitting residential buildings in such a way that while implementing energy-efficiency measures at the same time a healthy indoor environment is accomplished, especially in cities, where the ambient air and noise pollution is the highest.*

## 9.2.1 Mechanisms

### Residential buildings

People spend most of their life indoors, at home: on average nearly 16 hours per day during the week and 17 hours per day during the weekend. For children and elderly people, these figures are even higher: in the range of 19 to 20 hours a day. Research shows an obvious trend for time spent in the home to be increasing, especially with younger and older ages (Bonnefoy *et al.*, 2004).

It is estimated that in the EU 60 per cent of the 210 million buildings are residential buildings (E²APT, 2010), representing 75 per cent of the square metres involved. In square metre percentages, 64 per cent are single family houses and 36 per cent apartment blocks (BPIE, 2011; see Figure 9.2).

The indoor environment in a large part of these residential buildings is subjected to increased health effects, energy-efficiency improvements and possible climate change effects.

### Health effects

Findings of a pan-European housing survey by the World Health Organization clearly indicated a link between indoor environmental housing conditions, including the immediate environment, and human health and wellbeing (Bonnefoy *et al.*, 2004). These data confirmed that the indoor dwelling characteristics that most affect human health are connected to the indoor environmental parameters

a

b

**Figure 9.2** *(a) Neighbourhood with apartment and housing blocks in Delft, The Netherlands; (b) Residential area 'La Mottaz' from the air, Switzerland*

Source: (a) Bluyssen; (b) Claude-Alain Roulet.

thermal comfort, lighting, moisture, mould and noise (see also Section 1.3). Sleep disturbance, linked to a multitude of indoor physical parameters, increased the risk of household accidents by at least 46 per cent (some 350 million Europeans complain regularly about sleeping problems); and obesity and mental disorder seem to have a relation with the conditions of homes and neighbourhoods people are living in together with the increasing time they spend indoors (Bonnefoy et al., 2004).

The WHO estimated that the exposure-risk relationship of *new-onset asthma* in children attributable to mould in their home environment is about 2.4 while for dampness this is 2.2. Put another way, about 12 per cent of new childhood asthma in Europe can be attributable to indoor mould exposure, while for indoor dampness this is about 15 per cent (WHO, 2011h and i). In cold climates 15 per cent of dwellings have signs of dampness problems in general and 5 per cent have signs of mould problems. In warm climates the corresponding estimates are 20 per cent for dampness and 25 per cent for mould. The estimate for water damage is 10 per cent regardless of climate.

Excess *winter deaths* (number of deaths in winter above the average for the previous and subsequent four-month seasons) due to cold homes account for 30 per cent of all excess winter deaths in 11 countries of the WHO European region (WHO, 2011h and i). Deaths from cardiovascular diseases are directly linked to exposure to excessively low indoor temperatures for long periods (50–70% of excess winter deaths are attributed to cardiovascular conditions, and some 15–33% to respiratory disease).

It has been shown that 2–12 per cent of all *lung cancer deaths* in the EU can be attributed to radon exposure, besides the well-known sources, such as (WHO, 2011h and i):

- *Smoking*: exposure to residential second-hand smoke affects non-smokers through respiratory infections and asthma in children and lung cancer and coronary heart disease in adults.
- *Combustion*: carbon-based fuel combustion such as gas and solid fuels can lead to household carbon monoxide poisoning; solid fuel use has been linked to COPD and lung cancer in adults and pneumonia in children.
- *Lead dust*: low-level exposure to lead has cognitive, developmental, neurological, behavioural, cardiovascular and other effects and higher exposure levels can result in acute poisoning.

While increases in the incidence of asthma and allergies worldwide has led to an increase in research on indoor air exposures in residences, outdoor air pollution and noise exposure from traffic, while being indoors, have been pointed out as large contributors to the living environment of cities in particular. As the proportion of the world's population living in urban areas increases (while in 2011 52% of the world population was living in urban areas, in 2050 67% is expected to live in cities; UN, 2012) more attention is required to the quality of urban environment conditions as well. Researchers have realized that most of our exposure to particles of outdoor origin occurs while we are indoors. Consequently, indoor conditions cannot only alter our exposures, but also the associated health risks (e.g. Bell et al., 2009).

## Energy consumption and retrofitting

More than 40 per cent of our residential buildings were constructed before the 1960s, when energy building regulations were very limited. Due to their age most of them require retrofitting or refurbishment. Retrofitting has been identified as the most immediate and cost effective mechanism to reduce energy consumption and carbon emissions in the building and construction sector (E²APT, 2010). Moreover, to meet EU energy performance targets set by the 2007 Energy Action Plan and the 20–20–20 targets adopted, it is necessary to double or triple the current retrofitting rate of 1.2 to 1.4 per cent to reach the short- and long-term goals of an energy reduction of 20 per cent by 2020 and a $CO_2$ emissions reduction of between 80–95 per cent by 2050 (EU, 2010b). The European households consumed in 2009 68 per cent of the total final energy use in buildings, of which space heating is the dominant factor for energy-use (up to 70%) in households, followed by cooling, hot water, cooking and appliances (BPIE, 2011). This is caused by intrinsic building characteristics, such as poor insulation, windows with very high U-values (U = heat transfer coefficient (W/m°C)) as well as the existence of thermal bridges (close to windows and balconies, or between the concrete structure and the brick walls). Improved insulation and increased air tightness of the building envelope can, however, lead to problems of dampness (Adan and Samson, 2011). Moreover, the insulation products applied could cause severe health problems, such as in the case of polyurethane foam (PUR). PUR, a polymer composed of a chain of organic units joined by urethane links, is formed by mixing an isocyanate (in general very reactive) with a polylol and a catalyst. When the forming reaction between those two components is fully established, the PUR is chemically inert. On the other hand, if the reaction is not completed or vapour escapes during the process, problems can arise. While isocyanates are known skin and respiratory sensitizers, the health implications of the also present amines, glycols and phosphates are still to be discovered (CDC-NIOSH, 2012).

## Climate change and indoor environment

Climate changes will affect different aspects of the indoor environment as well as the occupants of that indoor environment (Bluyssen, 2009a):

- The average rise in outdoor temperature is likely to lead to a rise in demand for air conditioning systems on the one hand and a decrease in energy demand for heating systems on the other. The increased change of temperature and humidity conditions caused by the increased precipitation (rain falls), causing changes of humidity conditions indoors, will most likely additionally lead to an increase in the need for highly adaptable buildings (including air conditioning systems).
- As a result of the increased storms and wind speed, air pollution comprising dust particles (from fine to the more heavy) are most likely to be transported more easily from one area to the other. For example, the orange dust coming from the Sahara covering cars after a rainfall will be a more frequent sight in European countries.

- Ozone is another pollutant of concern because elevated ozone outdoor concentrations can lead to an increase in secondary pollution indoors. High temperatures and UV-radiation stimulate the production of photochemical smog as well as ozone precursor biogenic VOCs (Wilby, 2007).
- The primary pollution of building products indoors can be influenced by an increase in indoor air temperature, but this effect is with increases of a few degrees most likely not very important. On the other hand, higher precipitation, both in summer and winter, could result in higher relative humidities and thus better and more living environments for moulds and bacteria.

One of the consequences of the climate change effects might be that people stay inside even more than they already do. Additionally, in an attempt to shut the sun out, blinds and curtains will be used. As a result the daylight exposure is decreased and alternatives have to be introduced to compromise that effect. The effect of storms and cloud forming may also influence the quality of daylight and indirectly the need for artificial forms of lighting that can adapt to changes rapidly.

Increased wind speeds and frequency of storms influence the acoustical quality indoors through the noise and vibrations perceived. Vibrations can be a cause of nuisance and health effects. The increase in use of air conditioning systems may cause more people to complain about noise originating from air conditioning systems.

## 9.2.2 Assessment

With respect to noise it has been acknowledged that a variety of factors influence noise annoyance and thus the effect of noise on health (see Section 3.3.3). In a review of 136 surveys on noise annoyance in residential areas, Fields (1993) reported that annoyance to noise is related to the amount of isolation from sound at home and to five attitudes (fear of danger from the noise source, noise prevention beliefs, general noise sensitivity, beliefs about the importance of the noise source, and annoyance with non-noise impacts of the noise source). Annoyance was not affected to an important extent by ambient noise levels, the amount of time residents are at home, or any of the nine demographic variables (age, sex, social status, income, education, home ownership, type of dwelling, length of residence or receipt of benefits from the noise source).

While health effects from exposure to ambient air and noise from road traffic was already well on the way, studies on health related to housing conditions probably started off in the 1990s when the consequences of the first energy crisis became more tangible in the identified association between dampness, moisture and mould with health effects in population studies in Europe, North America and elsewhere (WHO, 2009). Studies on homes and health performed in the last decades have been focused on:

- Air quality and health, both in homes and cities.
- Energy efficiency and indoor environment quality.

## Air quality and health

In the European study EXPOLIS (1994–98), personal exposure (home indoor and outdoor and workplace levels of $PM_{2.5}$, VOCs and CO) was assessed of approximately 500 subjects representing the adult populations of eight selected cities (Jantunen et al., 1998).

A pan-European housing and health survey was undertaken from 2002 to 2003 in eight European cities, named the LARES Survey (Large Analysis and review of European Housing and Health Status), in which the quality of 3,373 dwellings and the health status of 8,519 inhabitants was assessed in a *holistic* way, using questionnaires and inspection forms (Bonnefoy et al., 2004; WHO, 2007b). The outcome of this European study inspired further studies on homes and health in Europe, national and international, such as the European study AIRMEX (European Indoor Air Monitoring and Exposure Assessment study (2003–08); Kotzias et al., 2009). From AIRMEX, it was concluded that personal exposure concentrations were higher or similar to indoor as well as significantly higher than outdoor ones, home indoor concentrations dominated personal exposures indicating presence of strong *indoor sources* at home.

At the national level, surveys have been performed to collect data on population exposure to indoor pollutants in various indoor environments (dwellings, schools, offices etc.), such as in Germany (www.umweltbundesamt. de/gesundheit-e/survey/index.htm), in the UK (Coward et al., 2001), in The Netherlands (van Dongen and Vos, 2007) and in France (Kirchner et al., 2008). But also in other parts of the world, data has been collected on the indoor air quality in homes (see review in Mendell, 2007).

In a number of national and international studies it was found that the indoor residential risk factors of primary interest for asthma, allergies and respiratory health include (e.g. Norbäck et al., 1995; Øie et al., 1999; Bornehag et al., 2004a; Franchi et al., 2004; Mendell, 2007; see Figure 9.3):

- Allergens such as dust-mites, cockroaches and pet dander.
- Moisture, mould and endotoxin.
- Combustion products from appliances, tobacco or other combustion sources (e.g. outdoors).
- Indoor chemical emissions or emission-related materials or activities including formaldehyde or particleboard, phthalates of plastic materials, and recent painting; renovation and cleaning activities, new furniture, and carpets or textile wallpaper.

Several studies investigated associations between *ventilation rate* and *health effects*. Findings have been inconsistent, partly explained by the complexity of the processes involved (Mendell, 2007): while outdoor air generally dilutes the indoor concentrations of pollutants emitted by indoor sources, this is less true for semi-volatile compounds and some volatile compounds such as formaldehyde, for which lower air concentrations can lead to increased emissions.

In a study of 390 residences in Sweden it was found that buildings with mechanical ventilation have higher air exchange rates than the buildings with natural ventilation (Bornehag et al., 2005b). Multi-family houses had higher mean

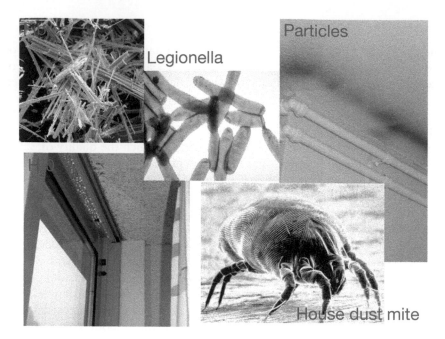

**Figure 9.3** *Some indoor residential risk factors*
Source: Bluyssen.

ventilation rates than other types of homes. Buildings from the 1960s and 1970s had lower ventilation rates than buildings from earlier or later construction periods. However, the type of ventilation system did not affect the risk for recurrent wheezing in very young Swedish children (one to two years old) in a study performed by Emenius *et al.* (2004).

Dimitroulopoulou (2012) identified 96 journal papers on air exchange rates, including specific data from 13 European countries, and he reviewed the current status of residential ventilation standards and regulations in Europe and compared them with the measured ventilation rates. The review showed that ventilation rates are in general lower than 0.5 h$^{-1}$, which is currently a standard in many countries.

Approximately 20 per cent of the European population is allergic to mites and fungi, and the prevalence of asthma and allergies in domestic buildings is increasing (Institute of Medicine, 2000). In the past 30 years, there has been an apparent change in the percentage of asthmatics sensitized to mould. A recent analysis of fungal allergy suggests these percentages are rising (Simon-Nobbe *et al.*, 2008). Asthma can be triggered by exposure to a variety of allergens such as pollen, animal dander, cockroaches, dust mites and moulds as well as various pollutants and irritants, such as environmental tobacco and other factors (e.g. cold air and exercise; Adan and Samson, 2011). According to the WHO, 235 million people suffered from asthma in 2011 (WHO, 2011c).

Residents in approximately 25 per cent of the European social housing stock, i.e. more than 45 million people, are exposed to increased (i.e. 30–50%) health risks associated with exposure to *moulds* (Bonnefoy *et al.*, 2004). A recent investigation of the housing stock shows that some 15 per cent of Dutch dwellings have fungal problems, Belgium reported 20 per cent in social housing, Germany about 30 per cent and the UK earlier indications were that 20–25 per cent was severely affected. Mould growth frequently occurs in indoor environments that are considered relatively dry (on average < 50 per cent relative humidity). In modern dwellings about 60 per cent of problem cases suffer from disfigurement in bathrooms, 40 per cent in kitchens and only few per cent in other rooms (Adan and Samson, 2011).

The fear of mould problems are sometimes justified and sometimes not. According to Adan and Samson (2011) mould growth seems to depend on moisture load and therefore neither thermal nor ventilation forms are a guarantee for a mould-free environment.

High levels of airborne viable fungi have been related to wheezing and persistent coughing among infants with family history of asthma, while high levels of 1,3 beta-glucan in dust protected from wheezing and low levels increased risk for wheezing. Findings with bacterial endotoxin have also been inconsistent: high levels have both been found protective from asthma and a risk for wheezing (Hyvaerinen *et al.*, 2009).

In a review, Mendell (2007) summarized 21 studies in the epidemiologic literature on associations between indoor residential chemical emissions, or *emission-related materials or activities*, and respiratory health or allergy in infants or children. Composite wood materials that emit formaldehyde, flexible plastics that emit plasticizers and new paint have all been associated with increased risks of respiratory and allergic health effects in children (e.g. Diez *et al.*, 2000; Bornehag and Nanberg, 2010). Elevated risks were also reported for renovation and cleaning activities, new furniture and carpets or textile wallpaper. Common residential practices such as using pressed wood furnishings in children's bedrooms, repainting infant nurseries, and encasing mattresses and pillows with vinyl for asthmatic children raised questions.

## Energy efficiency and IEQ

In the *energy efficiency* and *IEQ studies*, such as the European projects EPIQR (Energy Performance, Indoor Air Quality, Retrofit; Bluyssen, 2000) and HOPE (Health Optimisation Protocol for Energy-efficient Buildings) (Roulet *et al.*, 2006b), thermal comfort, light, noise and air quality problems were all considered next to the energy efficiency of the buildings investigated.

In HOPE, which ran from 2002 to 2004, 97 apartment buildings and 59 office buildings from 9 countries, of which approximately 75 per cent were designed to be energy efficient, were studied with questionnaires, checklists and (for a selection) environmental monitoring techniques (see Study F, Box 8.4). From the results it was concluded that low-energy buildings with good indoor environment quality and healthy occupants exist and vice versa (Roulet *et al.*, 2006b). In the HOPE project, 2,750 valid questionnaires from 97 apartment buildings were collected. Building average comfort marks given by occupants on a scale from 1 (satisfactory) to 7 (unsatisfactory) comprised: 3.37 for lighting, 3.09 for comfort overall, 2.95 for air quality, 2.87 for thermal comfort and 2.67 for acoustics (Roulet *et al.*, 2006b). In an earlier study, EPIQR, 26 apartment buildings were investigated with a questionnaire and a checklist (Bluyssen, 2000) for which a similar order of grouping of complaints was found. Highest complaints were found on noise (47.3%) and control of noise (49.1%), followed by thermal comfort problems (draught 45.1%; quality heating 42.5% and control of temperature 42.1%), and air quality complaints (stuffy/bad smell 39.9%; condensation on windows 39.5%; control of ventilation 38.2%). Not being able to control your environment seems an important complaint to be taken seriously. The challenge is to optimize energy efficiency of buildings for the health and comfort of occupants. Next to that occupant behaviour (e.g. opening windows, washing, cooking, sleeping, working (office at home), studying, watching television) is an important aspect as well as occupant needs and requirements, in this road towards energy-efficient and healthy buildings.

### 9.2.3 Analysis

Tables 9.1 and 9.2 show for a number of studies performed for which building-related indicators and dose-related indicators, respectively, associations have been found with the occupant-related indicators studied. From the studies performed, it is clear that associations between dose-related indicators and occupant-related indicators have been difficult to obtain, and/or are often inconsistent. For dampness and mould different characteristics (or conditions) of the built environment are used instead of indoor air measurements of fungi and/or bacteria. But also with that approach difficulties occur. Different definitions of indoor dampness and mould, but also different assessment methods, sometimes make comparison between studies difficult, and other exposures that may induce the same health effect such as asthma are not well known. For example, in the 'Dampness in Buildings and Health' (DBH) study, four dampness indices were associated with the higher prevalence of airway, nose and skin symptoms among 10,851 children between one and six years old (Bornehag *et al.*, 2005c):

- *Water leakage*: flooding and/or water leakage during the previous year or earlier during the child's lifetime, in the child's or parent's bedroom or in the kitchen or bathroom.
- *Floor moisture*: loosening or bubbled or discoloured flooring material in the child's or parent's bedroom or in the bathroom.
- *Visible dampness*: visible mould or damp spots in the child's or parent's bedroom.
- *Condensation on windows*: more than 5 cm of condensation on the inside of the windowpane during the winter.

While in the WHO report (2011i) two types of indicators were defined:

- A *general indicator for 'dampness'*, which includes observations of high relative humidity, condensation on surfaces, moisture/water damage, signs of leaks and stained/discoloured surface materials.
- A *specific indicator for 'mould'* includes observations of visible microbial growth, especially visible mould, and mould odour.

Occupants' perceptions have been the basis for the assessment of dampness/ mould in most population studies (WHO, 2011h). In a study performed by

Tham *et al.* (2007) questionnaire data (from parents), including information on dampness and indoor mould in the rooms where the children sleep, from 4,759 children (age 1.5 to six year) attending 120 randomly selected day-care centres, was obtained. In a study performed by Naydenov *et al.* (2008) similar results were found for 216 pre-school children in the cities of Burgas and Sofia in Bulgaria when the questionnaire data from the parents were applied. However, the non-professional dwelling inspectors reported less visible mould, but more damp stains and a mouldy odour than the parents. Strong significant relationships between parental reports of dampness problems and the health of the children were found in a study among 500 Danish children, whereas analogous relationships were not found using the inspectors' observations (Toftum *et al.*, 2009).

And then there is the question of the other factors that are usually not asked for or looked for in questionnaire and checklists. There is a need to include behaviour and social dimensions in indoor environmental research. For example, in a survey of 40 homes, 70 per cent (28 homes) responded to a questionnaire

Table 9.1 *Framework of associations found between building-related indicators and selected occupant-related indicators for studies performed on homes*

| Occupant-related indicators → <br><br> Building-related indicators ↓ | Annoyance | Sleep disturbance | Cardiovascular diseases | Asthma, allergy and respiratory symptoms | Lung cancer, COPD | Excess winter death | Example references |
|---|---|---|---|---|---|---|---|
| *Air quality* | | | | | | | |
| • Combustion sources | | | | + | | | Mendell, 2007 |
| • Type of ventilation system (natural versus mechanical) | | | | +/− | | | Emenius *et al.*, 2003 |
| • Dampness and moulds (several indices) | | | | + | | | Bornehag *et al.*, 2005c; Toftum *et al.*, 2009 |
| • Furniture (formaldehyde) | | | | + | | | Diez *et al.*, 2000 |
| • New painting | | | | + | | | Diez *et al.*, 2000 |
| • Furnishings (PVC flooring) | | | | + | | | Bornehag and Nanberg, 2010 |
| • Cleaning tools and products | | | | + | | | Mendell, 2007 |
| • Renovation activities | | | | + | | | Mendell, 2007 |
| *Lighting* | | | | | | | |
| *Thermal comfort* | | | | | | | |
| • Control of heating system | + | | | | | | Bluyssen, 2000 |
| *Noise* | | | | | | | |
| • Outdoor sources | + | + | + | | | | See Section 3.3 |
| • Control | + | | | | | | Bluyssen, 2000 |
| *Other factors* | | | | | | | |
| • Energy efficiency | +/− | | | | | | Roulet *et al.*, 2006b |
| • Control of IEQ factors | + | | | | | | Roulet *et al.*, 2006b |

Note: +/− means inconclusive, while + means an association was found, positive or negative.

**Table 9.2** *Framework of associations found between dose–response-related indicators and selected occupant-related indicators for studies performed on homes*

| Occupant-related indicators → <br><br> Dose–response-related indicators ↓ | Annoyance | Sleep disturbance | Cardiovascular diseases | Asthma, allergy and respiratory symptoms | Lung cancer, COPD | Excess winter death | Example references |
|---|---|---|---|---|---|---|---|
| *Air quality* | | | | | | | |
| • ETS, combustion products | | | | + | + | | WHO, 2011h and i |
| • Ventilation rate | | | | +/– | | | Mendell, 2007 |
| • Fungi/endotoxin | | | | +/– | | | Hyvaerinen *et al.*, 2009 |
| • Formaldehyde | | | | + | | | Diez *et al.*, 2003 |
| • Phthalates | | | | +/– | | | See Section 3.4.3 |
| • VOCs | | | | | | | |
| • Radon | | | | | + | | WHO, 2011h and i |
| • Particles | | | | + | + | | See Section 3.5.4 |
| *Lighting* | | | | | | | |
| *Thermal comfort* | | | | | | | |
| • Indoor temperature | | | | | | | |
| • Outdoor temperature | | | | | | + | WHO, 2011h and i |
| • Relative humidity | | | | | | | |
| *Noise* | | | | | | | |
| • Noise level outdoors | – | | + | | | | Fields, 1993; Babisch, 2008 |

Note: +/– means inconclusive, while + means an association was found, positive or negative.

focused on occupants' overall satisfaction with their new low-energy houses, perceived indoor climate summer and winter, technical installations, regulation of the indoor climate summer and winter, practice of opening windows, availability and quality of information and heat consumption (Knudsen *et al.*, 2012). From the outcome it was concluded that besides from good information and communication on how to operate the technical installations, more robust and easy-to-use technical installations are required to enable the occupants to control the indoor climate and energy consumptions as intended.

Engvall *et al.* (2004) therefore developed a validated self-administered questionnaire for the occupants and a checklist for the building specifically for residential indoor investigations. This questionnaire is based on three categories of variables: population, indoor-environment-related complaints and symptoms, and health status and symptoms. In theory one could correct the results for the modifiers and/or confounders of interest. But even then the question remains how valid the outcome eventually is, when only one risk factor and one health outcome is considered (see Box 9.3).

---

**Box 9.3 Interrelationships**

The relationship of a single housing risk factor with one or more health outcomes is presented. . . . This approach is likely to underestimate the true extent of the relationships, as many dwellings have more than one health-threatening defect. Some conditions are directly linked; for example, energy-inefficient dwellings with low indoor temperatures will be prone to dampness and mould. For others, the presence of one condition may have a synergistic effect increasing the risk from another exposure, such as the parallel presence of tobacco smoke and radon, which strongly increases the risk of lung cancer. It is important to recognize the possible interrelationships among the various aspects of housing.

Source: WHO, 2011h, p. 2.

---

## 9.3 Children and school environment

*Children spend more time in schools than in any other place except at home. Unsatisfactory environmental conditions can have both short-term and long-term health effects, and can affect productivity or learning ability of the children.*

### 9.3.1 Mechanisms

#### Education and buildings

Since the first century AD, formal education compulsory for children from the age of six or seven already existed in the Western world with the formal Jewish education in instituted schools in cities, towns and villages (for more information see http://en.wikepdia.org/wiki/Compulsory_education). Nowadays in most countries all over the world children, from the age of 5 to 8 until the age of 16 to 18, are obliged to follow an education (Table 9.3). But in most countries, infants and young children already attend some kind of day-care or pre-school before the age of five.

Children spend more time in schools than in any other place except at home. While in the US more than 56 million children attend 125,000 public and private pre-school, primary and secondary schools (Geller *et al.*, 2007), in Europe more than 71 million children and almost 4.5 million teachers spend many hours per day inside schools (Rive *et al.*, 2011). The type of buildings these children and teachers stay in varies a lot: from one-level concrete or wooden buildings to several-floor brick stone square buildings (see Figure 9.4).

**Table 9.3** *Approximate categories of ages of children and types of schools they attend*

|  | 0–3 years | 4–5 years | 6–11 years | 12–18 years |
|---|---|---|---|---|
| Day-care/nursery/kindergarten | X |  |  |  |
| Pre-school |  | X |  |  |
| Elementary/primary school |  |  | X |  |
| Secondary/high school |  |  |  | X |

**Figure 9.4**
*A variety of school buildings: (a) east side of the Scuola dell'Infanzia 'G. Rodari' – Verona, Italy; (b) Delftsche School-vereeniging in Delft, The Netherlands; (c) Marquesa de Aloma School, Lisboa, Portugal*

Source: (a) Prof. Arch. Pietro Zennaro, colour designer, University Iuav of Venice; (b) Bluyssen; (c) João Pernão, Faculty of Architecture, Lisbon Technical University, Portugal.

a

b

c

## History

In the eighteenth century, in the Western world, the single-room house was perhaps the typical school classroom (Wu and Ng, 2003). Influenced by urbanization and industrialization, provision of separate classrooms for children with different abilities and age arose. From 1900 to 1930, an open-air school movement stressed aspects of health and welfare and placed emphasis on better ventilation and increased daylight. An open-air school required a garden site and classrooms that could be opened completely on one side. Additionally, after World War I, innovations of construction technology such as the use of steel framing made it possible to make use of maximum glazing area in schools. Large windows, however, have many drawbacks, such as glare, uncomfortable temperature conditions (both overheating in summer and undercooling in winter), which led to a reassessment of the use of daylighting in schools.

Reduction of window size, air conditioning and use of artificial lighting led to classrooms with very little daylight and even windowless schools. These school designs were referred to as 'exclusive', striving to exclude the effects of the environment, and are different from 'selective' designs, which seek to maximize the use of ambient energy in the form of solar gain and daylighting (Hawkes, 1982; see Figure 9.5). Scientists suggested that windowless schools should be used with caution, particularly since long-term effects were unknown (Wu and Ng, 2003).

Since the oil crisis in the 1970s, the passive-solar school design, a 'selective' model of environmental buildings, which permits direct sunlight penetration into the classrooms with careful controls, has been used (Wu and Ng, 2003). However, the large windows on the southern facade for maximizing solar gain for heating require a careful balancing of the visual aspects of glazing (both positive and negative, e.g. glare) and the interior thermal effects (solar gain and overheating).

Already by the mid 1990s, the US Department of Education reported that around 50 per cent of schools were experiencing unsatisfactory environmental conditions, such as poor ventilation, inadequate heating or lighting, or noise (Geller *et al.*, 2007). Those conditions can have both short-term and long-term *health effects*, and can affect the productivity or *learning ability* of the children.

**Figure 9.5**
*School classroom in Fransisco Amuda School, Lisboa, Portugal*
Source: João Pernão, Faculty of Architecture, Lisbon Technical University, Portugal.

Daisey *et al.* (2003) concluded, in a review of over 300 peer-reviewed publications, that there is a clear indication that classroom ventilation is typically inadequate and levels of specific allergens are sufficient to affect sensitive occupants. SBS and asthma were the most commonly reported health symptoms investigated in the schools. Of particular concern is the potential for increased risks of contracting certain communicable respiratory illnesses, such as influenza and common colds in classrooms with low ventilation rates (Fisk, 2001).

School size, class size and furniture arrangements influence the so-called crowded conditions or crowding. *Crowding* has an effect on privacy and on over-stimulation, influencing the school performance. A learning situation requires a high level of attention. Too much distraction from uncontrolled constant interaction with others, as a result of large group size or high spatial density, may influence this attention. Children who have attention-deficit hyperactivity disorder or autism-spectrum conditions are particularly vulnerable to crowding and thus overstimulation (Geller *et al.*, 2007).

But children are not only exposed at school, they spend a large part of their time at *home* as well. Risk factors identified most frequently include formalde-hyde or particleboard, phthalates or plastic materials, and recent painting (Mendell, 2007). Besides those risks, ETS (environmental tobacco smoke) exposure is an important risk factor (Crain *et al.*, 2002; see also Section 9.2.2).

The location or *site* at which the school is built is also an important issue. School buildings are often located on land that was or is available, preferably not too expensive. It is not uncommon that schools are built nearby major highways, increasing the exposure to traffic exhaust, or built in close proximity to landfills (contaminated; Geller *et al.*, 2007).

## Health effects

The prevalence of asthma and allergy among children has increased rapidly in the last decades. Respiratory allergies are very common and increasing throughout Europe affecting between 3 and 8 per cent of the adult population, and the prevalence being even higher in infants. Up to one out of five children in Europe has asthma, and this is expected to increase (WHO, 2007c). These diseases are the major causes of days lost from school. Asthma is a principal cause of school absences from chronic illness, responsible for 20 per cent of absences in elementary and high schools (Richards, 1986). According to Etzel (2007), this increase cannot be explained by genetic changes, and it is more likely to be due to changes in environmental exposures and/or in life style. The DBH study in Sweden in which a cross-sectional postal questionnaire was replied to by parents of 10,851 children, aged one to six years, found that attending day-care was associated with an increased risk of symptoms related to airways infections as well as eczema and allergic reactions to food (Hågerhed-Engman *et al.*, 2006). Children attending day care have an increased morbidity in airways infections and respiratory symptoms compared with children that stay at home (Nafstad *et al.*, 1999). Hågerhed-Engman *et al.* (2006) found that the higher risk for allergic diseases attending day-care centres should be explained by other factors than frequent infections. Building ventilation rate, quality and frequency of cleaning and other building-related factors were suggested. The presence of allergens from *furred pets* could be another possibility.

While poor air quality or rather air pollution can have a direct effect on health as well as performance, the other three indoor environment parameters (noise, poor lighting and unacceptable thermal comfort conditions) have a more indirect effect (via comfort related responses), but can have a clear effect on the learning performance and attendance of students as well (Mendell and Heath, 2005).

## Susceptibility

Children may be more susceptible than adults to the effects of toxic exposure because they are in the process of maturation and have higher metabolic demands as well as higher minute ventilation in early childhood (Mendell and Heath, 2005; Geller *et al.*, 2007). Additionally, infants and children in general have different activity patterns than adults. *Detailed age related health risk analysis* seems therefore necessary (WHO, 2006).

The highest susceptibility to chemical compounds can be observed in newborns and infants during the first weeks of life (Scheuplein *et al.*, 2002). Children have much higher lung area per body mass ratio than adults. On average, a one-year-old child inhales 7 ml per minute per kilogram of body weight (ml/min/kg), whereas an adult inhales 3–5 ml/min/kg. The amount of air passing through a child's lungs is therefore two times larger in volume. During maturity, the maturation process of reproductive and endocrine systems along with the nervous system occur very slowly and remain the most susceptible to pollutants' disruptive effect (Czernych *et al.*, 2012).

Another issue is the diagnosis of asthma. For young children (below the age of five) it is difficult to perform lung function testing and diagnose asthma. (Asthma is diagnosed with changes in lung function or response to treatment.) Young children have very small, narrow airways and often cough and wheeze with colds and chest infections, but this is not necessarily asthma (Etzel, 2007).

A 13-year follow-up study showed that children in large day-care centres suffered more often from common colds than children in home care, but during the early school years earlier day-care attendance was a protective factor for the common cold until about 13 years of age (Ball *et al.*, 2002).

Studies have also shown that young children (under about 13 years of age) are far more susceptible to poor acoustic conditions than adults (Shield and Dockrell, 2003). Children and adults who are hearing impaired (due to either permanent damage to their hearing or a temporary condition such as a cold or ear infection) are even more seriously affected by noise. It is estimated that at any one time up to 40 per cent of children in primary school class in the UK or US may have some form of hearing impairment (Shield and Dockrell, 2003).

## 9.3.2 Assessment

### Studies on schools

A number of problems related to the indoor school environment that are likely to have an effect on the health and performance of children and teachers have been identified in several mostly cross-sectional European studies, such as the HESE (Health Effect of School Environment) study (Simoni *et al.*, 2006) and the SEARCH project (Csobod *et al.*, 2010) – but also in diverse national projects, for example, BIBA in Belgium (VITO, 2010; Stranger *et al.*, 2008), a nationwide survey in France (Mercier *et al.*, 2011), EscoLAR in Portugal (Madureira *et al.*, 2012), in The Netherlands (Dijken *et al.*, 2006) as well as in Finland (Haverinen-Shaughnessy *et al.*, 2012). In these studies, health effects have been assessed by using self-administered questionnaires, and in a few also medical examination was applied (e.g. for assessing lung function, nasal patency or acoustic rhinometry). Indoor environment conditions were assessed by monitoring air pollutant concentrations (e.g. VOCs, $NO_2$, $O_3$, CO, $CO_2$, $SO_2$, PM, allergens, bacteria and moulds), inspecting the buildings with the use of a checklist (e.g. building characteristics and measures) and several physical measurements (e.g. temperature and relative humidity; see also Chapter 6). Indoor air quality (e.g. ventilation systems, dampness, indoor and outdoor air pollutants), thermal comfort, lighting conditions and performance, but also lack of awareness came out as important issues.

The exposure of children to indoor air pollutants and the effect on their health has been investigated more broadly, taking account of exposure in their homes as well as in their schools, in the Danish Indoor Environment and Children's Health (IECH) study (Clausen *et al.*, 2012), largely inspired by the Swedish DBH study (Bornehag *et al.*, 2004b; see Section 9.2.2). The original DBH study was also carried out in Bulgaria (Naydenov *et al.*, 2008) and Singapore (Tham *et al.*, 2007). A large epidemiological study on children's health was performed in Japan (Yoshino *et al.*, 2009), as well as in Australia (Marks G.B. *et al.*, 2010) and in China (Mi *et al.*, 2006). The purpose of those studies performed varied, as well as the age of the children involved.

### $CO_2$, ventilation and temperature

As explained in Section 8.2.1, with the exhalation of *people* indoors, $CO_2$ but also other compounds are exhausted into the indoor air. Many studies have found $CO_2$ concentrations above 1,000 ppm (concentration used as a 'dividing line' between adequate and inadequate ventilation (Shendell *et al.*, 2004)). But the findings with respect to the effects on health so far are inconclusive.

While Shendell *et al.* (2004) showed that a 1,000 ppm increase in $CO_2$ was associated with a 0.5–0.9 per cent decrease in annual average daily attendance, corresponding to a relative 10–20 per cent increase in student absence (of 436 classrooms from 22 primary and secondary schools in the states of Washington and Idaho). In a large study of asthma symptoms and exposure in 1,476 school children in 39 Swedish schools (Smedje and Norbäck, 2000), no statistically significant relationships were found (at 95 per cent confidence interval) between asthma and many commonly measured environmental factors, for example, $CO_2$, air exchange rates, humidity. No evidence suggested that elevated $CO_2$ levels were restricted to complaint schools.

And then Haverinen-Shaughnessy *et al.* (2011) found a linear association between classroom ventilation rates and student's academic achievement within the range of 0.9–7.1 l/s for fifth-grade classrooms of 100 elementary schools (one per school with an average of 21 students) of two school districts in the south-west United States. For every unit (1 l/s person) increase in the ventilation rate within that range, the proportion of students passing standardized test (i.e. scoring satisfactory or above) increased by 2.9 per cent for maths and 2.7 per cent for reading.

In a similar study performed in Finland with sixth grade students of 334 schools, maths achievement was associated with missed school days due to respiratory infections, headache, difficulties in concentration, and indoor temperatures perceived as too high in the classroom (Haverinen-Shaughnessy et al., 2012). From both studies, it was concluded that the proportion of gifted students in a classroom may be an important factor in group level analyses.

In a series of field experiments by Wargocki and Wyon (2007) the performance of schoolwork by 10- to 12-year-old children was measured during week-long experimental periods of improved classroom air quality (by increasing outdoor air supply rates from 3 to 9.5 l/s per person) and during weeks in which moderately elevated classroom temperatures were avoided (by cooling from about 24–25 to 20°C). The results showed that doubling the ventilation rate would improve school performance by 8–14 per cent while reducing the temperature by 1°C would improve it by 2–4 per cent, depending on the nature of the task.

It is clear that minimum ventilation rates based on mainly body odour (with $CO_2$ as an indicator) and to some extent on primary emissions from some building materials are not preventing occupants and visitors of a space developing health symptoms (asthma, etc.) and/or comfort complaints (odour, irritation). This is particularly the case for children and schools. It is also clear that the school environment, indoor air quality in particular, may affect the health of occupants of schools, school children and teachers. However, few comparable data are available.

## Pollution sources other than people

Besides children (students), several other sources of pollution can contribute to indoor air pollution, such as building and furnishing materials, outdoor air (site of school building), cleaning techniques, ventilation systems and maintenance.

In a study of 579 asthmatic Danish children, Hansen et al. (1987) reported an increased severity of asthma in schools with carpet compared with those with no carpet. Csobod et al. (2010) found an increased prevalence of children woken by wheezing at night and carpets on the floor in classrooms.

Textile materials, such as mattresses, pillows, curtains and carpets, found in day-care centres or nurseries, can act as a reservoir for dust mites and allergens (e.g. from children who have a cat and/or dog at home). Cat allergen has been linked to asthma and current asthma risk (Smedje and Norbäck, 2001), dog allergen with current wheeze and breathlessness during daytime (Kim et al., 2005).

Formaldehyde can be emitted from urea formaldehyde foam insulation, glues, fibreboard, pressed board, plywood, particle board, carpet backing and fabrics. Formaldehyde concentration has, for example, been related to cumulative asthma and nocturnal attack of breathlessness (Zhao et al., 2008), and to asthma among children without history of atopy (Smedje and Nörback, 2001).

The effects of *moisture-damage* repairs on microbial exposure and symptoms in over 1,300 school children aged 6 to17 of two moisture damaged and two non-damaged schools was studied by Meklin et al. (2005) by questionnaires before and after repairs. A significant decrease of symptoms was observed among the school children in one of the damaged schools. Haverinen-Shaughnessy et al. (2012) studied the occurrence of moisture problems in three countries from different climate regions of Europe. They found that the prevalence of moisture problems in school buildings was the highest in Spain (most common: moisture/water damage), but lower and similar in Finland (most common: mould odour) and The Netherlands (most common: dampness).

Epidemiological data, supported by experimental studies, point to a possible correlation between phthalate exposure (e.g. from PVC flooring) and *asthma and airway diseases* in children, indicating a role for phthalates in the early mechanisms of the pathology of allergic asthma (Bornehag and Nanberg, 2010).

In 60 schools (242 classes) in 6 participating countries (Albania, Bosnia and Herzegovina, Hungary, Italy, Serbia and Slovakia) health data from 5,242 children showed an association between heavy traffic and industry and decreased lung function (Csobod *et al.*, 2010). Mi *et al.* (2006) performed a cross-sectional study in 30 classes of 10 junior high-schools situated in 2 districts of Shanghai (in China there is a 6-year primary school followed by a 3-year junior high school). From the answers to the questionnaires of 1,414 pupils with a mean age of 13 years, and indoor and outdoor air pollution measurements, followed that ambient air pollution of $SO_2$, $NO_2$, $O_3$ and particles caused by industry or traffic are a major problem to the indoor environment in schools. Current asthma was associated with outdoor $NO_2$. Some studies on the effects of outdoor pollutants suggest that absence from school increased with higher outdoor concentrations of ozone, nitrogen oxides and carbon monoxide (e.g. in: Park *et al.*, 2002; Chen *et al.*, 2000), although Chen *et al.* (2000) found a negative correlation with $PM_{10}$.

Also applying *unhealthy or no cleaning techniques* can have an effect. For example broom use for cleaning classrooms was associated with increased prevalence of children with a chronic cough with phlegm, and mop use with bleach was associated with an increased prevalence of skin rash and eczema (Csobod *et al.*, 2010). Wällinder *et al.* (1999) showed that removal of settled dust by vacuum cleaning can also have an impact on children's health.

From studies on components of HVAC-systems (Bluyssen *et al.*, 2003), it is clear that HVAC systems can contribute to indoor air pollution, specifically bad maintenance, and therefore can also form a risk to the health and performance of children at schools. However, this has not been investigated yet in schools.

Because outdoor air pollutants can be a major problem indoors, air cleaning and air filtration have been applied as possible strategies to reduce the effects. For example, Wargocki *et al.* (2008) found in a field intervention study that electrostatic air cleaners had an effect on particle concentration but not on symptoms, environmental perceptions or school performance in the pupils tested. On the other hand, Mattsson and Hygge (2005) observed that pupils sensitive to pollen and pet allergen had less irritation when electrostatic air cleaners were active (during pollen season).

Very few studies have been performed on the relationships between type of ventilation system applied and performance, attendance or health effects in schools. For example, Norbäck *et al.* (2011) studied respiratory symptoms, perceived air quality and physiological signs in elementary school pupils in relation to *displacement and mixing ventilation system* (see Box 9.4).

In an intervention they changed from mixing ceiling ventilation to displacement ventilation in one classroom, while two others were kept unchanged. All

---

**Box 9.4  Displacement versus mixing ventilation**

- General advantage displacement over mixing: higher air-change efficiency (depending on the occupancy distribution; with increased movement in the room the plug flow can be disturbed and the concentration distribution moves towards a mixed situation).
- General disadvantage displacement over mixing: thermal discomfort (temperature gradient might lead to cold discomfort and draught around the feet because of lower air supply temperature).

---

classrooms had floor heating. The results showed that displacement ventilation improved perceived indoor air quality and dyspnoea was less compared to controls. There was no effect of displacement ventilation on thermal discomfort or draught, perhaps due to floor heating.

Zuraimi *et al.* (2007) observed an effect on health of different ventilation systems. There were more symptoms when there was hybrid ventilation. Marks *et al.* (2010) showed in a double-blind, cluster-randomized crossover study in 400 primary school students attending 22 schools in New South Wales (Australia) that exposure to unflued gas heaters compared with flued gas heaters was associated with an increased risk of symptoms (e.g. cough and wheeze). There was no evidence of an adverse effect on lung function. In New South Wales, unflued gas heaters have been the principal form of heating used in schools for many years.

## Lighting

*Lighting* is a key feature of the school environment. Since the 1990s, a couple of studies have been conducted.

In 1992, Küller and Lindsten found a significant correlation between patterns of daylight level, hormone pattern and students' behaviour. They studied eight-year old students in four classrooms over one year in Sweden, and concluded that 'work in classrooms without daylight may upset the basic hormone pattern, and this in turn may influence the children's ability to concentrate or cooperate, and also eventually have an impact on annual body growth and sick leave' (Küller and Lindsten, 1992, p. 316).

A direct correlation between student learning and lighting quality was demonstrated in 1999 by the Heschong Mahong Group: students with more daylight in their classrooms progressed more than 20 per cent faster in maths and reading skills than their counterparts in classrooms without daylight (Heschong Mahong Group, 1999; Heschong, 2002). Three large school districts with a range of day lighting conditions in their classrooms and different climates in the US were selected. From 8,000 to 9,000 students (second through fifth graders) per district, test scores and demographic information were collected. For a sample of the schools in each district day lighting conditions were classified by day lighting experts in more than 2,000 classrooms with the use of codes (see Table 9.4).

**Table 9.4** *Qualitative codes applied to assign day-lighting conditions for each classroom*

| 5 | Best day lighting | Classroom is adequately lit with day lighting for most of the school year. Adequate daylight available throughout classroom. |
|---|---|---|
| 4 | Good day lighting | Classroom has major daylight component, and could occasionally be operated without any electric lights. Noticeable gradient in illumination levels. |
| 3 | Average condition | Classroom has acceptable daylight levels directly next to windows or under skylights. Strong illumination gradient. Some electric lights could occasionally be turned off. |
| 2 | Poor day lighting | Illumination is always inadequate without electric lights. Glare a likely problem. |
| 1 | Minimal day lighting | Small, token windows or top lighting. |
| 0 | | Classroom has no windows or top-lighting. |

Source: Heschong Mahong Group, 1999.

In one of the school districts it was found that students in classrooms with the largest window areas were found to progress 15 per cent faster in maths and 23 per cent faster in reading that those with the least window areas. The other two school districts only had one standardized test at the end of the school year. For these districts, students in classrooms with the most window area or day lighting were found to have 7 per cent to 18 per cent higher scores on the standardized tests than those with the least window area or day lighting. What is important to mention is the fact that the models applied only explained about 25 per cent of the natural variation in student performance. The other 75 per cent of unexplained variation might be purely random or explained by other factors not included in their models, such as teacher quality, home, life, health, nutrition, motivation, and so on.

While Hathaway (1992) suggested there is a correlation between absenteeism and lighting, the Heschong Mahong Group (1999) found that student attendance was not affected by lighting.

Light has an effect on sight, but also on mood and general health (see Section 3.3.2). Appropriate lighting requires a certain brightness as well as even glare-free lighting of a balanced spectrum. Winterbottom and Wilkins (2009) investigated 90 classrooms across 11 secondary schools in the UK. They measured flicker, illuminance at desks and luminance at whiteboards. Results showed that 80 per cent of classrooms are lit with 100 Hz fluorescent lighting (which is said to be able to cause headaches and impair visual performance). Glare was induced by daylight and fluorescent lighting, interactive whiteboards and dry-wipe whiteboards, and patterns from venetian blinds. Excessive lighting (over 1,000 lux) was caused by daylight and fluorescent lighting.

## Noise

*Noise* can also affect learning. Noise adversely impacts the ability of teachers and students to communicate, as well as students' attention, memory and thus motivation and academic achievement (Geller *et al.*, 2007). In addition noise can cause annoyance and stress responses (see Section 3.2.4).

The majority of effects of noise on children's performance in the classroom has been focused on external (environmental) noise, of which much of the work has been concerned with exposure to aircraft noise. Both chronic and acute exposure may adversely affect performance (Shield and Dockrell, 2003). External noise has, for example, been shown to affect children's recall in a study of 1,358 children aged 12–14 years in their own classrooms using standard tests, but under different noise conditions (quiet: ambient noise; noisy: 55 and/or 66 dBA $L_{eq}$; Hygge, 2003). The children's recall and recognition of a text a week later was tested and a statistically significant decline in performance associated with noisy conditions caused by aircraft and road traffic was found. Train noise and verbal noise did not affect recall or recognition, while some of the pairwise combinations of aircraft noise with train or road traffic did interfere.

Development of reading skills when chronically exposed to aircraft noise was studied by Evans and Maxwell (1997). They concluded that the harmful effects of noise are related to chronic exposure rather than interference effects during the testing session itself. The adverse correlation of chronic noise with reading was partially attributable to deficits in language acquisition.

Besides external noise transmitted through the building facade to a classroom, noise inside a classroom may include noise from teaching equipment (computers, projectors and so on), noise from building services in the classroom, and noise transmitted through the walls, floor and ceiling from other parts of the school, but most importantly noise generated by the pupils themselves (Shield and Dockrell, 2003). Several recent studies have investigated the internal noise on children's performance. For example, Maxwell and Evans (2000) found in a study of pre-school children who had been exposed to levels in the classroom of 75 dB(A) compared to children in acoustically treated classrooms, thereby reducing background noise levels and reverberation times, that children performed better in the acoustically treated rooms. However, in a study of older children (aged 13 and 15), a poor correlation was found between sound level and standard of work during maths classes (Lundquist *et al.*, 2000).

Air conditioning and ventilation systems contribute to the level of classroom noise. However, when occupied no difference was detected in noise levels with classrooms without heating and ventilating noise (Shield and Dockrell, 2004). The effects of noise on the performance of children has been studied more than the annoyance experienced by noise. With adults, dose–response relationships between noise and annoyance have been established, but for children few studies have been concerned with annoyance (Shield and Dockrell, 2003).

## Other factors

Other factors studied that can have an effect on wellbeing and performance of children in classrooms are:

- *Class size* (Nye *et al.*, 2000): small classes appear to benefit all kinds of students in all kinds of schools.
- *Seating location* (Montello, 1988): although class participation seems to be influenced by seating location, seating location doesn't seem to influence course achievement of college students in lecture-style classrooms.

- *Arrangement of seats* (Woolner *et al.*, 2007): the arrangement of desks and chairs has been studied a lot, however, how it affects the performance of pupils depends on the type of task, the level of the pupils involved and required involvement.
- *School building condition* (Durán-Narucki, 2008): in run-down school facilities, students attended less days on average and had lower grades in a sample of 95 elementary schools in New York City.
- *Outdoor school environment* (Ozdemir and Yilmaz, 2008): health outcomes (e.g. BMI) of primary school students seem to be related to the landscape features and physical qualities of schoolyards.
- *Teacher–student interaction* (Baker, 1999): school satisfaction is affected by perceptions of a caring, supportive relationship with a teacher and a positive classroom environment.
- *Classroom furniture* (Panagiotopoulou *et al.*, 2004): ergonomic dimensions of chairs and desks affect the sitting posture of the children and thus their general wellbeing.
- *Colours* (Engelbrecht, 2003): depending on the age of children different colours are considered stimulating; younger children prefer bright colours and patterns, while adolescent prefer more subdued colours (see Section 3.2.4).

### 9.3.3 Analysis

It is difficult to come to strong conclusions about the impact of learning environments because of the multi-factorial nature and the diverse and disconnected nature of the research performed: which tends to focus on elements and fails to synthesize understandings (e.g. interactions between different factors). Other risk factors and confounding factors, such as age and exposure in the home environment of the children, make it even more complex (WHO, 2007c). Schools are systems in which the environment is just one of the many interacting factors (Woolner *et al.*, 2007).

In Tables 9.5 and 9.6, an overview of the building-related indicators and dose-related indicators, respectively, is presented, for which an association or correlation has been found with the occupant-related indicators mostly studied: 'asthma, allergy and respiratory symptoms', 'learning ability and performance', 'absence' and 'discomfort'. From Table 9.6 is clear that causal relations between dose-related indicators in classrooms and the performance of students have not been established or are inconclusive, except for temperature in the classroom. On the other hand, for both the dose-related indicators and the building-related indicators, in the air quality group, several associations have been established with asthma, allergy and respiratory symptoms. This is mostly related to the fact that the majority of studies performed have been focused on this association. However, it must be noted that the outcome of the selected studies performed are difficult to compare because the studies had a different set-up (different protocols, different measurement techniques and/or different factors indicators, other factors and end-points).

**Table 9.5** *Framework of associations found between building-related indicators and selected occupant-related indicators for studies performed on children and schools*

| Building-related indicators ↓ | Asthma, allergy and respiratory symptoms | Learning ability and performance | School absence | Discomfort | Examples of references |
|---|---|---|---|---|---|
| *Air quality* | | | | | |
| • Outdoor sources (traffic, landfills) | + | | | | Csobod et al., 2010 |
| • Dampness and moulds | + | | | | Meklin et al., 2005; Haverinen-Shaughnessy et al., 2012 |
| • Furnishings (carpeting – VOCs, PVC-flooring phthalates and dust) | + | | | | Hansen et al., 1987; Bornehag and Nanberg, 2010; Csobod et al., 2010 |
| • Flued vs unflued heating | + | | | | Marks et al., 2011 |
| • Displacement vs mixing system | + | | | | Nörback et al., 2011 |
| • Electrostatic vs bag filter | +/– | +/– | | | Wargocki et al., 2008; Mattson and Hygge, 2005 |
| • Cleaning tools and products | + | | | | Csobod et al., 2010; Wällinder et al., 1999 |
| *Lighting* | | | | | |
| • Window area | | + | | | Heschong, 2002 |
| • Patterns of light | | + | | | Küller and Lindsten, 1992 |
| • Excessive daylight | | | | + | Wu and Ng, 2003 |
| *Thermal comfort* | | | | | |
| • Overheating | | | | + | Haverinen-Shaughnessy et al., 2012 |
| *Noise* | | | | | |
| • Road traffic | | + | | | Hygge, 2003 |
| • Aeroplanes | | + | | | Hygge, 2003; Evans and Maxwell, 1997 |
| • Internal noise | | +/– | + | | Maxwell and Evans, 2000; Lundquist et al., 2000 |
| *Other factors* | | | | | |
| • Crowding (children per class or per m2) | + | | | | Nye et al., 2000; Geller et al., 2007 |
| • Condition of building | | + | + | | Duran-Narucki, 2008 |
| • Age of children | + | + | | | WHO, 2006 |
| • Colour of internal walls | | + | | | Engelbrecht, 2003 |

Note: +/– means inconclusive, while + means an association was found, positive or negative.

**Table 9.6** *Framework of associations found between dose-related indicators and selected occupant-related indicators for studies performed on children and schools*

| Occupant-related indicators → <br><br> Dose-related indicators ↓ | Asthma, allergy and respiratory symptoms | Learning ability and performance | School absence | Discomfort | Examples of references |
|---|---|---|---|---|---|
| *Air quality* | | | | | |
| • Particles | + | | +/− | | Chen *et al.*, 2000 |
| • Ventilation rate/child/$CO_2$ | +/− | +/− | | | Shendell *et al.*,2004; Smedje *et al.*, 2000; Haverinen-Shaughnessy *et al.*, 2011; Wargocki and Wyon, 2007 |
| • Fomaldehyde | + | | | | Zhao *et al.*, 2008; Smedje and Nörback, 2001 |
| • Allergens | + | | | | Smedje, 2001; Kim *et al.*, 2005 |
| • $NO_2$ | + | | + | | Mi *et al.*, 2006; Chen *et al.*, 2000; Park *et al.*, 2002 |
| • Phthalates | + | | | | Bornehag and Nanberg, 2010 |
| *Lighting* | | | | | |
| • Illuminance | | | + | | Winterbottom and Wilkins, 2009; Wu and Ng, 2003 |
| • Glare | | | + | | Winterbottom and Wilkins, 2009; Wu and Ng, 2003 |
| *Thermal comfort* | | | | | |
| • Temperature in classroom | | + | | | Wargocki and Wyon, 2007 |
| • Humidity | − | | | | Smedje *et al.*, 2000 |
| *Noise* | | | | | |
| • Noisy conditions | | +/− | | | Hygge, 2003 |
| • Background noise levels and reverberation times | | +/− | | | Maxwell and Evans, 2000; Lundquist *et al.*, 2000 |
| *Other factors* | | | | | |
| • ETS at home | + | | | | Crain *et al.*, 2002 |

Note: +/− means inconclusive, while + means an association was found, positive or negative.

## 9.4 Workers and office buildings

*Offices show an almost identical character no matter where you are in the world. Nowadays, offices usually are highly equipped with electronics, artificial lighting and more and more mechanical ventilation. Flexible open arrangements and high density of occupants seem to go hand in hand with increased health symptoms and productivity loss.*

### 9.4.1 Mechanisms

#### Office work and buildings

The first office buildings appeared in the mid-1800s in Europe, providing rooms for rent to small firms such as lawyers, brokers and bankers to conduct their business (Sundstrom and Sundstrom, 1986). With the growth of organizations into large enterprises during the late 1800s, the need for more office space resulted in building structures with many stories, first expanding sideways but soon going upwards due to lack of space. The highest office building, named a skyscraper, today is the 829.84-m-tall Burj Khalifa in Dubai, United Arab Emirates. Skyscrapers are increasingly common where land is expensive. Skyscrapers are considered symbols of a city's economic power and define the city's identity and skyline (see Figure 9.6). While the first skyscrapers (multi-storey building of at least 100 m) comprised steel frames and curtain walls of glass or polished stone, today they can be built almost entirely with reinforced concrete.

**Figure 9.6**
*Skyscrapers defining the skyline*
Source: painting Guusje Bluyssen.

**Figure 9.7**
*(a) (left) A skyscraper and (b) (opposite) a relatively old, storied building (built just after World War II), both situated on the same square, central station of Rotterdam*

Source: Bluyssen.

a

b

In many countries today, office buildings are not necessarily skyscrapers. Several types of office buildings are situated near to each other (see Figure 9.7). Location is an important factor.

## Health effects

With the development of the office building, the indoor environment conditions have changed considerably during the past 150 years. The first problems were related mainly to poor lighting, bad air (e.g. smoke and body odours) and temperature, influencing both productivity and sickness rates. Technology developments, such as the electric light bulb by Edison in 1879 and the ventilation and air conditioning systems in the 1930s, seem to resolve those issues. However, in combination with the personal computers and so-called Burolandschaft design (an entirely open-plan office arrangement which allowed no private offices) in the 1960s and 1970s, they also created 'new' problems (such as flickering light, reflections, limited view to the outside, Legionnaires' disease, draught and noise of co-workers). Anything that couldn't be explained was named Sick Building Syndrome. Building-related illnesses and productivity loss were more tangible. Additionally, psychosocial factors (e.g. privacy, control, identity and status) began to win ground (Mendell, 1993).

The open office required fewer square metres per person and was easier to maintain and rebuild. In the 1970s and the beginning of the 1980s reports on people's problems regarding lack of privacy and noise from colleagues started to emerge (Hedge, 1982), and by the 1980s architects began to use private offices again in combination with the open-plan office (Sundstrom and Sundstrom, 1986).

Total costs                                  Building (10% of total):

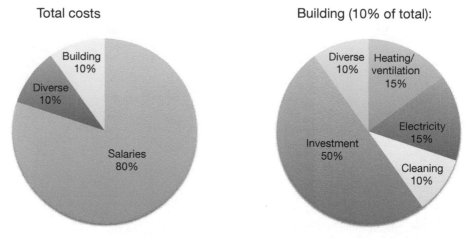

**Figure 9.8** *Example of running costs in an office building: it can be seen that energy costs represented by heating, ventilation and electricity costs only represent 3 per cent of the total costs, while salary costs for office workers represent 80 per cent. It is thus very important to keep our workers comfortable and healthy!*
Source: Bluyssen, based on Clements-Croome, 2002.

## 'New working'

Within the last 10 years more and more private companies and public institutions (again) have replaced the cellular offices with open-plan offices (Pejtersen *et al.*, 2006). With the introduction of ICT (Information and Communication Technology) and more flexible ways of organizing work processes, the work environment has changed accordingly. There are a growing number of organizations that allow office workers to work at home as a teleworker and/or move from fixed workplaces to shared work places (flex places) in more open and transparent offices (Croon *et al.*, 2005). The latter is also named 'new working'.

From a cost-efficiency point of view, the introduction of these office concepts seem advantageous. Organizations save office space, reduce service costs and increase flexibility of office use. However, these new office concepts may affect the health and comfort of office workers, as well as their work performance. It should not be forgotten that even though it seems costs are reduced by saving on service costs, salary costs are still the major part of the running costs of an office building (see Figure 9.8). When an office worker gets sick or his/her performance is affected, this will affect the overall costs much more.

Next to the recession, a side-effect of flex working is the decrease in the demand for office space. In 2011, more than 10 per cent of the available office space in 38 European cities was vacant: e.g. Amsterdam 18 per cent, Brussels 11.2 per cent, Milan 10.8 per cent and Athens 20 per cent (BNP Paribas Real Estate, 2012).

## 9.4.2 Assessment

### Studies

In the last two decades, many studies all over the world have been performed to identify and solve health and comfort problems of office workers. Studies such as the European Audit project (Bluyssen *et al.*, 1996), the BASE-study in the US (Apte *et al.*, 2000), the Building in Use studies in the UK (Leaman, 1996), the COPE (Cost-effective Open Plan) study in Canada (Veitch *et al.*, 2002) and more recently the European HOPE project (Bluyssen *et al.*, 2011a) and the CBE (Center for the Built Environment)-study in the US (Frontczak *et al.*, 2012) have all shown how complex the relationships between building conditions (thermal comfort, lighting, moisture, mould and noise) and human wellbeing in office buildings are.

By prohibiting smoking in public buildings in most countries at the beginning of the twenty-first century, one serious problem was at last eliminated. Other major problems in office buildings, specifically studied in the last decade, have been associated with: type of office arrangement (single, shared, cubicles and open-space offices); the amount of personal control over the indoor environment; thermal comfort; noise issues; daylight and view issues; ventilation system and air quality; and psychosocial and personal factors.

### Open plan versus single office space

Croon *et al.* (2005) defined three dimensions that can be used to describe office concepts:

- *The office location*: the place at which the office worker carries out his/her activities, in the conventional office or at home (teleworking).
- *The office layout*: the arrangement of workplaces and type of boundaries (e.g. open workplaces, cell offices; see Figure 9.9).
- *The office use*: the manner in which workplaces are assigned to office workers, a fixed workplace or desk sharing.

In a systematic review of the literature on the effect of office concepts on worker health and performance, Croon *et al.* (2005) found strong evidence that working in open workplaces reduces privacy and job satisfaction. That office arrangement could play a role in the prevalence of symptoms and sickness absence was already studied in the 1970s and 1980s (Oldham and Brass, 1979; Hedge, 1982; Mendell, 1993). With another wave of open-plan offices in the last decade, studies on the effect of open-plan offices on health and comfort of office workers again emerged (Pejtersen *et al.*, 2006, 2011). Exposure to viruses has been shown to be higher in open-plan offices (Li *et al.*, 2007).

Pejtersen *et al.* (2006) studied eleven naturally and eleven mechanically ventilated office buildings, in which nine had mainly cellular offices (comprising 1 and 2 occupants), five of the buildings had mainly open-plan offices (comprising 7 to 28 occupants or of more than 28 occupants; multi-person offices comprised 3 to 6 occupants), with a questionnaire survey. They found that occupants in open-plan offices are more likely to perceive thermal discomfort (too high temperature, varying temperature), poor air quality (dry air, stuffy air

a

b

d

**Figure 9.9**
*Types of office arrangement: (a) (top left) single, (b) (top
right) shared, (c) (left) open-space and (d) (above) cubicle*
Source: Bluyssen.

c

and being bothered by dust and debris) and noise (in the room) and they more frequently complain about central nervous system symptoms (e.g. headache) and throat and eyes irritation than occupants in multi-person and cellular offices. The strength of the observed association did not change after adjustment for psychosocial risk factors.

In a national survey of 2,202 Danish inhabitants between 18–59 years of age, self-reported sickness absence was found to be significantly related to the number of occupants in the office (also self-reported) when adjusted for confounders (age, gender, socioeconomic status, body mass index, alcohol consumption, smoking habits and physical activity during leisure time; Pejtersen *et al.*, 2011). The study represented more than 2,000 different offices.

Frontzack *et al.* (2011) performed a web-based survey of 52,980 occupants in 351 office buildings, in which the occupants were generally satisfied with their workspace and with the overall building, even if they registered high dissatisfaction with sound privacy, temperature, noise level and air quality. The lowest satisfaction level was observed for sound privacy, being lowest in cubicles, highest in single offices. Satisfaction with the workspace and with almost all indoor environmental parameters and building features was higher in private offices than in a shared offices and cubicles.

## Personal control

Preference for private offices may be related to greater freedom to organize the office space, ability to control the indoor environment to a greater extent in a private office and freedom from having to negotiate the conditions with co-workers. According to Evans and McCoy (1998), insufficient spatial resources, inflexible spatial arrangements, and lack of climatic or lighting controls can all threaten individual needs to effectively interact with the indoor environment (see Box 9.5). People's perception of control of their environment may affect their comfort and satisfaction.

---

### Box 9.5 Design elements for control

- *Spatial resources* include both design density and volume. The amount of available space, visual exposure, structural depth, openness of the perimeter, brightness, and extent of view can all moderate the effects of crowding on human behaviour.
- *Flexibility* can be provided by degree of perimeter openness, moveable partitions, and semi-fixed furniture.
- *Responsiveness*, the clarity and speed of feedback one receives when acting upon a setting or object, or rather unresponsiveness, can be a major factor in the development of helplessness.
- *Privacy*, the ability to regulate social interaction, can be influenced by spatial hierarchy. Size, location and degree of stimulus isolation of interiors influence the effectiveness of buildings to provide privacy. Deeper spaces afford more privacy and enhance ability to regulate social interaction.
- *Social interaction* is influenced by the visual or acoustical permeability of barriers, the functional distance between spaces and furniture arrangements.
- *Feeling of territoriality* can be influenced by clearly delineated and visibly marked boundaries, semi-public spaces and good surveillance opportunities.

Source: Evans and McCoy, 1998.

In a literature survey on how the indoor environment in buildings affects human comfort, it was found that creating a comfortable thermal indoor environment is often considered to be the most important factor for achieving overall satisfaction with IEQ and that providing people with the possibility to control the indoor environment improves thermal and visual comfort and overall satisfaction with IEQ (Frontczak and Wargocki, 2011). Huizinga *et al.* (2006) found, in a study of over 34,000 survey responses in 215 buildings in North America and Finland, that people with access to a thermostat and/or an operable window were statistically more satisfied with their workplace temperature. Personalized ventilation systems, which provide the occupant with a higher degree of control of their local microenvironment, may result in both higher satisfaction and performance (Kaczmarczyk *et al.*, 2002).

Having the opportunity to control your own environment is nice, but whether the control devices or mechanism also respond adequately is another point. Bordass *et al.* (1993) suggested that perceived comfort and control appeared to come not so much from the particular individual control device, but from systems that can respond quickly when people find conditions unsatisfying. They showed that the perceived degree of control with the temperature, ventilation and lighting decreases consistently with the increased number of people sharing an office.

Another point of attention is the automation of control given by the available control technology advances and strict energy demands. Due to the complexity of control, the user's control opportunities are decreased and even bypassed, especially in mechanically ventilated buildings. Leaman and Bordass (2001) concluded that it is a mistake to allow automation to remove occupants completely from the control loop. Toftum (2010) showed that the degree of control, as perceived by the occupants, is more important for the prevalence of adverse systems and building-related symptoms than the ventilation mode per se.

## Thermal comfort

While Wyon *et al.* (1979) found, in an experimental chamber study with 17-year old high school students, somewhat more evidence for adverse effects on performance from higher temperatures within the comfort range, Witterseh *et al.* (2004) showed that performance on tests with adults in a controlled office/laboratory setting did not change when the temperature increased from 22 to 26 and 30°C. Additionally, subjects reported decreased self-estimated performance and increased difficulty thinking and concentrating as well as increased severity of several self-reported symptoms.

Seppänen and Fisk (2005) re-analysed data including 150 assessments of performance from 26 studies obtained in office environments, factories, field laboratories and school classrooms. Considering the effect of temperature, the percentage change in performance per degree increase in temperature was calculated. The data were weighted by both sample size and relevance of the outcome (i.e. objectively reported work performance was assigned a higher weighting than simple visual tasks). The analysis showed that performance increased with temperature up to 20–23°C and decreased with temperature above 23–24°C. Maximum performance was predicted to occur at a temperature of 21.6°C.

## Noise

Associations between self-reported noise exposure and long-term sickness absence has been demonstrated (Kristiansen *et al.*, 2008). Office noise has been correlated with absenteeism and job satisfaction, but has also been shown to influence perceived physical wellbeing and cognitive performance (e.g. Donald and Siu, 2001; Leather *et al.*, 2003). Clausen *et al.* (2009) found a relation between self-reported noise exposure and long-term sickness absence for men, not for women. Jahncke *et al.* (2011) exposed students to two open-plan office noise conditions, high and low noise, during work in a simulated open-plan office. The results indicated that participants remembered fewer words, rated themselves as more tired, and were less motivated with work in high noise compared to low noise.

Many studies in office buildings have shown that, especially telephones, which keep ringing on vacant workplaces, and other people's conversation are generally the most disturbing noise sources in open-plan offices (Branbury and Berry, 2005). Ventilation noise has also been shown to influence performance (Hygge and Knez, 2001). While the nature of noise seems crucial for disrupting effects (Jones *et al.*, 1993), it is clear that effects also depend on the task being performed: the potential impact of office noise on performance is completely different depending on whether silent work at a computer workstation or a communication task has to be performed (Liebl *et al.*, 2012).

## Daylight and view

With the hypothesis that occupant access to daylight and views of the outdoors may have a direct link via circadian biochemistry to many improvements in occupant health and well being, and a likely reduction in the prevalence of SBS (Veitch, 2011; see also Section 3.3.2), several studies have been performed in the last decade.

Analysis of data from post-occupancy evaluation studies in 29 office buildings, 492 workstations, with a questionnaire including questions to report satisfaction on indoor environment qualities (thermal, air, acoustic, lighting, spatial and overall), showed that workstation location (near a window or not) and gender difference affect occupant satisfactions even though environmental conditions are similar (Choi *et al.*, 2009).

In a field study of 200 office workers it was found that better access to views was associated with fewer complaints of fatigue, headache, difficulty concentrating and other health complaints, along with fewer complaints about environmental comfort conditions in the building, such as air quality, thermal and acoustic conditions (Heschong and Roberts, 2009).

Aries *et al.* (2010) studied 10 office buildings in The Netherlands and found that window views that are rated as being more attractive are beneficial to building occupants by reducing discomfort (see Figure 9.10 for examples of views). On the other hand, being close to a window and rating the lighting as being of lower quality can result in thermal and glare problems.

**Figure 9.10** *Different views out of an office space*
Source: Bluyssen.

## Ventilation (system) and air quality

The relation between ventilation, air quality indoors and outdoors, and other sources of pollution has been a major topic in several large studies on office buildings (e.g. Sundell *et al.*, 1994). The first European-wide study focusing on the indoor air quality in office buildings was the European Audit project (Bluyssen *et al.*, 1996a). The outcome of this project, which covered 6,537 occupants representing more than 30,000 occupants in the 56 audited buildings, showed that even though ventilation rates were well above existing ventilation standards and average pollutant concentrations generally met the requirements of existing national standard, nearly 30 per cent of the occupants and 50 per cent of the visitors found the air unacceptable. It was concluded that besides the occupants, it is important to take into account other sources of pollution, comprising building and furnishing materials and HVAC-systems.

Several studies have pointed out that the hygienic state and design of HVAC-systems are associated with increased symptoms and complaints. In the European project AIRLESS (see Study E, Box 5.14), air pollution caused by and/or originating from HVAC-systems and its components (air filters, air ducts, air humidifiers, heating and cooling coils, and rotating heat exchangers) was studied (Bluyssen *et al.*, 2003). In normal comfort ventilation systems the filters (see Figure 9.11) and the ducts seem to be the most common sources of pollution, especially odours.

**Figure 9.11** *An example of a dirty filter: the colour of the bag filter used to be pink!*
Source: Bluyssen.

If humidifiers and rotating heat exchangers are used, they are also reasonable suspects as remarkable pollution sources especially if not constructed and maintained properly. The pollution load caused by the heating and cooling coils is in general less notable.

Several studies have been performed on the use of supply air filters in relation to complaints and symptoms. For example, Wargocki *et al.* (2004) monitored the performance of 26 call-centre operators before and after replacing a used filter with a clean filter. The replacing of the filter, while maintaining a high ventilation rate, improved the performance of the operators by 10 per cent (measured by talk-time). A positive effect of air filters has also been measured. Mendell *et al.* (2002) found that reduced concentrations of airborne particles in a double-blind cross over study of enhanced particle filtration in an office building, were associated with less air quality complaints, but not with reduced symptoms among 396 respondents.

In all of the 56 office buildings of the European Audit project, the air was perceived as being dry, and in a large number the occupants felt slightly warm (Bluyssen *et al.*, 1996a). According to Wolkoff (2013) ocular discomfort belongs to the top two reported symptoms in public office workers. In a follow-up of the BASE study (in which the indoor environment and occupant perceptions of 100 office buildings randomly were selected across the US), a one-year study of workers' health in four large office buildings, eye irritation was positively correlated with floor dust and reported lack of office cleanliness (Chao *et al.*, 2003).

Dry eye is a multifactorial disease of the tears and the ocular surface that results in symptoms of discomfort, visual disturbance and tear film instability with potential damage to the ocular surface (see also Section 3.5.4). Exposure to environmental pollution, visually high demanding (VDU) work and age have been shown to be related to ocular discomfort (dry air; dry, irritated and tired eyes). Other risk factors include the thermal climate (temperature, air velocity and humidity), the use of contact lenses, nutrition and personal care products.

### 9.4.3 Analysis

In Tables 9.7 and 9.8, an overview of the building-related indicators and dose-related indicators, respectively, is presented, for which an association or correlation has been found with the studied occupant-related indicators. While annoyance or satisfaction with the office environmental aspects has been studied mostly, ocular discomfort is an upcoming occupant-related indicator being associated with air quality, lighting and thermal aspects.

What is interesting to note is the fact that overall satisfaction with work space and with the building isn't linearly related with dissatisfaction of the individual factors noise, thermal comfort or air quality (Kim and de Dear, 2012; Frontczak *et al.*, 2012). As was seen in Part I of this book, besides the physical parameters, psychosocial and personal factors can influence wellbeing and even cause similar health effects. This has been of particular interest in several studies on office buildings (Brauer *et al.*, 2006; Marmot *et al.*, 2006; Runeson *et al.*, 2006).

In 1998 Jaakola already proposed an office environment model based on Sir Karl Popper's theory of three worlds (1972; see Figure 9.12). The model showed

**Table 9.7** *Framework of associations found between building-related indicators and selected occupant-related indicators for studies performed on workers and office buildings*

| Occupant-related indicators → Building-related indicators ↓ | Annoyance, satisfaction | Performance and productivity | Absence | Ocular discomfort | Headaches | Example references |
|---|---|---|---|---|---|---|
| *Air quality* | | | | | | |
| • Operable windows | + | | | | | Huizinga et al., 2006 |
| • HVAC systems | + | | | | | Bluyssen et al., 2003 |
| • Building and furnishing materials | | | | | | Bluyssen et al., 1996a |
| • Used vs clean filter | | + | | | | Wargocki et al., 2004 |
| • Particle filtration | + | | | | | Mendell et al., 2002 |
| • Office cleanliness | | | | + | | Chao et al., 2003 |
| *Lighting* | | | | | | |
| • Workstation location near window | + | | | | | Choi et al., 2009 |
| • Window view | + | | | | | Aries et al., 2010 |
| *Thermal comfort* | | | | | | |
| • Thermostat for control temp | + | | | | | Huizinga et al., 2006 |
| *Noise* | | | | | | |
| • Noise exposure indoors (self-reported) | | | + | | | Kristianen et al., 2008; Clausen et al., 2009 |
| • People's conversation and telephones | + | | | | | Branbury and Berry, 2005 |
| • Ventilation system | + | | | | | Hygge and Knez, 2001 |
| *Other factors* | | | | | | |
| • Open plan versus single office space | + | | + | + | + | Pjetersen et al., 2006 and 2011; Frontzack et al., 2011 |
| • Personal control versus automated control | + | | | | | Frontzack et al., 2011; Toftum, 2010 |
| • VDU work and age | | | | + | | Wolkoff, 2013 |

*Note:* + means an association was found, positive or negative.

that both the physical environment (world 1) and the social environment (world 3) affect and influence the 'personal' world (world 2). The 'personal' world deals with states of consciousness, mental states or behavioural actions (thinking, emotions and memories).

According to Sundstrom and Sundstrom (1986), three levels of analysis can be distinguished in office buildings:

- *Individual workers*: in relation to indoor environment conditions (e.g. lighting, air quality, noise, thermal comfort), workstations (e.g. equipment, chair, floor space) and supporting environment (e.g. hallways, restrooms, work areas), mediated by several processes (e.g. adaptation, arousal, overload, stress, fatigue and attitudes).

**Table 9.8** *Framework of associations found between dose-related indicators and selected occupant-related indicators for studies performed on workers and office buildings*

| Occupant-related indicators → Dose-related indicators ↓ | Annoyance, satisfaction | Performance and productivity | Absence | Ocular discomfort | Headaches | Example references |
|---|---|---|---|---|---|---|
| *Air quality* <br> • Ventilation rate/worker <br> • Floor dust <br> • Environmental pollution | +/– | | | + <br> (+) | | Bluyssen *et al.*, 1996a <br> Chao *et al.*, 2003 <br> Wolkoff, 2013 |
| *Lighting* <br> • Glare at desks/screen | | | | (+) | | Wolkoff, 2013 |
| *Thermal comfort* <br> • Temperature <br> • Air velocity, humidity and temperature | | + | | (+) | | Seppänen and Fisk, 2005 <br> Wolkoff, 2013 |
| *Noise* <br> • Noise level indoors | + | + | + | | | Donald and Siu, 2001; Leather *et al.*, 2003; Jahncke *et al.*, 2011 |

*Note*: +/– means inconclusive, while + means an association was found (positive or negative) and (+) that an indication of a relation has been given.

**Figure 9.12**
*The office environment model*
Source: Bluyssen; adapted from Jaakola, 1998.

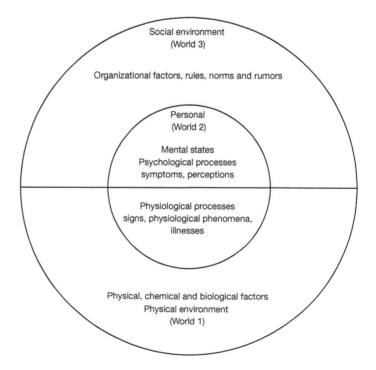

Social environment
(World 3)

Organizational factors, rules, norms and rumors

Personal
(World 2)

Mental states
Psychological processes
symptoms, perceptions

Physiological processes
signs, physiological phenomena,
illnesses

Physical, chemical and biological factors
Physical environment
(World 1)

- *Interpersonal relationships*: related to work spaces, room layout and building layout, mediated by interactions between people involving self-identity, status, choices in communication and regulation of interaction (privacy).
- *Organizations*: in the building (e.g. separation of work-units, differentiation of work-units) related to the organizational process.

Brauer and colleagues (2006) found that the perceived indoor environment was associated not only with the traditional Sick Building Syndrome (SBS) symptoms, but also with symptoms that cannot be physiologically linked to the indoor environment suggesting a risk of reporting bias when assessing nonspecific symptoms. The authors suggested that the associations found in previous cross-sectional studies on SBS symptoms and indoor environment factors may possibly be explained by information bias and reverse causality. Runeson *et al.* (2006) therefore recommended applying a multidisciplinary approach including psycho-social stress factors as well as personal factors, and indoor exposures, in studies on symptoms compatible with Sick Building Syndrome (SBS). Also other researchers have argued that a more complex research model than the conventional 'dose–response' model is required to explain symptoms and complaints in certain (exposure) situations (e.g. Jaakola, 1998; Berglund and Gunnarsen, 2000; Bluyssen and Adan, 2006). In a conventional 'dose–response' model it is assumed that the addition of the estimated effects of the measured concentrations or values of the parameters in the indoor environment (a person is exposed to) relates to the self-administered questions on the same parameters (i.e. air quality, thermal comfort, sound, light).

Berglund and Gunnarsson (2000) studied the role of personality in reporting Sick Building Syndrome (SBS) symptoms and found a significant influence of person-related factors in SBS. In an analysis of data from the US BASE-study, multiple personal factors correlated strongly with health and comfort symptoms (Mendell and Mirer, 2009) indicating that statistical control for those factors is required to evaluate possible relationships with other variables. The assumption in most guidelines that all people are approximately the same in their responses to certain stimuli may thus be incorrect. Jaakola (1998) stipulated that different determinants affect human health and comfort concurrently, and sometimes the effects may be synergistic. Frontczak *et al.* (2012) performed a literature survey on office building studies all over the world and identified a number of parameters influencing building occupants' satisfaction: office layout, office furnishing, thermal comfort, air quality, lighting, acoustic quality, cleanliness and maintenance, and personal workspace and overall building. In a review of environmental and societal influences acting on cardiovascular risk factors and diseases at a population level, Chow *et al.* (2009) drew similar conclusions. It is not only one aspect that determines the state of health, it is often the combination of more aspects.

## 9.5 IEQ analysis

*Depending on the scenario and the profile of the occupant of concern, patterns and interactions of cause–effect relationships need to be established, starting with the indicators of both causes and effects and the assessment protocols. To be able to perform such a situational analysis, the 'right' model and the 'right' data set or datasets are required: a model that is suitable for determining patterns and interactions, and that also takes account of dynamic behaviour. Current risk modelling uses mid-point or end-point models as the basis. In order to establish the algorithm for a scenario approach, 'new' (statistical) models should be studied with suitable datasets.*

### 9.5.1 IEQ modelling

#### Mid-point and end-point modelling

IEQ analysis is currently mostly based on mid-point or end-point modelling. The use of dose-related indicators (see Section 8.2), for example pollutant concentration indoors, is an example of mid-point modelling. A *mid-point indicator* can be defined as a parameter in a cause–effect chain or network (environmental mechanism) for a particular impact category that is between the inventory data and the category endpoints. The dose or environmental parameter, is used most frequently in guidelines and standards as well as in the commercial building assessment tools applied. Unfortunately, a discrepancy of current standards with end-users' needs is observed: indoor environmental quality as experienced by the occupants is often not acceptable and even unhealthy, even if standards and guidelines for individual environmental parameters are met.

For decision-making purposes, risks are usually determined by *end-point modelling*, using occupant-related indicators (Bare *et al.*, 2000). Endpoint characterization factors (or indicators) are calculated to reflect differences between stressors at an endpoint in a cause–effect chain and may be of direct relevance to society's understanding of the final effect. An example of endpoint modelling are human health impacts associated with climate change compared with those of ozone depletion using a common basis such as DALYs (disability adjusted life-years) (see Section 8.3.3.). With DALYs, originally developed by the WHO, years of life lost (YLL) and years lived with disability (YLD) are combined and standardized by means of severity weights. The DALY concept has been applied also at building level (WHO, 2011h and i), but the validity is questionable for two reasons: it is only valid when the exposure function for the risk factor (and sub-factors) and the disease under investigation is known and preferably valid for the population under investigation. And second, the interactions of the risk factors (and sub-factors) with other risk factors for the disease under investigation should be known. Additionally, people have different diseases, doctors hold different ideas about those diseases, and diseases carry different meanings in society (see Box 9.6).

Chow *et al.* (2009) conducted a convincing study on cardiovascular risk factors and disease, in which they emphasized the importance of:

- A better understanding of the impact of the environment, in which a first step is to develop better instruments to measure relevant factors. Most existing instruments focus on single risk factor or behaviour.
- Profiles instead of indices, which have limited value in indicating what should be done. Vlek (1996) introduces a multi-attribute characterization of risk, a risk profile, for capturing the multi-dimensionality as well as the multi-stage nature of risk, containing either objective assessments or subjective evaluations.

---

## Box 9.6  What is disease?

Disease has changed since 1812. While the top causes of death in 1900 were pneumonia (or influenza) and tuberculosis, in 2010 they were heart disease and cancer (Jones *et al.*, 2012). The definition of disease in Merriam-Webster's medical dictionary: 'an impairment of the normal state of the living animal or plant body'. As Jones *et al.* (2012, p. 2335) argue:

> But what is normal? What is impaired? Not every symptom constitutes a disease. Nor is it feasible simply to contrast 'disease', as diagnosed by doctors, with 'illness', as experienced by patients. A disease has characteristic signs and symptoms, afflicts particular people, and follows a characteristic course. Thus, a complex phenomenon. So it seems that diseases can never be reduced to molecular pathways, mere technical problems requiring treatments or cures. Disease is a complex domain of human experience, involving explanation, expectation and meaning.

---

## Box 9.7  Facts and problems identified with respect to indoor air quality

- The emission behaviour of sources is complex.
- Indoor and source surface chemistry create 'new' often unidentified compounds, not (yet) accounted for in current guidelines.
- The material constituents and moisture retention characteristics of a product determine the risk for microbial growth.
- The HVAC systems can be a source of pollution as well, which is not always acknowledged.
- To truly evaluate an exposure, all routes of exposure (both physiological and psychological) should be taken into account jointly.
- Individuals will react differently to the same exposure.
- It is difficult to relate symptoms to IAQ evaluation based on one situation. Other exposures in other environments, as well as time, play a role.
- Some compounds may have adverse effects on their own, while others, seemingly harmless, become harmful when they interact with each other or over time. Some compounds behave differently in a mixture rather than singly.
- Problems with IAQ are not only source related, but also building process related.

Source: Bluyssen, 2009b.

---

- Taking account of interactions, links, mediators and confounders. A particular challenge is how to disentangle the effects of multiple highly correlated factors.
- Taking account of how factors at multiple levels act (individual, household, community and macro-levels).

The cause–effect pathway is in many cases not clear at all. For example, for indoor air quality several facts and problems have been identified to explain this complexity (see Box 9.7).

### Situation or cause analysis

Another way of modelling is to focus on situations instead of mid-points or end-points. This type of analysis has been proposed by Köster's (2002) 'situational analysis', in which behaviour of people is observed and which is based on frequencies of occurrence of specific situations in people's daily lives; also by Ankley *et al.* (2010) in their so-called 'adverse outcomes pathway' concept, which addresses links between events at the molecular level with adverse outcomes at the population level; and in processes or systems with high quality and safety standards, in which root-cause analysis (RCA) is applied. RCA is a process designed to investigate and categorize the root causes of events, in order to prevent recurrence; RCA does not only help to identify what and how something happened, but also why it happened, so something can be done about it (Rooney and Vanden Heuvel, 2004).

Because not all the interactions or mechanisms taking place between the sources that produce the exposure parameters, the exposure parameters themselves and the exposure parameters and the exposed persons are known, *situation analysis* has been applied in the form of *short-cuts* (patterns of building-related indicators; Bluyssen *et al.*, 2011b; see Section 8.4.2). Patterns of those building-related indicators (and other indicators) including interactions among them have shown potential in both directions, negative and positive, stimulating or influencing health and comfort. An analytical model for determining those patterns and interactions is, however, required to validate this form of risk analysis, as well as large and complete datasets to work with.

## 9.5.2  Data modelling

### Complex systems approach

At different levels it has been acknowledged that the built environment and its indoor environment with occupants is a complex system: characterized by feedbacks, interrelations among agents and discontinuous non-linear relations as opposed to a deterministic system. A *deterministic system* is a system in which no randomness is involved in the development of future states of the system. A deterministic model will always produce the same output from a given starting condition or initial state.

At the *building level* Le Corbusier described buildings as complex machines for working and living (Le Corbusier 1925, p. 219): 'Une maison est une machine à habiter.'

At the *human level* Rea *et al.* (2006, p. 2) stated:

> Living organisms are better understood as complex adaptive systems character-ized by multiple participating agents, hierarchical organization, extensive interactions among genetic and environmental effects, nonlinear responses to perturbation, temporal dynamics of structure and function, distributed control, redundancy, compensatory mechanisms, and emergent properties.

Buildings, systems and people together form a dynamic system, and are composed of interacting elements which are embedded into feedback loops and

---

## Box 9.8 Interactions

Basically, the following interactions determine how well you feel, how healthy you are and how comfortable you are at a certain moment in time, and determine *your interaction with your environment* over time:

- *Interactions at human level*: Receiving information (sensations) can be looked upon from the physiology of the human body and/or from the psychological point of view. Interactions occur on both levels. Interactions between people should not be forgotten; those interactions can also have a significant effect on the physical and psychological state of a human being in the indoor environment.
- *Interactions at indoor environmental parameter level*: Important interactions are for example chemical reactions between pollutants in the air and microbiological growth at indoor surfaces, but also interactions between chemicals on (fine) dust and interactions between natural and artificial lighting.
- *Interactions at building level*: Interactions between elements of the building and between the building and the environment, such as interaction of the building with the ground it is built on (the foundation), interaction of outdoor environment with building (protection and transmission characteristics of the facade) and interaction of building with indoor environment (such as maintenance and emission of the indoor surfaces and the lighting, heating, cooling and ventilation systems that are integrated in the facade or not).

Source: Bluyssen, 2009a.

---

very sensitive to small changes or perturbations. Eliminating unwanted model details may cause a magnification of errors in model predictions. Predictions are useless after a certain amount of time (Lu *et al.*, 2010).

At the *human level*, the focus has been on isolating factors, primarily dose-related indicators, which may be causes of a particular disease state. But for many diseases there are numerous factors at different levels of influence, so a shift in methodological approach seems necessary (Galea *et al.*, 2010). At the *building level*, we have simplified all the complex design and management issues by using mainly linear models (Clements-Croome, 2011). Interactions (see Box 9.8), feedback loops and confounding and mediating factors have hardly been considered.

New models are needed to understand better how changes in environmental, socioeconomic and cultural factors affect building performance and this requires an integrated approach. There is a need for a significant paradigm shift from traditional reductionist and determinist approaches. A shift away from statistical association models focused on effect estimates to simulations in which we can test scenarios under different conditions, rather than simply observing associations within finite and specific datasets, is required (Galea *et al.*, 2010). The normally applied regression-based models (see Box 9.9) are concerned with assessing the relation between 'independent' variables and 'outcomes' of interest, and do little to take into account the dynamic and reciprocal relations between

---

### Box 9.9 Statistical models

A *statistical model* is a formalization of relationships between variables in the form of mathematical equations. It describes how one or more random variables are related to one or more variables. The model is statistical as the variables are not deterministically but stochastically related. Some statistical models are:

- *The general linear model* (restricted to continuous dependent variables): is a statistical linear model. For example, ANOVA (analysis of variance), linear regression, t-test and F-test.
- *The generalized linear model*: is a flexible generalization of ordinary linear regression that allows for response variables that have other than a normal distribution. For example logistic regression, linear regression and Poisson regression.
- *The multilevel model*: is a statistical model of parameters that vary at more than one level. Multi-level models are a subclass of hierarchical Bayesian models, which are general models with multiple levels of random variables and arbitrary relationships among the different variables. A *Bayesian network* or Bayesian model is a probabilistic graphical model that represents a set of random variables and their conditional dependencies via a directed acyclic graph.
- *Non-linear regression*: is a form of regression analysis in which observational data are modelled by a function, which is a nonlinear combination of the model parameters and depend on one or more independent variables. The data are fitted by a method of successive approximations. For example, Principal Component Analysis (e.g. applied in the re-analysis of the project HOPE data on office buildings (Description of Study F in Box 8.4); Bluyssen *et al.*, 2011a).
- *The structural equation model* (SEM): is a statistical technique for testing and estimating causal relations using a combination of statistical data and qualitative causal assumptions. Factor analysis, path analysis and regression all present special cases of SEM. The qualitative causal assumptions are represented by the missing variables in each equation, as well as vanishing co-variances among some error terms. These assumptions are testable in experimental studies and must be confirmed judgmentally in observational studies.

---

some 'exposures' and 'outcomes', discontinuous relations or changes in the relations between 'exposures' and 'outcomes' over time. On the other hand complex systems-dynamic analytical approaches make use of computer-based algorithms to model dynamic interactions between individual agents (e.g. persons, cells) or groups and their properties, within, and across levels of influence (Galea *et al.*, 2010).

### 'New' models

The *model applied* should thus be suitable for determining patterns and interactions, and take account of dynamic behaviour. Several techniques have been applied or considered for use, such as meta-analysis, Bayesian Networks, CARTs (classification and regression trees), fault tree analysis, Markov chains, SPNs (stochastic petri networks), structural equation models, neural networks and fuzzy logic. Some of those techniques are explained a little further below.

## Meta-analysis

The statistical analysis of aggregated data from published studies, meta-analysis, is becoming more and more popular in epidemiology (Blettner and Schattmann, 2005). Basically, one gathers results of previous studies following a strict protocol, calculates a pooled estimate (usually the odds-ratio) and the confidence interval, and performs an analysis of the heterogeneity of the study-specific effects as well as a sensitivity analysis. Babisch (2008) performed such a meta-analysis on studies in which the association between community noise and cardiovascular risk was investigated, in order to derive a common dose–effect curve. He performed the meta-analysis on both descriptive (cross-sectional) and analytical (case-control, cohort) studies.

## Bayesian networks

Bayesian networks make it possible to model a complex system integrating aspects of different levels (e.g. technical, human, organizational and environmental), different natures (e.g. qualitative and quantitative) and the temporal dimension (system dynamics), such as degradation, sequences in scenarios, evolution of symptoms and effects of operation conditions (Weber *et al.*, 2010). Bayesian networks are probabilistic graphical models, comprising nodes (i.e. the variables) and arcs (qualitative dependence relationships). An example of a simple Bayesian network is presented in Figure 9.13.

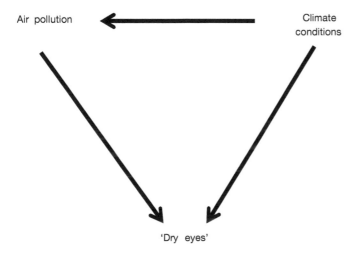

**Figure 9.13** *An example of a simple Bayesian network: certain air pollution may cause ocular discomfort (feeling of dry eyes), certain climate conditions (low humidity, high temperature, draft) may also cause the feeling of dry eyes (see Section 3.5.4). Certain climate conditions determine for some pollutants whether they are emitted more and/or stay airborne more than others (e.g. formaldehyde and particles)*
Source: Bluyssen.

Bell *et al.* (2009) applied Bayesian hierarchal modelling to explore whether air conditioning prevalence modified day-to-day associations between $PM_{10}$ and mortality, and between $PM_{2.5}$ and cardiovascular or respiratory hospitalizations, for those 65 years and older. They found that communities with higher air conditioning prevalence had lower PM effects.

## Structural equation models

Structural equation models are powerful analytical tools disentangling the effects of a specific variable on another high dimensional data with complex patterns of associations. These models generally consists of two parts (Brauer *et al.*, 2008):

- A *measurement model*: in which observed variables are considered reflections of a limited number of unobserved latent variables.
- A *structural model*: in which causal relationships among the latent variables and a set of co-variates are described.

Brauer *et al.* (2008) applied structural equation models to explore the temporal relationship and reversed effect between health and perception of the indoor environment. They showed in a two-phase prospective questionnaire study with a cross-lagged design including 1,740 adults a reversed causal relationship between health and complaints on the indoor environment.

## Neural networks and fuzzy logic

Neural networks are non-linear statistical data modelling tools, based on the biologically neural structure of the brain, and can be used to model complex relationships between inputs and outputs or to find patterns in data. Nodes (neurons) are connected together to form a network of nodes; algorithms are designed to alter the strength (weights) of the connections in the network to produce a desired signal flow (see an example of an artificial neuron in Figure 9.14). Fuzzy logic aims to model the imprecise form of human reasoning and decision-making. Durmisevic and Ciftcioglu (2010) used a fuzzy neural tree structure to estimate the overall design performance of multiple indoor environmental variables.

## 9.5.3 Data collection

A framework that makes the links and interactions between overall (system) requirements design and technical requirements for the different phases of a building clear to all stakeholders is needed (Bluyssen *et al.*, 2010b). Building, social and personal factors all can influence one's perceived health and comfort. It would be interesting to know whether or not it is possible to pinpoint the contribution of those factors (and sub-factors) and their potential combinations. If it is possible to identify relationships between one or combinations of those factors and sub-factors – whether these are building-, social- or personal-related – with perceived health and comfort, new ways of identifying causes of symptoms and complaints can be investigated.

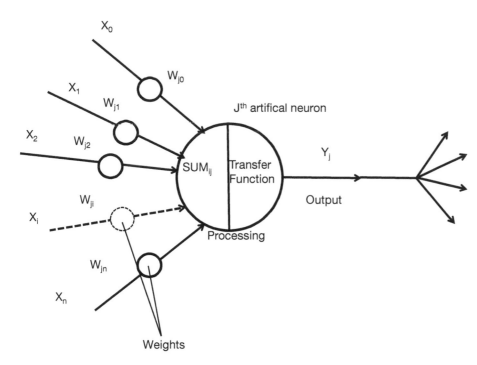

**Figure 9.14** *An artificial neuron with n number of inputs: comprises input channels, a processing element and an output channel. A certain weight is associated with the input or output connections, i.e. incoming information $x_n$ is multiplied by the corresponding weight $w_n$. The transfer functions transfer its input information to the output with a transformation (also named an activation function)*
Source: Bluyssen.

When a framework of indicators, interactions and dynamic behaviour over time per scenario is the ultimate goal, it is important to gather the 'right' package of information. Not only is enough data to be able to perform the analysis on important, but also the quality of the data collected (e.g. collected in a uniform way and preferably for more than one scenario). Errors, confounding factors, sensitivity and quality control are all important factors to take into account.

## Assessment protocol

The information gathered from previous studies and available information from different disciplines, shows that no consensus exists on which physical, physiological, psychological and social indicators and/or factors should be applied to explain responses of people to external stressors fully. For investigations of IEQ in an indoor environment, in principle three categories of measurement techniques are currently available: medical examination, a whole range of questionnaires and several observation and monitoring techniques (see also Chapter 6).

*Medical examination* can comprise measurement of responses and indicators that are part of the causal chain. However, no consensus exists on which of these indicators to apply in measuring the response to acute stress. The sensitivity, specificity and the biological intra- and interpersonal variability of each indicator may be important information, but also the rate of recovery, the peak during the day or perhaps an integrated measure (the total amount over time).

When an attempt is made to pinpoint the effects caused by exposure to a certain indoor environment (including different stressors), a *questionnaire* comprising questions related to all the relevant stressors, personal factors, other factors of influence and events may be required. Questionnaires developed and applied in other fields could be of help.

In addition, *observation and/or monitoring* may be a suitable way to inventory characteristics of the indoor and built environment and psychosocial environment, as well as subjects' behaviour and responses. A checklist, which should at least comprise two parts: one part concerned with the indoor and built environment and one part concerned with the psychosocial environment, which can be applied to inventory the possible stressors and factors of concern. Observation techniques can be helpful to assess subjects' behaviour and responses to certain situations (events). Also, new techniques are under development, such as automatic sensing and logging, which may be worthwhile to take into consideration in the search for *short-cuts* between stressors and effects in IEQ field investigations. These may help to identify relations between measurements of chemical and physical indoor environmental parameters and effects, as well as relevant relations between measurements of physiological indicators and effects.

## Study design

Assessment of health and comfort effects can be roughly performed in two settings:

1   *In a controlled (laboratory or semi-laboratory) environment*, in which the (direct) effect of certain changes of the environmental conditions on the health and comfort of a selected group of human subjects (with a control group or not) is monitored.
2   *In a real life situation*, in which a group of people is studied (longitudinal or cross-sectional) in their own habitat to assess in general the long-term effect of certain conditions. Interventions (changes) of certain conditions can be introduced as well: the effects before and after can be studied.

The pros and cons for both settings have been discussed in Chapter 5 for several fields of research. In general, laboratory settings provide few but high quality data and field situations, a lot but less qualitative data. In the field, often many factors are studied (because they cannot be controlled) at the same time, while in the laboratory often only one aspect is investigated under controlled conditions, with some exceptions.

To be able to define and fill a framework of indicators, interactions and dynamic behaviour over time per scenario, it seems necessary to apply both types of study design:

•   *In-the-field*: focused on interventions and long-term effects.
•   *In-the-lab*: focused on combined (direct) effects of environmental factors.

The combined effects of the environmental factors air quality, lighting, noise and thermal comfort have been conducted in several studies (Hygge and Knez, 2001; Witterseh *et al.*, 2004; Chen *et al.*, 2007; Clausen and Wyon, 2008; Liebl *et al.*, 2012). The most common combination of variables studied is that between noise and heat, and the second most common between noise and lighting. A compilation of the outcome of those studies is presented in Box 9.10 and Table 9.9. It was concluded that the effects of these parameters might simply add but also compensate each other (Lieble *et al.*, 2012). Such inhomogeneous results may result from the fact that a variety of performance measures were applied and people were exposed to different physical environments. Performance at different cognitive tasks might be variously sensitive to influences by different physical parameters, and thus results were not coherent.

For similar reasons, unfortunately, most of the studies performed so far (see Sections 9.2–9.4) are not suitable to be used for the 'pattern' recognition approach that could lead to the required framework with indicators and interactions. We are in need of enough (quantity) 'new' data and with the right quality.

---

## Box 9.10 Combined effect of environmental factors on performance

With respect to the first combination 'noise and heat', Witterseh *et al.* (2004) exposed 30 male subjects to combinations of 3 air temperatures (22, 26 and 30°C) and 2 acoustic conditions (quite (35 dBA) and open-plan office noise (55 dBA)), while performing office work for 3 hours under all 6 conditions. It was found that noise distraction and heat stress can sometimes counteract each other, as they both increase subjective distress and fatigue. In a study performed by Chen *et al.* (2007) on the combined effects of noise, heat and workload exposure, auditory fatigue (expressed in noise-induced temporary threshold shift recovery time) was enhanced by heat and workload.

'Noise and lighting' was studied by Lieble *et al.* (2012). They investigated the combined effects of acoustic and visual distraction on cognitive performance and wellbeing in mock-up offices. Visual distraction due to dynamic lighting caused significant complaints but did not impair performance. Veitch (1990) studied the effects of illuminance and intermittent office noise on reading comprehension (recognition), but found no interactions or main effects.

Nelson *et al.* (1984) varied air temperatures and illuminance, and reported productivity increases in cool air but no interaction with illuminance. While Wyon (2004) showed an increase in productivity of 6–9 per cent when common indoor air pollution sources (e.g. floor-coverings, used supply air filters and personal computers) were removed, Fang *et al.* (2004) found that perceived air quality interacts with thermal comfort.

In a study of three parameters, noise, heat and lighting, interactions between noise and heat on the long-term recall of a text, and between noise and light on the free recall of emotionally toned words, were found (Hygge and Knez, 2001).

**Table 9.9** *Studies of combined effects of the environmental factors air quality, lighting, noise and thermal comfort on performance/productivity*

|  | Noise | Thermal comfort | Air quality | Lighting |
|---|---|---|---|---|
| **Noise** | X | Interact (Witterseh et al., 2004; Chen et al., 2007) | — | No interaction (Lieble et al., 2012) Interact (Hygge and Knez, 2001) |
| **Thermal comfort** | Interact (Hygge and Knez, 2001) | X | Interact (Fang et al., 2004) | No interaction (Nelson et al., 1984) |
| **Air quality** | — | — | X | — |
| **Lighting** | No interaction (Veitch, 1990) | — | — | X |

*Note*: – means no studies performed.

# Appendices

# Appendix A: Questionnaires

## A.1 Demography and life style

An example of questions used to inventory demographic- and life-style-related information as applied in the European OFFICAIR study (Bluyssen *et al.*, 2012a). The actual format of the questionnaire differed from what is presented here.

### Personal data

|   | Question and sub-questions | Answers |
|---|---|---|
| 1 | Year of birth: Please enter a number between 1930 and 1995 | Year list |
| 2 | Gender: Please choose one of the following | Male<br>Female |

### Family status and home environment organization

|   |   | Question and sub-questions | Answers |
|---|---|---|---|
| 3 |   | Status: Please choose one of the following | Single<br>Married<br>In a relationship |
| 4 |   | Do you have any dependents (i.e. people you support financially, including children)? Please choose only one of the following | Yes<br>No |
| If yes |   |   |   |
|   | A | How many adult dependents do you have? Please choose only one of the following | Number list from 0 to 4, more than 4 |
|   | B | How many children dependents do you have? Please choose only one of the following | Number list from 0 to 4, more than 4 |

### Education

|   | Question and sub-questions | Answers |
|---|---|---|
| 5 | What is the highest level of education you have completed? Please choose only one of the following | Master, PhD or specialization<br>University or college or equivalent<br>Secondary school<br>Primary school or less<br>None |

## Commuting

| 6 | Please select each modality of commuting you usually reach your workplace with. Please choose all that apply | By car/cab<br>By subway<br>By train<br>By bus<br>By motorbike<br>By bike<br>On foot<br>Other |
|---|---|---|
| 7 | Please provide a time estimate for each modality (time per week); Please choose the appropriate response for each item you selected in question 9 | Less than 1 hour<br>Between 1 and 2 hours<br>Between 2 and 3 hours<br>Between 3 and 4 hours<br>More than 4 hours |
| 8 | If 'other' was chosen in question 9: Please specify which other modality you use | Please write you answer here |

## Life style

| 9 | Do you drink coffee? Please choose only one of the following | Yes<br>No |
|---|---|---|
| | If yes, how many per day? Please choose only one of the following | Number list from 0 to 5<br>More than 5 |
| 10 | Do you drink tea? Please choose only one of the following | Yes<br>No |
| | If yes, how many per day? Please choose only one of the following | Number list from 0 to 5<br>More than 5 |
| 11 | Have you ever smoked? Please choose only one of the following | Never<br>Former<br>Current |
| | **If current or former** | |
| A | When did you start to smoke (best estimate)? | Please enter a date |
| B | How many cigarettes per day (on average)? Please choose only one of the following | Number list from 0 to 40<br>More than 40 |
| C | How many cigars/pipes per day (on average)? Please choose only one of the following | Number list from 0 to 10<br>More than 10 |
| | **If former** | |
| D | When did you quit smoking (best estimate)? | Please enter a date |
| 12 | Do you drink alcohol? Please choose only one of the following | Yes<br>No |

|  | If yes | |
|---|---|---|
|  | Please specify how many glasses do you drink per week for each of the following? | Number list from 0 to 6<br>More than 6 |
|  | Beer<br>Wine<br>Other non-distillated beverages<br>Distillated beverages (e.g. whiskey, vodka, etc., also if drunk in cocktails) | |
| 13 | On average, how many days per week do you work out (sport, gym, commuting by bike or on foot, etc.)? Please choose only one of the following | Number list from 0 to 7 |
|  | If more than 0 | |
| 14 | How many minutes per work out session (on average)? Please choose only one of the following | Less than 30<br>31–60<br>More than 60 |

## Home

|  |  | |
|---|---|---|
| 15 | Does anybody smoke in your home?<br>Please choose only one of the following | Yes<br>No |
|  | If yes | |
| 16 | How many cigarettes and/or cigars and/or pipes per day (on average)? Please choose only one of the following | Number list from 0 to 40<br>More than 40 |
| 17 | Which of the following do you use at least once a week in your home? | |
| A | Air fresheners | Yes or No |
| B | Candles/incense | Yes or No |
| 18 | How far is your home from a major road (e.g. highway, multilane road, etc.).<br>Please choose only one of the following | Less than 100 metres<br>More than 100 metres |
| 19 | Which floor is your apartment/house located? If it is located on more than one floor, please indicate the bedroom floor | Number list from 0 (below ground) 1 (ground) to 10<br>Higher than 10 |

## A.2  Health and medical history

An example of questions used to inventory the health and medical history as applied in the European OFFICAIR study (Bluyssen *et al.*, 2012a).

### Personal medical history

|   | Regarding the following medical conditions | |
|---|---|---|
| 1 | Have you ever been told by a doctor that you have or had any? Please choose all that apply | Migraine<br>Asthma<br>Eczema<br>Allergy<br>High lipids in the blood (i.e. cholesterol, tryglycerides)<br>High blood pressure<br>Diabetes<br>Depression<br>Anxiety<br>Heart conditions<br>Other respiratory diseases<br>Psychiatric problems<br>None |
|   | **If yes** | |
| 2 | Year you were first told: please enter a number between 1930 and 2013 for each item. Only answer this question for the items you selected in question 1 | Year list |
| 3 | Are you still suffering from the above medical conditions? Please choose the appropriate response for each item. Only answer this question for the items you selected in question 1. | Yes<br>No |
| 3 | Have you received any medical treatment, including medications, for this in the past 12 months? Please choose the appropriate response for each item. Only answer this question for the items you selected in question 1. | Yes<br>No |
| 4 | Have you received any medical treatment, including medications, for this in the past 4 weeks? Please choose the appropriate response for each item. Only answer this question for the items you selected in question 1. | Yes<br>No |
| 5 | Please specify which kind.<br><br>Only answer this question for the items you selected in question 4. | Text (specification per item) |

## Family medical history

| 6 | Has somebody among your close relatives ever been told by a doctor that they have or had any of the medical conditions presented in question 1? Please choose all that apply | Selection of items<br>None |
|---|---|---|
| 7 | Please specify how many close relatives (parents, sisters/brothers, children) have ever been told by a doctor they have or had any of the medical conditions presented in question 1? Please choose the appropriate response for each item. Only answer this question for the items you selected in question 6. | Number list from 0 to 9<br>More than 9<br>Don't know |
| 8 | Please specify which kind.<br><br>Only answer this question for the items you selected in question 6. | Text (specification per item) |

# A.3 Work

An example of questions used to inventory work-related personal information as applied in the European OFFICAIR study (Bluyssen *et al.*, 2012a).

| 1 | | How would you describe the type of work you do (regardless of your job title)? Please choose only one of the following: | Managerial<br>Professional<br>Clerical/secretarial<br>Other |
|---|---|---|---|
| | A | If other, please specify. | Text |
| 2 | | When did you start working in this building (best estimate)? Please enter a date: | Date |
| 3 | | When did you start working at this workstation (best estimate)? Please enter a date: | Date |
| 4 | | Do you have a full-time or a part-time job? Please choose only one of the following: | Full-time<br>Part-time |
| 5 | | Which kind of contract do you have? Please choose only one of the following: | Permanent<br>Fixed-term |
| 6 | | On average, how many hours per week do you work in this building? Please choose only one of the following: | Number list from 1 to 60<br>More than 60 |
| 7 | | On average, how many hours per week do you work at this workstation? 'Your workstation' is the place (desk, cubicle, office, etc.) where you do the majority of your work. Please choose only one of the following: | Number list from 1 to 60<br>More than 60 |

| 8 | | Which best describes the space in which your workstation is located? Please choose only one of the following: | Single person, private office<br>Shared private office<br>Open space with partitions<br>Open space without partitions<br>Other |
|---|---|---|---|
| | A | If other, please specify. | Text |
| 9 | | How many people work in the room in which your workstation is located (including yourself)? Please choose only one of the following: | Number list from 1 to 100<br>More than 100 |
| 10 | | Does anybody smoke inside the room in which your workstation is located?<br>Please choose only one of the following: | Yes<br>No |
| | A | If yes: How many cigarettes and/or cigars and/or pipes per day (on average)?<br>Please choose only one of the following: | Number list from 1 to 40<br>More than 40 |
| 11 | | Which floor is your workstation located? | Below ground<br>Ground<br>Number list from 1 to 10<br>Higher than 10 |
| 12 | | How many windows are there in the room in which your workstation is located?<br>Please choose only one of the following: | 0<br>1<br>More than 1 |
| 13 | | Are any of the windows in your room openable?<br>Please choose only one of the following: | Yes<br>No |

## A.4 Work-related stress

An example of questions used to inventory work-related stress as applied in the European OFFICAIR study (Bluyssen et al., 2012a) based on Siegrist et al. (2004).

### Effort–reward imbalance

| 1 | | I am under constant time pressure due to heavy workload. Please choose only one of the following: | Yes<br>No |
|---|---|---|---|
| | A | If yes, this distresses me. Please choose only one of the following: | Not at all<br>Moderately<br>Considerably<br>Very much |
| 2 | | I have many interruptions and disturbances in my job. Please choose only one of the following: | Yes<br>No |
| | A | If yes, this distresses me. Please choose only one of the following: | Not at all<br>Moderately<br>Considerably<br>Very much |

| 3 | | I have a lot of responsibility in my job.<br>Please choose only one of the following: | Yes<br>No |
|---|---|---|---|
| | A | If yes, this distresses me. Please choose<br>only one of the following: | Not at all<br>Moderately<br>Considerably<br>Very much |
| 4 | | I am often pressured to work overtime.<br>Please choose only one of the following: | Yes<br>No |
| | A | If yes, this distresses me. Please choose<br>only one of the following: | Not at all<br>Moderately<br>Considerably<br>Very much |
| 5 | | My job is physically demanding.<br>Please choose only one of the following: | Yes<br>No |
| | A | If yes, this distresses me.<br>Please choose only one of the following: | Not at all<br>Moderately<br>Considerably<br>Very much |
| 6 | | Over the past few years, my job has become<br>more and more demanding. Please choose<br>only one of the following: | Yes<br>No |
| | A | If yes, this distresses me. Please choose only<br>one of the following: | Not at all<br>Moderately<br>Considerably<br>Very much |
| 7 | | I receive the respect I deserve from my<br>superiors. Please choose only one of<br>the following: | Yes<br>No |
| | A | If no, this distresses me. Please choose only<br>one of the following: | Not at all<br>Moderately<br>Considerably<br>Very much |
| 8 | | I receive the respect I deserve from my<br>colleagues. Please choose only one of the<br>following: | Yes<br>No |
| | A | If no, this distresses me. Please choose only<br>one of the following: | Not at all<br>Moderately<br>Considerably<br>Very much |
| 9 | | I experience adequate support in difficult<br>situations. Please choose only one of the<br>following: | Yes<br>No |
| | A | If no, this distresses me. Please choose<br>only one of the following: | Not at all<br>Moderately<br>Considerably<br>Very much |

| 10 | | I am treated unfairly at work.<br>Please choose only one of the following: | Yes<br>No |
|---|---|---|---|
| | A | If yes, this distresses me. Please choose<br>only one of the following: | Not at all<br>Moderately<br>Considerably<br>Very much |
| 11 | | My job promotion prospects are poor.<br>Please choose only one of the following: | Yes<br>No |
| | A | If yes, this distresses me.<br>Please choose only one of the following: | Not at all<br>Moderately<br>Considerably<br>Very much |
| 12 | | I have experienced or I expect to experience<br>an undesirable change in my work situation.<br>Please choose only one of the following: | Yes<br>No |
| | A | If yes, this distresses me.<br>Please choose only one of the following: | Not at all<br>Moderately<br>Considerably<br>Very much |
| 13 | | My job security is poor.<br>Please choose only one of the following: | Yes<br>No |
| | A | If yes, this distresses me.<br>Please choose only one of the following: | Not at all<br>Moderately<br>Considerably<br>Very much |
| 14 | | My current occupational position adequately<br>reflects my education and training.<br>Please choose only one of the following: | Yes<br>No |
| | A | If no, this distresses me.<br>Please choose only one of the following: | Not at all<br>Moderately<br>Considerably<br>Very much |
| 15 | | Considering all my efforts and achievements,<br>I receive the respect and prestige I deserve at<br>work. Please choose only one of the following: | Yes<br>No |
| | A | If no, this distresses me.<br>Please choose only one of the following: | Not at all<br>Moderately<br>Considerably<br>Very much |
| 16 | | Considering all my efforts and achievements,<br>my work prospects are adequate.<br>Please choose only one of the following: | Yes<br>No |
| | A | If no, this distresses me.<br>Please choose only one of the following: | Not at all<br>Moderately<br>Considerably<br>Very much |

| 17 | Considering all my efforts ad achievements, my salary/income is adequate. Please choose only one of the following: | Yes No |
|---|---|---|
| A | If no, this distresses me. Please choose only one of the following: | Not at all Moderately Considerably Very much |

## Over-commitment

| 18 | Please indicate to what you personally agree with these statements. Please choose the appropriate response for each item: | |
|---|---|---|
| A | I get easily overwhelmed by time pressures at work. | Strongly disagree Disagree Agree Strongly agree |
| B | As soon as I get up in the morning I start thinking about work problems. | Strongly disagree Disagree Agree Strongly agree |
| C | When I get home, I can easily relax and 'switch off' from work. | Strongly disagree Disagree Agree Strongly agree |
| D | People close to me say I sacrifice too much for my job. | Strongly disagree Disagree Agree Strongly agree |
| E | Work rarely lets me go, it is still on my mind when I go to bed. | Strongly disagree Disagree Agree Strongly agree |
| F | If I postpone something that I was supposed to do today I'll have trouble sleeping at night. | Strongly disagree Disagree Agree Strongly agree |

## Scoring of ERI and over-commitment (Bluyssen *et al.*, 2012b)

### ERI

Each ERI item could be scored from 1 (corresponding to 'Agree') to 5 ('Disagree'), with 0 corresponding to 'Not Applicable'.

The *extrinsic effort* score will be calculated using the 5-item version (*Effort scale*, eri1 + eri2 + eri3 + eri4 + eri6), excluding physical load, that has been found psychometrically appropriate in samples characterized predominantly by white collar jobs. The higher the score, the higher the effort (range 5 to 25).

*Reward* is calculated on general scale (*Reward scale*, eri7 + eri8 + eri9 + eri10 + eri11 + eri12 + eri13 + eri14 + eri15 + eri16 + eri17) and three sub-scales (*Esteem scale*, eri7 + eri8 + eri9 + eri10 + eri15; *Job security scale*, eri12 + eri13; *Job promotion/salary scale*, eri11 + eri14 + eri16 + eri17).

The *Efford/Reward ratio* (without eri5) is calculated as the *Effort scale/Reward scale* * 5/11.

### Over-commitment

Each over-commitment item could be scored from 1 (corresponding to 'Strongly disagree') to 4 ('Strongly agree').

Over-commitment scale is calculated by summing the six items.

## A.5 PANAS

An example of a questionnaire with which general states can be determined is the international PANAS (Positive–Negative Affective Scale) – Short form (Thompson, 2007)), applied in the European OFFICAIR study (Study B, Section 5.4.3; Bluyssen *et al.*, 2012a):

**Think about yourself and how you normally feel, to what extent do you generally feel?**

|  | Very slightly or not at all (1) | A little (2) | Moderately (3) | Quite a bit (4) | Extremely (5) |
|---|---|---|---|---|---|
| Upset |  |  |  |  |  |
| Hostile |  |  |  |  |  |
| Alert |  |  |  |  |  |
| Ashamed |  |  |  |  |  |
| Inspired |  |  |  |  |  |
| Nervous |  |  |  |  |  |
| Determined |  |  |  |  |  |
| Attentive |  |  |  |  |  |
| Afraid |  |  |  |  |  |
| Active |  |  |  |  |  |

The long version comprises:

- *Negative effect*: afraid, ashamed, distressed, guilty, hostile, irritable, jittery, nervous, scared and upset.
- *Positive effect*: active, alert, attentive, determined, enthusiastic, excited, inspired, interested, proud and strong.

An elaborated version (PANAS-X) contains 60 items. It measures also 11 specific affects: fear, sadness, guilt, hostility, shyness, fatigue, surprise, joviality, self-assurance, attentiveness and serenity.

## A.6 Health symptoms and comfort problems

Example of questions on health symptoms and comfort problems applied among others in the European HOPE project (Roulet *et al.*, 2006a).

### Your health during the past month (*or at this point in time*)

| | | | |
|---|---|---|---|
| 1 | During the past month, on how many days did you experience DRY EYES when you were at work in this building? | None<br>1 to 5<br>6 to 10<br>More than 10 | ❑<br>❑<br>❑<br>❑ |
| | If symptom experienced: was it better on days away from the office? | Yes<br>No | ❑<br>❑ |
| 2 | During the past month, on how many days did you experience WATERING or ITCHY EYES when you were at work in this building? | None<br>1 to 5<br>6 to 10<br>More than 10 | ❑<br>❑<br>❑<br>❑ |
| | If symptom experienced: was it better on days away from the office? | Yes<br>No | ❑<br>❑ |
| 3 | During the past month, on how many days did you experience a BLOCKED OR STUFFY NOSE when you were at work in this building? | None<br>1 to 5<br>6 to 10<br>More than 10 | ❑<br>❑<br>❑<br>❑ |
| | If symptom experienced: was it better on days away from the office? | Yes<br>No | ❑<br>❑ |
| 4 | During the past month, on how many days did you experience a RUNNY NOSE when you were at work in this building? | None<br>1 to 5<br>6 to 10<br>More than 10 | ❑<br>❑<br>❑<br>❑ |
| | If symptom experienced: was it better on days away from the office? | Yes<br>No | ❑<br>❑ |
| 5 | During the past month, on how many days did you experience a DRY/IRRITATED THROAT when you were at work in this building? | None<br>1 to 5<br>6 to 10<br>More than 10 | ❑<br>❑<br>❑<br>❑ |
| | If symptom experienced: was it better on days away from the office? | Yes<br>No | ❑<br>❑ |
| 6 | During the past month, on how many days did you experience CHEST TIGHTNESS OR BREATHING DIFFICULTY when you were at work in this building? | None<br>1 to 5<br>6 to 10<br>More than 10 | ❑<br>❑<br>❑<br>❑ |
| | If symptom experienced: was it better on days away from the office? | Yes<br>No | ❑<br>❑ |

| 7 | During the past month, on how many days did you experience FLU-LIKE SYMPTOMS when you were at work in this building? | None<br>1 to 5<br>6 to 10<br>More than 10 | ❏<br>❏<br>❏<br>❏ |
|---|---|---|---|
| | If symptom experienced: was it better on days away from the office? | Yes<br>No | ❏<br>❏ |
| 8 | During the past month, on how many days did you experience DRY SKIN when you were at work in this building? | None<br>1 to 5<br>6 to 10<br>More than 10 | ❏<br>❏<br>❏<br>❏ |
| | If symptom experienced: was it better on days away from the office? | Yes<br>No | ❏<br>❏ |
| 9 | During the past month, on how many days did you experience a RASH OR IRRITATED SKIN when you were at work in this building? | None<br>1 to 5<br>6 to 10<br>More than 10 | ❏<br>❏<br>❏<br>❏ |
| | If symptom experienced: was it better on days away from the office? | Yes<br>No | ❏<br>❏ |
| 10 | During the past month, on how many days did you experience HEADACHES when you were at work in this building? | None<br>1 to 5<br>6 to 10<br>More than 10 | ❏<br>❏<br>❏<br>❏ |
| | If symptom experienced: was it better on days away from the office? | Yes<br>No | ❏<br>❏ |
| 11 | During the past month, on how many days did you experience LETHARGY when you were at work in this building? | None<br>1 to 5<br>6 to 10<br>More than 10 | ❏<br>❏<br>❏<br>❏ |
| | If symptom experienced: was it better on days away from the office? | Yes<br>No | ❏<br>❏ |
| 12 | During the past month, on how many days did you experience ANY OTHER SYMPTOMS when you were at work in this building? | None<br>1 to 5<br>6 to 10<br>More than 10 | ❏<br>❏<br>❏<br>❏ |
| | If symptom experienced: was it better on days away from the office? | Yes<br>No | ❏<br>❏ |

Please describe symptoms

Consider any symptoms that you have experienced at work (but only those that are better when you are away from the office).

13  In which season do such symptoms tend to be at          Spring                  ❑
    their worst?                                            Summer                  ❑
                                                            Autumn                  ❑
    *Please tick one box only.*                             Winter                  ❑
                                                            No particular season    ❑

14  And during which part of the day do such symptoms       Morning                 ❑
    tend to be at their worst?                              Afternoon               ❑
                                                            Evening                 ❑
                                                            No particular time      ❑
    *Please tick one box only.*

15  How many days have you been absent from work            None                    ❑
    in the last month due to the working conditions in      Half a day              ❑
    your building?                                          1 to 2 days             ❑
                                                            > 2 days                ❑
                                                            3 to 5 days             ❑
                                                            Don't know              ❑
    *Please tick one box only.*

## Environmental conditions during the past month (*or at this point in time*)

16  How would you describe the typical indoor conditions in this office during the past
    month? Please tick one box per scale. The boxes with a shaded star represent the ideal
    point on each scale.

| | | | | | | | |
|---|---|---|---|---|---|---|---|
| Temperature | Comfortable | ★ | | | | | Uncomfortable |
| | Cold | | | ★ | | | Hot |
| | Not enough variation | | | ★ | | | Varies too much during the day |
| Air movement | Still | | | ★ | | | Draughty |
| Air quality | Dry | | | ★ | | | Humid |
| | Fresh | ★ | | | | | Stuffy |
| | Odourless | ★ | | | | | Smelly |
| | Satisfactory | ★ | | | | | Unsatisfactory |
| Light | Satisfactory | ★ | | | | | Unsatisfactory |
| | No glare | ★ | | | | | Glare |
| Noise | Satisfactory | ★ | | | | | Unsatisfactory |
| Vibration | Satisfactory | ★ | | | | | Unsatisfactory |
| Overall comfort | Comfortable | ★ | | | | | Uncomfortable |
| | | **1   2   3   4   5   6   7** | | | | | |

17   If you rate your overall comfort in the last month more than 4, what could be the reason for this? *You can tick more than one box.*

Not enough light/too much light/too much glare from sun and sky/glare from artificial light/noise from outside the building/noise from building systems (e.g. heating, ventilation, air conditioning, plumbing etc)/noise (other than from building systems) from within the building in winter/etc.

18   Please estimate how you think your productivity at work is influenced by the environmental conditions in the building in the winter.

| Productivity increased by | +30% or more | +20% | +10% | 0 | –10% | –20% | –30% or more | Productivity decreased by |
|---|---|---|---|---|---|---|---|---|

## Other aspects of your office environment

19   How much control do you personally have over the following aspects of your working environment?

| Temperature | None at all | | | | | | Full control |
|---|---|---|---|---|---|---|---|
| Ventilation | None at all | | | | | | Full control |
| Shading from sun | None at all | | | | | | Full control |
| Lighting | None at all | | | | | | Full control |
| Noise | None at all | | | | | | Full control |
| | | 1  2  3  4  5  6  7 | | | | | |

20   How would you describe the following in your office?

| Amount of privacy | Satisfactory | | | | | | Unsatisfactory |
|---|---|---|---|---|---|---|---|
| Layout | Like very much | | | | | | Do not like at all |
| Decoration | Like very much | | | | | | Do not like at all |
| Cleanliness | Satisfactory | | | | | | Unsatisfactory |
| View from window | Like very much | | | | | | Do not like at all |
| | | 1  2  3  4  5  6  7 | | | | | |

21   Have you or your colleagues ever made requests for improvements to the heating, ventilation or air conditioning in your office? Yes No [If No, go to question 22]

Give details [     ]

How satisfied were you with the following?

| Speed of response | Satisfactory | 1 | 2 | 3 | 4 | 5 | 6 | 7 | Unsatisfactory |
|---|---|---|---|---|---|---|---|---|---|
| Effectiveness of response | Satisfactory | 1 | 2 | 3 | 4 | 5 | 6 | 7 | Unsatisfactory |

22    Have you or your colleagues ever made requests for improvements to other aspects of your office environment? Yes No [If No, go to question 23]

Give details [    ]

How satisfied were you with the following?

| | | | | | | | | | |
|---|---|---|---|---|---|---|---|---|---|
| Speed of response | Satisfactory | 1 | 2 | 3 | 4 | 5 | 6 | 7 | Unsatisfactory |
| Effectiveness of response | Satisfactory | 1 | 2 | 3 | 4 | 5 | 6 | 7 | Unsatisfactory |

23    Are any of the windows in your room openable? Yes ❑    No ❑

# A.7 Personality test

An example of a brief personality test from Rammstedt and John (2007) (Appendix A the English version):

## How well do the following statements describe your personality?

| I see myself as someone who | Disagrees strongly | Disagrees a little | Neither agrees nor disagrees | Agrees a little | Agrees strongly |
|---|---|---|---|---|---|
| . . . is reserved. | (1) | (2) | (3) | (4) | (5) |
| . . . is generally trusting. | (1) | (2) | (3) | (4) | (5) |
| . . . tends to be lazy. | (1) | (2) | (3) | (4) | (5) |
| . . . is relaxed, handles stress well. | (1) | (2) | (3) | (4) | (5) |
| . . . has few artistic interests. | (1) | (2) | (3) | (4) | (5) |
| . . . is outgoing, sociable. | (1) | (2) | (3) | (4) | (5) |
| . . . tends to find fault with others. | (1) | (2) | (3) | (4) | (5) |
| . . . does a thorough job. | (1) | (2) | (3) | (4) | (5) |
| . . . gets nervous easily. | (1) | (2) | (3) | (4) | (5) |
| . . . has an active imagination. | (1) | (2) | (3) | (4) | (5) |

Scoring the BFI-10 scales (R = item is reversed-scored):

- Extraversion: 1R, 6
- Agreeableness: 2, 7R
- Conscientiousness: 3R, 8
- Neuroticism: 4R, 9
- Openness: 5R, 10

# Appendix B: Checklists

Based on checklists developed and applied in the European projects: HOPE, OFFICAIR and SINPHONIE.

## B.1 Description of outdoor environment

**1.1 What is the geographical location of the building?**        **Additional comments**

Seacoast ❑
North of the country ❑
South of the country ❑
East of the country ❑
West of the country ❑

**1.2 Where is the building situated?**        **Additional comments**

Industrial area ❑
Mixed industrial/residential area ❑
Commercial area ❑
Mixed commercial/residential area ❑
City centre, densely packed housing ❑
Town, with or without small gardens ❑
Suburban, with larger gardens ❑
Village in a rural area ❑
Rural area with no or few other homes nearby ❑

**1.3 Are there any nearby (within 100 metres) potential sources of outdoor air pollution that might influence the indoor environment?**        **Additional comments**

None ❑
Car parking close to the building ❑
Attached garage ❑
Direct access from basement or roof car park ❑
Busy road (at least part of the day) ❑
Highway ❑
Power plant for the building ❑
Other power plant (up to 1 km) ❑
Oil or coal-burning plants ❑
Industry (up to 10 km) ❑
Cooling towers ❑
Built on a landfill site ❑
Waste management site (tip or garbage dumpsters) ❑
   (up to 3 km)
Agricultural sources (up to 3 km) ❑
Other (specify) _____ ❑

**1.4 Are there any nearby (within 100 metres) noise sources outside the building that might influence the indoor environment?**

Additional comments

| | |
|---|---|
| None | ❑ |
| Car parking with minimum 50 places close to the building | ❑ |
| Busy road (at least part of the day) | ❑ |
| Highway | ❑ |
| Railway or station | ❑ |
| Subway | ❑ |
| Air traffic (up to 3 km) | ❑ |
| Sea, river or canal traffic | ❑ |
| Building, construction, etc. | ❑ |
| Sports events | ❑ |
| Other entertainment or leisure | ❑ |
| Factories or works | ❑ |
| Commercial premises | ❑ |
| Forestry, farming, etc. | ❑ |
| Community buildings (halls, churches, etc.) | ❑ |
| Other (specify) _____ | ❑ |

**1.5 Are there any nearby potential electromagnetic or 'electric noise' sources outside the building that might influence the indoor environment?**

Additional comments

| | |
|---|---|
| None | ❑ |
| Source | ❑ |
| Pylon | ❑ |
| Mobile phone network antenna | ❑ |
| Radio or television transmitter | ❑ |
| Other (specify) _____ | ❑ |

**1.6 What is the distance to the potential electromagnetic or 'electric noise' sources selected in question 1.5?**

| | Horizontal distance [m] | Vertical distance [m] |
|---|---|---|
| Source | _____ | _____ |
| Pylon | _____ | _____ |
| Mobile phone network antenna | _____ | _____ |
| Radio or television transmitter | _____ | _____ |
| Other | _____ | _____ |

# B.2  Description of building

**2.1 What is the ownership/tenancy status?**

Additional comments

| | |
|---|---|
| Owned by occupant | ❑ |
| Rented by occupant | ❑ |
| Rented by multiple occupants (of which survey area covers just one) | ❑ |
| Rented by multiple occupants (of which survey area covers more than one) | ❑ |

**2.2 Who is the maintainer of the building?**                    Additional comments

Municipality                                                    ❏
Foundation/Institution                                          ❏
Church                                                          ❏
Private                                                         ❏
Other (specify) _____                          ❏

**2.3 Has the building been certified by any program**          Additional comments
**(Legislation, Regulation (Energy Performance, IAQ,**
**Sustainability))?**

No                                                              ❏
Yes                                                             ❏
Which?_____

**2.4 When was the building completed?**              Year      Additional comments

Year of construction if purpose-built or year of
    construction of original building before conversion.    _____

**2.5 Has the building been converted?**                         Additional comments

No                                                              ❏
Yes                                                             ❏
If yes, when was conversion completed (year)?           _____

**2.6 Has the building been refurbished?**                       Additional comments
**(Refurbished does not include new furniture.)**

No                                                              ❏
Yes                                                             ❏
If yes, when was the last refurbishment completed (year)? _____

**2.7 Have there been any modifications in the last**           Additional comments
**year (as percentage)?**                              %

Floor structure                                      _____

Insulation                                           _____

Wall                                                 _____

Ceiling lining                                       _____

Heating system                                       _____

Ventilation system                                   _____

Windows                                              _____

**2.8 Number of storeys**                          Storeys      Additional comments

Occupied above ground                              _____

Unoccupied above ground                            _____

Occupied below ground                              _____

Unoccupied below ground                            _____

**2.9 Is the building investigated partially or entirely?**          **Additional comments**

Partially                                                              ❑
Entirely                                                               ❑

Total number of rooms (of the investigated part)          _____

Total number of occupants (of the investigated part)      _____

Total floor area (of the investigated part) in m$^2$       _____

**2.9A If the building is a school building: total number of rooms?**          **Additional comments**

Classrooms                                    _____

Dining rooms                                  _____

Gymnasiums/Sport hall                         _____

Teachers' rooms/Offices                       _____

Kitchen                                       _____

Library                                       _____

Bathrooms/Toilets                             _____

Garages                                       _____

Other _____

**2.9B If the building is an office building: total number of rooms?**          **Additional comments**

Open-plan offices                             _____

Single person offices                         _____

2–4 person offices                            _____

Meeting rooms                                 _____

Kitchen                                       _____

Library                                       _____

Bathrooms/Toilets                             _____

Garages                                       _____

Service rooms                                 _____

Other _____

**2.9C If the building is a home: total number of rooms?**          **Additional comments**

Sleeping rooms                                _____

Kitchen/living room together                  _____

Separate living room                          _____

Separate kitchen                              _____

Bathrooms/Toilets                             _____

Garages                                       _____

Other _____

| 2.10 What are the external walls constructed of (massive means made of solid bricks; lightweight means made of wood)? | Additional comments |
|---|---|
| Single wall | _____ |
| Double wall | _____ |
| Mixture of single and double | _____ |
| Massive structure (high thermal inertia) | _____ |
| Lightweight structure (low thermal inertia) | _____ |
| Mixture of massive and lightweight | _____ |
| Without insulation | _____ |
| With insulation | _____ |
|    External insulation thickness (mm) _____ | _____ |
|      All walls | _____ |
|      Some walls | _____ |
|    Cavity insulation thickness (mm) _____ | _____ |
|      All walls | _____ |
|      Some walls | _____ |
|    Internal insulation thickness (mm) _____ | _____ |
|      All walls | _____ |
|      Some walls | _____ |
| Type of insulation | _____ |
|    Mineral wool | _____ |
|    Glass wool | _____ |
|    Fibreglass | _____ |
|    Polystyrene | _____ |
|    Polyurethane | _____ |
|    Cork | _____ |
|    Other (specify) _____ | |

| 2.11 What is the type of foundation/ground floor? | Additional comments |
|---|---|
| Basement | |
| Slab on grade | |
| Crawl space | |
| Other (specify) _____ | |

| 2.12 How is the structure of the roof? | Additional comments |
|---|---|
| Flat roof | _____ |
| Ridge roof | _____ |
| Massive structure | _____ |
| Lightweight structure | _____ |
| Mixture of massive and lightweight | _____ |

Without insulation          _____

With insulation             _____

    External insulation thickness (mm) _____   _____

    Cavity insulation thickness (mm) _____   _____

    Internal insulation thickness (mm) _____   _____

Type of insulation          _____

    Mineral wool             _____

    Glass wool               _____

    Fibreglass               _____

    Polystyrene              _____

    Polyurethane             _____

    Cork                     _____

    Other (specify) _____

## 2.13 Type of glazing                          Additional comments

Single glazing                                   _____

Double glazing                                   _____

Double clear glazing with filling (argon or other)   _____

Double clear glazing with coating                _____

Double glazing with tinted internal pane         _____

Triple glazing                                   _____

Other (specify) _____

## 2.14 Are there solar shading devices present?

| | South side only | One or more other facades | Not present |
|---|---|---|---|
| External vertical blinds | ❏ | ❏ | ❏ |
| External shutters | ❏ | ❏ | ❏ |
| External roller shutters | ❏ | ❏ | ❏ |
| External louvres | ❏ | ❏ | ❏ |
| External screens | ❏ | ❏ | ❏ |
| External window films | ❏ | ❏ | ❏ |
| External horizontal blinds | ❏ | ❏ | ❏ |
| External awnings/canopies | ❏ | ❏ | ❏ |
| External overhangs | ❏ | ❏ | ❏ |
| External vertical fins | ❏ | ❏ | ❏ |
| Blinds between glazing | ❏ | ❏ | ❏ |
| Internal vertical blinds | ❏ | ❏ | ❏ |
| Internal louvres | ❏ | ❏ | ❏ |
| Atrium | ❏ | ❏ | ❏ |
| Double facade | ❏ | ❏ | ❏ |
| Other (specify) _____ | ❏ | ❏ | ❏ |

## 2.15 How are the solar shading devices controlled?          Additional comments

| | |
|---|---|
| No control (fixed) | ❏ |
| Individual | ❏ |
| Central down, individual up | ❏ |
| Automatic | ❏ |

**2.16 Has there been any major water leakage or flooding in the last five years? If yes, where from?**     Additional comments

No water leakage or flooding     ❏
Roof     ❏
Window     ❏
Facade     ❏
Basement     ❏
Water storage tanks     ❏
Water pipes     ❏
Other (specify) _____     ❏

**2.17 Are there materials containing asbestos in the building?**     Additional comments

Yes, flocculate     ❏
Yes, but compact     ❏
Yes, but sealed     ❏
No     ❏

**2.17A If yes to 2.16: Is there an asbestos management plan?**     Additional comments

Yes     ❏
No     ❏

**2.17B If yes to 2.16: Are materials containing asbestos all in a good state of maintenance?**     Additional comments

Yes     ❏
No     ❏

**2.18 Are there any lead components in the building?**     Additional comments

No lead components     ❏
Lead water pipes     ❏
Lead paint     ❏
Other (specify where) _____     ❏

**2.19 If in a radon-affected zone, is there proper construction of foundation and ventilation (control of pressure difference), or other measures to control migration of radon? Answer yes or no according to national/regional guidelines or local radon data**     Additional comments

Not designated as a radon-affected zone     ❏
Not known if a radon-affected area     ❏
Radon zone     ❏
    Migration controlled     ❏
    Migration not controlled     ❏
    Unknown     ❏

**2.20 Does the building contain radon bearing construction materials? (e.g. by product gypsum, alum shale, granites and volcanic tuffs)?**     Additional comments

Yes     ❏
No     ❏

**2.21 How is the density of nearby obstructions?**

1  Very dense       2  Moderately dense       3  Few buildings       4  Free standing

| **2.22 Facades with adjacent buildings** | **Additional comments** |
|---|---|
| One facade | ☐ |
| Two facades | ☐ |
| Three facades | ☐ |

| **2.23 Height of surrounding buildings** | **Additional comments** |
|---|---|
| Higher | ☐ |
| Same | ☐ |
| Lower | ☐ |
| Variable | ☐ |

| **2.24 Building shape** | **Additional comments** |
|---|---|
| Square | ☐ |
| Rectangular | ☐ |
| Square with atrium | ☐ |
| Other shape | ☐ |

| **2.25 Do neighbouring buildings with glass facades or light-coloured facades cause glare in the building?** | Yes | A little | No |
|---|---|---|---|
| Summer | ☐ | ☐ | ☐ |
| Winter | ☐ | ☐ | ☐ |

| **2.26 How many hours per week is the building occupied?** | Hours | Additional comments |
|---|---|---|
| At normal occupancy levels | _____ | |

| **2.27 Has there been fire damage (if yes, specify the date)?** | **Additional comments** |
|---|---|
| No | ☐ |
| Yes ____/____/____ | ☐ |
| Extent of the fire damage | |
|  Building wide | ☐ |
|  Limited spaces | ☐ |
|  Floors damaged | ☐ |

| **2.28 Are there visible air leaks (cracks in the construction) in the structure?** | **Additional comments** |
|---|---|
| No | ☐ |
| Yes | ☐ |

| 2.29 Where are the printers/copy machines in general located? | Additional comments |
|---|---|
| In the room | ❑ |
| In a separate printing room | ❑ |
| In the corridor | ❑ |

# B.3  Description of building services

Depending on the type of building investigated one can choose the questions that apply.

## Lighting

| 3.1 Has anything been done to optimise the use of daylight in this building? | Additional comments |
|---|---|
| No | ❑ |
| Internal light shelves | ❑ |
| External light shelves | ❑ |
| External light scoops | ❑ |
| Light pipes (with or without heliostat) | ❑ |
| Fixed louvres | ❑ |
| Horizontal blinds | ❑ |
| Reflecting blinds | ❑ |
| Mirrors (including mirrored louvres and light shelves) | ❑ |
| Holographic film | ❑ |
| Prismatic film | ❑ |
| Prismatic glazing | ❑ |
| Interior painted white | ❑ |
| Clear glazing | ❑ |
| High windows (adjacent to ceiling) | ❑ |
| Light coloured window frame | ❑ |
| 'Skylights' in bay windows | ❑ |
| Movable solar protection | ❑ |
| No fixed solar protections | ❑ |
| Automatic control of artificial lighting | ❑ |

## Heating and cooling

| 3.2 Are there heating systems present? | Additional comments |
|---|---|
| No | ❑ |
| Yes: In the whole building | ❑ |
| Yes: In some parts of the building | ❑ |

| 3.3 Are there cooling systems present? | Additional comments |
|---|---|
| No | ❑ |
| Yes: In the whole building | ❑ |
| Yes: In some parts of the building | ❑ |

**3.4 Are there non-electric heaters (for heating and/or hot water) present?**

**Additional comments**

No ❏
Outside building ❏
Inside building ❏

**3.5 Do all gas boilers or heaters or other equipment have proper flue gas pipes or chimneys?**

**Additional comments**

Yes ❏
No ❏
Not known ❏

**3.6 What is the cooling production plant?**

**Additional comments**

No cooling ❏
Heat exchanger on ground river or lake water ❏
Package air cooled chiller ❏
Water cooled chiller + cooling tower ❏
Water cooled chiller + dry cooler ❏
Heat pump (heating + cooling) ❏
Absorption type chiller + cooling tower ❏
Other (specify) _____ ❏

**3.7 What is the type of heating and cooling distribution network?**

**Additional comments**

Two pipe system ❏
Three pipe system ❏
Four pipe system ❏
Refrigerant distribution system ❏

**3.8 What are the heating and cooling terminal units?**

**Additional comments**

Hot water radiators or convectors ❏
Electrical radiators or convectors ❏
Two-pipe fan coil units ❏
Four-pipe or two-pipe/wire fan coil units ❏
Induction units ❏
Heating or cooling floor ❏
Heating or cooling ceiling ❏
Individual heat pumps or water loop ❏
Window units ❏
Split system ❏
Air conditioning cabinets ❏
Direct electric heaters ❏
Heated walls ❏
Other (specify) _____ ❏

**3.9 How is the room temperature controlled?**

**Additional comments**

Manual radiator valve ❏
Local thermostat at radiator/heating unit ❏
Local thermostat (e.g. on wall) ❏
Central sensor ❏
Facade sensor(s) – i.e. outside temperature ❏
Zone sensor(s) ❏

Manual control in room(s)     ❑
According to occupancy     ❑
Other (specify) _____     ❑

| 3.10 How many hours per week is the system running? | Hours | Additional comments |
|---|---|---|
| At levels designed for normal occupancy levels | _____ | |

**3.11 Are heaters located below windows to prevent draught in winter?**     **Additional comments**

Yes     ❑
Unnecessary – high performance glazing     ❑
No     ❑

**3.12 Are there any large glass surfaces (glazing on external walls) without thermal conditioning?**     **Additional comments**

Yes     ❑
No     ❑

**3.13 Is the temperature controlled by the system?**     **Additional comments**

During summer     ❑
During winter     ❑
No     ❑

| 3.13A If yes on question 3.13: What is the temperature set point and dead band range (°C)? | Set point | Range Min | Range Max |
|---|---|---|---|
| In summer | _____ | _____ | _____ |
| In winter | _____ | _____ | _____ |

**3.14 Is the relative humidity controlled by the system?**     **Additional comments**

During summer     ❑
During winter     ❑
No     ❑

| 3.14A If yes on question 3.14: What is the relative humidity set point and dead band range (°C)? | Set point | Range Min | Range Max |
|---|---|---|---|
| In summer | _____ | _____ | _____ |
| In winter | _____ | _____ | _____ |

## Hot water

**3.15 Is the sanitary hot water production located indoors?**     **Additional comments**

Yes     ❑
No     ❑
Not applicable     ❑

**3.16 Is hot water stored temperature known?**     **Additional comments**

Yes     ❑
No     ❑
Not applicable     ❑

**3.16A If yes on question 3.16:**       Additional comments

What temperature is hot water stored at (°C)?     _____

What temperature is hot water delivered at (°C)?     _____

**3.17 In the last year, has there been any reported case of legionella, aspergilla or humidifier-related fever related to the building?**      Additional comments

Yes      ❑
No      ❑

**3.18 Is there a documented procedure for managing the risk of legionella in the water supply?**      Additional comments

Yes      ❑
No      ❑

## Ventilation

**3.19 Are the windows openable?**      Additional comments

Yes      ❑
Yes, some      ❑
Yes, but occupants are not allowed to open them      ❑
No      ❑

**3.20 How is the building ventilated?**      Additional comments

Openable windows      ❑
Other natural ventilation (e.g. passive stack)      ❑
Mechanical ventilation      ❑
Hybrid/Mixed mode      ❑

**3.21 In naturally ventilated, or exhaust-only ventilated buildings, are there air transfer openings between rooms?**      Additional comments

Yes      ❑
Varies      ❑
No      ❑

## Mechanical ventilation

**3.22 Type of mechanical ventilation**      Additional comments

Supply system only      ❑
Both exhaust and supply      ❑
Exhaust system only for:      ❑
    Toilets/other polluted rooms only      ❑
    Other rooms      ❑
    Permanent      ❑
    Non-permanent      ❑
      Days per week      ❑
      Hours per day      ❑

| 3.23 What type of control system is there for mechanical ventilation? | | Additional comments |
|---|---|---|
| Central – manual (on/off) | ❏ | |
| Central – clock | ❏ | |
| Central – demand control (temperature, $CO_2$, other pollutant, relative humidity) | ❏ | |
| Local – manual (on/off) | ❏ | |
| Local – clock | ❏ | |
| Local – demand control (temperature, $CO_2$, other pollutant, relative humidity) | ❏ | |
| Recirculation control | ❏ | |

| 3.24 Air handling units (AHUs) | | Additional comments |
|---|---|---|
| 100% fresh air | ❏ | |
| AHU with recirculating air | ❏ | |
| AHU with recirculating fan, free cooling system | ❏ | |
| AHU for dual duct system and recirculating air | ❏ | |
| Other (specify) _____ | ❏ | |

| 3.25 Heating/cooling in AHU | Yes | No |
|---|---|---|
| Heating | ❏ | ❏ |
| Cooling | ❏ | ❏ |

| 3.26 What type of humidification? | | Additional comments |
|---|---|---|
| None | ❏ | |
| Spray | ❏ | |
| Evaporative | ❏ | |
| Steam | ❏ | |
| Ultrasonic | ❏ | |
| Infrasonic | ❏ | |
| Other (specify) _____ | ❏ | |

| 3.27 What type of water purification? | | Additional comments |
|---|---|---|
| None | ❏ | |
| Ozone | ❏ | |
| Biocide | ❏ | |
| High voltage | ❏ | |
| UV | ❏ | |
| Other (specify) _____ | ❏ | |

| 3.28 Is the system equipped with water droplet eliminators? | | Additional comments |
|---|---|---|
| Yes | ❏ | |
| No | ❏ | |

| 3.29 Is the system designed and maintained to collect and drain condensed water from cooling coils adequately? (Look for drainage panels, clear drainage holes, signs of overflow or spillage etc.) | | Additional comments |
|---|---|---|
| Not applicable | ❏ | |
| Yes | ❏ | |
| No | ❏ | |

| 3.30 What type of outdoor air filter is used? (Give type and class (filter grade according to EN779).) | Type | Class |
|---|---|---|
| Pre-filter | _____ | _____ |
| Main filter | _____ | _____ |

**3.31 What does the duct material comprise of?**          Additional comments

| | |
|---|---|
| Asbestos cement | ❏ |
| PVC | ❏ |
| Galvanized steel | ❏ |
| Other (specify) _____ | ❏ |

**3.32 Is the duct material in a good state of maintenance?**          Additional comments

| | |
|---|---|
| Yes | ❏ |
| No | ❏ |

**3.33 Is the duct insulated?**          Additional comments

| | |
|---|---|
| None | ❏ |
| Internally | ❏ |
|    Mineral fibre | ❏ |
|    Other (specify) _____ | ❏ |
| Externally | ❏ |
|    Mineral fibre | ❏ |
|    Other (specify) _____ | ❏ |

**3.34 What type of heat recovery is used?**          Additional comments

| | |
|---|---|
| None | ❏ |
| Fixed plate exchanger | ❏ |
| Rotating wheel exchanger | ❏ |
| Heat pipes | ❏ |
| Two-coil glycol water exchanger | ❏ |
| Other (specify) _____ | ❏ |

**3.35 How is the system operating at the time of inspection?**          Additional comments

| | |
|---|---|
| Full performance | ❏ |
| Reduced performance | ❏ |
| Stopped system | ❏ |
| 100% recirculation | ❏ |

**3.36 Is the system operating normally for the time of year at the time of inspection?**          Additional comments

| | |
|---|---|
| Yes, operating normally for the time of year | ❏ |
| No, operating too high for the time of year | ❏ |
| No, operating too low for the time of year | ❏ |

---

**3.37 What is the position of ventilation system intake?**　　　　**Additional comments**

| | |
|---|---|
| None | ❏ |
| Roof | ❏ |
| Facade | ❏ |
| Ground | ❏ |
| Other (specify) _____ | ❏ |

---

**3.38 What is the height of the ventilation system intake above ground level (m)? (If the building has individual intakes by storey or by zone, give height for each storey.)**　　m　　**Additional comments**

| | |
|---|---|
| Intake 1 | _____ |
| Intake 2 | _____ |

---

**3.39 What is the shortest distance of system intake from:**　　　　**Vertical [m]**　　　　**Horizontal [m]**

| | | |
|---|---|---|
| Exhaust outlets? | _____ | _____ |
| Cooling towers? | _____ | _____ |

---

**3.40 Are there any other potential pollutant sources close to the system intake?**　　　　**Additional comments**

| | |
|---|---|
| Yes | ❏ |
| No | ❏ |

---

**3.40 A If yes on question 3.40, give the shortest distance from intake for all that apply**　　m　　**Additional comments**

| | |
|---|---|
| Car parking close to the building | _____ |
| Attached garage | _____ |
| Direct access from basement or roof car park | _____ |
| Busy road | _____ |
| Power plant for the building | _____ |
| Other power plant | _____ |
| Industry | _____ |
| Cooling towers | _____ |
| Built on a landfill site | _____ |
| Waste management site (e.g. tip or dump) | _____ |
| Agricultural sources | _____ |
| Other (specify) _____ | _____ |

---

**3.41 What is the design outdoor flow rate? (Use whatever unit is available)**　　　　**Additional comments**

| | |
|---|---|
| $m^3/h$ | _____ |
| $ach^{-1}$ | _____ |
| $m^3/h$ person | _____ |

| 3.42 Is the exhaust ventilation of toilets, etc. running continuously to provide the basic ventilation for the building? | Additional comments |
|---|---|
| Yes | ☐ |
| No | ☐ |
| No exhaust ventilation of toilets | ☐ |

## Maintenance

| 3.43 How often are the air filters replaced? | Additional comments |
|---|---|
| No regular period for replacement | ☐ |
| Twice a year or more often | ☐ |
| Once a year | ☐ |
| Once every two years | ☐ |
| Less often than once every two years | ☐ |

| 3.44 The following components are | Cleaned per year | Intended to be cleaned per year | Date last time cleaned |
|---|---|---|---|
| Supply air ducts | ☐ | ☐ | ___/___/___ |
| Supply air devices | ☐ | ☐ | ___/___/___ |
| Exhaust air devices | ☐ | ☐ | ___/___/___ |

| 3.45 Does the building have an alarm system for malfunctioning of HVAC main components? | Additional comments |
|---|---|
| Yes | ☐ |
| No | ☐ |

| 3.46 Is the HVAC system managed by an external company? | Additional comments |
|---|---|
| Yes, managed by an external company | ☐ |
| No, managed internally | ☐ |
| Other (specify) _____ | ☐ |

# B.4  Description of room(s) and interior

Depending on the type of building investigated one can choose the questions and items that apply.

## General

| 4.1 Type of room | Additional comments |
|---|---|
| Classroom | ☐ |
| Office room – cellular | ☐ |
| Office room – landscape | ☐ |
| Meeting room | ☐ |
| Kitchen | ☐ |
| Living room | ☐ |
| Bedroom | ☐ |
| Other (specify) _____ | ☐ |

## 4.2 Dimensions and location of room — Additional comments

| | |
|---|---|
| Storey number | _____ |
| Floor area [m²] | _____ |
| Ceiling height [m²] | _____ |
| Windows area [m²] | _____ |

## 4.3 Number of occupants present/maximum possible (e.g. number of chairs, desks. . . .)

| | Now | Max | Additional comments |
|---|---|---|---|
| Children 0–4 | _____ | _____ | |
| Children 5–12 | _____ | _____ | |
| Children 13–18 | _____ | _____ | |
| Adults women 19–65 | _____ | _____ | |
| Adults men 18–65 | _____ | _____ | |
| Adults > 65 | _____ | _____ | |

## 4.4 Occupation room: Hours per day room is occupied

| | Hours | Additional comments |
|---|---|---|
| Monday | _____ | |
| Tuesday | _____ | |
| Wednesday | _____ | |
| Thursday | _____ | |
| Friday | _____ | |
| Saturday | _____ | |
| Sunday | _____ | |

**Lighting**

## 4.5 Type of lighting — Additional comments

| | |
|---|---|
| Natural | ❑ |
| Artificial | ❑ |
| Mixture | ❑ |

## 4.6 What is the type of artificial lighting? (Indicate all present) — Additional comments

| | |
|---|---|
| Fluorescent | ❑ |
| Compact fluorescent | ❑ |
| Incandescent | ❑ |
| Halogen | ❑ |
| Other (specify) _____ | ❑ |

**4.7 What is the location of the artificial lighting?**
**(Indicate all applicable)**

Additional comments

| | |
|---|---|
| Ceiling | ❑ |
| Walls | ❑ |
| Uplighters | ❑ |
| Individual lighting for occupants | ❑ |
| Low level/floor lighting | ❑ |
| Other (specify) _____ | ❑ |

**4.8 Percentage of glass in the different facades**

% Additional comments

| | |
|---|---|
| North | _____ |
| South | _____ |
| East | _____ |
| West | _____ |

**4.9 Contrast of window frames**

Additional comments

| | |
|---|---|
| Light-coloured window frames with light-coloured wall | ❑ |
| Light-coloured window frames with dark-coloured wall | ❑ |
| Dark-coloured window frames with light-coloured walls | ❑ |
| Dark-coloured window frames with dark-coloured walls | ❑ |

**4.10 What are the window frames constructed from?**

Additional comments

| | |
|---|---|
| Metal | ❑ |
| Wood | ❑ |
| PVC | ❑ |
| Aluminium | ❑ |
| Combination | ❑ |
| Other (specify) _____ | ❑ |

**4.11 How are main lights (e.g. ceiling or wall) controlled?**

Additional comments

| | |
|---|---|
| Automatic by time (building/floor/zone) | ❑ |
| Automatic with manual end control (building/floor/zone) | ❑ |
| Demand control: Daylight (photocells) | ❑ |
| Demand control: Occupants (motion sensors) | ❑ |
| Manual | ❑ |

**4.12 Solar shading devices**

Additional comments

| | |
|---|---|
| None | ❑ |
| South side only | ❑ |
| Other facades _____ | ❑ |
| External | ❑ |
| Internal | ❑ |

**4.13 Do solar shading devices hamper the use of**
**windows or decrease the ventilation capacity?**

Additional comments

| | |
|---|---|
| No | ❑ |
| Yes | ❑ |

**4.14 Control of the shading devices**    **Additional comments**

No control (fixed)    ❏
Individual    ❏
Central down, individual up    ❏
Automatic    ❏
Other (specify) _____    ❏

## Ventilation

**4.15 How is the room ventilated? (Indicate all present)**    **Additional comments**

Openable windows    ❏
Exhaust only    ❏
Supply and exhaust devices    ❏
Air conditioning cabinets    ❏
Other (specify) _____    ❏

**4.16 How is the ventilation controlled? (Indicate all present)**    **Additional comments**

None    ❏
Manual (on/off)    ❏
Central    ❏
Other    ❏

**4.17 Estimate of the percentage of office area with openable windows**    %    **Additional comments**

_____

**4.18 Are the windows opened?**    **Additional comments**

Yes    ❏
No    ❏

**4.19 Is the door to circulation area open?**    **Additional comments**

Yes    ❏
No    ❏

**4.20 Location of air supply devices (multi-code)?**    **Additional comments**

None    ❏
Floor    ❏
Windowsill    ❏
Ceiling    ❏
High on wall    ❏
Low on wall    ❏
Desks    ❏
Other (specify) _____    ❏

**4.21 Location of air exhaust devices (multi-code)?**    **Additional comments**

None    ❏
High    ❏
Low    ❏

**4.22 Designed air distribution principle**                    **Additional comments**

Displacement                                                    ❏
Mixing                                                          ❏
Other (specify) _____                        ❏

## Heating/cooling

**4.23 What type of heating system is present?**                **Additional comments**
**(Indicate all present)**

None                                                            ❏
Radiators or convectors                                         ❏
Heated floor                                                    ❏
Heated ceiling                                                  ❏
Heated walls                                                    ❏
Air conditioning cabinets                                       ❏
Air supply in ceiling                                           ❏
Air supply in wall                                              ❏
Air supply in floor                                             ❏

**4.24 What type of cooling system is present?**                **Additional comments**
**(Indicate all present)**

None                                                            ❏
Cooled floor                                                    ❏
Cooled ceiling                                                  ❏
Cooled walls                                                    ❏
Air conditioning cabinets                                       ❏
Air supply in ceiling                                           ❏
Air supply in wall                                              ❏
Air supply in floor                                             ❏

**4.25 How is the room temperature controlled?**                **Additional comments**
**(Tick all that apply)**

Manual radiator valve                                           ❏
Local thermostat at radiator/heating unit                       ❏
Local thermostat (e.g. on wall)                                 ❏
Manual control in room                                          ❏
Other (specify) _____                        ❏

**4.26 Are heaters located below windows to prevent**           **Additional comments**
**draught in winter?**

Yes                                                             ❏
Unnecessary – high performance glazing                          ❏
No                                                              ❏

## Furnishing

| 4.27 What is the main type of floor covering in the room? | | Additional comments |
|---|---|---|
| Carpet | ❏ | |
| Wood | ❏ | |
| Synthetic smooth floor covering | ❏ | |
| Exposed concrete | ❏ | |
| Stone/ceramic | ❏ | |
| Other (specify) _____ | ❏ | |

| 4.28 What is the main type of wall covering? | | Additional comments |
|---|---|---|
| Wallpaper | ❏ | |
| Enamel/gloss paint | ❏ | |
| Dispersion/emulsion paint | ❏ | |
| Wood/sealed cork | ❏ | |
| Porous fabrics including textiles | ❏ | |
| Stone/tile | ❏ | |
| Exposed concrete/plaster | ❏ | |
| Other (specify) _____ | ❏ | |

| 4.29 What is the main type of ceiling surface? | | Additional comments |
|---|---|---|
| Exposed concrete | ❏ | |
| Paint | ❏ | |
| Wallpaper | ❏ | |
| Synthetic material | ❏ | |
| Mineral fibre tiles | ❏ | |
| Wood fibre tiles; cork tiles | ❏ | |
| Wood | ❏ | |
| Gypsum/plaster | ❏ | |
| Other (specify) _____ | ❏ | |

| 4.30 Is there a suspended ceiling? | | Additional comments |
|---|---|---|
| No | ❏ | |
| Yes | ❏ | |

| 4.31 What material is the furniture (desks, tables, chairs, shelves/closet, etc.) made of? (Tick all that apply) | | Additional comments |
|---|---|---|
| Wood | ❏ | |
| Wood veneer | ❏ | |
| Plywood | ❏ | |
| Metal | ❏ | |
| Plastic laminate or composite | ❏ | |
| Other (specify) _____ | ❏ | |

| 4.32 What percentage of the furniture is | % | Additional comments |
|---|---|---|
| Less than one year old and made of particleboard or medium density fibreboard (MDF)? | _____ | |

**4.33 Are there any partitions within the rooms? If yes, give the material?**    Additional comments

No partitions ❏
Fabric ❏
Particleboard ❏
Metal ❏
Plastics ❏
Wood/wood laminate ❏
Other (specify) _____ ❏

## Moulds and dampness

**4.34 Is there visible mould growth in the office? (Answer 'yes' even if it appears only in small areas)**    Additional comments

No ❏
Yes ❏
    Where _____

| 4.35 Other damp/mould problems? | No | Yes | Additional comments |
|---|---|---|---|
| Noticeable mould odour | ❏ | ❏ | |
| Visible damp spots on walls, ceiling or floor | ❏ | ❏ | |
| Bubbles or yellow discolouration of plastic floors | ❏ | ❏ | |
| Blackened wood floor | ❏ | ❏ | |

**4.36 Does condensation tend to form on windows?**    Additional comments

No ❏
Yes ❏
    Inside ❏
    On the frame ❏
    Between glazing ❏
    Outside ❏

## Air pollution sources

**4.37 Which electronic IAQ sources are present in the room?**    Additional comments

Audiotape ❏
Computers/printers/photocopiers ❏
Data/video projector/TV/video conference ❏
Slide projector ❏
Servers ❏
Other (specify) _____ ❏

**4.38 Which non-electronic IAQ sources are present in the room (how many)?**    Additional comments

Blackboard with chalk _____

Whiteboard with markers _____

Electronic interactive board _____

Flip chart _____

Rugs – natural textile _____

Rugs – synthetic textile _____

Cushions – natural textile _____

Cushions – synthetic textile _____

Curtains – natural textile _____

Curtains – synthetic textile _____

Open shelves with gouaches, inks, etc. for graphic arts _____

Air fresheners (permanent: passive or electric plugged) _____

Air fresheners (occasionally: spray or other) _____

Plants in pots _____

Stuffed animals/pets _____

Other (specify) _____ _____

---

**4.39 Which other apparatus are present in the room?**      **Additional comments**

Air cleaners (specify type) _____  ☐

Space heaters  ☐

Humidifiers  ☐

Dehumidifiers  ☐

Other (specify) _____  ☐

---

**4.40 Are there any nearby (within 50 m from the windows) potential sources of outdoor air pollution that might influence the indoor environment?**      **Additional comments**

None  ☐

Car parking close to the building  ☐

Attached garage  ☐

Highway  ☐

Busy road  ☐

Low-traffic road  ☐

Power plant  ☐

Industry  ☐

Agricultural sources  ☐

Other (specify) _____  ☐

## Noise situation

**4.41 Are there any major indoor sources of noise?**      **Additional comments**

No indoor sources of noise  ☐

Occupants (distracting conversations)  ☐

Neighbours  ☐

Machines (photocopiers, computers, printers)  ☐

Vibration from fans, ducts, supply grilles or vents  ☐

Elevators  ☐

Other (specify) _____  ☐

---

**4.42 Is there any acoustic insulation applied?**      **Additional comments**

No  ☐

Yes (specify) _____  ☐

| 4.43 Can you hear outdoor noise sources? | | Additional comments |
|---|---|---|
| No outdoor noise sources | ❑ | |
| Traffic | ❑ | |
| Train | ❑ | |
| Airplane | ❑ | |
| People | ❑ | |
| Yes (specify) _____ | ❑ | |

## (Office) desk

| 4.44 How is the monitor of the PC or laptop in general positioned? | | Additional comments |
|---|---|---|
| Parallel to windows, with window in back | ❑ | |
| Parallel to windows, with window in front | ❑ | |
| Windows on the side, perpendicular to window | ❑ | |
| None of the above | ❑ | |

| 4.45 Can light from outside interfere with PC work? | | Additional comments |
|---|---|---|
| No | ❑ | |
| Yes | ❑ | |

| 4.46 How is the surface of the desk? (Tick all that apply) | | Additional comments |
|---|---|---|
| Glass | ❑ | |
| Reflecting light | ❑ | |
| Not reflecting light | ❑ | |
| Bright colour | ❑ | |
| Dark colour | ❑ | |

# B.5  Description of building use

Depending on the type of building investigated one can choose the questions and items that apply.

## General

| 5.1 Is there a documented complaints procedure for occupants who have a problem with the indoor environment? | | Additional comments |
|---|---|---|
| Yes | ❑ | |
| No | ❑ | |

## Activities

---

**5.2 What activities are carried out in the building?**   **Additional comments**

---

Office work ❑
Cooking ❑
Showering ❑
Sleeping ❑
Garage ❑
Underground car park ❑
Kitchen/restaurant ❑
Print shop ❑
Shop ❑
Industry ❑
Computer rooms ❑
Smoking allowed rooms ❑
Gymnasium ❑
Swimming pool ❑
Sauna ❑
Trash storage or trash separation room ❑
Dry-cleaning ❑
Other (specify) _____ ❑

---

**5.3 What types of general office machines are in occupied spaces?**   **Additional comments**

---

Laser printers ❑
Photocopiers ❑
Hot drinks machines ❑
Portable humidifiers ❑
Portable ionisers ❑
Portable air cleaners ❑
Other (specify) _____ ❑

---

**5.4 Is smoking permitted?**   **Additional comments**

---

No ❑
Only outside the building ❑
Only in separately-ventilated rooms ❑
    How many? _____
Yes ❑

---

**5.5 Are doors between rooms and circulation areas (corridors, landings, walkways, etc.) generally left open?**   **Additional comments**

---

Yes ❑
No ❑

---

## Pets, plants and pesticides

---

**5.6 Are there any pets present?**   **Additional comments**

---

Yes ❑
No ❑

---

| **5.7 How often are these types of animals present?** | Always | Sometimes | Rarely | Never |
|---|---|---|---|---|
| Furred | ❑ | ❑ | ❑ | ❑ |
| Feathered | ❑ | ❑ | ❑ | ❑ |
| Ants | ❑ | ❑ | ❑ | ❑ |
| Other (specify) _____ | ❑ | ❑ | ❑ | ❑ |

| **5.8 Are there any natural decorative plants present in the rooms?** | Additional comments |
|---|---|
| Yes | ❑ |
| No | ❑ |

| **5.9 In how many rooms?** | Additional comments |
|---|---|
| In every office | ❑ |
| Half of the offices | ❑ |
| Only a few of the offices | ❑ |

| **5.10A In the last year, have there been incidents of any of the following pests in or near the building?** | No incidents | Occupational space | Basement/ Outside |
|---|---|---|---|
| Rats | ❑ | ❑ | ❑ |
| Mice | ❑ | ❑ | ❑ |
| Cockroaches | ❑ | ❑ | ❑ |
| Ants | ❑ | ❑ | ❑ |
| Other insects | ❑ | ❑ | ❑ |
| Other (specify) _____ | ❑ | ❑ | ❑ |

| **5.10B In the last year, have you had to use a pesticide treatment for any of the following pests in or near the building?** | No incidents | Occupational space | Basement/ Outside |
|---|---|---|---|
| Rats | ❑ | ❑ | ❑ |
| Mice | ❑ | ❑ | ❑ |
| Cockroaches | ❑ | ❑ | ❑ |
| Ants | ❑ | ❑ | ❑ |
| Other insects | ❑ | ❑ | ❑ |
| Other (specify) _____ | ❑ | ❑ | ❑ |

| **5.11 Is there a pesticide treatment plan for the building?** | Additional comments |
|---|---|
| Yes | ❑ |
| No | ❑ |

## Cleaning

| **5.12 Is there a cleaning schedule for the communal parts of the building?** | Additional comments |
|---|---|
| Yes | ❑ |
| No | ❑ |
| Yes, some communal parts | ❑ |

**5.13 How often are cleaning activities carried out in the communal areas of the building?**

| | Daily | Twice or 3–4 times a week | Once a week | 1–3 times a month | Once a month | 2–4 times per year | Once a year | More than yearly | N/A |
|---|---|---|---|---|---|---|---|---|---|
| Floors/carpets swept/vacuumed | ❏ | ❏ | ❏ | ❏ | ❏ | ❏ | ❏ | ❏ | ❏ |
| Smooth floors washed | ❏ | ❏ | ❏ | ❏ | ❏ | ❏ | ❏ | ❏ | ❏ |
| Smooth floors waxed | ❏ | ❏ | ❏ | ❏ | ❏ | ❏ | ❏ | ❏ | ❏ |
| Smooth floors polished | ❏ | ❏ | ❏ | ❏ | ❏ | ❏ | ❏ | ❏ | ❏ |
| Walls dry-wiped/vacuumed | ❏ | ❏ | ❏ | ❏ | ❏ | ❏ | ❏ | ❏ | ❏ |
| Walls washed | ❏ | ❏ | ❏ | ❏ | ❏ | ❏ | ❏ | ❏ | ❏ |
| Ceilings dry-wiped/vacuumed | ❏ | ❏ | ❏ | ❏ | ❏ | ❏ | ❏ | ❏ | ❏ |
| Ceilings washed | ❏ | ❏ | ❏ | ❏ | ❏ | ❏ | ❏ | ❏ | ❏ |
| Surfaces dusted | ❏ | ❏ | ❏ | ❏ | ❏ | ❏ | ❏ | ❏ | ❏ |
| Surfaces polished | ❏ | ❏ | ❏ | ❏ | ❏ | ❏ | ❏ | ❏ | ❏ |
| Surfaces cleaned | ❏ | ❏ | ❏ | ❏ | ❏ | ❏ | ❏ | ❏ | ❏ |
| Other items (e.g. doors, banisters) dusted | ❏ | ❏ | ❏ | ❏ | ❏ | ❏ | ❏ | ❏ | ❏ |
| Other items polished | ❏ | ❏ | ❏ | ❏ | ❏ | ❏ | ❏ | ❏ | ❏ |
| Trash cans emptied | ❏ | ❏ | ❏ | ❏ | ❏ | ❏ | ❏ | ❏ | ❏ |

**5.14 How often are cleaning activities carried out in the rooms of the building?**

| | Daily | Twice or 3–4 times a week | Once a week | 1–3 times a month | Once a month | 2–4 times per year | Once a year | More than yearly | N/A |
|---|---|---|---|---|---|---|---|---|---|
| Floors/carpets swept/vacuumed | ❏ | ❏ | ❏ | ❏ | ❏ | ❏ | ❏ | ❏ | ❏ |
| Smooth floors washed | ❏ | ❏ | ❏ | ❏ | ❏ | ❏ | ❏ | ❏ | ❏ |
| Smooth floors waxed | ❏ | ❏ | ❏ | ❏ | ❏ | ❏ | ❏ | ❏ | ❏ |
| Smooth floors polished | ❏ | ❏ | ❏ | ❏ | ❏ | ❏ | ❏ | ❏ | ❏ |
| Walls dry-wiped/vacuumed | ❏ | ❏ | ❏ | ❏ | ❏ | ❏ | ❏ | ❏ | ❏ |
| Walls washed | ❏ | ❏ | ❏ | ❏ | ❏ | ❏ | ❏ | ❏ | ❏ |
| Ceilings dry-wiped/vacuumed | ❏ | ❏ | ❏ | ❏ | ❏ | ❏ | ❏ | ❏ | ❏ |
| Ceilings washed | ❏ | ❏ | ❏ | ❏ | ❏ | ❏ | ❏ | ❏ | ❏ |
| Surfaces dusted | ❏ | ❏ | ❏ | ❏ | ❏ | ❏ | ❏ | ❏ | ❏ |
| Surfaces polished | ❏ | ❏ | ❏ | ❏ | ❏ | ❏ | ❏ | ❏ | ❏ |
| Surfaces cleaned | ❏ | ❏ | ❏ | ❏ | ❏ | ❏ | ❏ | ❏ | ❏ |
| Other items (e.g. doors, banisters) dusted | ❏ | ❏ | ❏ | ❏ | ❏ | ❏ | ❏ | ❏ | ❏ |
| Other items polished | ❏ | ❏ | ❏ | ❏ | ❏ | ❏ | ❏ | ❏ | ❏ |
| Trash cans emptied | ❏ | ❏ | ❏ | ❏ | ❏ | ❏ | ❏ | ❏ | ❏ |

**5.15 When generally are the rooms cleaned?**          **Additional comments**

| | |
|---|---|
| In the morning before the arrival of workers/students/ children | ❏ |
| During occupation (e.g. work hours or school hours) | ❏ |
| In the evening (e.g. after work or school hours) | ❏ |
| Other | ❏ |
| Not known | ❏ |

**5.16 How often does a deep clean of the floors of the building take place?**          Additional comments

| | |
|---|---|
| Once a month or more often | ❏ |
| Once every three months or more often | ❏ |
| Once every six months or more often | ❏ |
| Once every year or more often | ❏ |
| Less often | ❏ |
| Never | ❏ |

**5.17 If any, when does this deep clean of the floors of the building take place?**          Additional comments

| | |
|---|---|
| In the morning before the arrival of workers/students/ children | ❏ |
| During occupation (e.g. work hours or school hours) | ❏ |
| In the evening (e.g. after work or school hours) | ❏ |
| Other | ❏ |
| Not known | ❏ |

**5.18 How often does a deep clean of other furniture in the building (e.g. chairs, furniture in communal areas, etc.) take place?**          Additional comments

| | |
|---|---|
| Once a month or more often | ❏ |
| Once every three months or more often | ❏ |
| Once every six months or more often | ❏ |
| Once every year or more often | ❏ |
| Less often | ❏ |
| Never | ❏ |

**5.19 If any, when does this deep clean of the furniture of the building take place?**          Additional comments

| | |
|---|---|
| In the morning before the arrival of workers/students/ children | ❏ |
| During occupation (e.g. work hours or school hours) | ❏ |
| In the evening (e.g. after work or school hours) | ❏ |
| Other | ❏ |
| Not known | ❏ |

**5.20 Are there any special cleaning requirements for the building?**          Additional comments

| | |
|---|---|
| Yes | ❏ |
| No | ❏ |

If yes, please specify which special requirements      _____

**5.21 Easy to clean**

| | Very difficult 1 | 2 | 3 | 4 | 5 | 6 | Very easy 7 |
|---|---|---|---|---|---|---|---|
| Does the layout of the rooms make it easy to clean? | ❏ | ❏ | ❏ | ❏ | ❏ | ❏ | ❏ |
| Are desks and storage shelves easy to clean? | ❏ | ❏ | ❏ | ❏ | ❏ | ❏ | ❏ |

**5.22 Are chemicals used for cleaning desks, shelves or cabinets in rooms?**          Additional comments

| | |
|---|---|
| Yes | ❏ |
| No | ❏ |
| Sometimes | ❏ |

**5.23 Are chemicals used for cleaning floors, ceilings, walls in rooms?**   Additional comments

| | |
|---|---|
| Yes | ❏ |
| No | ❏ |
| Sometimes | ❏ |

**5.24 If yes to question 5.22 or 5.23: Type of chemical used?**

| | Spray | Liquid | Additional comments |
|---|---|---|---|
| For floor cleaning or conservation | | | |
| Bleach or detergent with bleach | ❏ | ❏ | |
| Detergent without bleach | ❏ | ❏ | |
| Polish | ❏ | ❏ | |
| Other category relevant _____ | ❏ | ❏ | |
| For wall cleaning or conservation | | | |
| Bleach or detergent with bleach | ❏ | ❏ | |
| Detergent without bleach | ❏ | ❏ | |
| Polish | ❏ | ❏ | |
| Other category relevant _____ | ❏ | ❏ | |
| For windows cleaning | | | |
| Detergent with ammonia | ❏ | ❏ | |
| Detergent without ammonia | ❏ | ❏ | |
| Other category relevant _____ | ❏ | ❏ | |
| For furniture cleaning or conservation | | | |
| Detergent | ❏ | ❏ | |
| Polish | ❏ | ❏ | |
| Other category relevant _____ | ❏ | ❏ | |

**5.25 Are the windows open during cleaning of the classroom?**   Additional comments

| | |
|---|---|
| No | ❏ |
| Yes | ❏ |

# Appendix C: Indicators

## Indicators commonly used

Indicators commonly used for health and comfort issues as inventoried in the European project SuperBuildings (SuperBuildings, 2010a).

**Table C.1** *Visual comfort, sub-issues and indicators commonly used for health and comfort in the different existing building evaluation tools*

| Sub-issue | Indicator |
|---|---|
| Access to daylight | Access to daylight: yes or no<br>Intensity of daylight, expressed as the calculated daylight factor in lux and compared to reference values, regulation or standards<br>Percentage of floor area that is adequately day lit<br>Uniformity of daylight, expressed as a uniformity factor or in lux/h in comparison to reference values |
| Artificial lighting | Internal age and external lighting levels or average luminance, measured, expressed in lux and compared to regulation or standards<br>Uniformity of artificial lighting (calculated)<br>Presence of manual lighting control by users<br>Lighting zones<br>Measures to limit visual nuisance<br>Comfortable quality of the light emitted or colour index expressed in K (colour temperature) or IRC<br>Luminance of luminaries (cd/m$^2$)<br>The lighting system is conceived in order to avoid stroboscopic effects |
| Adequate view to the outside for all users | View to the outside: yes or no<br>% of users with adequate view out<br>Presence of view out, expressed as the availability, number and orientation of windows and openings and proved by photographs |
| Glare and reflections control | Presence of an (occupant-controlled) shading system (e.g. internal or external blinds), quality and type of control (e.g. automatic or manual)<br>Presence of measures against glare from daylight and/or artificial light<br>Glare rating luminaries (calculated)<br>Glare from light fixtures (horizontal)<br>Reflexion coefficients of walls, ceilings and floors |

Source: adapted from SuPerBuildings, 2010a.

**Table C.2** *Thermal comfort, sub-issues and indicators commonly used for health and comfort in the different existing building evaluation tools*

| Sub-issue | Indicator |
|---|---|
| Thermal comfort in winter and summer | Monitoring of room air and surface temperature |
| | Temperature control by occupants |
| | Take into account thermal comfort when planning and designing the building |
| | Setting of thermal comfort level during winter and summer |
| | Vertical temperature gradient |
| | Maximum temperature and time of this maximum temperature |
| | Presence of solar protection and protection to heat radiation from sun |
| | Design ventilation and heating based on thermal comfort in winter and summer |
| | Infiltration of heat to the interior or annual heating and cooling load and thermal transmission loss and summer insulation acquisition coefficient [$MJ/m^2$ or $W/m^2K$] |
| | Presence of thermal bridges |
| Draughts | Presence or not |
| Radiation | Presence or not |
| Relative humidity | Monitoring of humidity |
| | Humidity control |
| | Condensation |
| Ventilation | Limits for air speed in the space occupancy area for a set point temperature of 26°C |
| | Air conditioning presence or not |
| | Window openings for natural ventilation |

Source: adapted from SuPerBuildings, 2010a.

**Table C.3** *Acoustic comfort, sub-issues and indicators commonly used for health and comfort in the different existing building evaluation tools*

| Sub-issue | Indicator |
|---|---|
| Acoustic performance | Ambient or background noise level, measured or calculated and expressed in dB and compared to regulation or standards |
| | Acoustic pressure levels |
| | Reverberation intervals |
| Identification of different sources of noise, both indoors and outdoors | |
| Conduction of an acoustic study | |
| Measures to limit noise nuisance from indoors and outdoors through | Floor planning |
| | Acoustic insulation levels of openings, floors, walls |
| | Presence of sound-absorbing materials |
| | Noise control |
| | Use of silent equipment, measured and expressed in dB |

Source: adapted from SuPerBuildings, 2010a.

**Table C.4** *Indoor air quality, sub-issues and indicators commonly used for health and comfort in the different existing building evaluation tools*

| Sub-issue | Indicator |
| --- | --- |
| Ventilation | Ventilation rate, as number of air changes or volumes of air ($m^3/h$ or $m^3/h$/person)<br>Natural ventilation, expressed as percentage of windows opening<br>Mechanical ventilation evaluated by type and quality of the system<br>Control by users<br>Pollution prevention (e.g. $CO_2$ concentrations)<br>Purity of incoming air<br>Measures to optimize ventilation and air quality<br>Prevent air leaks |
| Emissions from materials | Indoor emissions of VOC, formaldehyde, fibres, carcinogens, chemicals, bacteria, wood treatment products, waterborne and airborne legionella contamination. Evaluation is based on laboratory analyses ($\mu g/kg$), certificates, labels or % of wall, floor and ceiling surface in contact with the indoor air |
| Other sub-issues | Moist control/humidity<br>Microbial contamination<br>Odours/olfactory comfort: identify sources of odours, provide a pleasant olfactory environment<br>Smoking prohibition<br>Outdoor emission of radon, based on in situ measurements |

Source: adapted from SuPerBuildings, 2010a.

# References

Adams, H. 'Patch' (1998) Memorable quotes for Patch Adams, www.imdb.com/title/tt0129290/quotes (accessed 21 December 2012).

Adan, O.C.G. (1994) *On the fungal defacement of interior finishes*, doctoral thesis, Technical University of Eindhoven, Wageningen, The Netherlands.

Adan, O.C.G., Samson, R.A. (eds) (2011) *Fundamentals of mold growth in indoor environments and strategies for healthy living*, The Netherlands, Wageningen Academic Publishers.

Adibi, J., Whyatt, R.M., Wiliams, L.P. Calafat, A.M., Camann, D., Herrick, R., Nelson, H., Bhat, H., Perera, F.P., Silva, M.J., Hauser, R. (2008) Characterization of phthalate exposure among pregnant women assessed by repeat air and urine samples, *Environmental Health Perspectives*, 116: 467–73.

Afshari, A., Matson, U., Ekberg, L.E. (2005) Characterization of indoor sources of fine and ultrafine particles: a study conducted in a full-scale chamber, *Indoor Air*, 15: 141–50.

Afshari, C.A., Hamadeh, H.K., Bushel, P.R. (2011) The evolution of bioinformatics in toxicology: advancing toxicogenomics, *Toxicological Sciences*, 120: S225–37.

Akimoto, T., Sasaki, M., Nakagawa, Y., Tanabe, S. (2009) Evaluation of workplace environment in creative workplace – monitoring during actual work and workers' behavior, *Conference Proceedings of Healthy Buildings 2009*, Syracuse, NY, paper 790.

AltTox.org (2009) Toxicity Testing Overview, http://alttox.org/ttrc/tox-test-overview (accessed 4 April 2012).

Alwaer, H., Sibley, M., Lewis, J. (2008) Factors and priorities for assessing sustainability of regional shopping centres in the UK, *Architectural Science Review*, 51: 391–402.

Andersson, J., Berggen, P., Gronkvist, M., Magnusson, S., Svensson, E. (2002) Oxygen saturation and cognitive performance, *Psychopharmacology*, 162: 119–28.

Ankley, G.T., Bennet, R.S., Erickson, R.J., Hoff, D.J., Hornung, M.W., Johnson, R.D., Mount, D.R., Nichols, J.W., Russom, C.L., Schmieder, P.K., Serrano, J.A., Tietge, J.E., Villeneuve, D.L. (2010) Adverse outcome pathways: a conceptual framework to support ecotoxicology research and risk assessment, *Environ. Toxicol. Chem.*, 29: 730–41.

Antelmi, I., De Paula, R.S., Shinzato, A.R., Peres, C.A., Mansur, A.J., Grupi, C.J. (2004) Influence of age, gender, body mass index, and functional capacity on heart rate variability in a cohort of subjects without heart disease, *The American Journal of Cardiology*, 93: 381–5.

Antonovsky, A. (1993) The structure and properties of the sense of coherence scale, *Soc. Sci. Med.*, 36: 725–33.

Apte, M.G., Fisk, W.J., Daisey, J.M. (2000) Associations between indoor $CO_2$ concentrations and sick building syndrome symptoms in US office buildings: an analysis of the 1994–1996 BASE study, *Indoor Air Journal*, 10: 246–57.

Apte, M.G. (2009) Response to 'Does filter media type really affect BRS?', *Indoor Air Journal*, 19: 526–8.

Aries, M.B.C., Veitch, J.A., Newsham, G.C. (2010) Window, view and office characteristics predict physical and psychological discomfort, *Journal of Environmental Psychology*, 30: 533–41.

Aries, M.B.C., Bluyssen, P.M. (2009) Climate change consequences for the indoor environment, *HERON*, 54: 49–69.

Arnetz, B.B., Berg, M. (1996) Melatonin and adrenocorticotropic hormone levels in video display unit workers during work and leisure, *J. Occup. Environm. Med.*, 38: 1108–10.

Arts, J.H.E., de Heer, C., Woutersen, R.A. (2006) Local effects in the respiratory tract: relevance of subjectively measured irritation for setting occupational exposure limits, *Int. Arch. Occup. Environ. Health*, 79: 283–98.

Ash, M. (2010) Dysregulation of the immune system: a gastro-centric perspective, Chapter 8 in: Nicolle, L., Woodriff Beirne, A. (eds) (2010) *Biochemical imbalances in diseases: a practitioner's handbook*, London and Philadelphia, PA: Singing Dragon.

ASHRAE (2004a) *ASHRAE Standard 55–2004: thermal environment conditions for human occupancy*, Atlanta, GA, American Society of Heating, Refrigerating and Air-Conditioning Engineers.

ASHRAE (2004b) *ASHRAE Standard 62.1–2004: ventilation for acceptable indoor air quality*, Atlanta, GA, American Society of Heating, Refrigerating and Air-Conditioning Engineers.

ASHRAE (2010a) *Guideline 10P, Interactions affecting the achievement of acceptable indoor environments*, second public review, Atlanta, GA, American Society of Heating, Refrigerating and Air-Conditioning Engineers.

ASHRAE (2010b) *Standard 55–2010: thermal environmental conditions for human occupancy* (ANSI approved), Atlanta, GA, American Society of Heating, Refrigerating and Air-Conditioning Engineers.

ASHRAE (2010c) *Performance measurements protocols for commercial buildings*, developed under the auspices of ASHRAE special project 115, www.ashrae.org (licensed 13 July 2010).

Aufderheide, M. (2008) An efficient approach to study the toxicological effects of complex mixtures, *Exp. Toxicol. Pathol.*, 60: 163–80.

Auliciems, A. (1981) Towards a psycho-physiological model of thermal perception, *Int. J. Biometeor*, 25: 109–22.

Ayres, J.G., Borm, P., Casee, P., Castranova, V., Donaldson, K., Ghio, A., Harrison, R.M., Hider, R., Kelly, F., Kooter, I.M., Marano, F., Maynard, R.L., Mudway, I., Nel, A., Sioutas, C., Smith, S., Baeza-Squiban, A., Cho, A., Duggan, S., Froines, J. (2008) Evaluating the toxicity of airborne particulate matter and nanoparticles by measuring oxidative stress potential: a workshop report and consensus statement, *Inhalation Toxicology*, 20: 75–99.

Babisch, W., Fromme, H., Beyer, A., Ising, H. (2001) Increased catecholamine levels in urine of subjects exposed to road traffic noise: the role of stress hormones in noise research, *Environment International*, 26: 475–81.

Babisch, W. (2002) The noise/stress concept, risk assessment and research needs, *Noise and Health*, 4: 1–11.

Babisch, W. (2006) Transportation noise and cardiovascular risk: updated review and synthesis of epidemiological studies indicate that the evidence has increased, *Noise and Health*, 8: 1–29.

Babisch, W. (2008) Road traffic noise and cardiovascular risk, *Noise and Health*, 10: 27–33.

Baker, D., Nieuwenhuijsen, M.J. (2008) *Environmental epidemiology, study methods and application*, New York: Oxford University Press.

Baker, J. (1999) Teacher–student interaction in urban at-risk classrooms: differential behaviour, relationship quality, and student satisfactions with school, *The Elementary School Journal*, 100: 57–70.

Bakke J.V., Norbäck D., Wieslander G., Hollund B.E., Florvaag E., Haugen E.N., Moen B.E. (2008) Symptoms, complaints, ocular and nasal physiological signs in university staff in relation to indoor environment: temperature and gender interactions, *Indoor Air*, 18: 131–43.

Balazova, I., Clausen, G., Rindel, J.H., Poulsen, T., Wyon, D.P. (2008) Open-plan office environments: a laboratory experiment to examine the effect of office noise and temperature on human perception, comfort and office work performance, *Proceedings of Indoor Air 2008*, Copenhagen, Denmark, paper 703.

Ball, T.M., Holberg, C.J., Aldous, M.B., Martinez, F.D., Wright, A.L. (2002) Influence of attendance at day-care on the common cold from birth through 13 years of age, *Arch. Pediatr. Adolesc. Med.*, 156: 121–6.

Bandura, A. (1982) Self-efficacy mechanism in human agency, *American Psychologist*, 37, 122–47.

Barabino, S., Chen, Y., Chauhan, S., Dana, R. (2012) Homeostatic mechanisms and their disruption in dry eye disease, *Progress in Retinal and Eye Research*, 31: 271–285.

Barclay, J.L., Miller, B.G., Dick, S., Dennekamp, M., Ford, I., Hillis, G.S., Ayres, J.G., Seaton, A. (2009) A panel study of air pollution in subjects with heart failure: negative results in treated patients, *Occup. Environ. Med.*, 66: 325–34.

Bare, J.C., Hostetter, P., Pennington, D.W., de Haes, H.A.U. (2000) State of the art: LCIA life-cycle impact assessment workshop, summary midpoint versus endpoints: the sacrifices and benefits, *Int. J. LCA*, 5: 319–26.

Barefoot, J.C., Dodge, K.A., Peterson, B.L., Dahlstrom, W.G., Williams, Jr. R.B. (1989) The Cook-Medley Hostility scale, item content and ability to predict survival, *Psychosomatic Medicine*, 5: 47–57.

Barett, P., Barett, L. (2010) The potential of positive places: senses, brain and spaces, *Intelligent Buildings International*, 2: 218–28.

Barett, P., Zhang, Y., Moffat, J., Kobbacy, K. (2013) A holistic, multi-level analysis identifying the impact of classroom design on pupils' learning, *Building and Environment*, 59: 678–89.

Bar-Haim, Y., Lamy, D., Pergamin, L., Bakermans-Kranenburg, M.J., van Ijzendoorn, M.H. (2007) Threat-related attentional bias in anxious and nonanxious individuals: a meta-analytic study, *Psychological Bulletin*, 133: 1–24.

Basanta, M., Jarvis, R.M., Xu, Y., Blackburn, G., Tal-Singer, R., Woodcock, A., Singh, D., Goodacre, R., Thomas, C.L., Fowler, S.J. (2010) Non-invasive metabolomic analysis of breath using differential mobility spectrometry in patients with chronic obstructive pulmonary disease and healthy smokers, *Analyst*, 135: 315–20.

Bassett, D.R., Ainsworth, B.E., Swartz, A.M., Strath, S.J., O'Brien, W.L., King, G.A. (2000) Validity of four motion sensors in measuring moderate intensity physical activity, *Medicine and Science in Sports and Exercise*, 32: S471–80.

Batty, G.D., Deary, I.J., MacIntyre, S. (2007) Childhood IQ in relation to risk factors for premature mortality in middle-aged persons: the Aberdeen children of the 1950s study, *Journal of Epidemiology and Community Health*, 61: 241–7.

Baumeister, R.F., Leary, M.R. (1995) The need to belong: desire for interpersonal attachments as a fundamental human motivation, *Psychological Bulletin*, 117: 497–529.

Bayer, C.W., Heerwagen, J., Gray, W.A. (2012) Health-centred buildings: a paradigm shift in buildings design and operation, *Proceedings of Healthy Buildings 2012*, Brisbane, Australia, paper 1E.5.

Béchamps, A. (1912) *The blood and its third anatomical element: application of the microzymian theory*, translated by M.R. Leverson, London: John Quesley.

Beelen, R., Hoek, G., Houthuijs, D., van den Brandt, P.A., Goldbohm, R.A., Fischer, P., Schouten, L.J., Armstrong, B., Brunekreef, B. (2009) The joint association of air pollution and noise from road traffic with cardiovascular mortality in a cohort study, *Occupational and Environmental Medicine*, 66: 243–50.

Beil, L. (2008) Medicine's new epicenter? Epigenetics: new field of epigenetics may hold the secret to flipping cancer's 'off' switch, *CURE (Cancer Updates, Research and Education)*, Winter, www.curetoday.com/index.cfm/fuseaction/article.show/id/2/article_id/949 (accessed 12 December 2012).

Bell, M.L., Ebisu, K., Peng, R.D., Dominici, F. (2009) Adverse health effects of particulate air pollution: modification by air conditioning, *Epidemiology*, 20: 682–6.

Bennet, K.M. (2005) Social engagement as a longitudinal predictor of objective and subjective health, *European Journal of Ageing*, 2: 48–55.

Berg, M., Arnetz, B.B., Liden, S., Eneroth, P., Kallner, A. (1992) Techno-stress. A psychophysiological study of employees with VDU-associated skin complaints, *J. Occup. Med.*, 34: 698–701.

Berglund, B., Lindvall, T., Schwela, D.H. (1999a) *Guidelines for community noise*, Geneva, WHO.

Berglund, B., Bluyssen, P.M., Clausen, G., Garriga-Trillo, A., Gunnarsen, L., Knoeppel, H., Lindvall, T., MacLeod, P., Mølahve, L., Winneke, G. (1999b) *Sensory evaluation of indoor air quality*, report no. 20, European Collaborative Action, Indoor Air Quality and Its Impact on Man, Joint Research Centre, Environment Institute, Ispra, Italy.

Berglund, B., Gunnarsson, A.D. (2000) Relationships between occupant personality and the Sick Building Syndrome explored, *Indoor Air Journal*, 10: 152–69.

Berglund, L.G., Cain, W.S. (1989) Perceived air quality and the thermal environment, *Proceedings of IAQ 89*, ASHRAE, Atlanta, GA, pp. 93–9.

Bernard, C. (1865), *An introduction to the study of experimental medicine*, original in French (1865), translated by H.C. Greene in 1957, New York, Dover Publications.

Bieling P.J., Antony, M.M., Swinson, R.P. (1998) The state-trait anxiety inventory, trait version: structure and content re-examined, *Behaviour Research and Therapy*, 36: 777–88.

Billings, J.S., Mitchell, S.W., Bergey, D.H. (1898) The composition of expired air and its effects upon animal life, *Smithsonian Contributions to Knowledge*, Washington DC, pp. 389–412.

Bjerregaard, P. (1983) Mean 24 hour heart rate, minimal heart rate and pauses in healthy subjects 40–79 years of age, *Eur. Heart J.*, 4: 44–51.

Blakemore, S-J., Burnett, S., Dahl, R.E. (2010) The role of puberty in the developing adolescent brain, *Human Brain Mapping*, 21: 926–33.

Blanchard, B.S. (2004) *System engineering management*, 3rd edition, Hoboken, NJ: John Wiley & Sons.

Blettner, M., Schlattmann, P. (2005) Meta-analysis in epidemiology, Chapter 7 in: Wolfgang, A. (ed.) *Handbook of epidemiology*, Springer: Berlin.

Bloemen, K., Hooyberghs, J., Desager, K., Witters, E., Schoeters, G. (2009) Non-invasive biomarker sampling and analysis of the exhaled breath proteome, *Proteomics Clin. Appl.*, 3: 498–504.

Bluyssen, P.M. (1990) *Air quality evaluated by a trained panel*, dissertation, October, Laboratory of Heating and Air Conditioning, Technical University of Denmark, Lyngby, Denmark.

Bluyssen, P.M. (1992) Indoor air quality management, a state of the art review and identification of research needs, *Indoor Environment*, 52: 326–34.

Bluyssen, P.M., Cox, C.W., van Drunen, Th.S. (1992) Evaluation of the indoor environment in problem buildings, *Proceedings of the International Conference Quality of the Indoor Environment*, Athens, Greece.

Bluyssen, P.M., Cox, C., Foradini, F., Dickson, D., Valbjørn, O. (1995) Identification of pollution sources by calculation of pollution loads, *Proceedings of Healthy Buildings 1995*, Milan, Italy, 3: 1335–40.

Bluyssen, P.M., Elkhuizen, P.A. (1996) Performance of trained and untrained panels in Europe, *Proceedings of Indoor Air 1996*, Nagoya, Japan, 1: 1053–8.

Bluyssen, P.M., de Oliveira Fernandes, E., Groes, L., Clausen, G.H., Fanger, P.O., Valbjørn, O., Bernhard, C.A., Roulet, C.A. (1996a) European Audit project to optimize indoor air quality and energy consumption in office buildings, *Indoor Air Journal*, 6: 221–38.

Bluyssen, P.M., Elkhuizen, P.A., Roulet C.A. (1996b) The Decipol method: a review, *Indoor Air Bulletin*, 3(6): 2–12.

Bluyssen, P.M., Cox, C., Plokker, W., Nicolaas, H., Soethout, L., Weerdenburg, M. (1999) EPIQR and IEQ: indoor environment quality investigation of 26 apartment buildings in Europe with a new tool, *Proceedings of Indoor Air 1999*, Edinburgh, Scotland, 4: 690–5.

Bluyssen, P.M. (2000) EPIQR and IEQ: indoor environment quality in European apartment buildings, *Energy and Buildings*, 31: 103–10.

Bluyssen, P.M. (2001) *State-of-the-art on performance concepts and tools for buildings*, TNO report 2001-GGI-R100, Delft, The Netherlands.

Bluyssen, P.M., Cox, C., Seppänen, O., de Oliveira Fernandes, E., Clausen, G., Müller, B., Roulet, C.A. (2003) Why, when and how do HVAC-systems pollute the indoor environment and what to do about it? *Building and Environment*, 38: 209–25.

Bluyssen, P.M., Adan, O.C.G. (2006) Marketing the indoor environment: standardization or performance on demand? *Proceedings of Healthy Buildings 2006*, Lisboa, Portugal, 5: 275–280.

Bluyssen, P.M. (2008) Management of the indoor environment: from a component related to an interactive top-down approach, *Indoor and Built Environment*, 17: 483–95.

Bluyssen, P.M. (2009a) *The indoor environment handbook: how to make buildings healthy and comfortable*, London: Earthscan.

Bluyssen, P.M. (2009b) Towards an integrative approach of improving indoor air quality, *Building and Environment*, 44: 1980–9.

Bluyssen, P.M., Aries, M.B.C. (2009) *Analysis of DALY as a health impact assessment indicator for the indoor environment*, TNO report, Delft, The Netherlands.

Bluyssen, P.M. (2010a) Towards new methods and ways to create healthy and comfortable buildings, *Building and Environment*, 45: 808–18.

Bluyssen, P.M., Oostra, M.A.R., Böhms, H.M. (2010b) A top-down system engineering approach as an alternative to the traditional over-the-bench methodology for the design of a building, *Intelligent Buildings International*, 2: 98–115.

Bluyssen, P.M. (2010c) *Indoor sources and health effects: background information and ways to go*, background document, Product Policy and Indoor Air Quality, 23 and 24 September, Directorate general for the Environment of the Belgian Federal Public Service of Health, Belgian Presidency of the Council of the European Union 2010, Brussels, Belgium.

Bluyssen, P.M., de Richemont, S., Crump, D., Maupetit, F., Witterseh, T., Gajdos, P. (2010d) Actions to reduce the impact of construction products on indoor air: outcomes of the European project Healthy Air, *Indoor and Built Environment*, 19: 327–39.

Bluyssen, P.M., Aries, M., van Dommelen, P. (2011a) Perceived comfort in office buildings: the European HOPE project, *Building and Environment*, 46: 280–8.

Bluyssen, P.M., Janssen, S., van den Brink, L.H., de Kluizenaar, Y. (2011b) Assessment of wellbeing in an office environment, *Building and Environment*, 46: 2632–40.

Bluyssen, P.M., Fossati, S., Mandin, C., Cattaneo, A., Carrer, P. (2012a) Towards a new procedure for identifying causes of health and comfort problems in office buildings, *Proceedings Healthy Buildings 2012*, Brisbane, Australia, paper 7.G7.

Bluyssen, P.M., de Kluizenaar, Y., Fossati, S., Mandin, C., Cattaneo, A., Carrer, P. (2012b) OFFICAIR Report D4.1 and D7.1 First Survey Results, TNO, Delft, The Netherlands.

Bluyssen, P.M., Alblas, M.J., Tuinman, I.L. (2013) In vitro exposure of human lung cells to emissions of several indoor air sources created in a climate chamber, *Sustainable Environment Research*, 23: 101–12.

BNP Paribas Real Estate (2012) *European office market*, www.realestate.bnpparibas. com/upload/docs/application/pdf/2012–11/propreport_office_uk-filigrane.pdf (accessed 19 November 2012).

Bonnefoy, X.R., Annesi-Maesona, I., Aznar, L.M., Braubachi, M., Croxford, B., Davidson, M., Ezratty, V., Fredouille, J., Ganzalez-Gross, M., van Kamp, I., Maschke, C., Mesbah, M., Moisonnier, B., Monolbaev, K., Moore, R., Nicol, S., Niemann, H., Nygren, C., Ormandy, D., Röbbel, N., Rudnai, P. (2004) Review of evidence on housing and health, background document for the Fourth Ministerial Conference on Environment and Health, Copenhagen, WHO Regional Office for Europe.

Bordass, B., Bromley, K., Leaman, A. (1993) User and occupant controls in office buildings, *Proceedings of ASHRAE Conference Building Design, Technology and Occupant Well-being in Temperate Climates*, Brussels, Belgium.

Bornehag, C.-G., Sundell, J., Bonini, S., Custovic A., Malmberg, P., Skerfving, S., Sigsgaard, T., Verhoeff, A. (2004a) Dampness in buildings as a risk factor for health effects, EUROEXPO: a multidisciplinary review of the literature (1998–2000) on dampness and mite exposure in buildings and health effects, *Indoor Air*, 14: 243–57.

Bornehag, C.-G., Sundell, J., Sigsgaard, T. (2004b) Dampness in buildings and health (DBH): report from an ongoing epidemiological investigation on the association between indoor environmental factors and health effects among children in Sweden, *Indoor Air* 14 (suppl. 7): 59–66.

Bornehag, C.-G., Sundell, J., Lundgren, B., Weshler, C.J., Sigsgaard, T., Hägerhed-Engmann, I. (2005a) Phthalates in indoor dust and their association with buildings characteristics, *Environmental Health Perspective*, 113: 1399–404.

Bornehag, C.-G., Sundell, J., Hägerhed-Engman, L., Sigsgaard, T. (2005b) Association between ventilation rates in 390 Swedish homes and allergic symptoms in children, *Indoor Air*, 15: 275–80.

Bornehag, C.-G., Sundell, J. Hägerhed-Engman, L., Sigsgaard, T., Janson, S., Aberg, N., the DBH Study group (2005c) Dampness at home and its association with airway, nose, and skin symptoms among 10851 preschool children in Sweden: a cross-sectional study, *Indoor Air*, 15 (suppl. 10): 48–55.

Bornehag, C.-G. (2009) Modern exposures and modern diseases, *Proceedings of Healthy Buildings 2009*, Syracuse, NY, plenary I1–082.

Bornehag, C.-G., Nanberg, E. (2010) Phthalates exposure and asthma in children, *International Journal of Andrology*, 33: 333–45.

Boschi N. (2002) Quality of life: meditations on people and architecture, *Proceedings of Indoor Air 2002*, Monterey, CA, 00.6405–6412.

Boyce, P.R. (2003) *Human factors in lighting*, 2nd edition, London and New York: Taylor & Francis.

BPIE (2011) *Europe's buildings under the microscope*, published by Buildings Performance Institute Europe, www.europeanclimate.org/documents/LR_CbC_study.pdf (accessed 29 October 2012).

Bradburn, N.M., Rips, L.J., Shevell, S.K. (1987) Answering autobiographical questions: the impact of memory and inference on surveys, *Science*, 236: 157–61.

Bradley, M.M., Lang, P.J. (1994) Measuring emotion: the self-assessment manikin and the semantic differential, *Journal of Behaviour Therapy and Experimental Psychiatry*, 25: 49–59.

Bradley, M.M., Miccoli, L., Escrig, M.A., Lang, P.J. (2008) The pupil as a measure of emotional arousal and autonomic activation. *Psychophysiology*, 45: 602–7.

Brainard, G.C., Hanifin, J.P., Greeson, J.M., Byrne, B., Glickman, G., Gerner, E., Rollag, M.D. (2001) Action spectrum for melatonin regulation in humans: evidence for a novel circadian photoreceptor, *Journal of Neuroscience*, 21: 6405–12.

Branbury, S.P., Berry, D.C. (2005) Office noise and employee concentration: identifying causes of disruption and potential improvements, *Ergonomics*, 48: 25–37.

Brantley, P.J., Jones, G.N. (1993) Daily stress and stress-related disorders, *Annals of Behavioral Medicine*, 15: 17–25.

Braucr, C., Mikkelsen, S. (2003) The context of a study influences the reporting of symptoms, *Int. Arch. Occup. Environ. Health*, 76: 621–4.

Brauer, C., Kolstad, H., Ørbæk, P., Mikkelsen, S. (2006) The sick building syndrome: a chicken and egg situation? *Int. Arch. Occup. Environ. Health*, 79: 465–71.

Brauer, C., Budtz-Jørgensen, E., Mikkelsen, S. (2008) Structural equation analysis of the causal relationship between health and perceived indoor environment, *Int. Arch. Occup. Environ. Health*, 81: 769–76.

van den Brink, L.H., Spiekman, M.E. (2009) *Interactie tussen gebruikers en installaties* (in Dutch) [Interaction between end-users and systems], TNO report, Delft, The Netherlands.

van den Brink, L.H., Attema, A.R., Kort, J., Spiekman, M.E. (2010) *Ontwikkelen van meetmethode: Waarom we de verwarming (niet) lager zetten* (in Dutch) [Development of a measurement method: why we can (not) turn down the heating system], TNO report, Delft, The Netherlands.

Brokelman, L., Vermande, H. (2005) *Faalkosten in de bouwwereld* (in Dutch) [Failure costs from the construction industry], SBR, Rotterdam, The Netherlands.

Brouwers, M.M., van Tongeren, M., Hirst, A.A., Bretveld, R.W., Roeleveld, N. (2009) Occupational exposure to potential endocrine disruptors: further development of a job exposure matrix, *Occupational and Environmental Medicine*, 66: 607–14.

Brown, R.J. (2004) Psychological mechanism of medically unexplained symptoms: an integrative conceptual model, *Psychol. Bull.*, 130: 793–812.

Brunekreef, B., Holgate, S.T. (2002) Air Pollution And Health, *The Lancet*, 360: 1253–42.

Bryan, P.A., Trinder, J., Curtis, N. (2004) Sick and tired: does sleep have a vital role in the immune system? *Nat. Rev. Immunol.*, 4: 457–567.

Bucchioni E., Kharitonov, S.A., Allegra, L., Barnes, P.J. (2003) High levels of interleukin-6 in the exhaled breath condensate of patients with COPD, *Respiratory Medicine*, 97: 1299–302.

Buchanan, I. Mendell, M., Mirer, A., Apte, M. (2008) Air filter materials, outdoor ozone and building-related symptoms in the BASE study, *Indoor Air Journal*, 18: 144–55.

Bulsing, P.J., Smeets, M., van den Hout, M. (2009) The implicit association between odors and illness, *Chem. Senses*, 34: 111–19.

Burge, S., Hedge, A., Wilson, S., Bass, J.H., Robertson, A. (1987) Sick Building Syndrome: a study of 4373 office workers, *Ann. Occup. Hyg.*, 31(4A): 493–504.

Burgess, H.J., Trinder, J., Kim, Y., Luke, D. (1997) Sleep and circadian influence on cardiac autonomic nervous system activity, *Am. J. Physiol. Heart Circ. Physiol.*, 273: 1761–8.

Burke, M.J., Brief, A.P., Georges, J.M. (1993) The role of negative affectivity in understanding relations between self-reports of stressors and strains: a comment on the applied psychology literature, *Journal of Applied Psychology*, 78: 402–12.

Caccavelli, D., Balaras, C., Gügerli, H., Allehaux, D., Witchen, K., Rasmussen, M.H., Bluyssen, P.M., Flourentzous, F. (2000) EPIQR-TOBUS: a new generation of decision-aid tools for selecting building refurbishment strategies, *Proceedings of Second International Conference on Decision Making in Urban and Civil Engineering*, Lyon, France.

Cain, W.S., Leaderer, B.P., Isseroff, R., Berglund, L.G., Huey, R.J., Lipsitt, E.D., Perlman, D. (1983) Ventilation requirements in buildings: control of occupancy odor and tobacco smoke odor, *Atmospheric Environment*, 17: 1183–97.

Cajochen, C., Zetzer, J.M., Czeisler, C.A., Dijk D.-J. (2000) Dose-response relationship for light exposure and ocular and electroencephalographic correlates of human alertness, *Behav. Brain Res.*, 115: 75–83.

Calabrese, E.J., Kenyon, E.M. (1991) *Air toxics and risk assessment*, Chelsea, MI: Lewis Publishers.

Cannon, W.B. (1963 [1932]) *The wisdom of the body*, first published in the Norton library 1963, New York: W.W. Norton and Company.

Capra, F. (1996) *The web of life: a new scientific understanding of living systems*, New York: Anchor books.

Carrer, P., Fanetti, A.C., Forastiere, F., Holcatova, I., Mølhave, L., Sundell, J., Viegi, G., Simoni, M. (2008) *EnVIE, Co-ordination Action on Indoor Air Quality and Health Effects*, WP1 final report Health effects, University of Milan, Italy.

Carrothers, T.J., Graham, J.D., and Evans, J. (1999) Putting a value on health effects of air pollution, *IEQ Strategies-Managing Risk*, October, 3(10).

Carslaw N., Langer S., Wolkoff P. (2009) Where is the link between reactive indoor air chemistry and health effects? *Atmospheric Environment*, 43: 3808–9.

Carter, N., Henderson, R., Lai, S., Hart, M., Booth, S., Hunyor, S. (2002) Cardiovascular and autonomic response to environmental noise during sleep in night shift workers, *Sleep*, 25: 457–64.

CDC-NIOSH (2012) *Science blog* – help wanted: spray polyurethane foam insulation research, http://blogs.cdc.gov/niosh-science-blog/2012/03/sprayfoam/ (accessed 28 January 2013).

Celermajer, D.S. (1997) Endothelial dysfunction: does it matter? Is it reversible? *JACC*, 30: 325–33.

CEN (2002a) *EN 12464–1 Light and lighting: lighting of work places* – Part 1: indoor work places, Brussels, Belgium.

CEN (2002b) *EN 12665 light and lighting: basic terms and criteria for specifying lighting requirements*, Brussels, Belgium.

CEN (2005) *prEN 15251 Criteria for the indoor environment including thermal, indoor air quality, light, and noise*, Brussels, Belgium.

CEN (2008) *prEN 15643–1 and 2: sustainability of construction works: sustainability assessment of buildings* – Part 1: general framework, and Part 2: framework for the assessment of environmental performance, AFNOR, France.

CEN (2010) *prEN 482 Workplace exposure: general requirements for the performance of procedures for the measurement of chemical agents*, Brussels, Belgium.

Chao, H.J., Schwartz J., Milton D.K., Burge, H.A. (2003) The work environment and workers' health in four large office buildings, *Environmental Health Perspectives*, 111: 1242–8.

Chapman, K., Robinson, S. (2007) *Challenging requirement for acute toxicity studies in the development of new medicines: a workshop report*, National Centre for the Replacement, Refinement and Reduction of Animals in Research (NC3Rs), London.

Chen, L., Jennison, B.L., Yang, W., Omaye, S.T. (2000) Elementary school absenteeism and air pollution, *Inhalation Toxicology*, 12: 997–1016.

Chen, C.J. Dai, Y.T., Sun, Y.M., Lin, Y.C., Juang, Y.J. (2007) Evaluation of auditory fatigue in combined noise, heat and workload exposure, *Industrial Health*, 45: 527–34.

Chen, Q., Hildemann, L.M. (2009) The effects of human activities on exposure to particulate matter and bioaerosols in residential homes, Environ. Sci. Technol., 43: 4641–6.

Choi, J.H., Aziz, A., Loftness, V. (2009) Decision support for improving occupant environmental satisfaction in office buildings: the relationship between sub-set of IEQ satisfactions and overall environmental satisfaction, *Proceedings of Healthy Building 2009*, Syracuse, NY, paper 747.

Choi, J.H., Aziz, A., Loftness, V. (2010) Investigation of the impacts of different genders and ages on satisfaction with thermal environments in office buildings, *Building and Environment*, 45: 1529–35.

Chou, J.W., Zhou, T., Kaufman, W.K., Paules, R.S., Bushel, P.R. (2007) Extracting gene expression patterns and identifying co-expressed genes from microarray data reveals biologically responsive processes, BMC genes, *BMC Bioinformatics*, 8: 427.

Chouraqui, P. Schnall, R.P., Dvir, I., Rozanski, A., Qureshi, E., Arditti, A., Saef, J., Feigin, P.D., Sheffy, J. (2002) Assessment of Peripheral Artery Tonometry in the Detection of Treadmill Exercise-Induced Myocardial Ischemia, *Journal of the American College of Cardiology*, 40: 2195–2200.

Chow, C.K., Lock, K., Teo, K., Subramanian, S.V., McKee, M., Yusuf, S. (2009) Environmental and societal influences acting on cardiovascular risk factors and disease at a population level: a review, *International Journal of Epidemiology*, 38: 1580–94.

CIB (1982) *Working with the performance approach in building*, CIB Report Publication 64, Rotterdam, The Netherlands, Council for Research and Innovation in Building and Construction.

CIOMS (2002) *International ethical guidelines for biomedical research involving human subjects*, Council for International Organizations for Biomedical Research, Geneva, Switzerland.

Clark, J.M., Thom, S.R. (2003) Oxygen under pressure, in: Brubakk, A.O., Neuman, T.S. (eds) *Bennett and Elliott's physiology and medicine of diving* (5th ed.), London: Saunders, pp. 358–418.

Clark, L.A., Watson, D. (1991) Tripartite model of anxiety and depression: psychometric evidence and taxonomic implications, *Journal of Abnormal Psychology*, 100: 316–36.

Clausen, G., Wyon, D.P. (2008) The combined effects of many different indoor environmental factors on acceptability and office work performance, *International Journal of Heating, Ventilation and Refrigeration Research*, 14: 103–13.

Clausen, G., Beko, G., Corsi, R.L., Gunnarsen, L., Nazaroff, W.W., Olesen, B.W., Sigsgaard, T., Sundel, J., Toftum, J., Weschler, C.J. (2011) Reflections on the state of research: indoor environmental quality, *Indoor Air*, 21: 219–30.

Clausen, G., Høst, A., Toftum, J., Bekö, G., Weschler, C., Callesen, M., Buhl, S., Ladegaard, M.B., Langer, S., Andersen, B., Sundell, J., Bornehag, C.-G., Sigsgaard, T. (2012) Children's health and its association with indoor environments in Danish homes and daycare centres: methods, *Indoor Air*, 22: 467–75.

Clausen, T., Christensen, K.B., Lund, T., Kristiansen, J. (2009) Self-reported noise exposure as a risk factor for long-term sickness absence, *Noise and Health*, 11: 93–7.

Clements-Croome, D. (2002) (ed.) *Creating the productive workplace*, London and New York: E&FN Spon.

Clements-Croome, D. (2004) *Intelligent buildings: design, management and operation*, London: Thomas Telford.

Clements-Croome, D. (2011) Sustainable intelligent buildings for people: a review, *Intelligent Buildings International*, 3: 67–86.

Clougherty J.E., Kubzansky L.D. (2009) A framework for examining social stress and susceptibility to air pollution in respiratory health environmental, *Health Perspectives*, 117: 1351–8.

Cohen, S., Kessler, R.C., Underwood, G.L. (1995) Measuring stress: a guide for health and social scientists, New York: Oxford University Press.

Cohen, S.M., Arnold, L.L. (2011) Chemical carcinogenesis, *Toxicological Sciences*, 120 (S1): S76–92.

Coldborn, T., Dumanoski, D., Myers, J.P. (1997) *Our stolen future*, New York: Plume.

Cook, W., Medley, D. (1954) Proposed hostility and pharisaic-virtue scales for the MMPI, *J. Appl. Psychol.*, 38: 414–18.

Corradi, L., Vinodkumar, A.M., Stefanini, A.M., Fioretto, E., Prete, G. (2002) Light and heavy transfer products in 58Ni+208Pb at the Coulomb barrier, *Physics Reviews*, C 66.

Coward, S.K.D., Llewelyn, J.W., Raw, G.J., Brown, V.M., Crump, D.R., Ross, D. (2001) *Indoor air quality in homes in England*, BRE report, BR433, London, UK.

Cox, C. (ed.) (2005) *Health optimisation protocol for energy-efficient buildings*, final report, 25 February 2005, TNO, Delft, The Netherlands.

Crain, E.F., Walter, M., O'Connor, G.T., Mitchell, H., Gruchalla, R.S., Kattan, M., Malindzak, G.S., Enright, P., Evans, R. 3rd, Morgan, W., Stout, J.W. (2002) Home and allergic characteristics of children with asthma in seven U.S. urban communities and design of an environmental intervention: the inner-city asthma study. *Environmental Health Perspectives*, 110: 939–45.

de Croon, E.M., Sluiter, J.K., Kuijer, P.P.F.M., Frings-Dresen M.H.W. (2005) The effect of office concepts on worker health and performance: a systematic review of the literature, *Ergonomics*, 48: 119–34.

Csobod, E., Rudnai, P.,Vaskovi, E. (2010) *School Environment and Respiratory Health of Children (SEARCH)*, International research project report within the programme 'Indoor air quality in European schools: preventing and reducing respiratory diseases', Regional Environmental Center for Central and Eastern Europe, Hungary, www.rec.hu/SEARCH (accessed 29 August 2012).

Culp, T.M. (2010) The metabolic syndrome: insuline resistance, dysglycaemia and dyslopideamia, Chapter 5 in: Nicolle, L., Woodriff Beirne, A. (eds) (2010) *Biochemical imbalances in diseases: a practioner's handbook*, London and Philadelphia: Singing Dragon.

Czernych, R., Wilma, K., Wolska, L. (2012) Age-related over- and underestimation of children health risk as a common mistake in risk assessment procedure, *Proceedings of Healthy Buildings 2012*, Brisbane, Australia, paper 4C5.

Dahl, M., Bauer, A.K., Aredouani, M., Soininen, R., Tryggvason, K., Kleeberger S.R., Kobzik, L. (2007) Protection against inhaled oxidants through scavenging of oxidized lipids by macrophage receptors MARCO and SR-AI, *Journal of Clinical Investigation*, 117: 757–64.

Daisey, J.M.A., Apte, W.J. (2003) Indoor air quality, ventilation and health symptoms in schools: an analysis of existing information, *Indoor Air*, 13: 53–64.

Darrat, I., Ahmad, N., Seidman, K., Seidman, M.D. (2007) Auditory research involving antioxidants, Curr. Opin. Otolaryngol, *Head Neck Surg.*, 15: 358–63.

de Dear, R. (2004) Thermal comfort in practice, *Indoor Air*, 14 (S7): 32–9.

de Dear, R. (2009) Towards a theory of adaptive thermal comfort: the pleasure principle, *Proceedings of Healthy Buildings 2009*, Syracuse, NY, Paper 804.

Deary, I.J., Batty, G.D., Pattie, A., Gale, C.R. (2008) More intelligent, more dependable children live longer, *Psychological Science*, 19: 874–80.

Deary, I.J., Liewald D., Nissan J. (2011) A free, easy-to-use, computer-based simple and four-choice reaction time programme: the Deary-Liewald reaction time task, *Behavior Research Methods*, 43: 258–68.

Delfino, R.J., Sioutas, C., Malik S. (2005) Potential role of ultrafine particles in associations between airborne particle mass and cardiovascular health, *Environmental Health Perspectives*, 113: 934–46.

Delongis A., Coyne J., Daof G., Folkman S., Lazarus, R.S. (1982) The relationship of hassles, uplifts, and major life events to health status, *Health Psychology*, 1: 119–36.

Demir, E., Desmet, P., Hekkert, P. (2006) Experiential concepts on design research: a (not too) critical review, in: Karlsson, M.A., Desmet, P.M.A., van Erp, J. (eds) *Proceedings of the 5th conference on design and emotion*, Göteborg, Sweden.

Desmet, P.M.A., Overbeeke, C.J., Tax, S.J.E.T. (2001) Designing products with added emotional value: development and application of an approach for research through design, *The Design Journal*, 4: 32–47.

Desmet, P.M.A., Hekkert, P. (2007) Framework of product experience, *International Journal of Design*, 1: 1–10.

Desmyter, J., Garvin, S., Lefebvre, P.-H., Stirano, F., Vaturi, A. (2010) *Perfection T1.4 A review of safety, security accessibility and positive stimulation indicators*, final report, 30 August, www.ca-perfection.eu/media/files/Perfection_D14_final.pdf (accessed 23 June 2012).

Deuble, M., de Dear, R. (2009) Do green buildings need green occupants? *Proceedings of Healthy Buildings 2009*, Syracuse, NJ, paper 229.

Diamanti-Kandarakis, E., Bourguignon, J-P., Ciudice, L.C., Hauser, R., Prins, G.S., Soto, A.M., Zoeller, R., T., Gore, A.C. (2009) Endocrine-disrupting chemicals: an endocrine society scientific statement, *Endocrine Reviews*, 30: 293–342.

Diener, E., Emmons, R.A., Larsen, R.J., Griffin, S. (1985) The satisfaction with life scale, *Journal of Personality Assessment*, 49: 71–5.

Diener, E. (2000) Subjective well-being, the science of happiness and a proposal for a national index, *American Psychologist*, 55: 34–43.

Diener, E., Seligman, M.E.P. (2002) Very happy people, *Psychological Science*, 13: 80–3.

Diez, U., Kroessner, T., Rehwagen, M., Richter, M., Wetzig, H., Schulz, R., Borte, M., Metzner, G., Krumbiegel, P., Herbarth, O. (2000) Effects of indoor painting and smoking on airway symptoms in atopy children in the first year of life: results of the LARS-study, Leipzig Allergy High-Risk Children Study, *Int. J. Hyg. Environ. Health*, 203: 23–8.

van Dijken, F., van Bronswijk, J.E.M.H., Sundell, J. (2006) Indoor environment in Dutch primary schools and health of the pupils, *Building Research and Information*, 34: 437–46.

Dimitroulopoulou, C. (2012) Ventilation in European dwellings: a review, *Building and Environment*, 47: 109–25.

Dockery, D.W., Pope, C.A., Xu, X., Spenglr, J.D., Ware, J.H., Fay, M.E., Ferris, B.G., Speizer, F.E. (1993) An association between air pollution and mortality in six US cities, *The New England Journal of Medicine*, 329: 1753–9.

Dolecek, T.A., Stamler, J., Caggiula, A.W., Tillotson, J.L., Buzzard, I.M. (1997) Methods of dietary and nutritional assessment and intervention and other methods in the multiple risk factor intervention trial, *American Journal of Clinical Nutrition*, 5: 196S–210S.

Donald, I., Siu, O. (2001) Moderating the stress impact of environmental conditions: the effect of organizational commitment in Honk Kong and China, *Journal of Environmental Psychology*, 21: 353–68.

van Dongen, J., Vos, H. (2007) *Health aspects of dwellings in the Netherlands* (in Dutch: *Gezondheidsaspecten van woningen in Nederland*), Delft, TNO report.

Dormann, C. (2003) Affective experiences in the home: measuring emotion, *Proceedings of Home Oriented Informatics and Telematics (HOIT)*, Irvine, CA.

Duffy, J.F., Wright, K.P. (2005) Entrainment of the human circadian system by light, *Journal of Biological Rhythms*, 20: 326–38.

Duffy, J.F. (2008) Forum: linking theory back to practice, *Building Research and Information*, 36: 655–8.

Duffy, J.F., Czeisler, C.A. (2009) Effect of light on human circadian physiology, *Sleep Med. Clin.*, 4: 165–77.

Durán-Narucki, V. (2008) School building condition, school attendance, and academic achievement in New York City public schools: a mediation model, *Journal of Environmental Psychology*, 28: 278–86.

Durmisevic, S., Ciftcioglu, O. (2010) Knowledge modelling for evidence-based design, *Health Environments Research and Design Journal*, 3: 101–23.

E²APT (2010) *The fundamental importance of building in future EU energy savings policies*, Energy Efficiency Action Plan Taskforce of the Construction Sector, http://euroace.org/LinkClick.aspx?fileticket=IYFmSEm7faM%3D&tabid=159 (accessed 29 October 2012).

EC (2012) *Guide to financial issues relating to FP7 indirect actions*, 16 January, ftp://ftp.cordis.europa.eu/pub/fp7/docs/financialguide_en.pdf.

Eder, W.E., Ege, M.J., von Mutius, E. (2006) The asthma epidemic, *N. Engl. J. Med.*, 355: 2226–35.

Edwards, P., Roberts, I., Clarke, M., Diguiseppi, C., Pratap, S., Wentz, R., Kwan, I. (2002) Increasing response rates to postal questionnaires: systematic review, *British Medical Journal*, 324: 1183–5.

Egloff, B., Schmuckle, S., Burns, L.R., Kohlman, C., Hock, M. (2003) Facets of dynamic positive affect: differentiating joy, interest, and activation in the positive and the negative affect schedule (PANAS), *Journal of Personality and Social Psychology*, 8: 528–40.

Elder, A., Gelein, R., Silva, Feikert, T., Opanashuk, L., Carter J., Potter, R., Maynard, A., Ito, Y., Finkelsei, J., Oberdörster, G. (2006) Translocation of inhaled ultrafine managese oxide particles to the central nervous system, *Environmental Health Perspectives*, 114: 1172–8.

Emenius, G., Svartengren, M., Korsgaard, J., Nordvall, L., Pershagen, G., Wickman, M. (2004) Building characteristics, indoor air quality and recurrent wheezing in very young children, BAMSE, *Indoor Air*, 14: 34–42.

Encyclopaedia Britannica (1991) Metabolism, in: *Macropaeda Knowledge in Depth*, vol. 23, pp. 975–1010.

Engelbrecht, K. (2003) *The impact of colour on learning*, http://sdpl.coe.uga.edu/HTML/W305.pdf (accessed 6 December, 2012).

Engvall, K., Norrby, C., Sandstedt, E. (2004) The Stockholm indoor environment questionnaire: a sociologically based tool for the assessment of indoor environment and health in dwellings, *Indoor Air*, 14: 24–33.

EPA (2003) A standardized EPA protocol for characterizing indoor air quality in large office buildings, www.epa.gov/iaq/base/methodology.html (accessed 12 December 2010).

Etzel, R.A. (2007) Indoor and outdoor air pollution: tobacco smoke, mould and diseases in infants and children, *Int. J. Hyg. Environ. Health*, 210: 611–16.

EU (2001) Directive 2001/95/EC of the European Parliament and of the Council of 3 December 2001 on *general product safety*, Brussels, Belgium.

EU (2004) Report from the commission to the European Parliament and the council concerning *existing community measures relating to sources of environmental noise*, pursuant to article 10.1 of Directive 2002/49/EC relating to the assessment and management of environmental noise, COM 160 final, Brussels, Belgium.

EU (2005) *Development of horizontal standardised assessment methods for harmonised approaches relating to dangerous substances under the construction products directive (CPD)*, Emission to indoor air, soil, surface water and ground water, M/366, Brussels, Belgium.

EU (2007a) *Together for health: a strategic approach for the EU: 2008–2013*, white paper, COM(2007) 630 final, Brussels, Belgium.

EU (2007b) *Improving quality and productivity at work: community strategy 2007–2012 on health and safety at work*, COM(2007) 62 final, Brussels, Belgium.

EU (2008) Communication from the Commission to the European Parliament, the Council, the European Economic and Social Committee and the Committee of the regions, *COM(2008) 400 final*, Brussels, Belgium.

EU (2010a) Directive 2010/31/EU of the European Parliament and the Council of 19 May 2010 on the *energy performance of buildings (recast)*, Belgium, Brussels.

EU (2010b) Towards a new energy strategy for Europe 2011–2010, council document, http://ec.europa.eu/energy/strategies/consultations/doc/2010_07_02/2010_07_02_energy_strategy.pdf (accessed 29 October 2012).

EU (2011) Regulation (EU) No. 305/2011 of the European Parliament and of the Council of 9 March 2011, Brussels, Belgium.

Evans, G., Maxwell, L. (1997) Chronic noise exposure and reading deficits, the mediating effects of language acquisition, *Environment and Behaviour*, 29: 638–56.

Evans, G.W., McCoy J.M. (1998) When buildings don't work: the role of architecture in human health, *Journal of Environmental Psychology*, 18: 85–94.

Evans, P., Halliwell, B. (1999) Free radicals and hearing: cause, consequence, and criteria, *Annals of the New York Academy of Sciences*, 884: 19–40.

Eysenck, H.J., Eysenck, M.W. (1985) *Personality and individual differences, a natural science approach*, New York: Plenum Press.

Fang, L., Clausen, G., Fanger, P.O. (1998) Impact of temperature and humidity on perception of indoor air quality during immediate and longer whole-body exposures, *Indoor Air*, 8: 276–84.

Fang, L., Wyon, D.P., Clausen, G., Fanger, P.O. (2004) Impact of indoor air temperature and humidity in an office on perceived air quality, SBS symptoms and performance, *Indoor Air* 14(suppl. 7): 74–81.

Fanger, P.O. (1982) *Thermal comfort*, Malabar, FL: Robert E. Krieger Publishing Company.

Fanger, P.O., Berg-Munch, B. (1983) Ventilation requirements for the control of body odor, *Proceedings of Engineer Foundation Conference Management of Atmospheres in Tightly Enclosed Space*, ASHRAE, Atlanta, GA.

Farmer, R., Sundberg, N.D. (1986) Boredom proneness: the development and correlates of a new scale, *Journal of Personality Assessment*, 50: 4–17.

Ferrie, J.E., Kivimaki, M., Head, J., Shipley, M.J., Vahtera, J., Marmot, M.G. (2005) A comparison of self-reported sickness absence with absences recorded in employers' registers: evidence from the Whiterhall II study, *Occup. Environ. Med.*, 62: 74–9.

Fibiger, W., Singer, G., Miller, A.J., Armstrong, S., Datar, M. (1984) Cortisol and catecholamines changes as functions of time-of-day and self-reported mood, *Neurosci. Biobehav. Rev.*, 8: 523–30.

Fiehn, O. (2001) Combining genomics, metabolome analysis, and biochemical modelling to understand metabolic networks, *Comp. Funct. Genomics*, 2: 155–68.

Fields, J.M. (1993) Effect of personal and situational variables on noise annoyance in residential areas, *J. Acoust. Soc. Am.*, 93: 2753–63.

Figuerio, M.G., Bierman, A., Plitnick, B., Rea, M.S. (2009) Preliminary evidence that both blue and red light can induce alertness at night, *BMC Neuroscience*, 10: 105.

Finalayson-Pitts, B.J., Pitts, J.N. (2000) *Chemistry of the upper and lower atmosphere*, San Diego, CA: Academic Press.

Fisk, W.J. (2000) Review of health and productivity gains from better IEQ, *Proceedings of Healthy Buildings 2000*, Helsinki, Finland, 4 August: pp. 22–34.

Fisk, W.J. (2001) Estimates of potential nationwide productivity and health benefits from better indoor environments: an update, in: Spengler, J. Samet, J.M., McCarthy, J.F. (eds) *Indoor Air Quality Handbook*, New York: McGraw-Hill, pp. 1–36.

Fisk, W.J., Lei-Gomez, Q., Mendell, M.J. (2007) Meta-analysis of the associations of respiratory health effects with dampness and mold in homes, *Indoor Air*, 17: 284–96.

Fowler, J.H., Christakis, N.A. (2009) Dynamic spread of happiness in a large social network: longitudinal analysis over 20 years in Framingham heart study, *British Medical Journal*, 337: 1–9.

Fraga, C.G., Litterio, M.C., Prince, P.D., Calabro, V., Piotrkowski, B., Galleano, M. (2011) Cocoa flavanols: effects on vascular nitric oxide and blood pressure, *Journal of Clinical Biochemistry and Nutrition*, 48: 63–7.

Frampton, M.W. (2011) *Health effects endpoints: oxidative stress and inflammation*, University of Rochester Medical Center, www.healtheffects.org/ACES/Frampton.pdf (accessed 2 November 2011).

Franchi, M., Carrer, P., Kotzias, D., Rameckers, E.M.A.L., Seppanen, O., van Bronswijk, J.E.M.H., Viegi, G. (2004) THADE report, *Towards healthy air in dwellings in Europe*, EFA, Naples, Italy, http://ec.europa.eu/health/ph_projects/2001/pollution/fp_pollution_2001_frep_02.pdf (accessed 4 November 2012).

Frank, S.M., Raja, S.N., Bulcao, C., Goldstein, D.S. (2000) Age-related thermoregulatory differences during core cooling in humans, *Am. J., Physiol. Regulatory Integrative Comp. Physiol.*, 279: R349–R354.

Frankel, M., Timm, M., Hansen, E.W., Madsen, A.M. (2012) Comparison of sampling methods for the assessment of indoor microbial exposure, *Indoor Air*, 22: 405–14.

Freathy, R.M., Kazeem, G.R., Morris, R.W., Johnson, P.C.D., Paternoster, L., Ebrahim, S., Hattersley, A.T., Hill, A., Hingorani, A.D., Holst, C., Jefferis, B.J., Kring, S.I.I., Mooser, V., Padmanabhan, S., Preisig, M., Ring, S.M., Sattar, N., Upton, M.N., Vollenweider, P., Waeber, G., SØrensen, T.I.A., Fraying, T.M., Watt, G., Lawlor, D.A.,

Whincup, P.H., Tozzi, F., Smith, G.D., Munafo, M. (2011) Genetic variation at CHRNA5-CHRNA3-CHRNB4 interacts with smoking status to influence body mass index, *International Journal of Epidemiology*, 40: 1617–28.

Fromme, H., Körner, W., Shahin, N., Wanner, A., Albrecht, M., Boehmer, S., Parlar, H., Mayer, R., Liebl, B., Bolte, G. (2009) Human exposure to polybrominated diphenyl ethers (PBDE), as evidenced by data from a duplicate diet study, indoor air, house dust, and biomonitoring in Germany, *Environment International*, 35: 1125–35.

Frontczak, M., Wargocki, P. (2011) Literature survey on how different factors influence human comfort in indoor environments, *Building and Environment*, 46: 922–37.

Frontczak, M., Schiavon, S., Goins, J., Arens, E., Zhang, H., Wargocki, P. (2012) Quantitative relationships between occupant satisfaction and aspects of indoor environmental quality and building design, *Indoor Air*, 22: 119–31.

deFur, P.L., Evans, G.W., Cohen Hubal, E.A., Kyle, A.D., Morello-Frosch R.A., Williams, D.R. (2007) Vulnerability as a function of individual and group resources in cumulative risk assessment, *Environmental Health Perspectives*, 115: 817–24.

Galea, S., Riddle, M., Kaplan, G.A. (2010) Causal thinking and complex system approaches in epidemiology, *International Journal of Epidemiology*, 39: 97–106.

Gee, G., Payne-Sturges, D. (2004) Environmental health disparities: a framework for integrating psychosocial and environmental concepts, *Environmental Health Perspectives*, 112: 1645–53.

Geiser, M., Rothen-Rutishauser, B., Kapp, N., Schürch, S., Kreyling, W., Schulz, H., Semmler, M., Im Hof, V., Heyder, J., Gehr, P. (2005) Ultrafine particles cross cellular membranes by nonphagocytic mechanisms in lungs and in cultured cells, *Environmental Health Perspectives*, 113: 1555–60.

Geldard, A. (1972) The sense of smell, Chapter 5 in: *The human senses*, 2nd edition, New York: John Wiley & Sons.

Geller R.J., Rubin, I.L., Nodvin, J.T., Teague, W.G., Frumkin, H. (2007) Safe and healthy school environments, *Pediatric Clinics of North America*, 54: 351–73.

Giannattasio, C., Ferrari, A.U., Mancia, G. (1994) Alterations in neural cardiovascular control mechanisms with ageing, *J. Hypertens. Suppl.*, 12: S13–17.

Goldberg, D.P. (1972) *The detection of psychiatric illness by questionnaire*, London: Oxford University Press.

Goldberg, L.R. (1992) The development of markers for the big-five factor structure, *Psychological Assessment*, 4: 26–42.

Goonewardena, S.N., Prevette, L.E., Desai, A.A. (2010) Metabolomics and atherosclerosis, *Curr. Artheroscler. Rep.*, 12: 267–72.

Goor, D.A., Sheffy, J., Schnall, R.P., Arditti, A., Caspi, A., Bragdon, E.E., Sheps, D.S. (2004) Peripheral arterial tonometry: a diagnostic method for detection of myocardial ischemia induced during mental stress tests: a pilot study, *Clinical Cardioology*, 27: 137–41.

Gosling, S.D., Rentfrow, P.J., Swann, W.B. (2003) A very brief measure of the big five personality domains, *Journal of Research in Personality*, 37: 504–28.

Goyer, R.A. (1995) Nutrition and metal toxicity, *American Journal of Clinical Nutrition*, 61: 646S-650S.

Graham, J.M.A., Janssen, S.A., Vos, H., Miedema, H.M.E. (2009) Habitual traffic noise at home reduces cardiac parasympathetic tone during sleep, *International Journal of Psychophysiology*, 72: 179–86.

Grandjean, P., Poulsen, L.K., Heilmann, C., Steuerwald, U., Weihe, P. (2010) Allergy and sensitization during childhood with prenatal and lactational exposure to marine pollutants, *Environmental Health Perspectives*, 118: 1429–33.

Graveland, H.V.R., Rensen, S.A., Brunekreef, W.M., Gehring, B. (2011) Air pollution and exhaled nitric oxide in Dutch schoolchildren, *Occup. Environ. Med.*, 68: 551–6.

Green, P.J., Kirby, R., Suls, J. (1996) The effects of caffeine on blood pressure and heart rate: a review, *Annals of Behavioral Medicine*, 18: 201–16.

Grös, D.F., Antony, M.M., Simms, L.J., McCabe, R.E. (2008) Psychometric Properties of the State–Trait Inventory for Cognitive and Somatic Anxiety (STICSA): comparison to the State–Trait Anxiety Inventory (STAI), *Psychology Assessment*, 4: 369–81.

Grundy, S.M., Brewer, H.B., Cleeman, J.I., Smith, S.C., Lenfant, C. (2004) Definition of metabolic syndrome, report of the national heart, lung, and blood institute/American heart association conference on scientific issues related to definition, Journal of the American Heart Association, *Circulation*, 109: 433–8.

Haase, A., Rott, S., Mantion, A., Graf, P., Plendll, J., Thünemann, A.F., Meier, W.P., Taubert, A., Luch, A., Reiser, G. (2012) Effects of silver nanoparticles on primary mixed neural cell cultures: uptake, oxidative stress and acute calcium responses, *Toxicological Sciences*, 126: 457–68.

Hågerhed-Engman, L., Bornehag, C.G., Sundell, J. (2006) Day-care attendance and increased risk for respiratory and allergic symptoms in preschool age, *Allergy*, 61: 447–53.

Hågerhed-Engman, L., Sigsgaard, T., Samuelson, I., Sundell, J., Janson, S., Bornehag, C.-G. (2009) Low home ventilation rate in combination with moldy odor from the building structure increase the risk for allergic symptoms in children, *Indoor Air*, 19: 184–92.

Halliwell, B., Gutteridge, J.M. (2007) *Free radicals in biology and medicine*, 4th edition, Oxford: Oxford University Press.

Hanciles, S., Pimlott, Z. (2010) Polyunsaturated fatty acid (PUFA) imbalances, Part 2: PUFAs in the brain, Chapter 4 in: Nicolle, L., Woodriff Beirne, A. (eds) (2010) *Biochemical imbalances in diseases: a practitioner's handbook*, London and Philadelphia: Singing Dragon.

Hänninen, O., Asikainen, A., Bischof, W., Hartmann, T., Carrer, P. Seppänen, de Oliveira Fernandes, E., Leal, V., Malvik, B., Kephalopoulos, S., Braubach, M., Wargocki, P. (2012) Health implications of alternative potential ventilation guidelines, *Proceedings of Healthy Buildings 2012*, Brisbane, Australia, paper 3A.3.

Hansen, A.M., Meyer, H., Guntelberg, F. (2008) Building-related symptoms and stress indicators, *Indoor Air*, 18: 440–6.

Hansen, L., Bach, E., Kass Ibsen, E., Osterballe, O. (1987) Carpeting in schools as an indoor pollutant, *Proceedings of Indoor Air 1987*, Berlin, Germany, vol. 2: 727–31.

Hartley, T., Maguire, E.A., Spiers, H.J., Burgess, N. (2003) The well-worn route and the path less traveled: distinct neural bases of route following and wayfinding in humans, *Neuron*, 37: 877–88.

de Hartog, J.J., Lanki, T., Timonen, K.L., Hoek, G., Janssen, N.A.H., Ibald-Mulli, A., Peters, A., Heinrich, J., Tarklainen, T.H., van Grieken, R., van Wijnen, J.H., Brunekreef, B., Pekkanen, J. (2009) Associations between PM2.5 and heart rate variability are modified by particle composition and beta-blocker use in patients with coronary heart disease, *Environmental Health Perspectives*, 117: 105–11.

Harvey P.W., Everett D.J., Springall C.J. (2007) Adrenal toxicology: a strategy for assessment of functional toxicity to the adrenal cortex and steroidogenesis, *J. Appl. Toxicol.*, 27: 103–15.

Haskell, E.H., Palca, J.W., Walker, J.M., Berger, R.J., Heller, H.C. (1981) The effects of high and low ambient temperatures on human sleep stages, *Electroencephalography and clinical neurophysiology*, 51: 494–501.

Hathaway, W.E. (1992) A study into the effects of types of lighting on children. A case of daylight robbery, www.advancedglazings.com/expSolera/researchPaper_Daylight_Robbery.pdf (accessed 6 December 2012).

Haverinen-Shaughnessy, U., Moschandreas, D.J., Shaughnessy, R.J. (2011) Association between substandard classroom ventilation rates and students' academic achievement, *Indoor Air*, 21: 121–31.

Haverinen-Shaughnessy, U., Turunen, J., Palonen, J., Putus, T., Kurnitski, J., Shaughnessy, R. (2012) Health and academic performance of sixth grade students and indoor environmental quality in Finnish elementary schools, *British Journal of Educational Research*, 2: 42–58.

Hawkes, D. (1982) The theoretical basis of comfort in the 'selective' control of environments, *Energy and Buildings*, 5: 127–34.

Hays, R.D., Hayashi, T., Stewart, A.L. (1989) A five-item measure of socially desirable response set, *Educational and Psychological Measurement*, 49: 629–36.

Hays, R.D., Sherbourne C.D., Mazel, R.M. (1995) User's manual for the medical outcomes study (MOS) Core measures of health-related quality of life, retrieved from the RAND Corporation web site www.rand.org/health/surveys_tools.html (accessed 22 May 2012).

HCN (Health Council of The Netherlands) (2004) *The influence of night-time noise on sleep and health*, report no. 2004/14E, HCN, The Hague, The Netherlands.

Headey, B., Wearing, A. (1992) *Understanding happiness: a theory of subjective well-being*, Melbourne: Longman Cheshire.

Health Council of The Netherlands (2000) *RSI*, publication no. 2000/22, The Hague, The Netherlands.

Healthpages.org (2011) Guide to cumulative trauma disorders (CTDs), http://healthpages.org/health-a-z/guide-cumulative-trauma-disorders/ (accessed 4 September 2011).

Hedge, A. (1982) The open-plan office: a systematic investigation of employee reactions to their work-environment, *Environment and Behaviour*, 14: 519–42.

Hedge, A., Erickson, W.A., Rubin, G. (1992) Effects of personal and occupational factors on sick building syndrome reports in air conditioned offices, in: Quick, J.C., Murphy, L.R., Hurell, J.J. (eds) *Work and well being: assessment and interventions for occupational mental health*, Washington DC: American Psychological Association, pp. 286–98.

Hellhamer, D.H., Hellhammer, J. (2008) *Stress the brain–body connection*, Key issues in mental health, editors: A. Riecher-Rössler, M. Steiner, vol. 174, Basel, Switzerland: Karger.

Henderson, D., Bielefeld E.C., Carney Harris, K., Hua Hu, B. (2006) The role of oxidative stress in noise-induced hearing loss, *Ear and Hearing*, 27: 1–19.

Henry, I.P. (1986) Neuroendocrine patterns of emotional response, in: R. Plutchik, H. Kellerman (eds) *Emotion: theory research and experience*, San Francisco, CA: Academic Press, pp. 37–60.

Henvinet (2009) The Henvinet policy briefs: 'Expert elicitation on health implications of decaBDE', 'Expert elicitation on health implications of HBCD', 'Expert elicitation on health implications of phthalates', and 'Expert elicitation on neurodevelopmental implications of CPF', henvinet.nilu.no/Home/tabid/339/language/en-US/Default.aspx (accessed 24 May 2013).

Hepple, B., Peckham, C., Baldwin, T., Brazier, M., Bronsword, R., Calman, K., Harper, P., Harries, R., Lipton, P., Southwark, P., Plant, R., Raff, M., Ross, N., Sewell, H., Smith, P., Strathern, D.M., Williamson, A. (2005) *The ethics of research involving animals*, London: The Nuffield Council on Bioethics.

van Herk, H., Porting, Y.H., Verhallen, T.M.M. (2004) Response styles in rating scales: evidence of method bias in data from six EU countries, *Journal of Cross-cultural Psychology*, 35: 346–60.

Herz, R.S. (2002) Influence of odors on mood and affective cognition, Chapter 10 in: Rouby, C., Schaal, B., Dubois, R., Gervais, R., Holley, A. (eds) *Olfaction, taste, and cognition*, Cambridge: Cambridge University Press.

Heschong, L. (2002) Day lighting and student performance, *ASHRAE J.*, 44: 65–7.

Heschong, L., Roberts, J.E. (2009) Linking daylight and views to the reduction of sick building syndrome, *Proceedings of Healthy Building 2009*, Syracuse, NY, paper 797.

Heschong Mahong Group (HMG) (1999) *Daylighting in schools: an investigation into the relationship between daylighting and human performance*, Pacific Gas and Electric Company, HMG project no. 9803, San Francisco, CA.

Heudorf, U., Mersch-Sundermann, V., Angerer, J. (2007) Phthalates: toxicology and exposure, *Int. J. Hyg. Environ. Health*, 210: 623–34.

Hinson, J., Raven, P., Chew, S. (2010) *The endocrine system*, 2nd edition, Systems of the body, China: Churchill Livingstone Elsevier.

Hoem, H.K., Bjelland, H. (2006) Making users talk about product experiences: exploring the three levels of human processing in a product design context, in: Karlsson, M.A., Desmet, P.M.A., van Erp, J. (eds) *Proceedings of the 5th conference on design and emotion*, Göteborg, Sweden.

Höfler, W. (1968) Changes in regional distribution of sweating during acclimatization to heat, *J. Appl. Physiol.*, 25: 503–6.

Hofman, W.F., Kumar, A., Tulen, J.H. (1995) Cardiac reactivity to traffic noise during sleep in man, *J. Sound Vib.*, 179: 577–89.

Hollander, A.G.M., Melse, J.M. (2006) Valuing the health impact of air pollution, in: Ayres, J., Maynard, R., Richards, R. (eds), *Air pollution and health*, Air Pollution Reviews, vol.3, London: Imperial College Press.

Holmes, E., Loo, R.L., Stamler, J., Bictash, M., Yap, I.K., Chan, Q., Ebbels, T., De Iorio, M., Brown, I.J., Veselkov, K.A., Daviglus, M.L., Kesteloot, H., Ueshima, H., Zhao, L., Nicholson, J.K., Elliot, P. (2008) Human metabolic phenotype diversity and its association with diet and blood pressure, *Nature*, 453: 396–400.

Holmes, T.H., Rahe, R.H. (1967) The social readjustment scale, *Journal of Psychosomatic Research*, 11: 213–18.

Hölscher, B.H.J., Jacob, B., Ritz, B., Wichmann, H.-E., Hölscher, B. (2000) Gas cooking, respiratory health and white blood cell counts in children, *Int. J. Hyg. Environ. Health*, 203: 29–37.

van Hoof, J., Mazej, M., Hensen, J.L.M. (2010) Thermal comfort: research and practice, *Frontiers in Bioscience*, 15: 765–88.

Houle, J.N., Staff, J., Mortimer, J.T., Uggen, C., Blackstone, A. (2011) The impact of sexual harassment on depressive symptoms during the early occupational career, *Society and Mental Health*, 2: 89–105.

van den Hout, F. (1989) Kleur en uitzicht [Colour and view], Technische Universiteit Eindhoven (in Dutch).

Houtman, I., Douwes, M., de Jong, T., Meeuwsen, J.M., Jongen, M., Brekelmans, F., Nieboer-Op de Weegh, M., Brouwer, D., van den Bossche, S., Zwetsloot, G., Reinert, D., Neitzner, I., Hauke, A., Flaspoler, E., Zieschang, H., Kolk, A., Nies, E., Bruggemann-Prieshoff, H., Roman, D., Karpowicz, J., Perista, H., Cabrita, J., Corral, A. (2008) *New forms of physical and psychological health risks at work*, European Parliament, Policy Department Economic and Scientific Policy, IP/A/EMPF/ST/2007-19, PE 408.569, Brussels, Belgium.

Hsu, N.Y., Lee, C.C., Wang, J.Y., Li, Y.C., Chang, H.W., Chen, C.Y., Bornehag, C.G., Wu, P.C., Sundell, J., Su, H.J. (2012) Predicted risk of childhood allergy, asthma and reported symptoms using measured phthalate exposure in dust and urine, *Indoor Air*, 22: 186–99.

Huizinga, C., Abbaszadeh, S., Zagreus, L., Arens, W. (2006) Air quality and thermal comfort in office buildings: results of a large indoor environmental quality survey, *Proceedings of Healthy Buildings 2006*, Lisboa, Portugal, 3: 393–7.

Hummel, T., Heilman, S., Murphy, C. (2002) Age-related changes in chemosensory functions, Chapter 27 in: Rouby, C., Schaal, B., Dubois, R., Gervais, R., Holley, A. (eds) *Olfaction, taste, and cognition*, Cambridge: Cambridge University Press.

Humphreys, M.A., Nicol J.F. (1998) Understanding the adaptive approach to thermal comfort, *ASHRAE Trans.*, 104: 991–1004.

Humphreys, M.A. (2005) Quantifying occupant comfort: are combined indices of the indoor environment practicable? *Building Research and Information*, 33: 317–25.

Humphreys, M.A., Nicol F.J. (2007) Self-assessed productivity and the office environment: monthly surveys in five European countries, *ASHRAE Trans.*, *113*: 606–16.

Hygge, S., Knez, I. (2001) Effects of noise, heat and indoor lighting on cognitive performance and self-reported affect, *Journal of Environmental Psychology*, 21: 291–9.

Hygge, S. (2003) Classroom experiments on the effects of different noise sources and sound levels on long-term recall and recognition in children, *Applied Cognitive Psychology*, 17: 895–914.

Hyvaerinen, A., Karakainen, P., Meklin T., Rintala, H., Päivi Kärkkäinen, Korppi, M., Putus, T., Pekkanen, J., Nevalainen, A. (2009) Airborne microbial levels: associations with childhood asthma and moisture damage, *Proceedings of Healthy Building 2009*, Syracuse, NY, paper 168.

Institute of Medicine (2000) *Cleaning the air, asthma and indoor exposures*, Committee on the Assessment of Asthma and Indoor Air, Washington DC: National Academy Press, 438.

Intille S.S., Rondoni J., Kukla C., Iacono I., Bao, L. (2003) A context-aware experience sampling tool. *CHI 2003*, Lauderdale, FL.

Irwin, M., Thompson, J., Miller, C., Gillin, J.C., Ziegler, M. (1999) Effects of sleep and sleep deprivation on catecholamine and interleukin-2 levels in humans: clinical implications, *J. Clin. Endocrinol. Metab.*, 84: 1979–85.

Isaksen, I.S.A., Dalsøren, S.B. (2011) Getting a better estimate of an atmospheric radical, *Science* 331 (6013): 38–9.

ISEE (2012) *Ethics guidelines for environmental epidemiologists*, www.iseepi.org/about/ethics.htm (accessed 5 May 2012).

ISO (2005a) *EN ISO, 7730 Moderate thermal environments: determination of the PMV and PPD indices and specification of the conditions for thermal comfort*, International Organization for Standardization, Geneva, Switzerland.

ISO (2005b) *ISO 17025 General requirements of the competence of testing and calibration laboratories*, International Organization for Standardization, Geneva, Switzerland.

ISO (2006) *ISO/TS21929 Building construction: sustainability in building construction – sustainability indicators* – Part I: framework for the development for indicators for buildings, International Organization for Standardization, Geneva, Switzerland.

ISO (2007) *ISO 20988: 2007 Air quality: guidelines for estimating measurement uncertainty*, International Organization for Standardization, Geneva, Switzerland.

ISO (2008a) *ISO/TS 21931-1 Sustainability in building construction: framework for methods of assessment for environmental performance of construction works* – Part 1: buildings, Revision of ISO/TS 21931-1: 2006, International Organization for Standardization, Geneva, Switzerland.

ISO (2008b) *ISO 9001 Quality management systems: requirements*, International Organization for Standardization, Geneva, Switzerland.

ISO (2008c) *ISO/EC 15288:2008 International standard, systems and software engineering: system life cycle processes*, Piscataway, NJ.

Izard, C.E., Dougherty, F.E., Bloxom, B.M., Kotsch, W.E. (1974) *The differential emotions scale: a method of measuring the subjective experience of discrete emotions*, Vanderbilt University, Nashville, TN.

Jaakola, J.J.K., Tuomaala, P., Seppanen, O. (1994) Air recirculation and sick building syndrome: a blinded crossover trial, *Am. J. Public Health*, 84: 422–8.

Jaakola, J.J.K. (1998) The office environment model: a conceptual analysis of the Sick Building Syndrome, *Indoor Air Journal*, 4 (suppl.): 7–16.

Jamison, R.N., Raymond, S.A., Levine, J.G., Slawsby, E.A., Nedeljkovic, S.S., Katz, N.P. (2001) Electronic diaries for monitoring chronic pain: 1-year validation study, *Pain*, 91: 277–85.

Jahncke, H., Hygge, S., Halin, N., Green, A.M., Dimberg, K. (2011) Open-plan office noise: cognitive performance and restoration, *Journal of Environmental Psychology*, 31: 373–82.

Janssen, S.A. (2002) Negative affect and sensitization of pain, *Scandanavian Journal of Psychology*, 43: 131–7.

Janssen S.A., van Dongen, J.E.F. (2007) *Persoonlijke en situationale factoren bij hinderbeleving van industrie* – literatuurstudie (in Dutch) (Personal and situational factors with annoyance of industry – literature study), TNO-report, Delft, Nederland.

Jantunen, M.J., Hänninen, O., Katsouyanni, K., Knöppel, H., Keunzli, N., Lebret, E., Maroni, M., Saarela, K., Sram, R., Zmirou, D. (1998) Air pollution exposure in European cities: the Expolis study, *JEAEE*, 8: 495–518.

Jantunen, M.J., Oliveira Fernandes, E., Carrer, P., Kephalopoulos, S. (2011) *Promoting actions for healthy indoor air (IAIAQ)*, European Commission Directorate General for Health and Consumers, Luxembourg, http://ec.europa.eu/health/healthy_environ ments/docs/env_iaiaq.pdf (accessed 7 August 2012).

Jenkins, S., Brown, R., Rutterford, N. (2009) Comparing thermographic, EEG, and subjective measures of affective experience during simulated production interactions, *International Journal of Design*, 3: 53–65.

Johnson, F., Mavrogianni, A., Ucci, M., Vidal-Puig, A., Wardle, J. (2011) Could increased time spent in a thermal comfort zone contribute to population increases in obesity? *Obes. Rev.*, 12: 543–51.

Jones, D.M., Macken, W.J., Murray, A.C. (1993) Disruption of visual short-term memory by changing-state auditory stimuli: the role of segmentation, *Mem. Cogni.*, 21: 318–28.

Jones, D.S., Podolsky, S.H., Greene, J.A. (2012) The burden of disease and the changing task of medicine, *The New England Journal of Medicine*, 366 (25): 2333–8.

Jørgensen, B. (2006) *Integrating lean design and lean construction; processes and methods*, Ph.D. thesis, Technical University of Denmark.

Kaczmarczyk, J., Zeng, Q., Melikov, A., Fanger, P.O. (2002) The effect of a personalized ventilation system on air quality prediction, SBS symptoms and occupant's perform-ance, *Proceedings of Indoor Air 2002*, Monterey, CA, 4: 1042–7.

Kang, K.-S., Trosko, J.E. (2011) Stem cells in toxicology: fundamental biology and practical considerations, *Toxicological Sciences*, 120: S269–89.

Kapit, W., Elson, L.M. (2002) *The anatomy coloring book*, 3rd edition, San Francisco, CA: Benjamins/Cummings Science Publishing.

Kapit, W., Macey, R.I., Meisami, E. (2000) *The physiology coloring book*, 2nd edition, San Francisco, CA: Benjamins/Cummings Science Publishing.

Karahanglu, A. (2008) *A study of consumer's emotional responses towards brands and branded products*, master's thesis, Middle East Technical University, Turkey.

Karasek, R., Brisson, C., Kawakami, N., Houtman, I., Bongers, P., Amick, B. (1998) The job content questionnaire (JCQ): an instrument for internationally comparative assessments of psychosocial job characteristics, *Journal of Occupational Health Psychology*, 3: 322–55.

Kelly, F.J. (2003) Oxidative stress: its role in air pollution and adverse health effects, *Occup. Environ. Med.*, 60: 612–16.

Kendall S., Teicher, J. (2000) *Residential open building*, London and New York: E & FN Spon.

Khan, V.J., Markopoulos, P., Eggen, B. (2009) Features for the future Experience Sampling Tool. *MobileHCI 2009*, Bonn, Germany.

Khaneman D., Krueger A.B., Schkade A.A., Schwarz N., Stone A.A. (2004) A survey method for characterizing daily life experience, the day reconstruction method, *Science*, 306: 1776.

Kiecolt-Glaser, J.K., Preacher, K.J., MacCallum, R.C., Atkinson, C., Malarkey, W.B., Glaser, R. (2003) Chronic stress and age-related increases in the proinflammatory cytokine IL-6, *Proceedings of the National Academy of Sciences*, 100: 9090–5.

Kim, J.L.E., Mi, L., Johansson, Y., Smedje, M., Nörback, G. (2005) Current asthma and respiratory symptoms among pupils in relation to dietary factors and allergens in the school environment, *Indoor Air*, 15: 170–82.

Kim, J., de Dear, R. (2011) Nonlinear relationships between individual IEQ factors and overall workspace satisfaction, *Building and Environment*, 49: 33–40.

Kirchner, S., Derbez, M., Duboudin, C., Elias, P., Gregoire, A., Lucas, J.-P., Pasquier, N., Ramalho, O., Nathalie Weiss, N. (2008) Indoor air quality in French Dwellings, *Proceedings of Indoor Air 2008*, Lyngby, Denmark, paper 574.

Kjaergaard, S.K., Hodgson, M. (2001) The assessment of irritation using clinical methods and questionnaires, *AIHAJ*, 62: 711–16.

Klerman, E.B. (2005) Clinical aspects of human circadian rhythms, *Journal of Biological Rhythms*, 20: 375–86.

Kluizenaar, Y., Janssen, S., Lenthe, F.J., Miedema, H.M.E. (2009) Long term road traffic noise exposure is associated with an increase in morning tiredness, *J. Acoust. Soc. Am.*, 126: 626–33.

Knol, A.B., Staatsen, B.A.M. (2005) *Trends in the environmental burden of disease in the Netherlands 1980–2020*, RIVM report 500029001/2005, Bilthoven, The Netherlands.

Knudsen, H.N., Jensen, O.M., Kristensen, L. (2012) Occupant satisfaction with new low-energy houses, *Proceedings of Healthy Buildings Conference 2012*, Brisbane, Australia, paper 2E.2.

Kolarik B., Lagercrantz, L., Sundell, J. (2009) Nitric oxide in exhaled and aspirated nasal air as an objective measure of human response to indoor air pollution, *Indoor Air*, 19: 145–52.

Köster, E. (2002) The specific characteristics of the sense of smell, Chapter 3 in: Rouby, C., Schaal, B., Dubois, R., Gervais, R., Holley, A. (eds) *Olfaction, taste, and cognition*, Cambridge: Cambridge University Press.

Kotzias D., Koistinen K., Kephalopoulos S., Schlitt C., Carrer P., Maroni, M., Jantunen M., Cochet C., Kirchner S., Lindvall T., McLaughkin J., Mølhave L., de Oliveira Fernandes E., Seifert B. (2005) The INDEX project: critical appraisal of the setting and implementation of indoor exposure limits in the EU, Joint Research Centre, Ispra, Italy.

Kotzias, D., Geiss, O., Tirendi, S., Barrero-Moreno, J., Reina, V., Gotti, A., Cimino-Reale G., Casati, B., Marafante, E., Sarigiannis, D. (2009) Exposure to multiple air contaminants in public buildings, schools and kindergartens the European monitoring and exposure assessment (AIRMEX) study, *Fresenius Environmental Bulletin*, 18: 670–81.

Kouba, E., Wallen, E.M., Pruthi, R.S. (2007) Uroscopy by Hippocrates and Theophilus: prognosis versus diagnosis, *J. Urol.*, 177: 50–52.

Kreyling, W., Semmler-Behnke, M., Moller, W. (2006) Ultra-fine particle-lung interactions: does size matter? *Journal of Aerosol Medicine*, 19: 74–83.

Kristenson, M., Eriksen, H.R., Sluiter, J.K., Starke, D., Ursin, H. (2004) Psychobiological mechanisms of socioeconomic differences in health, *Soc. Sci. Med.*, 58: 1511–22.

Kristiansen, J., Clausen, Th., Christensen, K.B., Lund, Th. (2008) Self-reported noise exposure as a risk factor for long-term sickness absence, *9th International congress on noise as a public health problem* (ICBEN), Foxwoods, CT, 302–6.

Kubey, R., Larson, R., Csikzentmihalyi, M. (1996) Experience sampling method applications to communication research questions, *Journal of Communication*, 46: 99–120.

Kubzansky, L.D., Kawachi, I., Sparrow, D. (2000) Socioeconomic status, hostility, and risk factor clustering in the Normative aging study: any help from the concept of allostatic load? *Annals of Behavioral Medicine*, 24: 330–8.

Kuhbander, C., Hanslmayr, S., Maier, M.A., Pekrun, R., Spitzer, B., Pastoetter, B., Bauml, K.-H. (2009) Effects of mood on the speed of conscious perception: behavioural and electrophysiological evidence, *Social Cognitive and Affective Neuroscience*, April, 7: 1–8.

Küller, R., Lindsten, C. (1992) Health and behaviour of children in classrooms with and without windows, *J. Env. Psycholog.*, 12: 305–17.

Kuperman, Y., Isser, O., Regev, L., Musseri, I., Navon, I., Neufeld-Cohen, A., Gil, S., Chen, A. (2010) *Proceedings of the National Academy of Sciences* (PANAS), 107: 8393–8.

Kushner, M.G., Abrams, K., Borchardt, C. (2000) The relationship between anxiety disorders and alcohol use disorders: a review of major perspectives and findings, *Clin. Psychol. Rev.*, 20: 149–71.

Kuvin, J.T., Patel, A.R., Sliney, K.A., Pandian, N.G., Sheffy, J., Schnall, R.P., Karas, R.H., Udelson, J.E. (2003) Assessment of peripheral vascular endothelial function with finger arterial pulse wave amplitude, *American Heart Journal*, 46: 168–74.

Lan, L., Lian, Z., Pan, L. (2010) The effects of air temperature on office workers' well-being, workload and productivity evaluated with subjective rating, *Applied Ergonomics*, 42: 29–36.

Lan, L., Wargocki, P., Wyon, D.P., Lian, Z. (2011) Effects of thermal discomfort in an office on perceived air quality: SBS symptoms, physiological responses and human performance, *Indoor Air*, 21: 376–90.

Lang, I., Bruckner, T., Triebig, G. (2008) Formaldehyde and chemosensory irritation in humans: a controlled human exposure study, *Regul. Toxicol. Pharm.*, 50: 23–36.

Langdridge, D., Hagger-Johnson, G. (2009) *Introduction to research methods and data analysis in psychology*, 2nd edition, London: Pearson Education.

La Porte, R.E., Kuller, E.H., Kupfer, D.J., McPartland, R.J., Matthews, G., Caspersen, C. (1979) An objective measure of physical activity for epidemiologic research, *American Journal of Epidemiology*, 109: 158–68.

La Rovere, M.T., Pinna, G.D., Maestri, R., Mrtara, A., Capomolla, S., Febo, O., Ferrari, R., Franhini, M., Gnemmi, M., Opasich, C., Riccardi, P.G., Traversi, E., Cobelli, F. (2003) Short-term heart rate variability strongly predicts sudden cardiac death in chronic heart failure patients, *Circulation*, 107: 565–70.

Larsson, M. (2002) A memory systems approach, Chapter 14 in: Rouby, C., Schaal, B., Dubois, R., Gervais, R., Holley, A. (eds) *Olfaction, taste, and cognition*, Cambridge: Cambridge University Press.

Larsson, M., Weiss, B., Janson, S., Sundell, J., Bornehag, C-G. (2009) Associations between indoor environmental factors and parental-reported autistic spectrum disorders in children 6–8 years of age, *Neurotoxicology*, 30: 822–31.

de Laurentiis, G., Paris, D., Melck, D., Maniscalco, M., Marisco, S., Corso, G., Motta, A., Sofia, M. (2008) Metabonomic analysis of exhaled breath condensate in adults by NMR spectroscopy, *ERJ Express*, doi: 10.1183/09031936.00072408.

Layard, R., Glaister, S. (1994) *Cost–benefit analysis*, 2nd edition, Cambridge: Cambridge University Press.

Leaman, A. (1996) Building in use studies, in: Baird, G. (ed.), *Building Evaluation Techniques*, New York: McGraw-Hill.

Leaman, A., Bordass, A. (2001) Assessing building performance in use, 4: the Probe occupant surveys and their implications, *Building Research and Information*, 29: 129–43.

Leaman, A., Bordass, B., Stevenson, F. (2010) Building evaluation: practice and principles, *Building Research and Information*, 39: 436–49.

Leather, P., Beale, D., Sullivan, L. (2003) Noise, psychosocial stress and their interaction in the workplace, *Journal of Environmental Psychology*, 23: 213–22.

Le Corbusier (1925) *Urbanism*, Paris.

LEnSE partners (2007) *Stepping stone 2: development of a sustainability assessment methodology, framework and content*, Klomp Grafische Communicatie, Amersfoort, The Netherlands, March 2007.

van Lenthe, F.J., Schrijvers, C.T., Droomers, M., Joung, I.M., Louwman, M.J., Mackenbach, J.P. (2004) Investigating explanations of socio-economic inequalities in health: the Dutch GLOBE study, *European Journal of Public Health*, 14: 63–70.

Lewtas, J. (2007) Air pollution combustion emissions: characterization of causative agents and mechanisms associated with cancer, reproductive, and cardiovascular effects: the sources and potential hazards of mutagens in complex environmental matrices – Part II, *Mutation Research*, 636: 95–133.

LHRF (2002) *Proceedings of Symposium Healthy Lighting*, Light and Health Research Foundation, Eindhoven, The Netherlands.

Li, N., Sioutas, C., Cho, A., Schmitz, D., Mistra, C., Semf, J., Wang, M., Oberley, T., Froines, J., Nel A. (2003) Ultrafine particulate pollutants induce oxidative stress and mitochondrial damage, *Environmental Health Perspectives*, 111: 455–60.

Li, N., Xia, T., Nel, A.E. (2008) The role of oxidative stress in ambient particulate matter induced lung diseases and its implications in the toxicity of engineered nanoparticles, *Free Radical Biol. Med.*, 44: 1689–99.

Li, Y., Leung, G.M., Tang, J.W., Yang, X., Chao, C.Y.H., Lin, J.Z., Lu, J.W., Nielsen, P.V., Niu, J., Qian, H., Sleigh, A.C., Su, H.-J.J., Sundell, J., Wong, T.W., Yen, P.L. (2007) Role of ventilation in airborne transmission of infectious agents in the built environment: a multidisciplinary systematic review, *Indoor Air*, 17: 2–18.

Libert, J.P., Nisi, J.Di, Fukuda, H., Muzet, A., Ehrhart, J., Amoros, C. (1988) Effect of continuous heat exposure on sleep stages in humans, *Sleep*, 11: 195–209.

Liebl, A., Haller, J., Jödicke, B., Baumgartner, H., Schlittmeier, S., Hellbrück, J. (2012) Combined effects of acoustic and visual distraction on cognitive performance and well-being, *Applied Ergonomics*, 43: 424–34.

Lindberg, L., Ezra, A. (2008) Alcohol, wine and cardiovascular health, *Clinical Cardiology*, 31: 347–51.

Lindstrom, M. (2008) *Buyology: how everything we believe about why we buy is wrong*, Croydon: Random House Business Books.

Loft, S., Vistisen, K., Ewertz, M., Tjønneland, A., Overvad, K., Enghusen Poulsen, H. (1992) Oxidative DNA damage estimated by 8-hydroxydeoxyguanosine excretion in humans: influence of smoking, gender and body mass index, *Carcinogenesis*, 13: 2241–7.

Lorch, R. (2011) The relevance of time, Keynote SB11, 40: 40 Helsinki Looking back and looking forward, *Proceedings of Sustainable Buildings 2011 (SB11)*, Helsinki, Finland.

Lovibond, P.F., Lovibond, S.H. (1995) The structure of negative emotional states: comparison of the depression anxiety stress scales (DASS) with the Beck depression and anxiety inventories, *Behaviour Research and Therapy*, 33: 335–42.

Lu C., Ma Y., Lin J., Li C., Lin R.S., Sung, F. (2007) Oxidative stress associated with indoor air pollution and sick building syndrome-related symptoms among office workers in Taiwan, *Inhalation Toxicology*, 19: 57–65.

Lu, X., Clements-Croome, D.J., Viljanen, M. (2010) Integration of chaos theory and mathematical models in building simulation, Part I: literature review, *Automation in Construction*, 19: 447–51.

Lundquist, P., Holmberg, K., Landstrom, U. (2000) Annoyance and effects on work from environmental noise at school, *Noise and Health*, 2: 39–46.

Lupisek, A., Botsi, S., Hajek, P., Sakkas, N., Hodkova, J. (2009) Perfection D.1.1: an inventory of relevant standards, regulations and technologies, Final version, October 25, 2009, www.ca-perfection.eu (accessed 23 June 2012).

Madigan, M., Martinko, J., Parker, J. (eds) (2005) *Brock biology of microorganisms*, 11th edition, NJ: Prentice Hall.

Madureira, J., Paciencia, I., Stranger, M., Ventura, G., Oliveira de Fernandes, E. (2012) Field study on schoolchildren's exposure to indoor air in Porto, Portugal: preliminary results, *Proceedings of Healthy Buildings 2012*, Brisbane, Australia, paper 4C.4.

Mahnke, F.H., Mahnke, R.H. (1996) *Color, environment and human response*, New York: John Wiley & Sons.

van Marken Lichtenbelt, W.D., Vanhommerig, J.W., Smulders, N.M., Drossaerts, B.S., Kemerink, G.J., Bouvy, N.D., Schrauwen, P., Teule, G.J.J. (2009) Cold-activated brown adipose tissue in healthy men, *The New England Journal of Medicine*, 360: 1500–8.

Marks, G.B.E., Aust, W., Toelle, N., Xuan, B.G., Belousova, W., Cosgrove, E., Jalaludin, C., Smith, B. (2010) Respiratory health effects of exposure to low-NO$_x$ unflued gas heaters in the classroom: a double-blind, cluster-randomized, crossover study, *Environmental Health Perspectives*, 118: 1476–82.

Marmot, A.F., Eley, J., Stafford, S.A., Warrick, E., Marmot, M.G. (2006) Building health: an epidemiological study of 'sick building syndrome' in the Whitehall II study, *Occup. Environ. Med.*, 63: 283–9.

Martin, G.N., Carlson, N.R., Buskist, W. (2010) *Psychology*, 4th edition, Harlow UK: Pearson Education.

Marty, M.S., Carney, E.W., Rowlands, J.C. (2011) Endocrine disruption: historical perspectives and its impact on the future of toxicology testing, *Toxicological Sciences*, 120(S1): S93–108.

Maschke, C., Rupp, T., Hecht, K. (2000) The influence of stressors on biochemical reactions: a review of present scientific findings with noise, *Int. J. Hyg. Environ. Health*, 203: 45–53.

Matthews, A., Mackintosh, B. (1998) A cognitive model of selective processing in anxiety, *Cognit. Ther. Res.*, 22: 539–60.

Mattson, M., Hygge, S. (2005) Effects of particulate air cleaning on perceived health and cognitive performance in school children during pollen season, *Proc. of Indoor Air 2005*, Beijing, China, vol. 2: 111–5.

Maxwell, L., Evans, G. (2000) The effect of noise on pre-school children's pre-reading skills, *Journal of Environmental Psychology*, 20: 91–7.

McClellan, S., Hamilton, B. (2010) *So stressed: a plan for managing women's stress to restore health, joy and peace of mind*, London: Simon & Schuster.

McDonagh, D., Anne, B., Haslam, C. (2002) Visual product evaluation: exploring users' emotional relationships with products, *Applied Ergonomics*, 33: 231–40.

McEwen, B.S. (1998) Protective and damaging effects of stress mediators, *N. Engl. J. Med.*, 338: 171–9.

McEwen, B.S. (2004) Protective and damaging effects of the mediators of stress and adaptation: allostasis and allostatic load, Chapter 2 in: Schulkin (ed.) (2004) *Allostasis, homeostasis and the cost of physiological adaptation*, New York: Cambridge University Press.

McIntyre, D.A. (1980) *Indoor Climate*, Essex, UK: Applied Science Publishers.

McKinley, R.A., McIntire, L.K., Schmidt, R., Repperger, D.W., Caldwell, J.A. (2011) Evaluation of eye metrics as a detector of fatigue: human factors, *The Journal of the Human Factors and Ergonomics Society*, 53: 403–14.

McNair, D.M., Lorr, M., Droppleman, L.F. (1971) *Manual for the profile of mood states (POMS)*, San Diego, CA: Educational and Industrial Testing Service.

Meerdink, G., Rozendaal, E.Z., Witteveen, C.J.E. (1988) Daglicht en uitzicht in kantoorgebouwen (in Dutch) [Daylight and view in office buildings], Directoraat-Generaal van de Arbeid, S 51, Voorburg, The Netherlands.

Mehrabian, A. (1996) Pleasure–arousal–dominance: a general framework for describing and measuring individual differences in temperament, *Current Psychology*, 14: 261–92.

Meklin, T., Potus, T., Pekkanen, J., Hyvarinen, A., Hirvonen, M.-R., Nevalainen, A. (2005) Effects of moisture-damage repairs on microbial exposure and symptoms in schoolchildren, *Indoor Air*, 15 (suppl. 10): 40–7.

Melamed, S., Ugarten, U., Shirom, A., Kahana, L., Lerman, Y., Froom, P. (1999) Chronic burnout, somatic arousal and elevated salivary cortisol levels, *Journal of Psychosomatic Research*, 46: 591–98.

Mendell, M.J. (1993) Non-specific symptoms in office workers: a review and summary of the epidemiologic literature, *Indoor Air*, 12: 227–36.

Mendell, M.J., Fisk, W.J., Petersen, M.R., Hines, C.J., Dong. M., Faulkner, D., Deddens, J.A., Ruder, A.M., Sullivan, D.A., Boeniger, M.F. (2002) Indoor particles and symptoms among office workers: results from a double-blind crossover study, *Epidemiology*, 13: 296–304.

Mendell, M.J., Heath, G.A. (2005) Do indoor pollutants and thermal conditions in schools influence student performance? A critical review of the literature, *Indoor Air*, 15: 27–52.

Mendell, M.J. (2007) Indoor residential chemical emissions as risk factors for respiratory and allergic effects in children: a review, *Indoor Air*, 17: 259–77.

Mendell, M.J., Mirer, A.G. (2009) Can the US EPA BASE study be used to provide reference levels for building-related symptoms in offices? *Proceedings of Healthy Building 2009*, Syracuse, NY, paper 756.

Mercier, F., Raffy, G., Derbez, M., Glorennec, P., Blanchard, O., Bonvallot, N., Le Bot, B. (2011) Cumulative indoor exposures to semi-volatile organic compounds (SVOCs) in France: contamination levels in thirty French schools, *Proceedings of Indoor Air 2011*, Austin, TX, paper 554.

Merriam-Webster Dictionary (2011) 'Stress', www.merriam-webster.com/dictionary/stress (accessed 3 October 2011).

Meyer, H.W., Würtz, H., Suadicani, P., Valbjørn, Sigsgaard, T., Gyntelberg, F. (2005) Molds in floor dust and building-related symptoms among adolescent school children: a problem for boys only? *Indoor Air*, 15 (suppl. 10): 17–24.

Meyer T.J., Miller M.L., Metzger R. L., Borkovec T.D. (1990) Development and validation of the Penn State Worry questionnaire, *Behaviour Research and Therapy*, 28: 487–95.

Mi, Y.H.N., Tao, D., Mi, J., Ferm, Y.L. (2006) Current asthma and respiratory symptoms among pupils in Shanghai, China: influence of building ventilation, nitrogen dioxide, ozone and formaldehyde in classrooms, *Indoor Air*, 16: 454–64.

Michael-Titus, A., Revest, P., Shortland, P. (2010) *The nervous system*, 2nd edition, Systems of the body, China: Churchill Livingstone Elsevier.

Miedema, H.M.E., Vos, H. (2003) Noise sensitivity and reactions to noise and other environmental conditions, *J. Acoust. Soc. Am.*, 113: 1492–1504.

Miedema, H.M.E., Vos, H. (2007) Associations between selfreported sleep disturbance and environmental noise based on reanalyses of pooled data from 24 studies, *Behav. Sleep. Med.*, 5: 1–20.

Miller, M.R., Hankinson, J., Brusasco, V., Burgos, F., Casaburi, R., Coates, A., Crapo, R., Enright, P., van der Grinten, C.P.M., Gustafsson, P., Jensen, R., Johnson, D.C., MacIntyre, N., McKay, R., Navajas, D., Pedersen, O.F., Pellegrino, R., Viegi, G., Wanger, J. (2005) Standardisation of spirometry, *Eur. Respir. J.*, 26: 319–38.

Miller, R.L., Ho, S.M. (2008) Environmental epigenitics and asthma: current concepts and call for studies, *Am. J. Respir. Crit. Care Med.*, 177: 567–73.

Milton, D.K., Glencross, P.M., Walters, M.D. (2000) Risk of sick leave associated with outdoor air supply rate, humidification and occupant complaints, *Indoor Air*, 10: 212–21.

Miyakawa, M., Matsui, T., Kishikawa, H., Murayama R., Uchiyama I., Itoh, T., Yoshida, T. (2006) Salivary chromogranin A as a measure of stress response to noise, *Noise and Health*, 8: 108–13.

Moher, D., Hopewell, S., Schulz, K.F., Montori, V., Goetzche, P.C., Deveraux, P.J., Elborune, D., Egger, M., Altman, D.G. (2010) CONSORT 2010 explanation and elaboration: updated guidelines for reporting parallel group randomised trials, *British Medical Journal*, 340: 869–98.

Mølhave, L., Liu, Z., Jørgensen, A.H., Pedersen, O.F., Kjaergaard, S. (1993) Sensory and physiological effects on humans of combined exposures to air temperatures and volatile organic compounds, *Indoor Air*, 3: 155–69.

Montello, D. (1988) Classroom seating location and its effect on course achievement, participation, and attitudes, *Journal of Environmental Psychology*, 8: 149–57.

Montuschi, P. (2002) Indirect monitoring of lung inflammation, *Natural Reviews Drug Discovery*, 1, 238–42.

Morawska, L., He, C., Johnson, G., Jayaratne, R., Salthammer, T., Wang, H., Uhde, E., Bostrom, T., Modini, R., Ayoko, G., McGarry, P., Wensing, M. (2009) Printers and copiers: how to use science of emissions to minimize human exposures, *Proceedings of Healthy Building 2009*, Syracuse, NY, paper 578.

Morello-Frosch, R., Shenassa, E.D. (2006) The environmental 'Riskscape' and social inequality: implications for explaining maternal and child disparities, *Environmental Health Perspectives*, 114: 1150–3.

Morrow, R.H., Bryant, J.H. (1995) Health policy approaches to measuring and valuing human life: conceptual and ethical issues, *American Journal of Public Health*, 85: 1356–60.

MRC, Medical Research Council (2000) A framework for development and evaluation of RCTs for complex interventions to improve health, www.mrc.ac.uk/Utilities/Documentrecord/index.htm?d=MRC003372 (accessed 3 April 2012).

Mucha, M., Skrzypiec, A.E., Schiavon, E., Attwood, B.K., Kucerova, E., Pawlak, R. (2011) Lipocalin-2 controls neuronal excitability and anxiety by regulating dendritic spine formation and maturation, *Proceedings of the National Academy of Sciences (PNAS)*, 108: 18436–41.

Müller, A., Yeoh, C. (2010) Compromised detoxification, Chapter 3 in: Nicolle, L., Woodriff Beirne, A. (eds) (2010) *Biochemical imbalances in diseases: a practitioner's handbook*, London and Philadelphia, PA: Singing Dragon.

Murray, C.J.L., Lopez, A.D. (eds) (1996) *The global burden of disease*, A comprehensive assessment of mortality and disability from disease, injury and risk factors in 1990 and projected to 2020, Vol I., Harvard, MA: Harvard University Press.

Muzet, A., Ehrhart, J., Candas, V., Libert, J.P., Vogt, J.J. (1983) REM sleep and ambient temperatures in man, *Int. J. Neurosci.*, 18: 117–26.

Muzet, A. (2007) Environmental noise, sleep and health, *Sleep Med. Rev.*, 11: 135–42.

Myatt, T.A., Johnston, S.L., Zuo, Z., Wand, M., Kebadze, T., Rudnick, S., Milton, D.K. (2004) Detection of airborne rhinovirus and its relation to outdoor air supply in office environments, *Am. J. Respir. Crit. Care Med.*, 169: 1187–90.

Mygind, N. (1986) *Essential allergy, an illustrated text for students and specialists*, Oxford: Blackwell Scientific Publications.

Nabkasom, C., Miyai, N., Sootmongkol, A., Junprasert, S., Yamamoto, H., Arita, M., Miyashita, K. (2005) Effects of physical exercise on depression, neuroendocrine stress hormones and physiological fitness in adolescent females with depressive hormones, *European Journal of Public Health*, 16: 179–84.

Nadler, B. (2007) Inflammation, free radicals, oxidative stress and antioxidants, www.beverlynadler.com/html/inflammation.html (accessed 2 November 2011).

Nafstad, P., Hagen, J.A., Oie, L., Magnus, P., Jaakola, J.J. (1999) Day care centres and respiratory health, *Pediatrics*, 103: 753–8.

Nater, U.M., Rohleder, N. (2009) Salivary alpha-amylase as a non-invasive biomarker for the sympathetic nervous system: current state of research, *Psychoneuroendocrinology*, 34: 486–96.

National Cancer Institute (2012) Cell proliferation, www.cancer.gov/dictionary?cdrid=46479 (accessed 20 June 2012).

Naydenov, K., Melikov, A., Markov, D., Stankov, P., Bornehag C.-G., Sundell, J. (2008) A comparison between occupants' and inspectors' reports on home dampness and their association with the health of children: the ALL-HOME study, *Building and Environment*, 43: 1840–9.

Nelson, T.M., Nilsson, T.H., Johnson, T.M. (1984) Interaction of temperature, illuminance and apparent time on sedentary work fatigue, *Ergonomics*, 27: 89–101.

Nenonen, S., Airo, K., Bosch, P., Fruchter, R., Koivisto, S., Gerberg, N., Rothe, P., Ruohomäki, Vartiainen, M. (2009) *Managing workplace resources for knowledge work*, www.proworkproject.com/prowork/final-report/ (accessed 31 July 2012).

Newbold, R.R., Padilla-Banks, E., Snyder, R.J., Philips, T.M., Jefferson, W.N. (2007) developmental exposure to endocrine disruptors and the obesity epidemic, *Reprod. Toxicol.*, April–May, 23: 290–6.

Nicol, J.F., Wilson, M. (2011) A critique of European Standard EN, 15251: strengths, weaknesses and lessons for future standards, *Building Research and Information*, 39: 183–93.

Nicolle, L., Hallam, A. (2010) Polyunsaturated fatty acid (PUFA) imbalances, Chapter 4 in: Nicolle, L., Woodriff Beirne, A. (eds) *Biochemical imbalances in diseases: a practitioner's handbook*, London and Philadelphia, PA: Singing Dragon.

Nicolle, L., Woodriff Beirne, A. (eds) (2010) *Biochemical imbalances in diseases: a practitioner's handbook*, London and Philadelphia: Singing Dragon.

Nodder, J. (2010) Compromised thyroid and adrenal function, Chapter 6 in: Nicolle, L., Woodriff Beirne, A. (eds) (2010) *Biochemical imbalances in diseases: a practitioner's handbook*, London and Philadelphia, PA: Singing Dragon.

Noejgaard, J.K., Christensen, K.B., Wolkoff, P. (2005) The effect on human eye blink frequency of exposure to limonene oxidation products and methacrolein, *Toxicology Letters*, 156: 241–51.

Nolan, J., Batin, P.D., Andrews, R., Lindsay, S.J., Brooksby, P., Mullen, M., Baig, W., Flapan, A.D., Cowley, A., Prescott R.J., Neilson, J.M.M., Fox, K.A.A. (1998) Prospective study of heart rate variability and mortality in chronic heart failure, *Circulation*, 98: 1510–6.

Norbäck, D., Björnsson, E., Janson, C., Widström, J., Boman, G. (1995) Asthmatic symptoms and volatile organic compounds, formaldehyde and carbon dioxide in dwellings, *Occup. Environ. Med.*, 52: 388–95.

Norbäck, D., Wålinder, R., Wieslander, G., Smedje, G., Erwall, C., Venge, P. (2000) Indoor air pollutants in schools: nasal patency and biomarkers in nasal lavage, *Allergy*, 55: 163–170.

Norbäck, D., Wieslander, G., Zhang, X., Zhao, Z. (2011) Respiratory symptoms, perceived air quality and physiological signs in elementary school pupil in relation to displacement and mixing ventilation system: an intervention study, *Indoor Air*, 21: 427–37.

Nöske, I., Brasche, S., Hellwig, R.T., Bischof, W., Popfinger, B., Gebhardt, H., Levchuk, I., Bux, K. (2011) Impact of elevated temperatures in a controlled office environment on skin moisture and skin temperature: the HESO Study, *Proceedings of Indoor Air 2011 Conference*, Austin, TX, paper A105.

NRC (2004) *Research priorities for airborne particulate matter*, IV, continuing research progress, report of the national research council of the national academies, www.nap.edu/openbook.php?record_id=10957&page=1 (accessed 18 December 2012).

Nye, B., Hedges, L., Konstantopoulos, S. (2000) The effects of small classes on academic achievement: the results of the Tennessee class size experiment, *American Educational Research Journal*, 37: 123–51.

Oberdörster, G., Oberdörster, E., Oberdörster, J. (2005) Nanatoxicology: an emerging discipline evolving from studies of ultrafine particles, *Environmental Health Perspectives*, 113: 823–39.

van Odijk, J., Kull, I., Borres, M.P., Brandtzaeg, P., Edberg, U., Hanson, L.A., Høst, A., Kuitunen, M., Olsen, S.F., Skorfving, S., Sundell, J., Wille, S. (2003) Breastfeeding and allergic disease: a multidisciplinary review of the literature (1966–2001) on the mode of early feeding in infancy and its impact on later atopic manifestations, *Allergy*, 58: 833–43.

Oldham, G.R., Brass, D.J. (1979) Employee reactions to an open-plan office: naturally occurring quasi experiment, *Admin. Sci. Q.*, 24: 267–84.

Øie, L., Nafstad, P., Botten, G., Magnus, P., Jaakola, J.J.K. (1999) Ventilation in homes and bronchial obstruction in young children, *Epidemiology*, 10: 294–9.

de Oliveira Fernandes, O., Jantunen, M., Carrer, P., Seppanen, O., Harrison P., Kephalopoulos, S. (2008) *EnVIE: Co-ordination Action on Indoor Air Quality and Health Effects* Project no. SSPECT-2004–502671, final report, http://paginas.fe.up.pt/~envie/finalreports.html (accessed 14 August 2012).

Ostir, G.V., Markides, K.S., Black, S.A., Goodwin, J.S. (2000) Emotional wellbeing predicts subsequent functional independence and survival, *J. Am. Geriarr. Soc*, 48: 473–8.

Otto T., Cousens, G., Herzog, C. (2000) Behavioural and neuropsychological foundations of olfactory fear conditioning, *Behavioural Brain Research*, 110: 119–28.

Ozdemir, A., Yilmaz, O. (2008) Assessment of outdoor school environments and physical activity in Ankara's primary schools, *Journal of Environmental Psychology*, 28: 287–300.

Palladio, A. (1570) *The four books of architecture*, translated by Ware, I., 1738, New York: Dover publications (1964 edition).

Panagiotopoulou, G., Christoulas, K., Papanckolaou, A., Mandroukas, K. (2004) Classroom furniture dimensions and anthropometric measures in primary school, *Applied Ergonomics*, 35: 121–8.

Parati, G. (2004) Assessing circadian blood pressure and heart rate changes: advantages and limitations of different methods of mathematical modelling. *J. Hypertens.*, 22: 2061–4.

Park, H., Lee, B., Ha, E.H., Lee, J.T., Kim, H., Hong, Y.C. (2002) Association of air pollution with school absenteeism due to illness, *Arch. Pediatr. Adolesc. Med.*, 156: 1235–9.

Parker, S. (2003) *Hormones, injury, illness and health*, Chicago, IL: Heineman Library.

Pasanen, P.O., Teijonsalo, J., Seppänen, O., Ruuskanen, J., Kalliokoski, P. (1994) Increase in perceived odor emission with loading of ventilation filters, *Indoor Air*, 4: 106–13.

Paules, R. (2003) Phenotypic anchoring: linking cause and effect, *Environmental Health Perspectives*, 111: A338–9.

Pearlin, L.I., Schooler, C. (1978) The structure of coping, *Journal of Health and Social Behaviour*, 19: 2–21.

Peccia, J., Hernandez, M. (2006) Incorporating polumerase chain reaction-based identification, population characterization, and quantification of microorganisms into aerosol science: a review, *Atmospheric Environment*, 40: 3941–61.

Pejtersen, J., Allermann, L., Kristensen, T.S., Poulsen, O.M. (2006) Indoor climate, psychosocial work environment and symptoms in open-plan offices, *Indoor Air*, 16: 392–401.

Pejtersen, J.H., Feveile, H., Christensen, K.B., Burr, H. (2011) Sickness absence associated with shared and open-plan offices: a national cross sectional questionnaire survey, *Scand. J. Work Environ. Health*, 37: 376–82.

Pereg, D., Gow, R., Mosseri, M., Lishner, M., Rieder, M., Van Uum, S., Koren, G. (2010) Hair cortisol and the risk for acute myocardial infarction in adult men, *Stress*, 14: 73–81.

Perner, L. (2007) Consumer behaviour: the psychology of marketing, www.consumerpsychologist.com (accessed 21 August 2007).

Pernot, C.E.E., Koren, L.G.H., Dongen, J.E.F., van Bronswijk, J.E.M.H. (2003) Relatie EPC-niveau en gezondheidsrisico's als onderdeel van het kwaliteitsniveau van gebouwen (in Dutch) (Relation EPC-level and health risks as part of the building quality level) TNO report, Delft, The Netherlands.

Peters, A., Veronesi, B., Calderon-Garciduenas, L., Gehr, P., Chen, L.C., Geiser, M., Reed, W., Rothen-Rutishauser, B., Schürch, S., Schulz, H. (2006) Translocation and potential neurological effects of fine and ultrafine particles a critical update, *Particle Fibre Toxicol.*, 3: 1–13.

Phull, S. (2010) Poor energy production in increased oxidative stress, Chapter 9 in: Nicolle, L., Woodriff Beirne, A. (eds) (2010) *Biochemical imbalances in diseases: a practitioner's handbook*, London and Philadelphia, PA: Singing Dragon.

Pieper, S. (2008) *Prolonged cardiac activation, stressful events and worry in daily life*, Ph.D. thesis, University of Leiden, The Netherlands.

Podsakoff, P.M., MacKenzie, S.B., Lee, J.Y., Podsakoff, N.P. (2003) Common method biases in behavioural research: a critical review of the literature and recommended remedies, *J. Appl. Psychol.*, 88: 879–903.

Pope, C.A. III, Dockery, D.W. (2006) Health effects of fine particulate air pollution: lines that connect, *J. Air Waste Manage Assoc.*, 56: 709–42.

Popper, K.R. (1972) *Objective knowledge: an evolutionary approach*, Oxford: Oxford University Press.

Postlethwait, E.M. (2007) Scavenger receptors clear the air, *Journal of Clinical Investigation*, 117: 601–4.

Power, M.L. (2004) Viability as opposed to stability: an evolutionary perspective on physiological regulation, in: Schulkin (ed.) *Allostasis, homeostasis and the cost of physiological adaptation*, New York: Cambridge University Press, pp. 343–64.

Preller, L., Zweers, T., Boleij, J.S.M., Brunekreef, B. (1990) Gezondheidsklachten en klachten over het binnenklimaat in kantoorgebouwen (in Dutch) (Health symptoms and complaints on indoor climate in office buildings), Directoraat-Generaal van de Arbeid, Ministerie van Sociale Zaken en Werkgelegenheid, S83, Voorburg, The Netherlands.

Pressman, S.D., Cohen, S. (2005) Does positive affect influence health? *Psychol. Bull.*, 131: 925–71.

Puri, K., Lynam, H. (2010) Dysregulated neurotransmitter function, Chapter 10 in: Nicolle, L., Woodriff Beirne, A. (eds) (2010) *Biochemical imbalances in diseases: a practitioner's handbook*, London and Philadelphia, PA: Singing Dragon.

Quirce S., Lemière, C., de Blay, F., del Pozo, V., Gerth Van Wijk, R., Maestrelli, P., Pauli, G., Pignatti, P., Raulf-Heimsoth, M., Sastre, J., Storaas, T., Moscato, G. (2010) Noninvasive methods for assessment of airway inflammation in occupational settings, *Allergy*, 65: 445–58.

Quirk, W.S., Seidman, M.D. (1995) Cochlear vascular changes in response to loud noise. *Am. J. Otol.*, 16: 322–5.

Radloff, L.S. (1977) The CES-D scale: a self-report depression scale for research in the general population, *Applied Psychological Measurement*, 1: 385–401.

Rammstedt, B., John, O.P. (2007) Measuring personality in one minute or less: a 10-item short version of the big five Inventory in English and German, *Journal of Research in Personality*, 41: 203–12.

Raw, G.J., Roys, M.S., Leaman, A. (1990) Further findings from the office environment survey: productivity, *Proceedings of Indoor Air 1990*, Toronto, Canada, 1: 231–6.

Raw, G.J., Roys, M.S., Tong, D. (1994) Questionnaire design for sick building syndrome, Part II: the effect of symptom list and frequency scale, *Proceedings of Healthy Buildings 1994*, Budapest, Hungary, 1: 481–6.

Raymond, C.J. (2011) Environmental issues past, present and future: changing priorities and responsibilities for building design, Keynotes: 40: 40 Looking backward and forward, *Proceedings of Sustainable Buildings 2011 (SB11)*, Helsinki, Finland.

Rea, M.S. (1982) An overview of visual performance, *Lighting Design and Application*, 12 (11): 35–41.

Rea, M.S., Bierman, A., Figuerio, M.G., Bullough, J.D. (2008) A new approach to understanding the impact of circadian disruption on human health, *Journal of Circadian Rhythms*, 6: 7.

Rea, T.J., Brown, C.M., Sing, S.F. (2006) Complex adaptive system models and the genetic analysis of plasma HDL-cholesterol concentration, *Perspect. Biol. Med.*, 49: 490–503.

Reiss, R., Andersson, E.L., Cross, C.E., Hidy, G., Hoel, D., Mcclellan, R., Moolgavkar, S. (2007) Evidence of health impacts of sulfate- and nitrate containing particles in ambient air, *Inhalation Toxicology*, 19, 419–49.

Reiter, R.J., Melchiorri, D., Sewerynek, E., Poeggeler, B., Barlow-Walden, L., Chuang, J., Oritz, G.G., Acufia-Castroviejo, D. (1995) Review of the evidence supporting melatonin's role as an antioxidant, *J. Pineal. Res.*, 18(10): 1–11.

Reiter, R.J., Carneiro, R.C., Oh, C.S. (1997) Melatonin in relation to cellular anti-oxidative defence mechanisms, *Horm. Metab. Res.*, 29(8): 363–73.

Richards, W. (1986) Allergy, asthma and school problems, *J. Sch. Health*, 56: 151–2.

Riether, C., Doenien, R., Pacheco-Lopez, G., Niemi, M., Engler, A., Engler, H., Schdelowski, M. (2008) Behavioural conditioning of immune functions: how the central nervous system controls peripheral immune responses by evoking associate learning processes. *Rev. Neurosci.*, 19: 1–17.

Rimbach, G., Egert, S., de Pascual-Teresa, S. (2011) Chocolate: (un)healthy source of polyphenols, *Genes Nutr.*, 6: 1–3.

Rive, S., Rudnai, P., Bruinen de Bruin, Y., Annesi-Maesano, I. (2011) *The indoor air quality in European schools: an overview of air pollutant concentrations and associated health effects*, SINPHONIE WP2 report, Paris, France.

Robertson, D.G., Watkins, P.B., Reily, M.D. (2011) Metabolomics in toxicology: preclinical and clinical applications, *Toxicological Sciences* 120 (S1): S146–70.

Rolando, M. (1984) Tear mucus ferning test in normal and keratoconjunctivitis sicca eyes, *Chibret Int. J. Ophtalomol.*, 2(4): 32–41.

Rooney, J.J., Vanden Heuvel, L.N. (2004) Root cause analysis for beginners, *Quality Progress*, July: 45–53.

Roosens, L., Abdallah, M.A.E., Harrad, S., Neels, H., Covaci, A. (2009) Factors influencing concentrations of polybrominated diphenyl ethers (PBDEs) in studens from Antwerp, Belgium, *Environ. Sci. Technol.*, 43: 3535–41.

Roponen, M., Seuri, M., Nevalainen, A., Randell, J., Hirvonen, M.R. (2003) Nasal lavage method in the monitoring of upper airway inflammation: seasonal and individual variation, *Inhalation Toxicology*, 15: 649–61.

Rotter, J.B. (1966) Generalized expectancies for internal versus external control of reinforcement, *Psychological Monographs*, 80 (1, Whole no. 609).

Rouby, C., Bensafi, M. (2002) Is there a hedonic dimension to odors, Chapter 9 in: Rouby, C., Schaal, B., Dubois, R., Gervais, R., Holley, A. (eds) *Olfaction, taste, and cognition*, Cambridge: Cambridge University Press.

Roulet, C.-A., Bluyssen, P.M., Cox, C., Foradini, F. (2006a) Relations between perceived indoor environment characteristics and well-being of occupants at individual level, *Proceedings of Healthy Buildings 2006*, Lisboa, Portugal, 3: 163–8.

Roulet, C.-A., Flourentzou, F., Foradini, F., Bluyssen, P., Cox, C., Aizlewood, C. (2006b) Multi-criteria analysis of health, comfort and energy-efficiency in buildings, *Building Research and Information*, 34: 475–82.

Rudel, R.A., Perovich, L.J. (2009) Endocrine disrupting chemicals in indoor and outdoor air, *Atmospheric Environment*, 43: 170–81.

Runeson, R., Norbäck, D., Stattin, H. (2003) Symptoms and sense of coherence: a follow-up study of personnel from workplace buildings with indoor air problems, *Int. Arch. Occup. Environ. Health*, 76: 29–38.

Runeson, R., Wahlstedt, K., Wieslander, G., Norbäck, D. (2006) Personal and psycho-social factors and symptoms compatible with sick building syndrome in the Swedish workforce, *Indoor Air*, 16: 445–53.

Russcl, J.A. (1980) A circumplex model of affect, *Journal of Personality and Social Psychology*, 39: 1161–78.

Rutten, P.G.S., Trum, H.M.G.J. (1998) Prestatiegericht ontwerpen en evalueren (in Dutch) (Performance based design and evaluation), Technische Universiteit Eindhoven, The Netherlands.

Sacks, J.D., Stanek, L.W., Luben, T.J., Johns, D.O., Buckley B.J., Brown, J.S., Ross, M. (2011) Particulate matter-induced health effects: who is susceptible? *Environmental Health Perspectives*, 119: 446–54.

Santos, A.M.B., Gunnarsen, L. (1999) *Indoor climate optimization with limited sources*, SBI report 314, Hoersholm, Denmark.

Saucier, G. (1994) Mini-markers: a brief version of Goldberg's unipolar big-five markers, *Journal of Personality Assessment*, 63: 506–16.

Sauvé, B., Koren, G., Walsh, G., Tokmakereijn, S., Van Uum, S.H. (2007) Measurement of cortisol in human hair as a biomarker of systemic exposure. *Clin. Invest. Med.*, 30: E183–91.

Scheuplein, R., Chamley, G., Dourson, M. (2002) Differential sensitivity of children and adults to chemical toxicity, *Regul. Toxicol. Pharm.*, 35: 448–67.

Schiffman, R.M., Christianson, M.D., Jacobsen, G., Hirsch, J.D., Reis, B.L. (2000) Reliability and validity of the Ocular Surface Disease Index, *Archives of Ophthal-mology*, 118: 615–21.

Schimmack, U. (2006) *Internal and external determinants of subjective well-being: Review and policy implication. Happiness and public policy: theory, case studies and implications*, New York: Palgrave Macmillan.

Schulkin, J. (2004) *Allostasis, homeostasis and the cost of physiological adaptation*, Cambridge: Cambridge University Press.

Schultz, D.P., Schultz, S.E. (2006) *Psychology and work today*, 9th edition, New Jersey: Pearson Education International.

Schulz, K.F., Altman D.G., Mohr, D. (2010) Consort 2010 statement: updated guidelines for reporting parallel group randomised trials, *British Medical Journal*, 340: 698–702.

Schwartz, G.E., Brown, S.L., Ahern, G.L. (1980) Facial muscle patterning and subjective experience during affective imagery: sex differences, *Psychophysiology*, 17: 75–82.

Seagrave, J., McDonald, J.D., Mauderly, J.L. (2005) In vitro versus in vivo exposure to combustion emissions, *Exp. Toxicol. Pathol.*, 57 (Suppl. 1): 233–8.

Seeger, K. (2009) Metabolic changes in autoimmune disease, *Curr. Drug Discov. Technol.*, 6: 256–61.

Seeman, T.E., Singer, B.H., Rowe, J.W., Horwitz, R.I., McEwen, B.S. (1997) Price of adaptation: allostatic load and its health consequences, *Arch. Intern. Med.*, 157: 2259–68.

Seeman, T., Epel, E., Gruenewald, T., Karlamangla, A., McEwen, B.S. (2010) Socio-economic differentials in peripheral biology: cumulative allostatic load, *Annals of the New York Academy of Sciences*, 1186: 223–39.

Seidman, M.D. Standring, R.T. (2010) Noise and quality of life, *International Journal of Environmental Research and Public Health*, 7: 3730–8.

Seppänen, O., Fisk, W.J. (2005) Some quantitative relations between indoor environmental quality and work performance or health, *Proceedings of Indoor Air 2005*, Beijing, China, pp. 40–53.

Serway, R.A. (1996) Physics for scientists and engineers with modern physics, 4th edition, Saunders College Publishing, USA.

Shendell, D.G., Prill, R., Fisk, W.J., Apte, M.G., Blake, D., Faulkner, D. (2004) Associations between classroom $CO_2$ concentrations and student attendance in Washington and Idaho, *Indoor Air*, 14: 333–41.

Sheth, J.N., Mittal, B., Newman, B.I. (1999) Researching customer behaviour, Chapter 13 in: *Customer behavior, consumer behavior and beyond*, Orlando, FL: The Dryden Press.

Shield, B., Dockrell, J. (2003) The effects of noise on children at school: a review, *Building Acoustics*, 10: 97–116.

Shield, B., Dockrell, J. (2004) External and internal noise surveys of London primary schools, *Journal of the Acoustical Society of America*, 115: 730–38.

Siegel, D.J. (2011) *Mindsight, the new science of personal transformation*, New York: Bantam Books Trade Paperbacks.

Siegrist J., Klein D., Voigt, K.H. (1997) Linking sociological with physiological data: the model of effort-reward imbalance at work, *Acta Physiol. Scand.*, 161: 112–6.

Siegrist J., Starke D., Chandola T., Godin, I., Marmot, M., Niedhammer, I., Peter, R. (2004) The measurement of effort-reward imbalance at work: European comparisons. *Soc. Sci. Med.*, 58(8): 1483–99.

Silbernagel, S., Despopoulos, A. (2008) *Atlas van de fysiologie* (in Dutch), Sesam, 15de Nederlandse druk, translation of 7th German version Taschenatlas der physiologie (2007).

Simoni, M., Annesi-Maesano, I., Sigsgaard, G., Norbäck, D., Wieslander, G., Lystad, W., Canciani, M., Viegi, G., Sestini, P. (2006) Relationships between school indoor environment and respiratory health in children of five European countries (HESE study), *Eur. Respir. J.*, 28: 837.

Simon-Nobbe, B., Denk, U., Pöll, V., Rid, R., Breitenbach, M. (2008) The spectrum of fungal allergy, *Int. Arch. Allergy Immunol.*, 145: 58–86.

Singer, B., Ryff, C.D., Seeman, T. (2004) Operationalizing allostatic load, Chapter 4 in: Schulkin (ed.) *Allostasis, homeostasis and the cost of physiological adaptation*, Cambridge: Cambridge University Press.

Sjödin, A., Päpke, O., McGahee, E., Focant, J.-F., Jones, R.S., Pless-Mulloli, T., Toms, L.-M.L., Herrmann, T., Müller, J., Needham, L.L., Patterson, D.G. (2008) Concentration of polybrominated diphenyl ethers (PBDEs) in household dust from various countries, *Chemosphere*, 73: S131–6.

Skov, O., Valbjørn, O., DISG (1987) The 'Sick' building syndrome in the office environment, The Danish town hall study, *Environment International*, 13: 339–49.

Slovic, P. (2010) *The feeling of risk: new perspectives on risk perception*, London: Earthscan.

Smedje, G.N., Norbäck, D. (2000) New ventilation systems at select schools in Sweden: effects on asthma and exposure, *Arch. Environ. Health*, 55: 18–25.

Smedje, G.N., Nörback, D. (2001) Incidence of asthma diagnosis and self-reported allergy in relation to the school environment: a four-year follow-up study in school-children, *Int. J. Tuberc. Lung Disc.*, 5: 1059–66.

Smith, A.D., Cowan J.O., Brassett K.P., Filsell, S., McLachian, C., Monti-Sheehan, G., Herbison, P.G., Taylor, R.D. (2005) Exhaled nitric oxide: a predictor of steroid response, *Am. J. Respir. Crit. Care Med.*, 172: 453–9.

Smith, K.R. (2003) The global burden of disease from unhealthy buildings: preliminary results from a comparative risk assessment, *Proceedings of Healthy Buildings 2003*, Singapore, pp. 118–26.

Snedecor, G.W., Cochran, W.G. (1980) *Statistical methods*, 7th edition, Ames, IA: The Iowa State University Press.

Sommers J., Vodanovich S.J. (2000) Boredom proneness: its relationship to psychological and physical health symptoms, *Journal of Clinical Psychology*, 56(10): 149–55.

Sormunen, P., Holopainen, R., Jokela, M., Laine, T., Dehlin, S., Heikkilä, K., Nummelin, O., Hirvonen, T., Sandesten, S., Fristedt, S., Benning, P., Åberg P., Olofsson, T., Schade, J., Matthyssen, A., Gerene, S., Fijneman, M., Bluyssen, P.M. (2009) *Stakeholder values, stakeholder preferences and requirements for the life cycle design process*, InPro Task 2.3: Capturing stakeholder values, Deliverable 14b, Helsinki, Finland.

Spiekman, M.E. (2012) *Social embodied agents to support elderly with mild dementia living alone*, master's thesis, TNO-DV 2011 S419, Soesterberg, The Netherlands.

Spreng, M. (2000) Central nervous system activation by noise, *Noise and Health*, 7: 49–57.

Stanojevic, S., Wade, A., Stocks, J., Hankinson, J., Coates, A.L., Pan, H., Rosenthal, M., Corey, M., Lebecque, P., Cole, T.J. (2008) Reference ranges for spirometry across all ages, A new approach, *Respir. Crit. Care Med.*, 177: 253–60.

Stephens, B., Siegel, J.A. (2012) Penetration of ambient submicron particles into single-family residences and associations with building characteristics, *Indoor Air*, 22: 501–13.

Steptoe, A., Cropley, M., Griffith, J., Kirschbaum, C. (2000) Job strain and anger expression predict early morning elevations in saliva cortisol, *Psychosomatic Medicine*, 62: 286–92.

Steptoe, A., Brydon, L. (2005) Association between acute lipid stress responses and fasting lipid levels 3 years later, *Health Psychology*, 24: 601–7.

Steptoe, A., Kunz-ebrecht, S.R., Wright, C., Feldman, P.J. (2005) Socioeconomic position and cardiovascular and neuroendocrine responses following cognitive challenge in old age, *Biol. Psychol.*, 69: 149–66.

Steptoe, A., Hamer, M., Chida, Y. (2007) The effects of acute psychological stress on circulating inflammatory factors in humans: a review and meta-analysis, *Brain, Behaviour and Immunity*, 21: 901–12.

Steptoe, A., Poole, L. (2010). Use of biological measures in behavioral medicine, in: Steptoe, A. (ed.) *Handbook of behavioral medicine research*, New York: Springer, pp. 619–32.

Sterling, P. (2004) Principles of allostasis: optimal design, predictive regulation, pathophysiology, and rational therapeutics, Chapter 1 in: Schulkin (ed.) *Allostasis, homeostasis and the cost of physiological adaptation*, New York: Cambridge University Press, pp.17–64.

Stevens, R.G., Rea, M.S. (2001) Light in the built environment: potential role of circadian disruption in endocrine disruption and breast cancer, *Cancer Causes Control*, 12: 279–87.

Stevens, R.G., Blask, D.E., Brainard, G.C., Hansen, J., Lockley, S.W., Provencio, I., Rea, M.S., Reinlib, L. (2007) Meeting report: the role of environmental lighting and circadian disruption in cancer and other diseases, *Environmental Health Perspectives*, 115: 1357–62.

Stevens, S.S. (1957) On the psycho-physical law, *Psychol. Review*, 64: 153–81.

Stevenson, R.J. (2010) An initial evaluation of the functions of human olfaction, *Chem. Senses*, 35: 3–20.

Stone, A.A., Neale, J.M. (1982) Development of a methodology for assessing daily experiences: decisions for the researcher, in: Baum, A., Singer, J.E. (eds) *Advances in environmental psychology: environment and health*, Hillsdale, NJ: Lawrence Erlbaum, 4: 49–83.

Straif, K., Baan, R., Grosse, Y., Secretan, B., Ghissassi, F.E., Bouvard, V., Altieri, A., Benbrahim-Tallaa, L., Coglino, V. (2007) Carcinogenicity of shift-work, painting, and fire-fighting, *The Lancet Oncology*, 8: 1065.

Stranger, M., Potgieter-Vermaakc, S., Van Grieken, R. (2008) Characterization of indoor air quality in primary schools in Antwerp, Belgium, *Indoor Air*, 18: 454–63.

Stroem-Tejsen, P., Wyon, D.P., Zukowska, D., Jama, A. (2009) Finger temperature as a predictor of thermal comfort for sedentary passengers in a simulated aircraft cabin, *Proceedings of Healthy Building 2009*, Syracuse, NY, paper 382.

Subramanian, J., Govindan, R. (2007) Lung cancer in never smokers: a review, *J. Clin. Oncol.*, 25: 561–70.

Sun, S., Schiller, J., Gazdar, A. (2007) Lung cancer in never smokers: a different disease, *Nature Rev. Cancer*, 7: 778–90.

Sundell, J., Lindvall, T., Stenberg, B. (1994) Associations between type of ventilation and airflow rates in office buildings and the risk of SBS-symptoms among occupants, *Environment International*, 20: 239–51.

Sundell, J., Levin, H., Nazaroff, W.W., Cain, W.S., Fisk, W.J., Grimsrud, D.T., Gyntelberg, F., Li, Y., Persily, A.K., Pickering, A.C., Samlet, J.M., Spengler, J.D., Taylor, S.T., Weschler, C.J. (2011) Ventilation rates and health: multidisciplinary review of the scientific literature, *Indoor Air*, 21: 191–204.

Sundstrom, E., Sundstrom M.G. (1986) *Work places: the psychology of the physical environment in offices and factories*, New York: Cambridge University Press.

Sung, E.J., Min, B.C., Kim, S.C. (2005) Effects of oxygen concentrations on driver fatigue during simulated driving, *Applied Ergonomics*, 35: 25–31.

SuPerBuildings (2010a) *Conclusions about the needs for development of sustainability indicators and assessment methods*, final report WP2 D2.1, http://cic.vtt.fi/super buildings/node/6 (accessed 31 May 2012).

SuPerBuildings (2010b) *Concept and Framework*, final report WP4.1, http://cic.vtt.fi/ superbuildings/node/6 (accessed 31 May 2012).

Swenberg, J.A., Lu., K., Moeller, B.C., Gao, L., Upton, P.B., Nakamura, J., Starr, T.B. (2011) Endogenous versus exogenous DNA adducts: their role in carcinogenesis, epidemiology and risk assessment, *Toxicological Sciences*, 120 (S1): S130–45.

Tahara, Y., Morito, N., Nishimiya, H., Yamagishi, H., Yamagichi, M. (2009) Evaluation of environmental and physiological factors of a whole ceiling-type air conditioner using a salivary biomarker, *Building and Environment*, 44: 1156–61.

Tallis, F., Eysenck, M., Mathews A. (1992) A questionnaire for the measurement of non-pathological worry, *Person. Individ. Diff.*, 13(2): 161–8.

Tamayo, P., Slonim, D., Mesirov, J., Zhu, Q., Kitareewan, S., Dmitrovsky, E., Lander, E.S., Golub, T.R. (1999) Interpreting patterns of gene expression with self-organizing maps: methods and application to hematopoietic differentiation, *Proc. Natl. Acad. Sci.*, 96: 2907–12.

Tamblyn, R.M., Menzies, R.I., Tamblyn, R.T., Farant, J.P., Hanley, J. (1992) The feasibility of using a double blind experimental cross-over design to study interventions for sick building syndrome, *Journal of Clinical Epidemiology*, 45: 603–12.

Tanabe, S., Nishihara, N. (2004) Productivity and fatigue, *Indoor Air*, 14: 126–33.

Tani, A., Hewitt, C.N. (2009) Uptake of aldehydes and ketones at typical indoor concentrations by houseplants, *Environ. Sci. Technol.*, 43: 8338–43.

Tarant, J.M. (2010) Blood cytokines as biomarkers of in vivo toxicity in preclinical safety assessments: considerations for their use, Review, *Toxicological Sciences*, 117: 4–16.

Taylor, J. (2006) *The mind, a user's manual*, Chichester, UK: John Wiley & Sons.

Taylor, S.E. (2006) Tend and befriend: biobehaviourial bases of affiliation under stress, *Current Directions in Psychological Science*, 15: 273–6.

Tham, K.W., Zuraimi, M.S., Koh, D., Chew, F.T., Ooi, P.L. (2007) Associations between home dampness and presence of molds with asthma and allergic symptoms among young children in the tropics, *Pediatr. Allergy Immunol.*, 18: 418–24.

Thompson, E.R. (2007). Development and validation of an internationally reliable short-form of the positive and negative affect schedule (PANAS), *Journal of Cross-cultural Psychology*, 38: 227–42.

Thompson, R., Keene, K. (2004) The pros and cons of caffeine, *The Psychologist*, 17: 698–701.

Thorgeirsson, T.E., Geller, F., Sulem, P., Rafnar, T., Wiste, A., Magnusson, K.P. *et al.* (2008) A variant associated with nicotine dependence, lung cancer and peripheral arterial disease, *Nature*, 452: 638–42.

Tochihara, Y., Ohnaka, T., Nagai, Y., Tokuda, T. (1993) Physiological responses and the thermal sensations of the elderly in cold and hot environments, *Journal of Thermal Biology*, 18: 355–61.

Toftum, J., Clausen, G., Bekö, G., Callesen, M., Weschler, C.J., Langer, S., Andersen, B., Høst, A. (2009) A case-base study of residential IEQ risk factors and parental reports of asthma and allergy symptoms among 500 Danish children: IECH, *Proceedings of Healthy Building 2009*, Syracuse, NY, paper 617.

Toftum, J. (2010) Central automatic control or distributed occupant control for better indoor environment quality in the future, *Build Environment*, 45: 23–8.

Trosko, J.E., Upham, B.L. (2005) The emperor wears no clothes in the field of carcinogen risk assessment: ignored concepts in cancer risk assessment, *Mutagenesis*, 20: 81–92.

Trueman, L., Bold, J., Gastro-intestinal imbalances, Part 1: The gastro-intestinal tract: use and abuse, Chapter 2 in: Nicolle, L., Woodriff Beirne, A. (eds) (2010) *Biochemical imbalances in diseases: a practitioner's handbook*, London and Philadelphia, PA: Singing Dragon.

van Tulder, M., Malmivaara, A., Koes, B. (2007) Repetitive strain injury, *The Lancet*, 369: 1815–22.

UN (2012) *World urbanization prospects*, the 2011 Revision, United Nations, New York, http://esa.un.org/unup/pdf/WUP2011_Highlights.pdf (accessed 2 November 2012).

Varghese, S., Gangamma, S., Patil, R., Sethi, V. (2005) Particulate respiratory dose to Indian women from domestic cooking, *Aerosol. Sci. Technol.*, 39: 1201–7.

Veitch, J.A. (1990) Office noise and illumination effects on reading comprehension, *Journal of Environmental Psychology*, 10: 209–17.

Veitch, J.A., Farley, K.M.J., Newsham, G.R. (2002) *Environmental satisfaction in open-plan environments: 1. Scale validation and method* (IRC-IR-844), Ottawa, National Research Council Canada, Institute for Research in Construction, http://nrc-cnrc.gc.ca/obj/irc/doc/pubs/ir/ir844/ir844.pdf (accessed 22 October 2010).

Veitch, J.A. (2011) Workplace Design contributions to mental healh and well-being, invited essay, *Healthcarepapers*, 11 (special issue): 38–46.

Verbrugge, L.M. (1980) Health diaries, *Medical Care*, 18: 73–95.

Vicente, P., Reis, E. (2010) Using questionnaire design to fight nonresponse bias in web surveys, *Social Science Computer Review*, 28: 251–67.

Vink, P. (2004) *Comfort and design: Principles and good practice*, Danvers, MA: CRC Press.

Virtanen, M., Singh-Manoux, A., Ferrie, J.E., Gimeno, D., Marmot, M.G., Elovianio, M., Jokela, M., Vahtera, J., Kivimäki, M. (2009) Long working hours and cognitive function, *American Journal of Epidemiology*, 169: 596–605.

Vischer, J.C. (1989) *Environmental quality in offices*, New York: Van Nostrand Reinhold.

Vischer, J.C. (2008) Towards an environmental psychology of workspace: how people are affected by environments for work, *Architectural Science Review*, 5(12): 97–108.

VITO (2010) *Indoor air quality in schools: influence of the ambient air, ventilation and classroom design, summary*, wwwb.vito.be/flies/flies_e_class.aspx (accessed 29 August 2012).

Vitrivius (100 BC) *Ten books of architecture*, translated by Morgan, M.H., New York: Dover Publications, 1961 edition.

Vlek, C.A.J. (1996) A multi-level, multi-stage and multi-attribute perspective on risk assessment, decision-making and control, *Risk Decision and Policy*, 1: 9–31.

Vonderheide, A.P., Mueller, K.E., Meija, J., Welsh, G.L. (2008) Polybrominated diphenyl ethers: causes for concern and knowledge gaps regarding environmental distribution, fate and toxicity, *Science of the Total Environment*, 400: 425–36.

Von Kempski, D. (2003) Air and well being: a way to more profitability, in: *Proceedings Healthy Buildings 2003*, Singapore, pp. 348–54.

Vrijkotte, T. (2010) *Work stress and cardiovascular disease risk*, Ph.D. thesis, Faculty of Psychology and Pedagogy, Vrije Universiteit, The Netherlands.

Vroon, P.A. (1990) Psychologische aspecten van ziekmakende gebouwen (in Dutch) [Psychological aspects of building related illness], published by Ministry of Housing, Planning and Environment, The Netherlands.

Wadsworth, E.J.K., Moss, S.C., Simpson, S.A., Smith, A.P. (2005) SSRIs and cognitive performance in a working sample, *Human Psychopharmacology*, 20: 561–72.

Wålinder, R., Norbäck, D., Wieslander, G., Smedje, G., Erwall, C. (1997) Nasal congestion in relation to low air exchange rate in schools, *Acta Otolaryngol. (Stockh)*, 117: 724–7.

Wälinder, R., Nörback, D., Wieslander, G., Smedje, G., Erwall, C., Venge, P. (1999) Nasal patency and lavage biomarkers in relation to settled dust and cleaning routines in schools, *Scand. J. Work Environ. Health*, 25: 137–43.

Wahlstedt, K.G.I., Eding, C. (1994) Psychosocial factors and their relations to psychosomatic complaints among postal workers, *European Journal of Public Health*, 4: 60–4.

Waldhauser, F., Dietzel, M. (1985) Daily and annual rhytms in human melatonin secretion: role in puberty control, *Annals of the New York Academy of Sciences*, 453: 205–14.

Walker, D.M., Gore, A.C. (2007) Endocrine-disrupting chemicals and the brain, in: Gore A.C. (ed.) *Endocrine-disrupting chemicals: from basic research to clinical practice*, Totowa, NJ: Humana Press, pp. 63–109.

Wallace, L. (2006) Indoor sources of ultrafine and accumulation mode particles: size distributions, size-resolved concentrations and sources strengths, *Aerosol Science and Technology*, 40: 348–60.

Wargocki, P., Wyon, D., Sundell, J., Clausen, G., Fanger, P.O. (2000) The effects of outdoor air supply rate in an office on perceived air quality, sick building syndrome (SBS) symptoms and productivity, *Indoor Air*, 10: 222–36.

Wargocki, P., Wyon, D.P., Fanger, P.O. (2004) The performance and subjective responses of call-center operations with and used supply air filters at two outdoor air supply rates, *Indoor Air* 14 (suppl. 8): 7–16.

Wargocki, P., Wyon, D.P. (2007) The effects of moderately raised classroom temperatures and classroom ventilation rate on the performance of schoolwork by children (RP-1257), *HVAC&R Research*, 13: 193–220.

Wargocki, P., Wyon, D.P., Jensen, K., Bornehag, C.G. (2008) The effects of electrostatic filtration and supply air filter condition in classrooms on the performance of schoolwork by children, *HVAC&R Research*, 14: 327–44.

Wargocki, P., Dalewski, M., Haneda, M. (2009) Physiological effects of thermal environment on office work, *Proceedings of Healthy Building 2009*, Syracuse, NY, paper 104.

Wargocki, P., Boschoff, W., Hanninen, O., Carrer, P., Hartmann, T. (2012) Principles of ventilation guidelines based on health (HealthVent project), *Proceedings of Healthy Buildings 2012*, Brisbane, Australia, paper 3A.1.

Waring, M.S., Siegel, J.A. (2011) The effect on an ion generator on indoor air quality in a residential room, *Indoor Air*, 21: 267–76.

Watson, J.D., Crick, F.H.C. (1953) Molecular structure of nucleic acids, *Nature*, 4356: 737–8.

Watson, J.D., Pennebaker, J.W. (1989) Health complaints, stress and distress: exploring the central role of negative affectivity, *Psychological Review*, 96: 234–54.

Watson, J.D., Tellegen, A. (1985) Towards a consensual structure of mood, *Psychological Bulletin*, 98: 219–35.

Watson, J.D., Wiese, D., Vaidya, J., Tellegen, A. (1999) The two general activation systems of affect: structural findings, evolutionary considerations, and psychobiological evidence, *Journal of Personality and Social Psychology*, 76: 820–38.

Weber, P., Medina-Oliva, G., Simon, C., Lung, B. (2010) Overview on Bayesian networks applications for dependability, risk analysis and maintenance areas, *Engineering Applications of Artificial Intelligence*, 25: 671–82.

Weichenthal, S., Dufresne, A., Infante-Rivard, C., Joseph, L. (2007) Indoor ultrafine particle exposures and home heating systems: a cross-sectional survey of Canadian homes during the winter months, *Journal of Exposure Science and Environmental Epidemiology*, 17: 288–97.

Wensing, M., Uhde, E., Salthammer, T. (2006) Plastics additives in the indoor environment: flame retardants and plasticizers, *Science of the Total Environment*, 339: 19–40.

Wensing, M., Delius, W., Omelan, A., Uhde, E., Salthammer, T., He, C., Wang, H., Morawska, L. (2009) Ultra-fine particles (UFP) from laser printers: chemical and physical characterization, *Proceedings of Healthy Building 2009*, Syracuse, NY, paper 171.

Weschler, C.J., Wells, J.R., Poppendieck, D., Hubbard, H., Pearce, T.A. (2006) Workgroup report: indoor chemistry and health, *Environmental Health Perspectives*, 114, 442–6.

Weschler, C.J., Nazaroff, W.W. (2008) Semivolatile organic compounds in indoor environments, *Atmospheric Environment*, 42: 9018–40.

Weschler, C.J. (2011) Chemistry in indoor environments: 20 years of research, *Indoor Air*, 21: 205–18.

Weschler, R.H., Nazaroff, W.W. (2012) SVOC exposure indoors: fresh look at dermal pathways, *Indoor Air*, 22: 356–77.

Westgaard, R.H., Winkel, J. (2011) Occupational musculoskeletal and mental health: significance of rationalization and opportunities to create sustainable production systems: a systematic review, *Applied Ergonomics*, 42: 261–96.

White, E., Armstrong, B.K., Saracci, R. (2008) *Principles of exposure measurement in epidemiology: collecting, evaluating and improving measures of disease risk factors*, 2nd edition, Oxford: Oxford University Press.

WHO (2000) *Guidelines for air quality*, 2nd edition, regional publication European series no. 91, World Health Organisation, Geneva, Switzerland.

WHO (2002) *Active ageing: a policy framework*, A contribution of the World Health Organization to the Second United Nations World Assembly on Ageing, Madrid, Spain.

WHO (2003) WHO technical meeting on exposure–response relationships of noise on health, 19–21 September 2002, Bonn, Germany.

WHO (2006) *Air quality guidelines: global update 2005*, Particulate matter, ozone, nitrogen dioxide and sulphur dioxide, WHO Regional office for Europe, Denmark.

WHO (2007a) *Health relevance of particulate matter from various sources*, Report on a WHO Workshop, Bonn, Germany, 26–27 March 2007, EUR/07/5067587.

WHO (2007b) *Large analysis and review of European housing and health status (LARES)*, Preliminary overview, WHO regional office for Europe, Denmark, www.euro.who.int/__data/assets/pdf_file/0007/107476/lares_result.pdf (accessed 6 November 2012).

WHO (2007c) *Children's health and the environment in Europe: a baseline assessment*, WHO Regional office for Europe, Denmark, www.euro.who.int/__data/assets/pdf_file/0009/96750/E90767.pdf (accessed 29 October 2012).

WHO (2008) *The global burden of disease 2004 update*, Geneva, Switzerland, www.who.int/healthinfo/global_burden_disease/GBD_report_2004update_full.pdf (accessed 7 November 2012).

WHO (2009) *Guidelines for indoor air quality: dampness and mould*, WHO regional office for Europe, Denmark, www.euro.who.int/__data/assets/pdf_file/0017/43325/E92645.pdf (accessed 6 February 2012).

WHO (2010) *WHO guidelines for indoor air quality: selected pollutants*, Denmark, WHO Regional office for Europe, www.euro.who.int/__data/assets/pdf_file/0009/128169/e94535.pdf (accessed 6 February 2012).

WHO (2011a) *Global status report on noncommunicable diseases 2010*, Geneva, www.who.int/nmh/publications/ncd_report2010/en/ (accessed 4 February 2012).

WHO (2011b) *Cardiovascular diseases (CVDs)*, Fact sheet no. 317, September 2011, www.who.int/mediacentre/factsheets/fs317/en/index.html# (accessed 1 February 2012).

WHO (2011c) *Asthma*, Fact sheet no. 307, May 2011, www.who.int/mediacentre/factsheets/fs307/en/index.html# (accessed 1 February 2012).

WHO (2011d) *Chronic obstructive pulmonary disease (COPD)*, Fact sheet no. 315, www.who.int/mediacentre/factsheets/fs315/en/index.html (accessed 1 February 2012).

WHO (2011e) *Obesity and overweight*, Fact sheet no. 311 (updated March 2011), www.who.int/mediacentre/factsheets/fs311/en/index.html# (accessed 1 February 2012).

WHO (2011f) *Diabetes*, Fact sheet no. 312, August 2011, www.who.int/mediacentre/factsheets/fs312/en/index.html# (accessed 4 February 2012).

WHO (2011g) *Depression*, December 23, www.who.int/mental_health/management/depression/en/ (accessed 4 February 2012).

WHO (2011h) *Environmental burden of disease associated with inadequate housing, Summary report*, Copenhagen, Denmark, www.euro.who.int/__data/assets/pdf_file/0017/145511/e95004sum.pdf (accessed 6 February 2012).

WHO (2011i) *Environmental burden of disease associated with inadequate housing*, Copenhagen, Denmark, www.euro.who.int/__data/assets/pdf_file/0003/142077/e95004.pdf (accessed 6 February 2012).

WHO (2012a) *Cancer*, Fact sheet no. 297, February, www.who.int/mediacentre/factsheets/fs297/en/index.html (accessed 4 February 2012).

WHO (2012b) *Disease and injury country estimates*, www.who.int/healthinfo/global_burden_disease/estimates_country/en/index.html (accessed 14 August 2012).

Wieslander, G., Norbäck, D., Nordstrom, K., Walinder, R., Venge, P. (1999) Nasal and ocular symptoms, tear film stability and biomarkers in nasal lavage, in relation to building-dampness and building design in hospitals, *Int. Arch. Occup. Environ. Health*, 72: 451–61.

Wiik, R. (2011) Indoor productivity measured by common response patterns to physical and psychosocial stimuli, *Indoor Air*, 21: 328–40.

Wilby, R.B. (2007) A review of climate change impacts on the built environment, *Built Environment*, 33: 31–45.

de Wilde, P., Coley, D. (2012) The implications of a changing climate for buildings, *Building and Environment*, 55: 1–7.

Wilson, S., Hedge, A. (1987) *The office environmental survey: a study of building sickness*, London: Building Use Studies.

Winterbottom, M., Wilkins, A. (2009) Lighting and discomfort in the classroom, *Journal of Environmental Psychology*, 29: 63–75.

Wisthaler, A., Weschler, C.J. (2010) Reactions of ozone with human skin lipids: sources of carbonyls, dicarbonyls and hydroxy carbonyls in indoor air, *Proc. Natl. Acad. Sci.*, 107: 6568–75.

Witterseh, T., Wyon, D.P., Clausen, G. (2004) The effect of moderate heat stress and open-plan office noise distraction on SBS symptoms and on the performance of office work, *Indoor Air*, 14: 30–40.

Wolkoff, P. (2010) Ocular discomfort by environmental and personal risk factors altering the precorneal tear film, *Toxicology Letters*, 199: 203–12.

Wolkoff, P. (2013) Indoor air pollutants in office environments: assessment of comfort, health, and performance, *Int. J. Hyg. Environ. Health*, 216: 371–94.

Woolner, P., Hall, E., Higgins, S., McCaughey, C., Wall, K. (2007) A sound foundation? What we know about the impact of environments on learning and the implications for building schools for the future, *Oxford Review of Education*, 33: 47–70.

World Medical Association (2008) *Declaration of Helsinki: ethical principle for medical research involving human subjects*, 59th WMA General Assembly, Seoul 2008, www.wma.net/ (accessed 1 April 2012).

Wright, C.E., Erblicht, J., Valdimarsdottir, H.B., Bovbjerg, D.H. (2007) Poor sleep the night before an experimental stressor predicts reduced NK cell mobilization and slowed recovery in healthy women, *Brain Behac. Immun.*, 21: 358–63.

Wu, C., Kuo, I., Su, T., Li Y., Lin L., Chan, C., Hsu, S. (2010) Effects of personal exposure to particulate matter and ozone on arterial stiffness and heart rate variability in healthy adults, *American Journal of Epidemiology*, 171: 1299–1309.

Wu, W., Ng, E. (2003) A review of the development of daylighting in schools, *Lighting Research Technology*, 35: 111–25.

Wüst, S., Federenko, I., Hellhammer, D.H., Kirschbaum, C. (2000) Genetic factors, perceived chronic stress and the free cortisol response to awakening, *Psychoneuro-immunology*, 25: 707–20.

Wyon, D.P., Andersen, I., Lundqvist, G.R. (1979) The effects of moderate heat stress on mental performance, *Scand. J. Work Environ. Health*, 5: 352–61.

Wyon, D.P. (1993) Healthy buildings and their impact on productivity, *Proceedings of Indoor Air 1993*, Helsinki, Finland, 6: 3–13.

Wyon, D.P. (1996) Individual microclimate control: required range, probable benefits and current feasibility, *Proceedings of Indoor Air 1996*, 1: 1067–72.

Wyon, D.P. (2004) The effects of indoor air quality on performances and productivity, *Indoor Air*, 14: 92–101.

Wyon, N.M., Wyon, D.P. (1987) Measurement of acute response to draught in the eye, *Acta Ophthalmologica*, 65: 385–92.

Xu, Y., Zhang, J. (2011) Understanding SVOCs, *ASHRAE Journal*, December: 121–5.

Yaglou, C.P.E., Riley, C., Coggins, D.I. (1936) Ventilation requirements, *ASHVE Trans.*, 42: 133–62.

Yamada, J., Stevens, B., de Silva, N., Gibbins, S., Beyene, J., Taddio, A., Newman C., Koren G. (2007) Hair cortisol as a potential biologic marker of chronic stress in hospitalized neonates, *Neonatology*, 92: 42–9.

Yao, Y., Lian, Z., Liu, W., Jiang, C., Liu, Y., Lu, H. (2009) Heart rate variation and electroencephalograph: the potential physiological factors for thermal comfort study, *Indoor Air*, 19: 93–101.

Yoshino, H., Hasegawa, K., Abe, K., Ikeda, K., Kato, N., Kumagai, K., Hasegawa, A., Mitamura, T., Yanagi, U., Ando, N., Matsuda, A. (2009) Investigation of children health problems in relation to indoor environmental factors in Japan, Part 1: Design of investigation and outcome of allergic symptoms, *Proceedings of Healthy Building 2009*, Syracuse, NY, paper 631.

Young, S.P., Wallace, G.R. (2009) Metabolomic analysis of human disease and its application to the eye, *J. Ocul. Biol. Dis. Infor.*, 2: 235–42.

Yu, I.T.S., Chiu, Y.-L., Au, J.S.K., Wong, T.-W., Tang, J.-L. (2006) Dose-response relationship between cooking fumes exposures and lung cancer among Chinese non-smoking women, *Cancer Res.*, 66: 4961–7.

Zagreus, L., Huizinga, C., Arens, E., Lehrer, E. (2004) listening to the occupants: a web-based indoor environmental quality survey, *Indoor Air*, 14(S8): 65–74.

Zald, D.H., Pardo, J.V. (1997) Emotion, olfaction and the human amygdala: amygdala activation during aversive olfactory stimulation, *Proc. Natl. Acad. Sci.*, 94: 4119–24.

Zhao, Z.S., Larsson, A., Wang, L., Zhang, Z., Norbäck, D. (2008) Asthmatic symptoms among pupils in relation to microbial dust exposure in schools in Taiyuan, China, *Pediatr. Allergy Immunol.*, 19: 455–65.

Zuckerman, M. (2005) *Psychobiology of personality*, 2nd edition, New York: Cambridge University Press.

Zuraimi, M.S.T., Chew, K.W., Ooi, F.T. (2007) The effect of ventilation strategies of child care centers on indoor air quality and respiratory health of children in Singapore, *Indoor Air*, 17: 317–27.

# Index

Milton Keynes UK
Ingram Content Group UK Ltd.
UKHW052030141024
449569UK00017B/757